PETROLEUM NANOBIOTECHNOLOGY

Modern Applications for a Sustainable Future

PETROLEUM NANOBIOTECHNOLOGY

Modern Applications for a Sustainable Future

Nour Shafik El-Gendy, PhD

Prof. of Petroleum & Environmental Biotechnology
Egyptian Petroleum Research Institute (EPRI)
October University for Modern Sciences & Arts (MSA)

Hussein Nabil Nassar, PhD

Researcher of Environmental Biotechnology
Egyptian Petroleum Research Institute (EPRI)
October University for Modern Sciences & Arts (MSA)

James G. Speight PhD, DSc

CD&W Inc., USA

First edition published 2022

Apple Academic Press Inc.
1265 Goldenrod Circle, NE,
Palm Bay, FL 32905 USA

4164 Lakeshore Road, Burlington,
ON, L7L 1A4 Canada

CRC Press
6000 Broken Sound Parkway NW,
Suite 300, Boca Raton, FL 33487-2742 USA

2 Park Square, Milton Park,
Abingdon, Oxon, OX14 4RN UK

Apple Academic Press exclusively co-publishes with CRC Press, an imprint of Taylor & Francis Group, LLC

Library and Archives Canada Cataloguing in Publication

Title: Petroleum nanobiotechnology : modern applications for a sustainable future / Nour Shafik El-Gendy, PhD, Hussein Nabil Nassar, PhD, James G. Speight, PhD.

Names: El-Gendy, Nour Shafik, author. | Nassar, Hussein Mohamed Nabil, author. | Speight, James G., author.

Description: First edition. | Includes bibliographical references and index.

Identifiers: Canadiana (print) 20210323000 | Canadiana (ebook) 20210323116 | ISBN 9781774630051 (hardcover) | ISBN 9781774638262 (softcover) | ISBN 9781003160564 (ebook)

Subjects: LCSH: Petroleum—Microbiology. | LCSH: Petroleum—Biotechnology. | LCSH: Nanobiotechnology. | LCSH: Nanoparticles.

Classification: LCC QR53.5.P48 E44 2022 | DDC 665.5—dc23

Library of Congress Cataloging-in-Publication Data

CIP data on file with US Library of Congress

ISBN: 978-1-77463-005-1 (hbk)
ISBN: 978-1-77463-826-2 (pbk)
ISBN: 978-1-00316-056-4 (ebk)

About the Authors

Nour Shafik El-Gendy, PhD

Nour Shafik El-Gendy, PhD, is a professor in the field of petroleum and environmental biotechnology. She is Head Manager of the Petroleum Biotechnology Laboratory and Former Acting and Vice Head of Process Design and Development Department, Egyptian Petroleum Research Institute (EPRI). Additionally, Dr. El-Gendy is Coordinator of Nanobiotechnology Program, Faculty of Nanotechnology for Postgraduate Studies, Cairo University, Sheikh Zayed Branch Campus. She is also Head of the Technology Innovation Support Center (TISC) office and Vice Head of Center of Excellence and member in Entrepreneurship Hub in October University for Modern Sciences and Arts, MSA University. She is a member of the Technical, Monitoring and Performance Evaluation Office and the coordinator of Water, Energy, and Environment Committee of the Egyptian Academy of Scientific Research and Technology (ASRT). She is the Former Advisor for the Egyptian Minster of Environment. She is the Vice Coordinator of the Scientific Research Committee and a member in the Rural Woman Committee, National Council for Women (NCW) of Egypt. Dr. El-Gendy was an executive committee member of the Egyptian Young Academy of Sciences (EYAS). She is also a member of many technical committees for the preparation of regulations of different faculties concerned with energy, environmental and biochemical engineering.

She is an expert in the assessment of environmental pollution and its remediation, wastewater treatment, biofuels, petroleum upgrading, green chemistry, nanobiotechnology, valorization of wastes, biocorrosion, and its mitigation. She has published 10 chapters, five books, and 116 research papers, and has supervised 27 MSc and PhD theses. Dr. El-Gendy is also an editor and reviewer for many international journals. She has also participated as PI, Co-PI, or research member in research projects concerned with bioethanol from lignocellulosic wastes, biofuels from algae, applications of nanobiotechnology in upgrading of petroleum, and bioremediation of polluted environment. Dr. El-Gendy participated in many international workshops and training courses and international conferences and seminars.

She is a member of many international associations and organizations concerned with petroleum industry and environmental health and sciences. She is also an active member in national committees concerning sustainable development, water desalination, wastewater treatment, biofuel standards, petroleum pollution, chemical engineering, and biotechnology. She is in international collaboration with different foreign international universities.

Dr. El-Gendy's teaching experience is with the Chemical Engineering Department, Faculty of Engineering, The British University in Egypt; Faculty of Biotechnology and Chemical Engineering Department, Faculty of Engineering, October University for Modern Sciences and Arts (MSA); and Faculty of Science, Monofiya University, Chemical Engineering Department, Faculty of Engineering, Cairo University and Biotechnology Program, Faculty of Science, Cairo University.

Dr. El-Gendy awarded professional diplomas for "Preparing Arab Women Leaders Internationally and Regionally," documented by the Egyptian Foreign Ministry and "Environmental Management". She honored in many scientific forums and awarded from India the International Scientist Award 2021 on Engineering, Science and Medicine. Her biography is recorded in *Who's Who in Science and Engineering.*

Hussein Nabil Nassar, PhD

Hussein Nabil Nassar, PhD, is a researcher in the field of petroleum and environmental biotechnology. He is a member of the Petroleum Biotechnology Laboratory, Process Design and Development Department at Egyptian Petroleum Research Institute (EPRI), Egypt. He is a lecturer in the Microbiology Department, Faculty of Pharmacy, October University for Modern Sciences and Arts (MSA) University, Cairo, Egypt. He is also a member of the Technology Innovation Support Center (TISC) office and the Center of Excellence, at the same university. He is also a scientific member of Liquid Chromatography and Water unit at Central Analytical Laboratories, Egyptian Petroleum Research Institution, as well as a lecturer in the Faculty of Nanotechnology for Postgraduate Studies, Cairo University, Sheikh Zayed Branch Campus.

Dr. Nassar received his PhD in the field of petroleum and environmental biotechnology from Al-Azhar University, Cairo, Egypt. He has published six book chapters, one book, and 44 research papers, and has supervised 4 PhD theses in the field of biofuels, biocorrosion, environmental pollution, bioremediation, valorization of agro-industrial wastes, nanobiotechnology, and wastewater treatment. Dr. Nassar is also an editor and reviewer in several international journals in the field of environmental biotechnology and pollution, bioenergy, nanobiotechnology, and petroleum microbiology and biotechnology. He participated as a research member in projects concerning bioethanol from lignocellulosic wastes, applications of nano-biotechnology in upgrading of petroleum, and bioremediation of petroleum polluted environment. He has participated in many international workshops and training courses and international conferences. Dr. Nassar is a member in many international associations concerned with petroleum industry, and environmental health and sciences.

James G. Speight, PhD, DSc

James G. Speight, PhD, DSc, has doctorate degrees in chemistry, geological sciences, and petroleum engineering and is the author of more than 85 books in petroleum science, petroleum engineering, biosciences, and environmental sciences as well as books relating to ethics and ethical issues. Dr Speight has more than 50 years of experience in areas associated with (1) the properties, recovery, and refining of reservoir fluids, (2) conventional petroleum, heavy oil, and tar sand bitumen, (3) properties and refining of natural gas, gaseous fuels, and crude oil quality, (4) production and properties of petrochemicals from various sources, (5) properties and refining of biomass, biofuels, biogas, and the generation of bioenergy, (6) reactor and catalyst technology, and (7) the environmental and toxicological effects of crude oil and its products. His work has also focused on safety issues, environmental effects, remediation, reactors associated with the production and use of fuels and biofuels, process economics, as well as ethics in science and engineering and in ethic in the universities.

In 1996, Dr. Speight was elected to the Russian Academy of Sciences and awarded the Gold Medal of Honor that same year for outstanding contributions to the field of petroleum sciences. In 2001, he received the Scientists without Borders Medal of Honor of the Russian Academy of Sciences and was also awarded the Einstein Medal for outstanding contributions and service in the field of geological sciences. In 2005, the Russian Academy of Sciences awarded Dr. Speight the Gold Medal—Scientists without Frontiers, in recognition of Continuous Encouragement of Scientists to Work Together Across International Borders. In 2007, Dr. Speight received the Methanex Distinguished Professor award at the University of Trinidad and Tobago in recognition of excellence in research. In 2018, he was awarded the American Excellence Award for Excellence in Client Solutions, United States Institute of Trade and Commerce, Washington, DC.

Contents

Abbreviations

AD	anno domini
ADS	adsorptive desulfurization
AFM	atomic force microscope
AgNPs	silver nanoparticles
ANA	anthranilic acid
API	American Petroleum Institute
ASTM	American Society for Testing and Materials
ATP	adenosine triphosphate
AuNPs	gold nanoparticles
AVLE	Artemisia vulgaris leaves' methanolic extract
BC	before christ
BCE	before the dommon era
BDN	biodenitrogenation
BDS	biodesulfurization
BET	Brunauer–Emmett–Teller
CAR	carbazole
CCD	central composite design
CE	common era
CMC	critical micelle concentration
CMNC	ceramic–matrix nanocomposites
CNTs	carbon nanotubes
COD	chemical oxygen demand
CPO	chloroperoxidase
CTAB	cetyl trimethyl ammonium bromide
DAF	dissolved air flotation
DLS	dynamic light scattering
DNA	deoxyribonucleic acid
EDX	energy dispersive X-ray
EDS	energy-dispersive X-ray spectroscopy
EOR	enhanced oil recovery
EPS	exopolysaccharide
EPS	extracellular polymeric substance
EY	eosin yellowish
fcc	face-centered cubic

FESEM	field emission scanning electron microscope
FFD	fractional factorial design
FTIR	Fourier transform infrared spectroscopy
GA	gum arabic
GC	gas chromatography
GCE	Glassy carbon electrode
GHG	greenhouse gas
GK	gum kondagogu
HDN	hydrodenitrogenation
HDS	hydrodesulfurization
HGO	heavy gas oil
HPLC	high-performance liquid chromatography
HRTEM	high-resolution transmission electron microscopy
ICP	inductively coupled plasma
IOs	iron oxides
IRB	iron-reducing bacteria
LMB	leucomethylene blue
MB	methylene blue
MCFF	mycelial cell free filtrate
MIC	microbiologically influenced corrosion
MMNC	metal–matrix nanocomposites
MNPs	magnetic nanoparticles
MO	methyl orange
MRI	magnetic resonance imaging
MS	mass spectrometry
MWCNTs	multiwalled carbon nanotubes
NAD	nicotinamide adenine dinucleotide
NCBI	national center for biotechnology information
NPs	nanoparticles
ODS	oxidative desulfurization
O/W	oil to water
OPs	onion peels
PAH	polyaromatic hydrocarbon
PANH	polycyclic aromatic nitrogen heterocyclic
PASH	polycyclic aromatic sulfur heterocyclic
PDI	polydispersity index
PdNPs	palladium nanoparticles
PEG	polyethylene glycol
PM	particulate matter

PNPs	polymeric NPs
PSB	phosphate sodium buffer
PVA	polyvinyl alcohol
PWW	petroleum wastewater
QDs	quantum dots
R&D	research and development
RhB	Rhodamine B
RNA	ribonucleic acid
ROS	reactive oxygen species
RSM	response surface methodology
SAED	selected area electron diffraction
SEM	scanning electron microscopy
SPR	surface plasmon resonance
SRB	sulfate-reducing bacteria
SSA	specific surface area
SWNTs	single-walled nanotubes
TEM	transmission electron microscopy
TOC	total organic carbon
TWW	tannery waste water
UV/Vis	ultraviolet/visible
VSM	vibration sample magnetometer
XPS	X-ray photoelectron spectroscopy
XRD	X-ray diffraction
ZVI	zero-valent iron

Preface

Research on nanotechnology application in the oil and gas industry has been growing rapidly in the past decade, as evidenced by the number of scientific articles published in the field. With conventional crude oil and natural gas reserves harder to find and more expensive to access, and produce, the pursuit of more game-changing technologies that can address the challenges of the industry has stimulated this growth.

Nanobiotechnology has the potential to revolutionize the petroleum industry both upstream and downstream, including exploration, drilling, production, and enhanced oil recovery (EOR), as well as the various processes within the refinery. Nanobiotechnology provides a wide range of alternatives for materials and processes to be utilized in a crude oil refinery. Nanoscale materials in various forms such as solid composites, complex fluids, and functional nanoparticle–fluid combinations are keys to the new technological advances.

The synthesis and application of reliable nanoscale materials is a progressive domain and the focus of modern nanotechnology. Conventional physicochemical approaches for the synthesis of metal nanoparticles have become obsolete owing to costly and hazardous materials. Thus there is an enormous revolution in the cost-effective and eco-friendly green (biological- and phyto-) synthesis of nanometal/metal oxides. Nanotechnology has successfully gained applications in many areas of life, thereby seen as the modern way of creating products, which results in high efficiency of use. In the petroleum-processing industries, this revolution is no exception. The efficiency of a number of conversion processes improves upon application of materials with the nanometer scale dimension, which is caused by improvements and developments of better material properties as the particle size decreases.

Nanobiotechnology is an interdisciplinary area of research involving synthesis, characterization, and applications of nanomaterials that are of use in the biotechnology field. It is one of the rapidly growing fields with significant applications in the refinery. Transition metal oxide nanoparticles are an important class of materials due to their attractive magnetic, electronic, and optical properties. This makes them preferable to be used in a variety of applications, such as catalysis, sensors, lithium-ion batteries, and environmental applications, etc. When compared with other classes of materials, transition-metal oxides have a variety of interesting properties and applications.

This book provides a state-of-the-art presentation on the history of nanotechnology and the worldwide development toward the green synthesis of nanotechnology, reaching to present and the future applications and aspects in petroleum industry. The advantages and drawbacks of microbial- and phytosynthesis of nanomaterials are discussed in detail. It points out and discusses the concept of reaching sustainability in nanotechnology, starting from the synthesis of metal/metal oxides and ends with their wide applications. There is a special emphasis in this book on some of the major challenges in petroleum industry, for example, microbial corrosion, bioupgrading of heavy crude oil and its fractions and the treatment of petroleum wastewater. Then, it describes the possible applications of nanobiotechnology to solve these problems. It ends with illustrations of challenges facing the green synthesis of nanomaterials and the possible recommended solutions. Toxicity of nanomaterials and its obstacles upon application are also covered. The challenges and opportunities of nanotechnology in petroleum industry are conferred. Finally, the further future aspects of application of nanobiotechnology in different sectors in petroleum industry are discussed. The book also debates how applications of nanobiotechnology with the beneficial properties of the green-synthesized nanomaterials in petroleum industry will open new sustainable industrial revolution in oil and gas industry with a real achievement for the three pillars of sustainability; social, environment, and economic. Moreover, the valorization of different available bio-wastes into nanomaterials, with the concept of reaching to zero waste is discussed from the point of achieving the 17 goals of sustainability, and overcome the problems of water scarce, waste management, and climate change.

CHAPTER 1

Petroleum Microbiology and Nanotechnology

1.1 INTRODUCTION

Nanoscience is the analysis of phenomena and material modification at atomic, molecular, and macromolecular scales, in which the properties of chemicals and materials differ significantly from the properties at a larger-than-molecular scale (McNaught and Wilkinson, 1997; Alemán et al., 2007; ASTM E2456). Nanotechnology is further defined as the understanding and regulating matter in dimensions from 1 to 100 nm (1 nm = 1×10^{-9} m = 3.2808×10^{-9} feet = 3.9370×10^{-8} inches). Nanoscience and nanotechnology thus involve the capability of detecting and controlling the actions of individual atoms and molecules (Nouailhat, 2008). Moreover, nanoobjects have (1) structures with new properties that have similar properties to those of a molecule and a solid; (2) specific qualities proving useful in different applications.

Nanotechnology is not an independent branch of science, but a multidisciplinary branch of science that incorporates disciplines such as chemistry, physics, biology (separately as botany and zoology), electrical engineering, biophysics, and material science to study particles with substantially new and enhanced physical, chemical, and biological properties, as well as functional properties (Fakruddin et al., 2012).

Biotechnology, on the other hand, is the wide field of science covering living systems and organisms for the creation or manufacture of products or any technical technology that uses biological systems, living organisms, or their derivatives for the production or alteration of products or processes for the particular use. It often overlaps with (related) areas, such as bioengineering, biomedical engineering, and molecular engineering, depending on the tools and applications. Biotechnology has been in use for millennia in many areas as (1) agriculture, (2) for production, and (3) medicine (Clark and Pazdernik, 2015; Douglas, 2016). In the late 20th century and early 21st century, biotechnology has expanded to include new and diverse science such as genomics, recombinant gene techniques, immunology as well as the development of pharmaceutical therapies and diagnostic test methods. Biotechnology, in particular, uses biological processes such as fermentation and harnesses biocatalysts such as enzymes, yeast, and other microbes to become microscopic plants in development, examples are: (1) 80% or more is streamlining the steps in chemical production processes, (2) improving manufacturing process efficiency, (3) reducing the use of and reliance on petrochemicals, (4) the use of biofuels to reduce greenhouse gas emissions, (5) a reduction in water use, (6) a reduction in waste generation, and (7) realizing the full potential of traditional biomass use.

It seems reasonable that the two disciplines will combine in the form of collaborative work terms in which each discipline can complement the other (Niemeyer and Mirkin, 2005). In this way, it is evidenced by merging the two fields (nanotechnology and biotechnology) that nanobiotechnology gives a discipline (nanobiotechnology) that has the potential to advance science through an understanding of chemical, physical, and biological processes and phenomena at the molecular level.

Nanobiotechnology, bionanotechnology, and nanobiology are words that refer to nanotechnology and biology intersection (Gazit, 2007). As the topic has only recently emerged, bionanotechnology and nanobiotechnology serve as broad words for different related technologies. This discipline tends to reflect the convergence of biological science into various nanotechnology fields. This computational biology approach enables scientists to visualize and build structures that can be used in biological science. Biologically influenced nanotechnology makes use of biological processes as the basis for uncreated technologies.

In the world of science and technology, chemists have been creating solutions, suspensions, and colloids for hundreds of years but observing the particles or mixtures they formed at the molecular level was difficult, if not impossible. Similarly, in the petroleum refining industry, the catalyst was synthesized to perform at the molecular level but identifying the individual reaction sites—and the ensuing reactions—was not always possible and often left to theoretical supposition. However, with the advances in various technologies, nanotechnology has come into being making it possible to check the output predictions and better understand the processes of chemical interaction.

Nanotechnology has been implemented successfully in many applications, such as nanoelectronics, nanobiomedicine, and nanodevices. But this technology has rarely been applied to the oil and gas industry, especially in upstream exploration and manufacturing. The oil and gas industry is in need to improve oil recovery and exploit unconventional resources. Nevertheless, the cost of exploration and oil production is under enormous pressure, and when the crude oil price is low and depressed it is more difficult to justify such expenditure. This comes along with the depletion of less viscous and high-quality low asphalt, sulfur, and nitrogen content world reserves of crude oil. Thus, there is a great need for new cost-effective, eco-friendly, sustainable, and energy-saving techniques to be applied in the production, refining, and processing of such heavy oil reserves. This chapter presents to the reader an outline of the science concepts on which nanobiotechnology is based and the potential uses of this technology in science and engineering as an introduction to its use in petroleum science and technology.

1.2 PETROLEUM MICROBIOLOGY

Petroleum microbiology is a state-of-the-art technology that recognizes the action of specific microbial species on crude oil. There are various beneficial and detrimental effects of microbes on crude oil. Scientifically, petroleum microbiology is a division of microbiology that applies microorganisms in different sectors of the petroleum industry from exploration, production, storage, handling, transport, refining, fractionations, petrochemical, and other petroleum products, also in the remediation of any petroleum pollution and upgrading of petroleum and/or its fractions. Petroleum microbiology research is multidisciplinary progressing via exploring the influences of microbial activities on crude oil composition and production, in addition to the microbial processes involved in hydrocarbon biodegradation. Thus, it can be applied in bioremediation of oil-polluted environments, microbial enhanced oil recovery, biodesulfurization, biodenitrogenation, and biodemetallization, bioupgrading of heavy crudes and refining residues.

Petroleum is a complex mixture of hydrocarbons and other organic compounds, including other organometallic elements, most notably vanadium and nickel complexes. Petroleum extracted from different reservoirs significantly differs in compositional and physical properties. Those hydrocarbons, long recognized as substrates that sustain microbial growth, are both a target and a result of microbial metabolism. Biodegradation by microorganisms can be helpful in modifying waxy crude oils, but conditions for downhole applications require the use of thermophiles, resistant to organic solvents, with heat-stable enzymes, and reduced oxygen requirements (Speight and El-Gendy, 2017).

Petroleum (also called crude oil) is by nature a combination of gaseous, liquid, and solid hydrocarbon compounds. Petroleum occurs in sedimentary rock deposits worldwide and also includes small quantities of compounds containing nitrogen, oxygen, and sulfur, as well as trace amounts of metallic constituents (Speight, 2014). On a molecular basis, the major constituents of conventional petroleum are derivatives of hydrocarbons (i.e., hydrogen and carbon compounds), which show great variation in their molecular structure. The simplest hydrocarbons are a large group of molecules in chain form, known as paraffins. This large series stretches from methane, which forms natural gas, to crystalline waxes through liquids refined into gasoline. A ring-shaped series of hydrocarbons, known as naphthenes, originate from volatile liquids, such as naphtha isolated to high molecular weight substances as a fraction of asphalt. Another group of ring-shaped hydrocarbons is known as aromatics; benzene, a common raw material for producing petrochemicals, is the principal compound in this category. At the other side, heavy oil, super heavy oil, and bitumen in tar sand contain fewer amounts of hydrocarbon derivatives than conventional petroleum. The constituents are more complex, higher boiling, and contain higher proportions of heteroatoms (nitrogen, oxygen, and sulfur) within the molecular structure of the constituents.

Thus, investigations of the character of petroleum need to be focused on the influence of its character on refining operations and the nature of the products that will be produced. Furthermore, one means by which the character of petroleum has been studied is through its fractional composition. However, the fractional composition of petroleum varies markedly with the method of isolation or separation, thereby leading to potential complications (especially in the case of the heavier feedstock) in the choice of suitable processing schemes for such feedstock. Crude oil can be fractionated into three or four general fractions: (1) the asphaltene fraction, (2) the resin fraction, (3) the aromatics fraction, and (4) the saturates fraction—the name of each fraction is not a true representation of the character of the fraction but more a name that has been applied as a guide to illustrate the means of separation. However, using this convenient nomenclature, interlaboratory investigations can be compared and the principle of predictability can be extended to each fraction to the ease or complexity of chemical transformation of the constituents and the character of the potential products.

In the current context, recent advances in molecular biology have extended the understanding of metabolic processes related to microbial petroleum hydrocarbon transformation. The physiological responses of microorganisms to the presence of hydrocarbons have been studied and characterized, including cell surface alterations and adaptive mechanisms for the uptake and efflux of these substrates. In addition, new molecular techniques have enhanced the ability of scientists to study the dynamics of microbial populations in petroleum-affected environments. Microbes are injected into partly depleted oil reservoirs to improve oil recovery but the effects of these microbes cannot be generalized and are, in fact, site-specific.

There is interest in replacing refinery desulfurization processes with methods of biodesulfurization by encouraging selective removal of sulfur without degradation of associated carbon molecules (El-Gendy and Speight, 2016; Speight and El-Gendy, 2017). Since microbes need an environment with some water, however, a two-phase oil–water system must be developed to maximize contact between the microbes and the hydrocarbon derivatives, and such an emulsion is not easily created with the higher density more viscous crude oils. Nevertheless, bacterial processes are being commercialized for the removal of hydrogen sulfide (H_2S) and sulfoxide derivatives from petrochemical waste streams. Microbes can also be used to extract nitrogen from crude oil, leading to reduced emissions of nitric oxide, provided that technical problems close to those faced in biodesulfurisation can be solved (El-Gendy and Speight, 2016; Speight and El-Gendy, 2017).

The collection of the available data related to petroleum microbiology is by no means exhaustive. But it does serve to illustrate the current understanding of some of the biotechnology applications. In addition, the potential of microbiology in petroleum recovery, petroleum refining, and the petroleum products and chemicals industries is enormous, and substantial progress in the field will be witnessed in the coming decades. New knowledge on biodegradation and biotransformation of hydrocarbons has been derived from the recent study, especially studies on specific biochemical processes, and the underlying genes and microorganisms. Ultimately, while petroleum has deep roots in the history of microbiology, it is evident that the topic continues to be of vital importance to industry and the environment and will remain an important field of fundamental and applied research for decades.

Therefore, a major challenge that has hindered the creation of microbiological applications for the petroleum industry in the past is the dynamic, heterogeneous, and dangerous existence of crude oil and various distillates, residues, heavy crude oil, extra-heavy oil, and tar sand bitumen, and petroleum-derived wastes (Speight, 2014). Furthermore, the successful application of microbiological techniques to petroleum is focused on the production of robust microbial biocatalysts that can adapt and tolerate the various hydrocarbon derivatives, and at commercially viable levels perform the desired biotransformation of the petroleum constituents.

A major issue related to the progress of the application of microbiological techniques to the transformation of petroleum constituents has been the variability and complexity of the constituents (Speight, 2104). Recognition that the behavior of petroleum is related to its composition has led to a multiplicity of attempts to create petroleum and its fractions as compositions of matter. As a consequence, different analytical techniques have been developed for the identification and quantification of any molecule in the lower boiling fractions of petroleum. It is now widely accepted that the term petroleum does not reflect a composition of matter but rather a mixture of different organic compounds that includes a broad variety of molecular weights and molecular forms that exist in equilibrium with each other (Speight, 2014). There may also be some concerns of the advisability (perhaps futility is a better word) of trying to identify every molecule in petroleum. The true emphasis would be to what ends such molecules can be used or the impact they have on manufacturing, including microbial transformation to other molecular compounds.

1.3 HISTORY OF NANOTECHNOLOGY

Nanotechnology is the science and technology of small things, that is, nano, this term originates from the Greek word "nanos" meaning "dwarf" in Greek (Kannan et al., 2010). Nanotechnology is classified as understanding and controlling matter in dimensions from 1 to 100 nm where specific phenomena allow for a new application (Hasan, 2015; Sriramulu and Sumathi, 2017). It is a multidisciplinary branch combining physics, chemistry, biology, electrical engineering, biophysics, material science, etc., to produce particles with novel and substantially improved physical, chemical, and/or biological properties and functionality due to their nanoscaled size which have many promising applications. For example, environmental perspectives, reducing the application of industrial chemicals, water disinfection and purification, renewable energy, delivering drugs and cancer treatment, food industry, packaging, etc. Nanoparticles (NPs) can be broadly classified into two groups, namely organic (which include carbon NPs) and inorganic (include magnetic, noble metal, and semiconductor NPs). They can be also categorized into natural, incidental, and engineered NPs.

NPs are widely used to generate energy from photoelectrochemical and electrochemical water splitting. Moreover, NPs can be utilized for electrochemical of CO_2 reduction to fuels precursors, solar cells, and piezoelectric generators which convert the mechanical energy into electricity. Some other examples for the application of nanotechnology in our daily life, magnetic recording tapes, computer hard drives, protective and glare—reducing eyeglass and window coatings, catalytic converters for automobiles, metal—grinding tools, dental bonding agents, longer—lasting tennis ball, burning and wound dressing, ink, etc.

Although the science of nanotechnology is relatively new; however, nanomaterials are discovered to be used for over a thousand years from painting to making steel (Table 1.1). It is believed to be prepared by chance without the consideration of its microscopic structure. In ancient history, NPs were used by the Damascenes to create swords with exceptionally sharp edges, the Romans to craft iridescent glassware. Moreover, the use of metallic NPs seems to have started with the beginning of glass-processing in Egypt and Mesopotamia back in the 14th and 13th centuries BCE. Silver has been used because of its antimicrobial activities since ancient times: Alexander the Great refused to drink water that had not been stored in silver vessels, and Paracelcus claimed the beneficial properties of silver toward health (Silver et al., 2006). The use of silver for its antimicrobial properties decreased when penicillin was discovered and the antibiotic period started (Klasen, 2000a). Since biocide-resistant strains emerged, the interest to use silver as an antimicrobial agent is rising again (Klasen, 2000b). Silver nanoparticles (AgNPs) have antimicrobial activities toward a wide versatile of Gram-negative and Gram-positive bacteria, and antifungal and antiviral activities (Sintubin et al., 2009).

Antimony was used as an opacifier and has been used as a reducing agent that forms copper NPs from cupric ions. The red color of some glasses is attributed to the presence of CuNPs or cuprous oxide NPs. Such pieces with CuNPs have been found in this region until the sixth century BCE (Brill and Cahill, 1988). There are a number of relatively famous examples of ancient artefacts that were created using nanocomposites.

Back in ancient Egyptian times, people used nanotechnology for cosmetics and hair dyes. Nanocrystals of PbS have been applied for hair blackening (Walter et al., 2016). Egyptian blue or cuprorivaite (Figure 1.1) which is a bright blue pigment that has been used from about 5000 years ago is composed of plentiful and low-cost elements, calcium, copper, silicon, and oxygen, in the form of monolayer nanosheets of calcium copper tetrasilicate, $CaCuSi_4O_{10}$, which have intense near-IR luminescence and are acquiescent to solution processing methods. When exposed to visible light, Egyptian blue emits near-infrared rays with extraordinary strength, with even single particles of the pigment noticeable from a distance of a few yards (Kakoulli et al., 2017). This suggests Egyptian blue could have a variety of modern applications and gives modern scientists hints to develop new nanomaterials. For example, remote controls, telecommunications, and security inks can be applied in advanced biomedical imaging because near-infrared radiation penetrates through tissue better than other wavelengths. Moreover, its abundant and inexpensive components compared to other near-infrared-emitting materials that contain rare earth elements, provide economic, and environmental benefits to future applications (Johnson-McDaniel et al., 2013).

TABLE 1.1 Notable Historical Milestones in the Evolution of Nanotechnology

Year	Event
400 BC	Reasoning about atoms and matter (Democritus)
500–1500 AD	Stained glass windows in European cathedrals used nanoparticles of gold chloride and other metal oxides
800–1700	Glowing, glittering luster ceramic glazes contained silver or copper or other metallic nanoparticles
1300–1800	Damascus saber blades contained carbon nanotubes and nanowires
1830s AD	Photography; Ag-nanoparticles
1857	Metal particles affects the color of church windows (Faraday)
1905	Calculate molecular diameter (Einstein)
1908	Explanation of dependence of color of glasses on metal size and kind
1931	Electron microscope
1936	Field emission microscope allowed near-atomic-resolution images of materials
1947	Semiconductor transistor
1951	Field ion microscope
1960	Ferrofluids
1968	Atomic layer growth
1974	Nanotechnology for fabrication methods below 1 μm
1981	Nobel Prize for scanning tunneling microscope
1985	Buckminster fullerenes (Bucky balls)
1986	Engines of creation
1991	Carbon nanotubes
1998	Carbon nanotube transistor
1999	Single molecule switch
2000	Construction of quantum corrals and quantum mirrors
2001	Soldering of nanotubes with e-beam
2004	Pentium IV processor based on 90 nm technology
2006	Smart nanomaterials with DNA molecules
2009	DNA-like robotic nanoscale assembly devices
2010	Nanoscale patterns and structures as small as 15 nm

FIGURE 1.1 Egyptian blue pigment.

The Middle East steel swords of Damascus (Figure 1.2) were made between AD300 and AD1700 and are known for their impressive strength, shattering resistance, and exceptionally sharp cutting edge. The steel blades contain nanoscale-oriented Fe_3C wires and carbon tubes, like structures that improved the material's properties. The thermal treatment during the shaping could explain the catalytic formation of these carbon nanostructures from hydrocarbons (Reibold et al., 2006).

FIGURE 1.2 Damascus steel swords.

From about AD400, the Roman Lycurgus Cup (it is known also as cage-cup), dating from the fourth century CE (Figure 1.3), is made of a glass that changes color when light is excelled through it. Since the glass contains gold, silver, and copper alloyed NPs in presence of chlorine (Barber and Freestone, 1990). Transmission electron microscopy (TEM) and X-ray analyses proved 7:3 silver–gold alloy, containing about 10% copper and 0.3% antimony which would have reduced the gold and silver salts (Freestone et al., 2007), which are distributed in such a way, making the glass looks like jade with an opaque greenish yellow color in reflected light or daylight (i.e., when illuminated from outside), but when the light passes and transmitted through the cup (i.e., when illuminated from inside), it appears as a brilliant red to ruby color (Freestone et al., 2007). Brill (1965) reported that the dichroism, which has been referred to as the Lycurgus effect, is due to the metallic colloidal dispersion, in particular to a combined effect of the absorption of gold which performs the red transmission, and the scattering of silver which performs the green reflectance. Thus, knowing the properties of the Lycurgus cup has led to new, potentially useful optical properties of nanocomposites. Researchers, for example, have created thin nanocomposite films that contain gold NPs which can absorb infrared while still transmitting light. Thus, in hot countries, these films can coat windows to reflect heat away, thus allowing light through the glass, thereby reducing the need for air conditioning.

FIGURE 1.3 The Roman Lycurgus cup.

Maya blue, which is an acid and corrosion-resistant azure pigment, was first produced in AD800 and discovered in Chichen Itza, a pre-Colombian Mayan city (Figure 1.4). This is a complex substance that contains the needle-shaped Palygorskite claycrystallites ($Mg_5(Si,Al)_8O_{20}(OH)_2$. $8H_2O$) with nanopores and superlattice nanostructure into which indigo dye molecules were combined chemically to create that environmentally stable pigment (José-Yacamán et al., 1996). The presence of some other metallic and metallic oxides NPs of iron, chromium, manganese, titanium or vanadium that would have come from the indigo colorant coming from a plant and Palygorskite clay, explained the extreme resistibility of that dye. The discovery of Maya blue helps with the exploration of a variety of nanoporous materials in which insertion and stabilization of organic dyes can be successfully achieved (Leitão and Seixas de Melo, 2013).

FIGURE 1.4 Maya blue pigment.

Pottery from all over the Mediterranean Renaissance during the period of 1450–1600AD (Figure 1.5) was often decorated with a shimmering metallic coating called lustre (metallic glazes). The color and shine gold and red colors are due to the incorporation of NPs of silver and copper (between 5 and 100 billionth of a meter) in the glazes, respectively, which act as arrays reflecting and diffracting the light (Padovani et al., 2003; Colomban, 2009).

FIGURE 1.5 Pottery from across the Renaissance Mediterranean world.

During the ninth century, metallic luster glazed ceramic decorations emerged in Mesopotamia. The Italian Deruta ceramicists are an example. Moreover, more than a thousand years ago, the Chinese were known to use gold NPs as an "inorganic dye" to create a red color in their ceramic porcelains (Hunt, 1976). Thus, the ancient lustreware offers a new procedure for generating metal–glass nanocomposites. Nowadays, glass factories use gold NPs as a red colorant, for example, the French crystal glass-making Baccarat (Louis, 2012) and the famous glass-makers from Murano (Moret, 2006). Example in Figure 1.6a and b displays iridescence under specular reflection with shiny blue and green colors. Analyzing the TEM showed a double layer of silver NPs in the outer layer with smaller sizes (510 nm) and larger ones (520 nm) in the inner layer. The distance between the two layers is constant at approximately 430 nm which results in interference effects (Figure 1.6c). The light dispersed by the second layer has a phase shift as opposed to the one dispersed by the first layer, and because the phase shift depends on the wavelength of the incoming light, each wavelength is dispersed differently.

While, in the medieval stained glass (Figure 1.7), NPs of gold and silver were stuck in the matrix of glass, generating the ruby red and deep yellow color, respectively. It is also called the "Ruby glass" and it is still used today in order to make cadmium-free intensely red glass (Delgado et al., 2011).

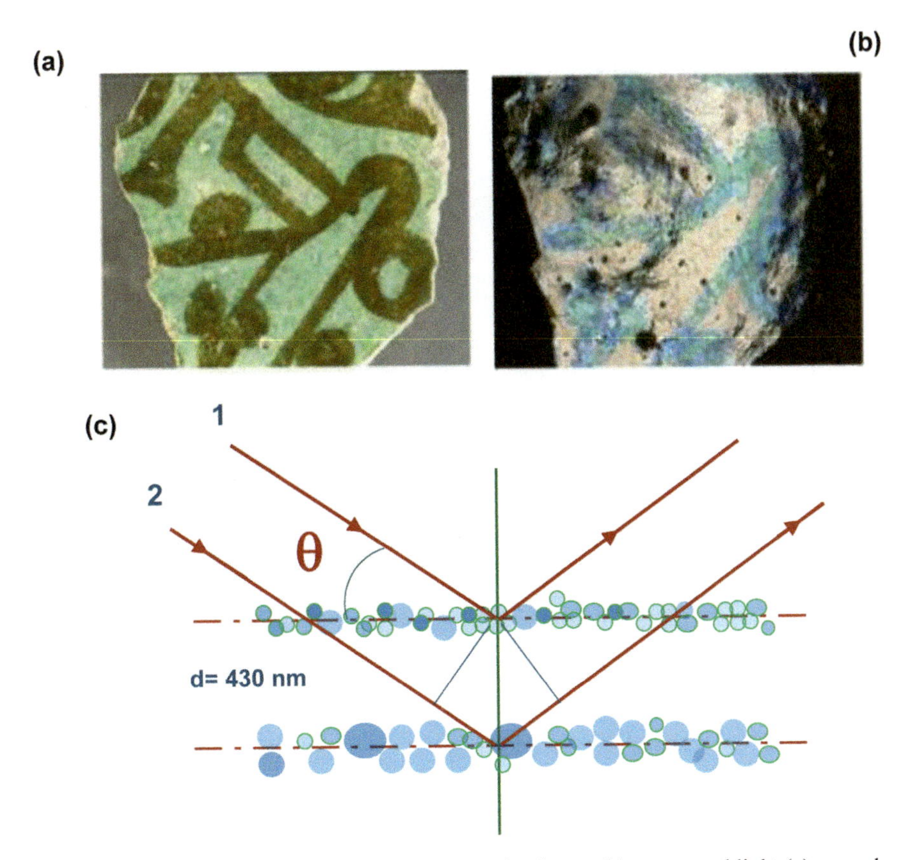

FIGURE 1.6 A medieval piece of a glazed ceramic observed by scattered light (a), specular reflection (b), and schematic diagram for interference phenomena due to the double layer (c).

During the ninth century CE in China, a very fine clay mineral named kaolin has also been used as raw material for the production of porcelains having a very fine thickness (<0.4 mm) (Rytwo, 2008).

However, nowadays, it has been discovered that the metal NPs size whatever gold or silver led to the variations in color (Table 1.2), which would be related to the dramatic change in material properties at the nanoscale.

The first synthesis of gold nanoparticles (AuNPs) was reported by Michael Faraday (Figure 1.8a), who was fascinated by the ruby color of colloidal gold and was the first to discover that the optical properties of gold colloids differ from those of the corresponding bulk metal (Faraday, 1857). That was probably the first reported observation of the effects of quantum size. Later, the specific colors of metal colloids were explained by Mie (1908).

FIGURE 1.7 Medieval stained glass.

TABLE 1.2 The Different Colors of Nanoparticles According to Sizes and Shapes

Metal Nanoparticles	Particles Size and Shape in Glass	Color Reflected
Gold	25 nm and spherical	Red
Gold	50 nm and spherical	Green
Gold	100 nm and spherical	Orange
Silver	100 nm and spherical	Yellow
Silver	40 nm and spherical	Blue
Silver	100 nm and prism	Red

The concept of a "nanometer" was first proposed in 1914 by Richard Zsigmondy (Figure 1.8b), the 1925 Nobel Prize Laureate in chemistry. He characterized the particle size in terms of nanometer and he was the first to use the microscope for measuring particle size, for example, gold colloids.

History and origin of modern nanoscience began by the father of modern nanotechnology Richard Feynman (Figure 1.8c), the 1965 Nobel Prize Laureate in physics. Within his historic lecture in 1959, entitled "there is plenty of room at the bottom" was presented during the American Physical Society meeting at the California Institute of Technology (Caltech). In which he outlined the idea of building objects from the bottom-up, and introduced the concept of manipulating matter at the atomic level (Feynman 1960; Schaming and Remita, 2015).

However, the word nanotechnology was firstly introduced by the Japanese scientist, Norio Taniguchi (Figure 1.8d) at Tokyo Science University. He was the first to use "nanotechnology" to define the processes of semiconductors that took place in a nanometer order. He supported the idea that nanotechnology consisted of the processing, separation, consolidation, and deformation of materials by one atom or one molecule (Taniguchi, 1974; Schaming and Remita, 2015). This brilliant suggestion did not gain much attraction until the mid-1980s when Eric Drexler (Figure 1.8e) published his book entitled "Engines of creation: The Coming Era of Nanotechnology" in 1986 which aided in understanding the potentiality of nanotechnology and proposed the idea of a nanoscale "assembler" which could create a copy of itself and other arbitrarily complex objects. Drexler's vision of nanotechnology is often called "molecular nanotechnology." Moreover, the carbon fullerenes were discovered in the 1980s by Kroto, Smalley, and Curl, who shared the Nobel Prize in Chemistry. Further, the science of nanotechnology was advanced when the Japanese physicist Sumio Iijima (Figure 1.8f) developed the carbon nanotubes (CNTs) (Morais et al., 2014; Khandel and Vishwavidyalaya, 2016).

Nearly all the industrialized nations developed nanotechnology programs in the late 1990s and early 2000s, leading to a global explosion of nanotechnology activities. Among the initiatives in Germany is the Nano-Initiative of the German Government, which includes (1) NanoMobil for the automobile industry, NanoLux for the optical industry, (2) NanoFab for the electronics industry, (3) nano for life science industries, and (4) nano in production for the production of nanomaterials production. Ministry of Education, Culture, Sports, Science, and Technology and Ministry of Economy, Trade, and Industry have led the activities in Japan. Among their other initiatives is the creation of the Nanotechnology Researchers Network, which promotes nanotechnology research by universities and private companies by making sophisticated and large-scale equipment owned by public institutions and other universities, such as high-voltage electron microscopes and nanofabrication facilities accessible to general researchers.

Nanotechnology contributes significantly to the growth of modern industries such as the electronics, biomedicine, materials, manufacturing, and energy industries. Nanotechnology has become one of the most

FIGURE 1.8 The six champions of nanotechnology.

important technologies that have tremendous applications. For example, in medical and pharmaceutical fields, the structure of deoxyribonucleic acid (DNA) can be considered nanoscience, diagnostic biosensors, drug delivery systems, and imaging probes. The use of nanomaterials has increased dramatically in the food and cosmetics industries for improvements in processing, packaging, shelf life, and bioavailability. Zinc oxide quantum dot NPs also show antimicrobial activity against foodborne bacteria, and NPs are now used as food sensors for food quality and health detection (Mo et al., 2014; Zhang et al., 2014; Dai et al., 2015; Han et al., 2015). In addition, the developments made by nanotechnology have now spread toward many areas in the oil and gas industry, including exploration and increased oil recovery, drilling, processing, refining, and petrochemical industries. But it is still rarely been applied in real fields, especially in upstream exploration and manufacturing. Nevertheless, there is a widespread belief that nanotechnology may be exploited to develop novel nanomaterials with enhanced performance to combat the technological barriers of applying nanotechnology in the petroleum industry. Rising governmental and global oil industry funding tools have been allocated for exploration, drilling, production, refining, and wastewater treatment. Nanosensors, for example, allow for accurate measurement of reservoir conditions. Nanofluids prepared using functional nanomaterials will show better performance in processes of oil production, and nanocatalysts have increased efficiency in oil refining and petrochemical processes. Nanomembranes enhance the separation of oil, water, and gas, purification of oil and gas, and elimination of impurities from wastewater. Functional nanomaterials can play a significant role in manufacturing smart, efficient, and more robust equipment (He et al., 2016; Peng et al., 2018).

In addition, nanotechnology has played an important role in studying the mineral composition, microporous structure, and rock physical properties of unconventional reservoirs; the reservoir nanosensor is still in the laboratory process, and the reservoir nanorobots still need a crucial advancement in technology; numerical simulation technology for oil and gas migration in micro-nanoporous media is expected to become an important means of unconventional oil and gas migration mechanism, distribution pattern and resource evaluation. Nanomaterials have size and surface effect, wetting characteristics, particle migration inhibition characteristics, shear thickening behavior, photocatalytic properties of the nanometer, nanofiltration, and wear resistance, and nanocorrosion. Those promote the broad applications of nanomaterials in the fields of enhanced oil recovery (EOR), water treatment, engineering, and anticorrosion (Liu et al., 2016; Sun et al., 2017; Etim et al., 2018; Fakoya et al., 2018).

1.4 SOME IMPORTANT DEFINITIONS

1.4.1 NANOSCIENCE VERSUS NANOTECHNOLOGY

Nanoscience is the analysis of phenomena and material modification at atomic, molecular, and macromolecular scales, where properties vary significantly from those at a wider scale (McNaught and Wilkinson, 1997; Alemán et al., 2007; ASTM E2456). Nanotechnologies are the design, characterization, manufacture and application of structures, devices and systems by manipulating shape and size at nanometre scale (Fakruddin et al., 2012). Thus, the focus is to generate NPs with significantly novel and improved physical, chemical, and/or biological properties and functionality. Consequently, nanotechnology is science and technology at the molecular level and is further defined as the understanding and regulating matter at dimensions from 1 to 100 nm (1 nm = 1×10^{-9} m = 3.2808×10^{-9} feet = 3.9370×10^{-8} inches). Consequently, nanoscience and nanotechnology involve the ability to see and control individual atoms and molecules.

Although the science of nanotechnology is relatively new, nanomaterials are nevertheless discovered to be used for more than a thousand years, from painting to steel production. Nanotechnology also covers chemistry, biology, several fields of engineering, material science, and medicine, in terms of scientific fields. There was a tendency when the terms nanoscience and nanotechnology were first used to view them as a distinct academic discipline or as a new industry. However, to be correct, nanoscience and nanotechnology are interrelated with other disciplines.

Researchers have known for centuries that properties of materials, such as hardness, electrical conductivity, elasticity, and adhesion, are (or are likely to be) reliant on the atomic or crystalline structure of these materials. Nanotechnology is an aid in understanding the mechanisms behind many previously explored phenomena in the fields of chemistry, physics, and biology. For instance, chemists have been producing solutions, suspensions, and colloids for hundreds of years but it was difficult, if not impossible, to analyze the particles or mixtures they produced at the molecular level. With the developments in various technologies, performance forecasts can be checked and the processes of chemical interaction better understood.

On a historical basis, modern nanoscience and nanotechnology are very recent, but nanoscale materials have been in use for decades. In Damascus, as previously mentioned, NPs were used to build swords with extraordinarily sharp edges, the Romans design iridescent glassware. In addition, the use of metallic NPs seems to have started with the beginning of glass-making in Egypt and 14th century Mesopotamia. In the stained glass windows of medieval churches, colors were formed by alternative-sized gold and silver particles. Antimony was used as a pacifier and has been used as a reducing agent. The copper NPs or cuprous oxide NPs were also used in ancient artefacts.

1.4.2 NANOMETER SCALE

The nanoscopic scale (or nanoscale) typically refers to nanotechnology-applicable structures with a length scale usually referred to as 1–100 nm. The nanometer (International spelling used by the International Bureau of Weights and Measures; SI symbol: nm) or nanometer (American spelling) is a metric scale unit of length equal to one billionth of a meter (0.000000001 m). The name combines the nano (from Ancient Greek, nanos, "dwarf") SI prefix with the term meter of the parent unit (from Greek, metrn, "measurement unit"). It can be written in scientific notation as 1×10^{-9} m, in engineering notation as 1 E−9 m, and which is just 1/1,000,000,000 m. One nanometer is tens of ångströms. In most solids the nanoscopic scale is (roughly) a lower bound to the mesoscopic scale. One nanometer is approximately the length of 10 hydrogens or 5 Si atoms arranged in a line.

1.4.3 NANOMATERIALS

Nanomaterial is classified as a material having at least one nanoscale; external dimension or internal structure or surface structure (Schaming and Remita, 2015). Nanomaterials can be categorized depending on the overall shape (i.e., the macroscopic dimensions of the material) to all three dimensions that are in the nanoscale (NPs, quantum dots (<10 nm), nanoshells, nanorings, and microcapsules), two dimensions are in the nanoscale (nanotube fibers and nanowires), and one dimension is in the nanoscale (thin films, layers, and coatings) (Tiwari et al., 2012). The term NP was described by the American Society for Testing and Materials (ASTM) in 2006 as particles with at least two or more dimensions ranging from 1 to 100 nm. In addition, the British Standards Institution defined NPs as particles with one or more nanoscale dimensions (with order dimensions of 100 nm or less).

The other more applicable classification is depending on the dimensionality of the nanoscale component with which the material is made. Thus, nanomaterials can be classified according to the dimension of the nanostructure used to make it: zero-dimensional (0D), one-dimensional (1D), two-dimensional (2D), and three-dimensional (3D). A nanostructure is said to have one dimension if it has a length larger than 100 nm in one direction only. Consequently, an NP is considered to be 0D, as it has no dimension with a length larger than 100 nm. Fullerenes are the example for the 0D nanostructures. Nanowire or a fiber is a 1D, a thin film is a 2D nanostructure, whereas 3D nanomaterials are nanophase materials consisting of equiaxed nanometer-sized grains (Figure 1.9).

Organic NPs, such as DNA, proteins, and virus, as well as inorganic NPs, such as iron oxyhydroxides, aluminosilicates, and metals, that can be produced during weathering, volcanic eruptions, forest fires, incinerators, or from fuel combustion, are known as nonintentionally produced nanomaterials.

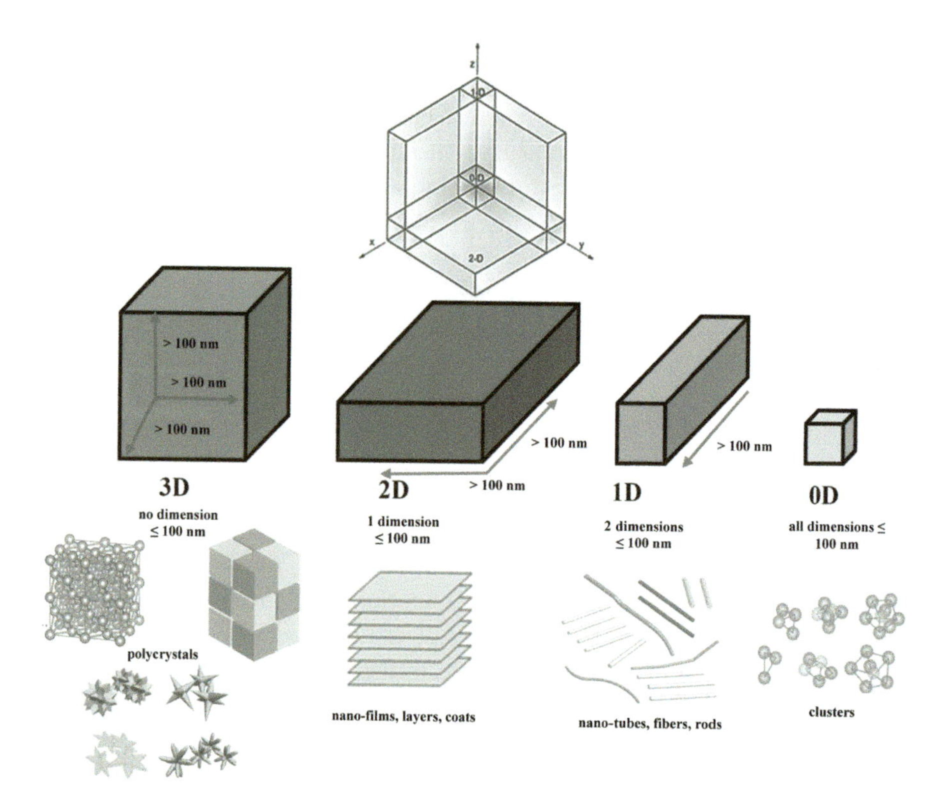

FIGURE 1.9 Classification of materials according to the nanostructural dimensions.

Nevertheless, nanotechnology does not generally include the nonintentionally produced nanomaterials. Thus, nanomaterials are bigger than single atoms but smaller than cells and bacteria (Table 1.3). Nanomaterials can be made up of different elements, such as metals (e.g., gold or silver NPs), metal oxides (e.g., zinc oxide or titanium dioxide NPs), semiconductors (e.g., Si), or carbon (e.g., CNTs). Bulk nanomaterials are therefore larger artifacts made of structures that have well-identified domains with an average size of <100 nm.

According to the composition, the nanomaterials can be classified into;

Carbon-based nanomaterials consisted mainly of carbon. That includes fullerenes, CNTs, graphene, etc. Until the late twentieth century, only two well-defined allotropes of carbon were discovered; diamond, which composed of a 3D crystalline array of carbon atoms, and graphite that composed of stacked sheets of 2D hexagonal arrays of carbon atoms. Then came, the fullerenes which contain nanomaterial that is made of the hollow globular cage, like allotropic, sphere, ellipsoid, tube, and many other shapes and forms of carbon. Spherical ones are known as buckyballs (named after R. Buckminster Fuller) and were discovered in 1985 and consist of 60 carbon atoms. A fullerene is a carbon allotrop in the shape of a hollow sphere, an ellipsoid, a line, and several other shapes, as well as a number of novel solubility properties (Beck and Mándi, G. 1997; Arikawa, 2006; Semenov et al., 2010).

TABLE 1.3 Examples to Figure Out Nanomaterials

Example	Approximate Size	Notes
The head of a pin	1,000,000 nm across	You can see these with your eyes unaided
The page of a book	100,000 nm thick	
Fingernail	Grows at a rate of 1 nm/s That means they grow 86,400 nanometres in a day, but that is still too small for you to notice a difference	
A human hair	40,000–80,000 nm diameter	
A red blood cell	7000 nm across	You can see these using a light microscope
Cancer cell	10,000 nm	
Bacteria	1000–10,000 nm	
A single particle of smoke	1000 nm	
Virus	100 nm	You need an electron microscope or other device to see these
DNA molecule	1–2 nm wide	
Antibody	10 nm	
Most atoms	0.1–0.2 nm	
10 hydrogen atoms side by side	1 nm long	
Glucose molecule	1 nm	
Transistor on latest computer chips	100 nm	

Buckyball is simply the common name for a family of totally carbon-composed molecules (in the form of a hollow sphere, ellipsoid, tube, or plane) forming a spherical shape. Allotropes are different structural forms of the same elements by way of description and explanation and can exhibit very different physical properties and chemical behavior. The allotropic forms change is triggered by the same forces which affect other structures, such as pressure, temperature, and light. Therefore, the stability of an allotrope depends on particular conditions. As an example of allotropes having different chemical behavior are (1) ozone (O_3) which is a much stronger oxidizing agent than (2) oxygen, sometimes referred to for clarification as dioxygen (O_2). The first fullerene molecule, that is, buckminsterfullerene (C_{60}) was discovered by Richard Smalley, Robert Curl, James Heath, Sean O'Brien, and Harold Kroto at Rice University in 1985, and won the Nobel Prize in Chemistry for its discovery in 1996. It has the structural shape of a soccer ball, a truncated icosahedron with 32 faces, 20 hexagonal, and 12 pentagonal rings, where each carbon is sp^2 hybridized. All the rings are fused and all double bonds are conjugated together. Given their intense conjugation, they behave chemically and physically as electron-deficient alkenes, rather than as aromatic systems rich in electrons (Bakry et al., 2007).

Figure 1.10 signifies the distinguished fullerenes consisting of C_{60} and C_{70} with the diameter of 7.114 and 7.648 nm, respectively, and other fullerenes consisting of C_{76} and C_{84}. Carbon "onions" were also found and consisting of carbon cages like Russian dolls inside each other (Figure 1.10e). Such carbon particles have millions of atoms and thousands of condensed shells. Carbon cages consisting not of pentagonal and hexagonal rings (like C60), but heptagonal (7 membered) rings and hypothesic structures were postulated. Fullerenes have different applications from electronics, sensors, renewable energy to medicine, due to their electrical conductivity, high strength, structure, electron affinity, and versatility. The cylindrical fullerenes are also called CNTs (buckytubes) of many thousands of times long as they are wide, have the same icosahedral structure as the fullerenes (Figure 1.10f), and were identified in 1991 by Iijima Sumio of Japan. Thus, a CNT can be as thin as a few nanometers (small as 2 nm in diameter) and yet be as long as hundreds of microns.

CNTs resemble a graphite sheet rolling upon itself. According to their diameter telicity and the helical orientation of the rows of hexagonal rings in the walls of the tubes, they can be classified as metallic or semiconducting. According to the number of rolled sheets, they can be classified into single-walled nanotubes (SWNTs), double-walled nanotubes, or multiwalled carbon nanotubes (MWCNTs) as many as 100 tubes (walls) or more, respectively (Figure 1.11). Applied quantum chemistry, specifically, orbital hybridization best explains chemical bonding in nanotubes. The chemical bonding of nanotubes includes carbon atoms that are fully sp^2-hybrid. Such bonds, close to those of graphite and stronger than those found in alkanes and diamonds (which employ sp^3-hybrid carbon atoms), give their special strength to nanotubes. CNTs are at least 100 times stronger than steel, but just one-sixth as heavy, so almost any material could be reinforced by nanotube fibers.

Such impressive structures have a fascinating array of electrical, magnetic, and mechanical properties. Researchers estimate that Young's MWNT modulus (a measure of their elasticity, or ability to rebound from stretching or compression) is greater than that of carbon fibers by a factor of 5–10. MWNTs are capable of absorbing loads readily through a series of reversible elastic deformations, such as buckling or kinking, in which the carbon atom bonds remain intact. Nanotubes are much better at generating heat and energy than copper. CNTs are already used in polymers for conductivity control or enhancement and are applied to antistatic packaging. Moreover, due to their unique physical, chemical, and mechanical characteristics, they can be used in the preparation of nanocomposites for many commercial applications such as fillers, adsorbents for environmental remediation, and as means of support for various inorganic and organic catalysts.

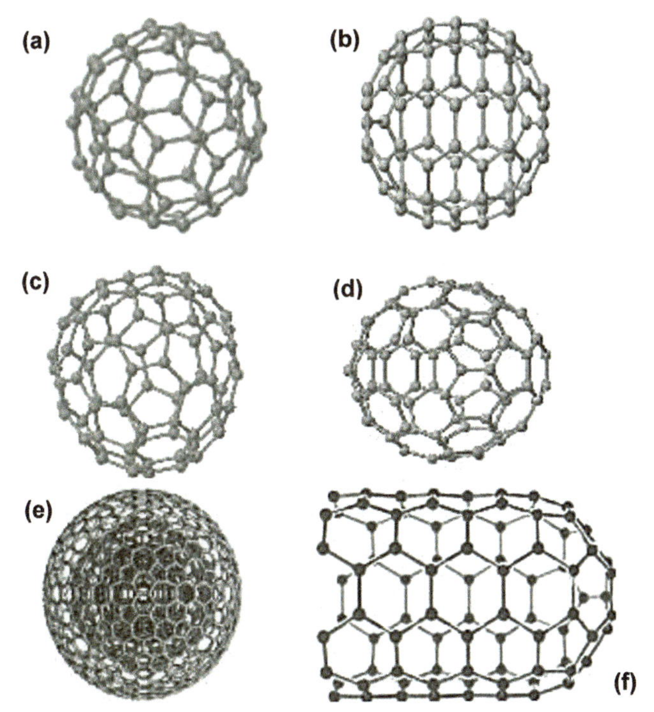

FIGURE 1.10 Different forms of fullerenes/buck balls (a) C_{60}, (b) C_{70}, (c) C_{76} and (d) C_{84}, and buckytubes (e).

Oxidation would decappe nanotubes and the resulting open tubes filled with metals such as lead or even buckyballs. Atoms of boron and nitrogen may be embedded in CNT walls. Therefore, by encapsulation in nanotube skins, microscopic metal particles that would otherwise be quickly oxidized can be stabilized in air. CNTs are also highly electrically conductive, which could potentially make metal wires extremely cost-effective to replace. CNTs exhibit faster phononon transport than diamond, previously recognized as the best thermal conductor, and nanotubes' electric current-carrying capacity is about four orders of magnitude greater than copper. CNTs' semiconductive properties make them candidates for next-generation computer chips. CNTs are peculiar in that they are lengthwise thermally conductive, but not through the tube itself. Thus, CNTs play a part in thermal insulation on both sides. To generate tiny insulated electrical wire, conducting CNTs can be coated with sheaths of metal sulfides.

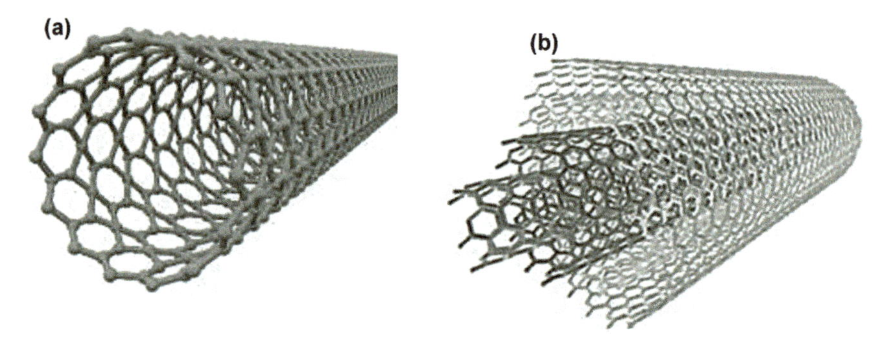

FIGURE 1.11 SWNTs are just like a regular straw (a), MWNTs are series of nesting tubes of ever growing diameters (b).

One of the most important nanoproducts in the world is carbon black, an amorphous powder of NPs ranging from 20 to 50 nm in thickness. More than 90% of this is used to strengthen rubber, primarily car tires, and is created by the incomplete combustion of heavy aromatic oil or gas. In parallel, studies of semiconductor nanocrystals led to the creation of quantum dots if the quantization of electronic energy levels occurs small enough (typically below 10 nm). A quantum dot is a semiconductor in which the propagation of electrons is limited in three dimensions (different from the quantum wires in which 2D propagation is controlled and the quantum wells in which propagation is controlled in one direction). A quantum dot's properties slip between the properties of bulk semiconductors and discrete molecules. These nanoscale particles are also used as drug carriers or imaging agents in biomedical applications and have the ability to be used in fluid dynamic applications. For example, semiconductor NPs (quantum dots) are lead sulfides with complete passivation by oleic acid, oleyl amine, and hydroxyl ligands (size ~5 nm). Additionally, semisolid and soft NPs were made. The liposome is a semisolid prototype nanoproduct of nature. Various types of liposome NPs are widely commonly used as anticancer drug delivery systems and vaccines.

For stabilizing emulsions, NPs with a half hydrophilic character and the other half hydrophobic character (Janus particles, amphoteric particles) are especially useful. Particle forms may be self-assembled at interfaces between water and oil, which act as solid surfactants. Internally, hydrogel NPs made from the core–shell of N-isopropylacrylamide hydrogel can be colored with affinity baits. These baits of affinity allow the NPs to isolate and extract unwanted proteins while improving the target analytes.

Metal-based nanomaterials are made of alkali and noble metallic NPs like copper, gold, silver, metal oxides, etc. There are mono- and bi-metallic NPs, which are composed of one and two different metals, respectively. The properties of these NPs are determined by the constituent metals and their nanometric size. In addition, in today's cutting-edge materials the facet, size, and shape-controlled synthesis of metal NPs are relevant. The metal NPs are characterized by localized surface plasmon resonance, thus, possess unique optoelectrical properties. Metal NPs have different applications, for example, for obtaining a high-quality scanning electron microscope (SEM images); AuNPs coating is used for sample preparation to enhance the electronic stream. Titanium dioxide (TiO_2) NPs are widely used in applications, like paint, sunscreen, and toothpaste. Moreover, copper, gold, and silver have a broad absorption band in the visible zone of the electromagnetic solar spectrum. ZnO NPs same as TiO_2 NPs can be used for sunscreen as they are colorless, transparent, and better reflect/scatter ultraviolet light than bigger particles. Thus, protect more human skin against UV-induced damage. Amorphous silica NPs are used in the food industry, as an anticaking mediator to retain the flow properties in powder foodstuffs (e.g., instant soups) and to condense pastes (Heiligtag and Niederberger, 2013). El-Gendy et al. (2015a, b) used acid TiO_2 hydrosol and TiO_2 NPs for photodegradation of Basic Blue-41 and Reactive Red-84, respectively. El-Gendy et al. (2016) used ZnO NPs for photodegradation of chlorophenols.

Dendrimers, which are nanosized polymers, are built from branched units (i.e., monomers are assembled into a tree structure around a central core, Figure 1.12). Dendrimer originates from the Greek word dendra that means a tree. It is a highly branched 3D structure that offers a great functionality and flexibility of the surface. They can work on the surface, and they can contain molecules in their cavities. Dendrimers are specifically implemented for drug delivery, diagnosis, and tumor therapy.

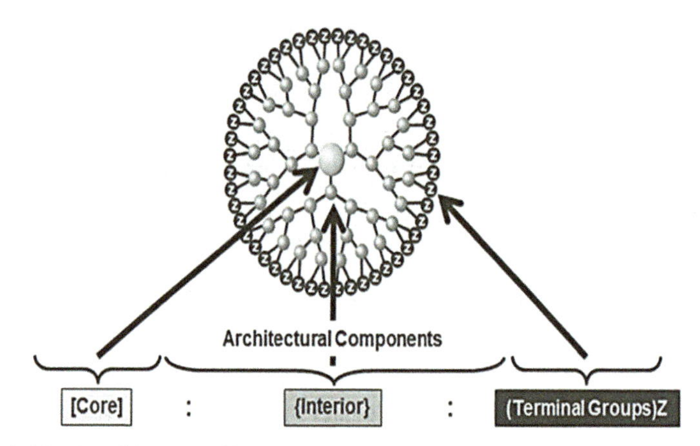

FIGURE 1.12 Dendrimers architecture.

Nanocomposite contains a mixture of simple NPs or compounds, such as nanosized clays within a bulk material. The NPs give better physical, mechanical, and/or chemical properties to the initial bulk material. A nanocomposite is a substance consisting of many phases in which one, two, or three dimensions of at least one phase is <100 nm. Materials made from nanocomposite can be multilayer structures. There are several ways to make multilayers, such as by gas-phase deposition, or from self-assembly of monolayers. In some cases, spinodal decomposition can be used to obtain multilayered structures in the bulk of mixed oxides. Magnetic multilayered materials used in storage media are an example of multilayer nanocomposites. Polymer–clay nanocomposites can improve significantly the properties of an initial polymer at low cost and have many successful applications. Moreover, the size of the nanostructures incorporated into the nanocomposite offers enormous applications. Nanostructure of <5 nm has some catalytic applications. Nanocomposites with nanostructures of <50 nm or 100 nm show refractive index changes and some mechanical strengthening or superparamagnetism, respectively. Nanocomposite itself can be classified into organic, inorganic, or inorganic/organic. It can be also classified into ceramic–matrix, metal–matrix, or polymer–matrix nanocomposites. The ceramic–matrix nanocomposites (CMNC) are materials where the ceramic (oxides, nitrides, silicides, etc.) is the main component followed by metals.

CMNC has better optical, electrical, magnetic, and/or corrosion resistance properties relative to the traditional materials. Metal-matrix nanocomposites (MMNC), for example, are the reinforced metal-matrix composites with CNTs. Another example is the highly energetic superthermite or nanothermite materials that contain an oxidizer and a reducing agent. The explosive reaction proceeds much faster than that in the case of thermite materials prepared from microparticles. This kind of material has military applications as explosives or propellants and in pyrotechnics. Researchers have created a method for producing anodes from a composite made with Si nanospheres and carbon NPs for the lithium-ion batteries. The anodes that are made of the Si–carbon nanocomposite make closer contact with the lithium electrolyte, allowing for quicker charging or power discharge. Thus, produces batteries with greater power output. For the development of structural components with a high strength to weight ratio, nanocomposite may be used. For instance, an epoxy that contains CNTs can be used to produce nanotube–polymer composite windmill blades. These blades are strong but lightweight. Thus longer windmill blades become practical and would increase the amount of electricity generated by each windmill. With a similar weight of CNTs, it results in stronger/stiffer parts when used with graphene than epoxy composites. As, graphene bonds are better for the epoxy polymers, allowing for a more effective coupling of the graphene into the composite structure. This property could result in the manufacture of components with higher strength-to-weight ratios for applications such as windmill blades or aircraft components and render nanocomposite lightweight sensors. A nanocomposite polymer–nanotube conducts electricity, how well it conducts depends on the spacing of the nanotubes. This property can be extended to the development of polymer–nanotube nanocomposite patches that serve as stress sensors on windmill blades.

The nanocomposite will bend when strong wind gusts bend the blades too. Bending changes the electrical conductance of the nanocomposite sensor which causes an alarm to be sounded. This alarm will allow the shutdown of the windmill before unnecessary damage occurs. For making a conductive paper, a nanocomposite of cellulosic materials and CNTs may be used. When an electrolyte soaks up this conductive material, a flexible battery is formed. Investigators are trying to integrate magnetic NPs and fluorescent NPs into a particulate nanocomposite. The magnetic properties of the particle in nanocomposite make the tumor more noticeable during a preoperative magnetic resonance imaging (MRI) treatment. While the nanocomposite particle's fluorescent property might help the surgeon better see the tumor while operating.

1.4.4 NANOPARTICLES

NPs which are atomic or molecular aggregates with a dimension between 1 and 100 nm and are viewed as the fundamental building blocks of nanotechnology (Yadav et al., 2017). They exhibit new or enhanced properties based on unique features such as scale, distribution, and morphology. The large surface-to-volume ratio of NPs compared to their bulk materials makes them attractive candidates for many applications (Schröfel et al., 2014; Palomo and Filice, 2016). NPs are not simple molecules themselves, they are made up of a surface layer that can be worked with a number of small molecules, metal ions, surfactants, and polymers. A shell layer is a chemically different material from the core in all aspects. And finally, the core is essentially the central portion of the NP and usually refers to the NP itself (Shin et al., 2016). The nanosized materials (e.g., metals NPs) have physical, chemical, electronic, electrical, mechanical, magnetic, thermal, dielectric, optical, and biological properties that differ from those in the bulk form (Da Costa et al. 2014; Shi et al., 2015; Zhou et al., 2015). For example bulk silver (Ag) is nontoxic, but AgNPs have antimicrobial activity and are capable of killing viruses. Besides, the same metal, in its nanosized level, can be transformed to be a semiconductor or an insulator. When all particle dimensions are within the range of nanometers, it is known as iso-dimensional NPs. The size of NPs can influence the physiochemical properties of a substance, for example, the optical properties. Dreaden et al. (2012) reported the changes in colors and properties of AuNPs with the variations of its size and shape (Table 1.4), which can be used for applications of bioimaging. The alteration in aspect ratio, nanoshells thickness, or percentage of Au concentration would affect the absorption properties of the NPs, consequently, different absorption colors are observed.

TABLE 1.4 Color Dependence of AuNPs on Size and Shape

Metal Nanoparticles	Particles Size and Shape in Glass	Change in Color With Different Parameters
Gold	50 nm nanorods	Aspect ratio From blue to red
Silica–gold core–shell nanoparticles	140 nm nanoshells	Shell thickness From pale purple to dark orange
Gold	50 nm and nanocages	Percentage of AuNPs From yellow to pale blue

Moreover, NPs have distinct mechanical properties different from those of microparticles and their bulk materials which make them have different applications in tribology, surface engineering, nanofabrication, and nano-manufacturing. Not only are these but the thermal conductivities of metal NPs also stronger than those of solid formed fluids. For example, at room temperature, the thermal conductivity of copper is about 700 times greater than that of water, and about 3000 times greater than that of engine oil. The NPs of alumina (Al_2O_3) have a thermal conductivity higher than that of water. Thus, the nanofluids that are prepared by dispersing the broad total surface area nanometric scales solid particles into liquids such as water, ethylene glycol, or oils have improved thermal conductivity than traditional heat transfer fluids. As heat transfer occurs at the particle surface. Besides, the larger the total surface area of the NPs is the higher the stability of the suspension is.

However, from the theoretical point of view, NPs are frequently called nanoclusters or simply clusters known as the merger of millions of atoms or molecules (Ferrando et al., 2008). Such atoms or molecules can be of the same or different type. For example, ceramic NPs are nonmetallic inorganic solids, synthesized through heat, and successive cooling. In fact, they can be amorphous or polycrystalline and their surfaces can serve as a carrier for liquid droplets or gases. They can be also found in dense, porous, or hollow forms. Therefore, ceramics NPs have different applications, such as catalysis, photocatalysis, photodegradation of dyes, and imaging applications.

Semiconductor NPs which have properties between metals and nonmetals possess wide band-gaps consequently, express significant changes in their properties with bandgap alteration. Semiconductor NP materials, therefore, have applications in photocatalysis, photo optics, and electronic devices, as well as in water splitting applications due to their suitable bandgap and bandedge positions (Hisatomi et al., 2014). Polymeric NPs (PNPs) are mostly functionalized nanospheres or nanocapsular shaped organic-based NPs (Mansha et al., 2017). The former are matrix particles whose overall mass is generally solid and the other molecules are adsorbed at the spherical surface's outer boundary. In the latter case, the solid mass is completely encapsulated inside the particle (Rao and Geckeler, 2011). Ahmed et al. (2016) used nano-polyaniline for the removal of anionic sulphonated Acid Red 14 dye from water. Polyethylene oxide and polylactic acid NPs are applied for the intravenous administration of the drug. PNPs are more advantageous than liposomes, as potential drug carrier, especially the biodegradable PNPs. The magnetic

NPs with size <10–20 nm have irregular electronic distribution and can be used for heterogeneous and homogeneous catalysis, biomedicine, magnetic fluids, MRI data storage, and environmental remediation, such as water decontamination. Different techniques are applied for their preparation, such as solvothermal, coprecipitation, microemulsion, thermal decomposition, and flame spray synthesis. Iron oxide particles such as magnetite (Fe_3O_4) or its oxidized form maghemite (Fe_2O_3) are the most commonly employed NPs for biomedical applications. Superparamagnetic iron oxide NPs with sufficient surface chemistry is widely used in various vivo applications such as MRI contrast enhancement, tissue repair and immunoassay, biological fluid detoxification hyperthermia, drug delivery, and cell separation. Furthermore, they can be used as an effective sorbent for toxic soft materials and heavy metals such as mercury, lead, thallium, cadmium, and arsenic from natural water. Ferrofluids are ultrastable suspensions of tiny magnetic NPs in a liquid, combining the usual liquid activity with superparamagnetic properties. In the presence of a magnetic field, the NPs' magnetic moment attempts to align with the direction of the magnetic field which leads to a macroscopic magnetization of the liquid. The regulation of a ferrofluid's flow properties and its location within a technological device leads to countless applications in devices such as dynamic seals and dampers. Moreover, ferrofluids can be used as heat transfer materials in loudspeakers (Heiligtag and Niederberger, 2013).

In general, NPs are named after the real-world forms they might represent (Benelmekki, 2015; Makhlouf and Barhoum, 2018). As examples, the names nanospheres, nanochains, nanoreefs, nanoboxes have been used. These particle morphologies may arise spontaneously as an effect of a templating or directing agent present in the synthesis, such as micellar emulsions or anodized alumina pores, or from the materials' own innate crystallographic growth patterns. Many of these morphologies, such as long CNTs used to bridge an electrical junction, can serve a function. Amorphous particles typically take on a spherical form (due to their microstructural isotropy), whereas anisotropic microcrystalline whisker shape corresponds to their specific crystal habit. NPs are sometimes referred to as clusters, at the small end of the size scale. Spheres, rods, fibers, and cups are examples of the mostly developed shapes.

The chemical processing and synthesis of high-performance technological components for the private, industrial, and military sectors require the use of high-purity ceramics, polymers, and composite materials, nitrides, nonmetals (graphite, CNTs), and layered materials. In condensed bodies formed from fine powders, the irregular particle sizes and shapes in a typical powder often result in nonuniform packing morphologies resulting in variations in the packing density in the compact powder. Uncontrolled powder agglomeration due to van der Waals forces may also lead to heterogeneity of the microstructures. Differential stresses resulting from nonuniform drying shrinkage are directly related to the rate at which the solvent can be extracted, and therefore highly dependent on the porosity distribution. These stresses have been associated with condensed bodies with a plastic-to-brittle transformation and may result in the spread of cracks in unfired materials. Furthermore, any variations in the material packing density as it is prepared for the kiln are often exacerbated during the sintering process, resulting in inhomogeneous densification. It has been shown that some pores and other structural defects associated with density variations play a detrimental role in the sintering process by increasing and thereby limiting end-point densities. It has also been shown that differential stresses resulting from inhomogeneous densification result in the propagation of internal cracks, thus being the flaws regulating the intensity. Due to the distillation, (cf. purification) aspect of the process and have enough time to form single-crystal particles, inert gas evaporation, and inert gas deposition are free of many of these defects, but even their nonaggregated deposits have a log-normal size distribution, which is characteristic of NPs. That aggregation can be avoided the reason why modern gas evaporation techniques can deliver a fairly narrow distribution of scale. However, even in this case, due to the combination of drift and diffusion, random residence times in the growth zone create a size distribution that appears lognormal. It is therefore desirable to process a material in such a way that it is physically uniform in terms of component distribution and porosity, rather than using particle size distributions that maximize the green density. Having a uniformly distributed assembly of closely interacting suspended particles requires complete control over the forces of the interparticle. Monodisperse NPs and colloids provide this potential.

For example, monodisperse powders of colloidal silica (SiO_2) can be sufficiently stabilized to ensure a high degree of order in the aggregated colloidal or polycristalline colloidal solid. The degree of order appears to be constrained by the time and space permitted to create correlations of a longer range. Such faulty polycrystalline colloidal structures may seem to be the basic elements of the science of colloidal submicrometer materials and thus provide the first step in gaining a more comprehensive understanding of the processes involved in microstructural evolution in high-performance materials and components.

Functionalization is the introduction of organic molecules or polymers on the NP surface. NPs' surface coating defines many of their physical and chemical characteristics, notably stability, solubility, and targeting. A multivalent or polymeric coating confers a high degree of stability. Functionalized catalysts based on nanomaterials can be used to catalyze many organic reactions.

1.5 NP SYNTHESIS APPROACHES (STRATEGIES)

In the first 20 decades, studies on metal NPs were more focused on chemically synthesized NPs. Furtherly, physical and biological approaches were proposed for their safe and green nature. Nowadays, many works are concentrated on biological (microbial- and phyto-) procedures and their applications in several areas. Researches on nanotechnology are widely explored in developed countries as these are associated with their national income, intellectual property policies, and human resources. However, developing countries are trying to catch the wave of nanotechnology, especially India and China. It has been reported by a research review performed by Syafiuddin et al. (2017) that the Department of Science and Technology, India, has invested $20 million for nanomaterials science and technology development and China ranks the third in the number of nanotechnology patent applications filed.

NPs are formed in many ways including gas condensation, chemical precipitation, ion implantation, pyrolysis, and hydrothermal synthesis. Macro- or micro-scale particles in attrition are ground in a ball mill or other mechanism that reduces their size. The resulting particles are graded in the air for the recovery of NPs. A vaporous precursor (liquid or gas) is pushed at high pressure through an orifice in pyrolysis, and burnt. The resulting solid (a soot version) is air graded to recover oxide particulate matter from by-product gases. Traditional pyrolysis, rather than single primary particles, often result in aggregates and agglomerates. In comparison, ultrasonic nozzle spray pyrolysis helps to avoid the formation of agglomerates.

Thermal plasma may provide the energy needed to vaporize small particles of micrometer size. The temperatures of the thermal plasma are in the range of 9750 °C (17,650 °F), so that solid powder evaporates quickly. At the exit of the plasma field, NPs are produced upon cooling. DC plasma stream, DC arc plasma, and radiofrequency induction plasma are the main types of thermal plasma torches used to manufacture NPs. In the arc plasma reactors, an electric arc formed between the anode and the cathode provides the energy required for evaporation and reaction. At high pressure, for example, silica sand can be vaporized using an arc plasma, or thin aluminum wires that can be vaporized using the process of explosion wire. The resulting mixture of plasma gas and silica vapor can be easily cooled by quenching with oxygen, thus ensuring the consistency of the generated fumed silica.

In plasma torches with radiofrequency induction, energy coupling to the plasma is accomplished through the electromagnetic field produced by the induction coil. The plasma gas does not come into contact with electrodes, thus removing possible sources of contamination and allowing these plasma torches to work with a wide variety of gases including inert, reduced, oxidizing, and other corrosive atmospheres. Usually, the working frequency varies between 200 kHz (kilohertz) and 40 megahertz (MHz). Given that the residence time in the plasma of the injected feed droplets is very short, it is critical that the droplet sizes are small enough to get complete evaporation. The radiofrequency plasma approach has been used to synthesize different NP materials, such as the synthesis of various ceramic NPs such as Ti and Si oxides, carbides, and nitrides.

The process of inert gas condensation is also used for producing NPs from metals with a low melting point. The metal is vaporized in a vacuum chamber and then supercooled with a stream of inert gas. The supercooled metal vapor condenses into particles of nanometer size that can be educated in the stream of inert gas and deposited on a substrate or examined in situ.

NPs can also be formed using radiation chemistry in which radiolysis from gamma rays can create free radicals' solution. This quite simple method consumes the minutest amount of chemicals, including water, a soluble metallic salt, a radical scavenger (often secondary alcohol), and a surfactant (organic capping agent). In this practice, reducing radicals will reduce metallic ions into the zero-valence state. A scavenger chemical will favorably react with oxidizing radicals to inhibit the reoxidation of the metal. Once the zero-valence state reached, the metal atoms start to conjoin into particles. A chemical surfactant surrounds the particle along the formation, regulating its growth. In adequate concentrations, the surfactant molecules keep attached to the particles. This protects it from dissociating or forming clusters with other particles. The formation of NPs via the radiolysis technique controls the particle size and shape by optimizing the precursor concentrations and gamma dose.

The sol–gel route is a wet-chemical method (also recognized as chemical solution deposition) and is used for the fabrication of materials (typically a metal oxide) starting from the precursor which is a chemical solution (i.e., the sol), to an integrated network (i.e., the gel) which is either discrete particles or network polymers. Ideal precursors are metal alkoxides and metal chlorides, that undergo hydrolysis and poly-condensation reactions producing either a network of elastic solid or a colloidal suspension (or dispersion); that is a system consisted of discrete (usually amorphous) submicrometer particles that are dispersed to various degrees in a host fluid. Development of a metal oxide encompasses attaching the metal centers with oxo (M–O–M) or hydroxo (M–OH–M) bridges, which consequently generate metal-oxo or metal-hydroxo polymers in solution. Thus, the sol progresses forward into the production of a gel-like diphasic system of both liquid and solid phases in which the morphology varies from certain detached particles to a unremitting polymer networks.

In the case of the colloid, the particle volume fraction (or particle density) can be so small that a large amount of fluid will need to be extracted initially for identification of the gel-like properties. This can be achieved in a variety of ways—the easiest way is to give time for sedimentation and then drain the remaining liquid off. The centrifugation method can also be used to speed up the phase separation cycle.

Removal of the remaining phase of the liquid (solvent) involves a drying process, usually causing shrinking and densification. After all, the rate at which the solvent can be extracted is determined by the porosity distribution in the gel. Changes introduced during this processing step would obviously strongly influence the ultimate microstructure of the final product. Subsequently, a thermal treatment or firing process is often needed to promote further polycondensation and improve mechanical properties and structural stability through final sintering, densification, and grain development. One of the distinct benefits of using this approach is that densification is mostly done at a much lower temperature, as compared to the more conventional processing techniques.

The precursor sol may either be deposited on a substratum to form a film (e.g., by dip-coating or spin-coating), poured into an appropriate container with the desired shape (e.g., to produce a monolithic ceramic, glass, fiber, membrane, or aerogel) or used to synthesize powders (e.g., microspheres or nanospheres). The sol–gel method is a low-temperature and cheap technique that allows for fine control of the chemical composition of the substance. Also, small quantities of dopants, such as organic dyes and rare earth metals, can be incorporated into the sol and end up distributed evenly in the final product. This can be used as an investment casting material in ceramics production and manufacturing, or as a way of producing very thin films of metal oxides for various purposes. The sol–gel derived materials have diverse applications in optics, electronics, energy, space, (biosensors, medicine (such as for controlled drug release), and separation technology (such as adsorption chromatography) technology.

Ion implantation can be used for the treatment of surfaces of dielectric materials such as silica and sapphire for producing composites with near-surface dispersions of metal or oxide NPs.

NPs can be synthesized using two main techniques (Figure 1.13; Table 1.5); a top-down approach that allows us to control the manufacture of smaller, more complex objects, as illustrated by micro- and nano-electronics and bottom-up methods, which allow us to control the manufacture of atoms and molecules, as illustrated by supramolecular chemistry.

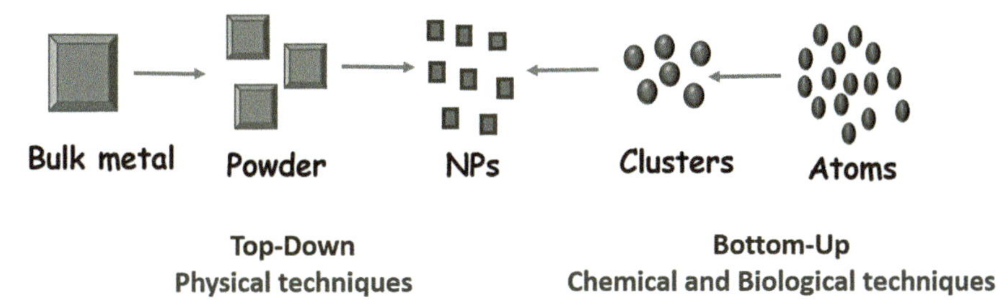

FIGURE 1.13 The two main approaches for nanoparticle synthesis.

TABLE 1.5 Comparison of the Bottom-Up and Top-Down Method of Synthesis

Method	Comment
Top-down	Begins with a material (chemical) generated on a larger scale
	Larger scale is then reduced to nanoscale
	May be (relatively) slow and not suitable for large scale production
	Physical forces used to combine units into larger stable structures
Bottom-up	Begins with atoms or molecules. Then, builds up to nanostructures
	Overall, the miniaturization of components (up to atomic level), followed by further self-assembly leading to the product formation
	Fabrication is often less expensive

1.5.1 TOP-DOWN SYNTHESIS

The top-down approach implies the use of larger (macroscopic) initial structures which can be managed externally in nanostructure processing. The method begins with a larger-scale pattern produced and then reduced to nanoscale. Whereas, the bulk materials are broken down into the NPs

(Balasooriya et al., 2017; Syafiuddin et al., 2017). The physical methods are top-down approaches. Generally, the top-down approach is slow and not suitable for large-scale production. Typical examples of this type of synthetic route include etching through the mask or chemical etching, ball milling, plasma arcing, attrition, lithography, application of severe plastic deformation, thermal evaporate, spray pyrolysis, ultra-thin films, pulsed laser desorption, thermal/laser ablation, sputtering, microwave, plasma arching, thermal evaporate, spray pyrolysis, sputter deposition, ultra-thin films, pulsed laser desorption layer by layer growth, molecular beam epistaxis, and diffusion flame synthesis of NPs (Joerger et al., 2000; Bello et al., 2017).

1.5.2 BOTTOM-UP SYNTHESIS

In the bottom-up approach, the NPs are built from atoms, molecules, and smaller particles/monomers (Balasooriya et al., 2017). In another word, the bottom-up synthetic approach begins with atoms or molecules and is much less costly to build up nanostructures and involves the miniaturization of material components (up to atomic level) with further self-assembly process leading to the creation of nanostructures. The physical forces working at the nanoscale are used during self-assembly to assemble simple units into larger, stable structures.

The synthesis of NPs via the bottom-up technique mainly requires metal precursors, solvent, reducing, and stabilization agents (Figure 1.14).

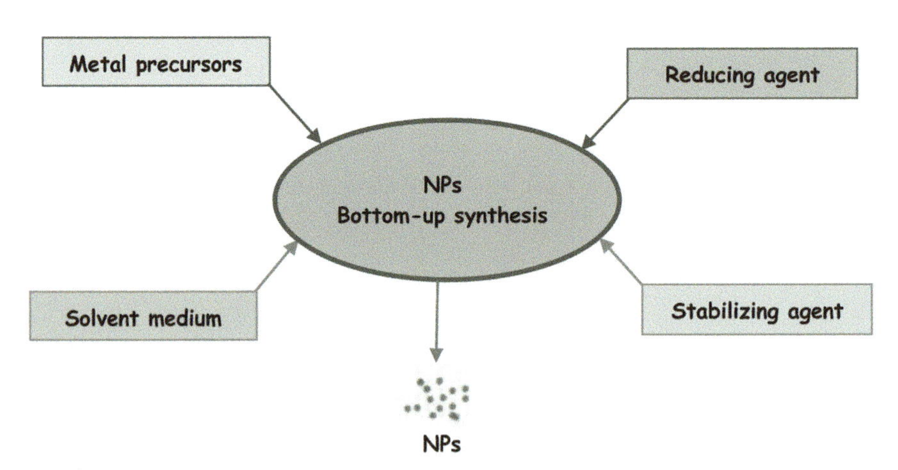

FIGURE 1.14 The main requirements for bottom-up synthesis of NPs.

The bottom-up approach is more convenient than the top-down approach because in the latter the surface structure is imperfect, the NPs are formed by the attrition have a relatively broad size distribution, in addition, chances of contamination with a significant amount of impurities are quite high (Sharma et al., 2017).

Typical examples of this process are the formation of quantum dots during epitaxial growth and the development of colloidal dispersal NPs. The bottom-up synthesis is typical of many processes that occur in nature. Examples are the spontaneous molecular organization into stable, structurally well-defined aggregates (nanometer-length scale). The molecules can be transported via liquids to surfaces to form self-assembled monolayers.

The chemical/electrochemical methods are considered as bottom-up approaches, for example, electrodeposition, sol–gel process, chemical solution deposition, chemical vapor deposition (Panigrahi et al., 2004; Oliveira et al., 2005), soft chemical method, Langmuir Blodgett method, catalytic route, hydrolysis (Pileni, 1997), coprecipitation method and wet chemical method (Gan et al., 2012), precipitation, atomic/molecular condensation, spray pyrolysis, laser pyrolysis, and aerosol pyrolysis (Haleemkhan et al., 2015). Most of the reducing and stabilizing agents have environmental concerns; however, some reducing agents such as borohydride, citrate, ascorbate, and glucose and other stabilizing agents such as trisodium citrate, carboxy methylcellulose, alkali lignin, mono- and di-saccahrides, and other benign surfactants have been applied to overcome this obstacle.

Biosynthesis of NPs is considered as one of the bottom-up approaches. It is eco-friendly, less energy, cost-effective, and single-step bioreduction method (Sathishkumar et al., 2009). Moreover, it depends on eco-friendly resources, such as plant extracts, agro-industrial wastes extracts, bacteria, fungi, micro- and macro-algae, and enzymes (Iravani, 2011).

1.6 CHARACTERIZATION METHODS

NPs have specific analytical criteria than traditional chemicals, for which chemical composition and concentration are sufficient metrics. NPs have certain physical features which must be measured for a complete depiction, for example, size, shape, surface morphology and properties, crystallinity, and dispersion state. In many applications, the homogeneity of those properties is essential. However, sampling and laboratory procedures can disrupt the distribution of other properties by their dispersion state or bias. An additional difficulty in an environmental sense is that certain approaches cannot detect low concentrations of NPs and can still have an adverse impact. NPs can be defined for certain applications in complex matrices such as water, soil, food, polymers, inks, complex mixtures of organic liquids, such as cosmetics, or blood. Chromatography (ion-exchange and high-performance liquid chromatography), centrifugation, ultracentrifugation, electrophoresis and filtration are practices that can be applied to separate NPs by size or other physical characteristics before or during the characterization step.

There are several techniques used to characterize NPs. Microscopy approaches generate micrographs and pictures of individual NPs to depict their shape, size, and location. Electron microscopy and scanning probe microscopy are the prevailing techniques. Since NPs have a size below the visible light diffraction limit, and traditional optical microscopy is not suitable. However, the latest generation of microscopes using laser light, which began in the 1980s, has allowed scientists to produce 3D images at various depth rates of the subject being examined using focalization and laser beam scanning. This type of microscope is known as a confocal microscope and is particularly suitable for use in the natural environment. Their ability to work with fluorescent markers leads to a very fascinating use of these microscopes. The laser beam excites a fluorescent material added to the sample for which the examiners are aware of the affinity of those molecular sites. Thus, the researcher can examine specific reactions selectively. Electronic sensors detect the fluorescent signals and a computer then amplifies and processes those signals.

Electron microscopes can be combined with spectroscopic methods which can carry out an elemental analysis. The method uses the wave properties of electrons but, since the particles need a vacuum in order to travel, the microscope is usually in the form of a metal vacuum chamber in which the following can be found: (1) the source of electrons, such as in cathode ray tubes used in television sets, and (2) the various elements of electronic optics, such as electromagnetic lenses that monitor the electrons' trajectories as well as the protection of the object to be studied.

Microscopy methods are destructive, and in the case of scanning probe microscopy may be sensitive to unwanted artifacts from sample preparation, or probe tip geometry. In addition, microscopy is based on the measurement of single particles, which means that large numbers of individual particles have to be described to estimate their bulk properties. Scanning electron microscopy (SEM), field emission scanning electron microscope, TEM, and high-resolution transmission electron microscopy are used for nanometer to micrometer scale morphological characterization (Omran et al., 2018a ,b).

The TEM has a 1000-fold higher resolution compared with the SEM (Eppler et al., 2000). In the TEM, the beam interacts with the crystalline sample and creates a diffraction figure or hologram. The beam interacts with the crystalline sample in a TEM and produces a diffraction image or hologram. The diffraction figure analysis allows an investigation to be made on the atomic structure of the analyzed sample. The final resolution is related to the electrons-associated wavelength, and hence to their energy. The most prevailing machines operate with voltages in the region of hundreds of thousands of volts. The surface of the sample under study is scanned in the SEM with an electron beam. The interaction between the electrons and the sample results in different signals (the emission of electrons and photons) which, when collected and analyzed, combine the image of the surface of the observed sample without using any mathematical method, as opposed to the transmission process of the electron microscope. The environmental SEMs helped researchers to observe objects in their natural state. The difference between the environmental SEM and the traditional ones—which involve a high vacuum at all column rates that make up the microscope—is that thanks to a differential diaphragm pump device, the sample is held at a specified pressure.

Spectroscopy, which tests particle interactions using electromagnetic radiation as a function of wavelength, is useful for characterizing concentration, scale, and shape in certain groups of NPs. NPs can use X-ray, ultraviolet-visible, and nuclear magnetic response spectroscopy. Light-scattering methods are used to determine particle size by using laser light, X-rays, or neutron scattering, with each approach appropriate for various size ranges and particle compositions. Other miscellaneous methods are surface charge electrophoresis, surface area surface Brunauer–Emmett–Teller test, and crystal structure X-ray diffraction (XRD), particle mass spectrometry, and particle number counters. The X-ray photon spectroscopy is an important technique for surface-sensitive analysis. It is useful for the identification of elements present and also their oxidation states. A widely used technique is UV/Visible spectroscopy (Pal et al., 2007). In general, light wavelengths in the 300–800 nm are used to characterize different metal NPs in the size range 2–100 nm (Feldheim and Foss, 2002). For example, spectrophotometric absorption extents in the wavelength ranges of 400–450 nm (Huang and Yang, 2004) and 500–550 nm (Shankar et al., 2004) are characteristic for silver and gold NPs, respectively. The size and shape of metal/metal oxide NPs and dielectric constant of the medium and surface adsorbed species affect the spectral position of plasmon band absorption as well as its width (Smitha et al., 2008; Omran et al., 2019). According to Mie's theory, only a single SPR band is expected in the absorption spectra of spherical NPs, whereas anisotropic particles could give rise to two or more SPR bands depending on the particles' shapes (Sosa et al., 2003; Das et al., 2010). The Fourier Transform Infrared Spectroscopy can be applied for identifying the organic functional groups (e.g., carbonyls, hydroxyls) attached to the surface of NPs, the other surface chemical residues, and the surface chemistry of biogenic NPs which would help in the presumptive elucidation of the mechanism involved in the biosynthesis of NPs (Chithrani et al., 2006; Ahmed et al., 2016). The dynamic light scattering (DLS) is applied to identify the surface charge and the size distribution of the suspended particles in a liquid (Jiang et al., 2009). Moreover, the techniques also used for characterization of size, surface charge of NPs are Zeta Sizer Nano ZS, utilizing DLS and electrophoretic light scattering, respectively. Zeta potential measured by DLS would help in determining of the stability of the biosynthesized NPs (Ahmed et al., 2016). Whereas, particles with a large negative or positive values of zeta potential are likely to repel each other and consequently are relatively more stable (Omran et al., 2018a,b; 2019a,b, 2020; Rajesh et al., 2018). The XRD is applied to identify the phase and crystalline structure of the NPs (Sun et al., 2000). X-rays penetrate the nanomaterial and the obtained diffraction pattern is matched with standards to acquire the structural information. Elemental composition of metal NPs is usually performed by energy dispersive spectroscopy (Strasser et al., 2010).

1.7 ORIGIN AND CONCEPT OF NANOBIOTECHNOLOGY

Nanomaterials are everywhere in nature around us and they are part of the environment since the creation of our planet, about 4.5 billion years ago. Although the nanosized fullerenes or graphene have been recently synthesized, they have been found in space. Nature has developed many ways of constructing macroscopic structures from nanobricks, such as seashells or bones. The synthesis is conducted at moderate temperature and pressure, using nanocrystals. Calcium carbonate or phosphate is involved in the biomineralization process. Biomimetic methods are also used nowadays. The iridescent colors of opals are due to packed silica nanospheres arranged in layers (Sanders, 1964). The unique and highly organized design at multiple length scales of nacre, the hierarchical biological nanocomposite found inside many mollusk shells (Nacre, also known as the mother of pearl, is an organic–inorganic composite material formed as an inner shell layer by some mollusks; it also constitutes the outer shelling of pearls. It is robust, buoyant, and shimmering) has inspired chemists to synthesize strong, tough, and stiff bioinspired ceramics formed by a dense packing of platelets presenting a long-range order (Bouville et al., 2014). The colors of butterflies are attributed to the fine structure of their wings, which, are multilayers of nanostructures that act like diffraction grids that make interferences, and consequently iridescences (Boulenguez et al., 2012). The feet nanostructures of geckos consist of a series of small ridges containing numerous hairs, each hair being subdivided into a thousand 200 nm wide projections. Thus, the total surface area of the feet of geckos is huge, leading to a strong surface adhesion, which is entirely due to van der Waals interactions (Moret, 2006). The self-cleaning superhydrophobia lotus leaves are due to the epicuticular wax crystalloids that cover its rough surface, leading to a water-repellent layer (Moret, 2006). The nanostructures of the gecko feet have inspired researchers to make some adhesive materials (Bogue, 2008) and the Lotus effect has been used to design new self-cleaning materials (Solga et al., 2007). Moreover, the natural molecular motors for many biological processes which involve a lot of motions at the nanoscale (e.g., animal and human cells) have inspired scientists to design molecular nanomachines (Moret, 2006). The DNA, ribonucleic acid (RNA), and ribosome constitute a nanoworld of extremely complex nanocomputers, nanosystems, and nanomachines. Moreover, proteins, membranes, and nucleic acids refer to giant natural nanostructures built as a result of self-assembling. Even wood is one of the most popular natural nanomaterials. It has a hierarchical scale structure. At the largest scale, wood contains soft fibers with a diameter of about 20–30 μm and a length typically between 2 and 5 mm. However, at an intermediate hierarchical scale, nanofibers are present with a diameter <100 nm and a length >1 μm. The smallest scale contains crystallites with a width <5 nm and a length <300 nm. Since the mechanical properties improve as the size of the structure decreases. It was found that the elasticity is multiplied by almost 12 and the strength by 100 as one goes from softwood structure to wood nanocrystals. Nanocellulosic materials are also present in nature. These can be collagen fibrils originating from animal sources; nanofibers originating from wood, plants, crops, or bacteria; crystals or whiskers deriving from wood, plants, and crops. Cellulose nanocrystals or whiskers can be extracted by mechanical and chemical processes for subsequent use in polymer nanocomposite materials (Nagô and Van de Voorde, 2014). The artificial growth of pearls inside mussels is another example. Lipid nanotechnology concerns with the preparation of lipid-based NPs, which contain lipid moieties. Lipid NPs are mostly characteristically spherical with a diameter ranging from 10 to 1000 nm. They are effectively applied in many biomedical applications, such as drug carriers and delivery and RNA release in cancer therapy (Mashaghi et al., 2013; Gujrati et al., 2014). Lipid NPs can be considered as PNPs that own a solid core composed of lipid and a matrix contains soluble lipophilic molecules, where surfactants or emulsifiers stabilize the external core of these NPs (Rawat et al., 2011). Needham et al. (2016) applied the solvent-exchange method (as an example for the bottom-up approach) to prepare limit-sized low-density lipoprotein NPs without using phospholipid and possessed high hydrophobicity, which is essential for medical cancer drug delivery purpose.

As stated above, nanotechnology is science and technology at the molecular level and is further defined as the understanding and control of matter at dimensions between 1 and 100 nm (Felix and Ahmed, 2104). In fact, as with many other fields, the nanotechnology techniques were in use centuries before the subject was formally established. In the simplest (and modern) sense, nanotechnology has been practiced for most of the 20th century in the form of many industrial processes. A particularly relevant example in the context of this book is the use of catalysts in the petroleum industry in which the catalyst is designed to contain molecular-specific sites that adsorb reactant molecules to provide specific products.

Biotechnology and nanotechnology are two of the 21st century's most promising technologies (Fakruddin et al., 2012). Recent words relating to the convergence of nanotechnology and biology are bionanotechnology, nanobiotechnology, and nanobiology (Gazit, 2007). The two terms bionanotechnology and nanobiotechnology are frequently interchangeably used. Bionanotechnology usually refers to the study of how the objectives of nanotechnology can be driven to develop existing processes and applications in biotechnology or to build new ones. Briefly, it is the application of nanotechnology in life sciences. For example, the application of NPs for the optimal dosage range medication delivery often contributes to increase the drug therapy effectiveness, reduce the side effects, and enhance patient compliance. The antimicrobial activity of some metal NPs, for example, Ag, TiO_2, ZnO, $BiVO_4$, Cu- and Ni-based NPs, make them used in textile, medicine, various households' products, water disinfection, and food packaging. Chitosan-coated silver nanotriangles (chitAgNPs) have been reported as photothermal agents against a line of human lung cancer cells (NCI-H460) (Boca et al., 2011). Moreover, AuNPs have been proved also as an important tool for hormone detection in a pregnant women's urine sample (Kuppusamy et al., 2014). Nanobiotechnology can be also considered as the integration of biotechnology and nanotechnology for developing bioactive, biosynthetic, and eco-friendly technology for the synthesis of nanomaterials (Gilaki, 2010). Physical and chemical methods for NPs preparation are complicated, outdated, expensive, and produce hazardous toxic waste that is harmful to the environment as well as human health (Kumar et al., 2016; Qj and Lam, 2017). To solve these problems, the bottom-up, green or biological approach is attracting many researchers due to its feasible, less toxic, mild reaction conditions, cost-effectiveness, and eco-friendly nature. Moreover, the biological method of synthesis of NPs has proved to be better than the chemical methods due to its slower kinetics, which offers a better control over crystal growth and reduced capital involved in the production (Sharma et al., 2009; Yadav et al., 2017). Furthermore, the biological synthesis of NPs finds extensive biomedical applications and is considered as the most commercialized NPs due to their effective antimicrobial potential including multidrug-resistant pathogens (Rai et al., 2012). Both prokaryotic and eukaryotic organisms have been reported to naturally produce inorganic nanomaterials either intra- or extra-cellularly by the process of biomineralization (Simkiss and Wilbur, 1989); magnetotactic bacteria act as nanofactory for Fe_3O_4 NPs (Matsunaga et al., 2007), diatoms synthesize silica nanospheres (Kröge et al., 1999), and many bacteria and archaea produce gypsum and calcium carbonate layers (Sleytr et al., 2007).

Primarily, as mentioned before, industry rely on expensive, energy-consuming, and toxic physicochemical nanosynthesis processes (microwave-assisted, sonochemical, pulsed laser deposition, electrochemical, photochemical, etc). Nevertheless, the green chemistry promoted the cost-effective and energy-saving alternative sustainable technologies that eliminate the hazardous substances, such as hydroxylamine, poly-N-vinylpyrrolidone, and sodium borohydride and tetrakis (hydroxymethyl) phosphonium chloride. However, water is the most widely studied solvent in green synthesis. To stabilize a NP in a dispersing medium, sufficient repulsive forces have to be developed to counteract the van der Waals interaction. Repulsive force is achieved through either electrostatic or steric stabilization. For steric stabilization, organic ligands are added to form a protective layer for example, polyvinyl alcohol, surfactants, ligands, or copolymers. The species present on the surface also play a role in the stability of NPs due to functionalization. For example, the addition of a dodecanethiol gold colloidal solution protected the nanostructures through the inherent nature of the stabilizing agent, which renders the NP system soluble in an organic solvent. Aliphatic and aromatic compounds contain a wide variety of functional groups, for example, cyano (–CN), mercapto (–SH), carboxylic acid (–COOH), and amino ($-NH_2$), known to have a high affinity for functionalization of noble metal NPs, and thus are useful as surface-protective functional groups. NPs have also been stabilized by the addition of charged species to the colloidal solution of noble metals. The efficiency of stabilization depended on the magnitude of adsorption of anions and cations on the surface of NPs. Stabilizing agent dosage determines the stability, reactivity, solubility, particle shape and size. The type of stabilizer also determines the catalytic properties of nanomaterials. Nevertheless, in biological methods, NP synthesis is carried out without the addition of any external reducing, stabilizing, or capping agents that are replaced by molecules produced by the living organisms, that is, bacteria, fungi, yeast, algae, plants (its extracts or bioproducts), etc., therefore, these living organisms are the nanobiofactories for the production of NPs. Synthesis of NPs with relatively high particle concentrations coupled with predefined shapes has been usually achieved using organic solvents.

In order to reveal even more favorable NPs functionality, some researchers look to the natural methods to produce NPs with novel properties or enhanced functions (Schröfel et al., 2014). Two bioroutes can be utilized for the synthesis of NPs, microbial via microorganisms, the eukaryotes (fungi and yeasts), and prokaryotes (bacteria and actinomycetes) and phytosynthesis via different plant extracts (Figure 1.15). Where the living cells act as operators at the nanolevel and generate targeted materials at very high efficiency. However, despite the stability and the green method of formation, the NP biosynthesis rate is not comparable with nonbiological synthesis methods. Nonetheless, biosynthesis is expected to be more commercially appropriate if the NPs could be synthesized faster and more efficiently on a large scale, with better control of size, morphology, stability, and the rate of preparation of NPs. Thus, this book will discuss the different routes of green- and bio-synthesis of nanomaterials.

The importance of exploring more environments and analyzing their microbial community to discover novel and/or better bioproducts will have a high impact on society and health, especially, with the increased antibiotic resistance in recent years and fewer new antibiotics emerging, work has begun to focus on these antibacterial NPs as possible new medical implements. Silver and gold NPs are good inhibitors for the growth of both Gram-positive and Gram-negative bacteria (Guzmán et al., 2009; Krishnaraj et al., 2010, Lima et al., 2013). Due to the fluorescence and surface plasmon resonance characteristics of AgNPs, it finds application as sensor in DNA sequences, mass spectrometry of peptides, colorimetric determination of histidine and ammonia, transport in microbial cells, and as biosensors (Vijayaraghavan and Kamala Nalini, 2010). Moreover, biosynthesized AgNPs are nontoxic (in

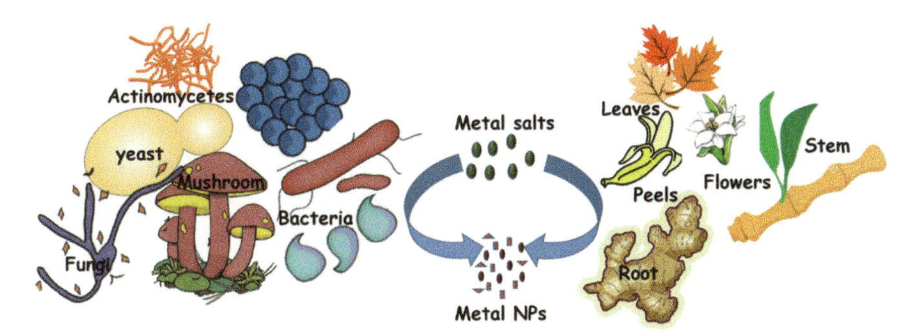

FIGURE 1.15 Biological routes for bottom-up synthesis of NPs.

low concentrations) to humans and are healthy inorganic antibacterial agents with high toxicity to a wide variety of microorganisms (Roy et al., 2013; Annamalai and Nallamuthu, 2016; Shanthi et al., 2016; Omran et al., 2018b). The mode of action of the AgNPs depends on monovalent ionic silver (Ag^+), which is released within the microbial cells and inhibits microbial growth by suppressing respiratory enzymes and components of electron transport (Li et al., 2006; Annamalai and Nallamuthu, 2016; Chen et al., 2016). The AgNPs have also been identified as affecting the cellular membranes (Chen et al. 2016; Durán et al. 2016). In addition, AgNP's antimicrobial activity is due to the formation of insoluble compounds through the inactivation of sulfhydryl groups in the cell wall and disruption of membrane-bound enzymes and lipids that lead to cell lysis (Dorau et al., 2004). Parashar et al. (2009) reported that AgNPs are efficient antimicrobial agents even at low concentrations and inhibit the growth of antibiotic-resistant bacteria, as they interact with bacterial membrane proteins and DNA, which possess sulfur and phosphorous compounds that have a high affinity toward silver ions. It has also been stated that the process may involve binding AgNPs to external proteins making pores, interfere with DNA replication, or form reactive oxygen species (ROS), such as hydrogen peroxide, superoxide anions, and hydroxyl radicals (Jung et al., 2008; Duncan, 2011; Durán et al., 2016). It is worth noting that the operation of AgNPs depends mainly on the size of NPs (Wu et al., 2014; Tamayo et al., 2014; Rahimi et al., 2016). Martinez-Castañón et al. (2008) stated that the antibacterial activity of Ag-NPs decreases as its particle size increases. It has also been demonstrated that 10 nm and smaller particles are more bioavailable by dissolving in the near vicinity of the cell surface or within the cells (Ivask et al., 2014). However, Patel et al. (2015) reported that the elongated-shaped 31.86 ± 1 nm Ag-NPs formed by the cyanobacterium *Limnothrix* sp. 37-2-1 did not express any biocidal activity against all the tested bacteria. Chernousova and Epple (2013) reported that the bactericidal effect of different AgNPs concentrations is bacterial class dependent. Nevertheless, Feng et al. (2000) observed that the Gram-positive and Gram-negative bacteria have different susceptibility to AgNPs, and attributed this to the differences in their membranes and cell walls. Generally, increasing the ratio between the surface area and the volume up to the nanoscale, the antimicrobial activity is improved, that is, more sites of interaction with biological receptors are available, providing a strong contact with microorganisms' surface, thus increasing their antimicrobial efficiency (Thakkar et al., 2010; Reidy et al., 2013). Where the antimicrobial activity of silver depends mainly on three common mechanisms: (1) absorption of free silver ions followed by disruption of ATP output and replication of DNA; (2) production of ROS by AgNPs and silver ion; and (3) direct disruption of cell membrane integrity via the creation of pits on the membrane surface leading to the death of cells (Marambio-Jones and Hoek 2010; Reidy et al. 2013). AgNPs have been tested for antifungal activity in biostabilized footwear materials against dermatophytes and other fungi (Falkiewicz-Dulik and Macura, 2008; Kim et al., 2008; Noorbakhsh, 2011; Marcato et al., 2012; Pereira et al., 2014).

The plants formed metallic NPs such as silver, platinum, and palladium NPs which effectively regulate the environmental malaria population (Kuppusamy et al., 2016). Biosynthesized gold, silver, and platinum NPs proved to achieve positive wounds have shown effective wound repair mechanisms and inflammatory tissue regeneration (Gurunathan et al., 2009). Moreover, biogenic and phytogenic NPs can be applied for cancer treatment. As these NPs act as novel control agents, the overexpression of cell growth in cancer cells will be halted and controlled with systemic cell cycle mechanisms (Akhtar et al., 2013). The phytogenic silver NPs were reported to regulate the cell cycle and enzymes in the bloodstream (Alt et al., 2004). Free radicals commonly induce cell proliferation and damage the normal cell function. Nevertheless, the phytogenic NPs relatively control the free radicals' formation from the cell. The moderate concentration of gold NPs was reported to induce the apoptosis mechanism in malignant cells (Dipankar and Murugan, 2012). Biogenic Ag and Au NPs can act as potent broad-spectrum antiviral agents to restrict virus-cell functions and are regularly inhibiting HIV-1 life cycle post-entry stages (Sun et al., 2005; Suriyakalaa et al., 2013). In some diabetes, mice treated AuNPs are documented to significantly reduce the number of liver enzymes such as alanine transaminase, alkaline phosphatase, serum creatinine, and uric acid. It also showed a decrease in the level of HbA (glycosylated hemoglobin) which is kept in the normal range (Daisy and

Saipriya, 2012). In another study, AuNPs are reported to inhibit α-amylase and acarbose sugar in diabetes-induced animal model (Manikanth et al., 2010). Not only was this but AgNPs reported to express higher antioxidant effects than those of ascorbic acids and other synthetic commercial standards (Kuppusamy et al., 2016). Table 1.6 summarizes some biological applications of heavy metals. For a biological method to be successfully compete with the chemical and physical synthesis of nanostructures, strict control over average particle size in the specific size range and uniform particle morphology is required (Quester et al., 2013).

1.8 APPLICATION OF NANOBIOTECHNOLOGY IN PETROLEUM INDUSTRY

Nanotechnology has played an important role in studying the mineral composition, microporous structure, and rock physical properties of unconventional reservoirs; the nanosensor reservoir is still in the laboratory process and the nanorobots reservoir will require a major advancement in technology; numerical simulation technology for oil and gas migration in micro-nano-porous media would become a key technology advancement for nonconventional oil and gas migration mechanism, distribution pattern, and resource evaluation. Nanomaterials have size and surface effect, wetting characteristics, inhibition characteristics of particle migration, shear thickening behavior, nanometer photocatalytic properties, nanofiltration, nanocorrosion, and wear resistance. Thus, nanomaterials have broad application prospects in the fields of EOR, water treatment, engineering, and anticorrosion (ShamsiJazeyi et al., 2014; Liu et al., 2016; Sun et al., 2017; Elnashaie et al., 2018; Etim et al., 2018; Fakoya et al., 2018). Nanotechnology has also enabled the production of nanoscale sensors for faster and more efficient measurement of key reservoir parameters, such as temperature and pressure (Krishnamoorti, 2006). Inclusion of synthetic NPs in low-volume fractions in the conventional fluid mixtures can modify wettability, viscosity, and corrosiveness. These nanofluids (sometimes referred to as "smart fluids") can provide significant improvements in sedimentation stability, transfer efficiency, and reduced erosion (Suleimanov et al., 2011).

Magnetic spherical-shaped 80 nm iron oxide NPs coated with hydrophobic oleylamine have been reported to have good application in oil reservoir (Esmaeilnezhad et al., 2019). It expressed efficient catalyzing empowerment in the initial oxygenation reactions, upgrading the efficiency of cracking step, and speed up the propagation of the combustion front. Moreover, its

TABLE 1.6 Examples for Some Biological Applications of the Green Synthesized Metal NPs

Biological Route	Metals	Size (nm)	Activity	References
Bacillus licheniformis Dahb1	Ag	18.69–63.42	Antibiofilm activity	Shanthi et al. (2016)
Deinococcus radiodurans R1 ATCC BAA-816	Ag	16.82	Antibacterial, antibiofouling and cytotoxicity activities	Kulkarni et al. (2015)
Pseudomonas aeruginosa strain BS-161R	Ag	8–24	Antimicrobial activity	Kumar and Mamidyala (2011)
Streptomyces albidoflavu	Ag	10–40	Antimicrobial activity	Prakasham et al. (2012)
Streptomyces sp. LK3	Ag	5	Antiparasitic activity	Karthik et al. (2014)
Streptomyces sp.09 PBT 005	Ag	98 –595	Antimicrobial, cytotoxicity	Kumar et al. (2015)
Alternaria sp.	Ag	10–30	Antibacterial activity	Singh et al. (2017)
Aspergillus oryzae MUM 97.19	Ag	14–76	Antifungal activity	Pereira et al. (2014)
F. oxysporum f. sp. lycopersici	Ag	5–13	Antibacterial and cytotoxicity activities	Husseiny et al. (2015)
Penicillium aculeatum Su1	Ag	4–55	Antimicrobial activity and cytotoxicity activity toward lung adenocarcinoma cells	Liang et al. (2017)
Trichoderma harzianum	Ag		–	Elamawi et al. (2018)
T. viride			–	
T. longibrachiatum		10	Antifungal	
Microcoleus sp	Ag	55	Antibacterial activity	Sudha et al. (2013)
Scenedesmus sp.	Ag	5–10	Antibacterial activity	Jena et al. (2014)
Piper longum leaf	Ag	18–41	Cytotoxicity activity toward HEp-2 cancer cells	Jacob et al. (2011)
Annona squamosa leaf	Ag	20–100	Cytotoxicity toward human breast cancer cell (MCF-7)	Vivek et al. (2012)

TABLE 1.6 *(Continued)*

Biological Route	Metals	Size (nm)	Activity	References
Eclipta alba leaf	Ag	310–400	Cytotoxicity toward mouse macrophage cells (RAW 254.7), human breast cancer cells (MCF-7), and human adenocarcinoma cells (Caco-2)	Premasudha et al. (2015)
Artemisia tournefortiana Rchb	Ag	22.89 ± 14.82	Cytotoxicity toward HT29 colon cancer cells	Baghbani-Arani et al. (2017)
Acalypha indica leaves	Ag and Au	20–30	Antibacterial	Krishnaraj et al. (2010)
Cassia fistula stem	Au	55–98	Antihypoglycemic	Daisy and Saipriya (2012)
Mentha piperita leaves	Ag and Au	90–150	Antibacterial	Mubarak Ali et al. (2011)
Mirabilis jalapa flowers	Au	100 nm	Antimicrobial	Vankar and Bajpa (2010)
Trigonella-foenum graecum seeds	Au	15–25	Catalytic	Aromal and Philip (2012)

magnetic properties added to its advantage as it can be recycled and reused since it can be easily recovered from the produced fluids upon applying external magnetic field. Thus, the dispersion of such magnetic hydrophobic iron oxide NPs near-wellbore would increase the quality of the produced oil, fuel availability, and propagation rate of the combustion front, via the decrement of the possible plugging risk of pore throats and the deactivation risk of the fixed bed catalysts (Esmaeilnezhad et al., 2019). Thus, it is recommendable in the Toe-to-Heel Air Injection, the thermal EOR method designated for the heavy oil reservoir. Not only this but the reusability of such magnetic iron oxide NPs would lower the cost of the materials and downstream processing and also reduce the possible negative environmental influences of the NPs, consequently, making the production procedure clean, green, and more effective. Nanocellulose fluid has been reported as a green flooding agent in enhancing oil recovery (Li et al., 2016). Nano-optical fiber has been reported as an oil microbe detection tool (Jahagirdar, 2008; El-Diasty and Ragab, 2013; Agista et al., 2018). MWCNTs fluid has been also reported as a good EOR agent (Alnarabiji et al., 2016). Nanofluids have been also reported for enhancing the microbial EOR (Amani, 2017; Rezk and Allam, 2019). ZnO NPs are reported to have efficient performance in EOR (Agista et al., 2018; Rezk and Allam, 2019). The synergetic effect of silica nanofluids and *Acintobacter calcoaceticus* (PTCC: 1318) biosurfactant in flooding tests in a glass micromodel has been examined (Khademolhosseini et al., 2015). Where, the nano/bio-material flooding expressed the highest heavy oil recovery relative to water, nanofluid, and biosurfactant flows (Khademolhosseini et al., 2015).

Nonetheless, NPs with significant improvements in optical, magnetic, and electrical properties compared with their bulk counterparts are excellent tools for sensor creation and image contrast agent formulation (Matteo et al., 2012). AgNPs were also used as optical sensors for the formation of adsorbents of small molecules (McFarland and Van Duyne, 2003). Hyperpolarized Si NPs have a novel tool for oil discovery measurement and imaging (He et al., 2016). To decrease shale permeability, NPs can be applied to the drilling fluid by physically plugging nanometer-sized pores and shutting off water leakage. NPs can, thus, provide possible alternatives for environmentally sensitive areas where oil-based drilling muds are used as a solution to problems related to instability in formation (Khalil et al., 2017). CuO and ZnO have been also reported to be used together with xanthan gum to improve rheological, thermal, and electrical properties of water-based drilling fluid (William et al., 2014). The choice of cementing materials as well as cementing techniques is important to achieve successful oil and gas exploration and production. The application of CNTs, TiO_2, Fe_2O_3, CuO, and ZnO NPs as additives in cementing helps in strengthening and increasing durability, resistance to water penetration, acceleration of hydration reaction, control calcium leaching, improves the impact and fracture toughness, and enhances the dry shrinkage and permeability resistance properties (Yu et al., 2010; Hurnaus and Plank, 2015). ZnO NPs have been reported for good stimulation and increasing the good productivity via the improvement of rheological characteristics of fracturing fluid (Nasr-El-Din et al., 2013).

Nanotechnology has also made important contributions to the refining industry over the last two decades. The performance of a variety of conversion processes increases the application of nanometer-scale catalytic materials, which is caused by enhancement and creation of better material properties as the particle size decreases. Examples of this are the applications of nanotechnology in many traditional processes of petroleum production, including catalytic cracking, oxidative dehydrogenation of alkanes, and desulfurization (Speight, 2014, 2017). The main advantages of refinery nanotechnology are the catalytic processes that are focused on the exposure of a broad surface area for reaction, thus reducing susceptibility to adverse and side reactions. To meet the rising demand for chemicals and fuels, the need for an improved catalyst with high activity, low deactivation, and low coke formation necessitates the increasing use of NPs as the catalyst. For example, anatase TiO_2 sandwich form polyoxometalate nanocomposite was used as an effective, reusable, and green nanocatalyst in the oxidative desulfurization of oil refineries (Rezvani et al., 2012). Khodadadi et al. (2012) reported the adsorptive desulfurization of diesel fuel using CuO NPs.

The production of mesoporous catalyst materials such as MCM-41 has changed the downstream refining process significantly. Nanofilters and NPs have the ability to extract hazardous radioactive compounds from industrial gas streams, such as nitrogen oxides, sulfur oxides, and related acids. Nanotechnology also offers options for carbon capture and long-term storage and has contributed to a new generation of nanomembranes to further separate gas streams and remove impurities from crude oil and crude oil products. In the catalytic converters of vehicles palladium, platinum and rhodium are stated to transform unburnt hydrocarbons and carbon monoxide into CO_2 and water (Capeness et al., 2015). Moreover, nickel NPs are reported as an efficient biocide, catalyst in the production of hydrogen and as a capture for the six platinum group metals are ruthenium (Ru), rhodium (Rh), palladium (Pd), osmium (Os), iridium (Ir), and platinum (Pt) in the refining process (Alonso et al., 2011; Crundwell et al., 2011; Ahamed and Alhadlaq, 2014; Capeness et al., 2015). PtNPs have also been applied in catalytic hydrogenation processes (Bratlie et al., 2007). PtNPs can be used in oxygen reduction reactions in fuel cells (Muthuswamy et al., 2013). Where, catalysts based on Pt NPs have been shown to perform enhanced activity for the electro-oxidation of formic acid (Waszczuk et al., 2002). Iron NPs are reported to be promising in catalysis of the Fischer–Tropsch synthesis and coal liquefaction (Huber, 2005; Galvis et al., 2012). AgNPs, AuNPs, and CuNPs have been stated for the reduction of nitro-aromatics into their corresponding amino-aromatics, in the presence of sodium borohydride ($NaBH_4$) (Ahmed et al., 2016; Qu et al., 2017; Zaheer, 2018). Furthermore, AgNPs are used as a catalyst in methanol and ethylene oxidation to formaldehyde and ethylene oxide, respectively (Nagy and Mestl, 1999). One of the most important applications of AgNPs and CuNPs is their efficient application as antibiofilm agents (LewisOscar et al., 2015). Nanomagnetic fluid is reported as efficient mitigation of corrosion and scale occurrence to carbon steel in acidic media, as ferromagnetic NPs form a good protective layer (Murugesan et al., 2016). Several metal oxide NPs, especially iron oxide NPs, have been reported for the hydrocracking of heavy petroleum residue obtained from the vacuum distillation unit (Khalil et al. 2017). AgNPs have been applied in the manufacturing of H_2 gas sensors (Mohamedkhair et al., 2019). That would have applications in H_2-fueled engines and hydrocracking process, decreasing the threat of fire and explosions. Adio et al. (2019) reported that activated carbon prepared from date-pits, which was functionalized with 25 mM AgNPs acted as an efficient adsorbent for mercury from Arabian gas-condensate, within only one hour. That was attributed to the higher active surface area of the sorbent material. Such cost-effective adsorbent can be regenerated and used four successive times without losing its activity (Adio et al., 2019). The surface treatment of porous Si with TiO_2, SnO_2, or Au NPs was reported to enhance the sensing capabilities for low concentration CO molecule (Alwan et al., 2017).

Upgrading extra-heavy crude oil tar sand bitumen has been another important challenge for refiners (Speight, 2014, 2017). Such feedstocks are not always convenient to be transported to the refinery owing to their high density and viscosity, where they can be processed into useful products. Nanocatalysts that however give a solution at the production site for partial (or even full) upgrading of extra-heavy crude oil and tar sand bitumen. NiO nanocrystals-functionalized SiO_2 NPs have been reported for viscosity reduction and deasphalting of heavy crude oil (Montes et al., 2018). TiO_2 and Fe_3O_4 NPs have been reported for deasphatling of heavy crude oil. It has been also proved that the asphaltenes adsorption is a metal-oxide-specific process, which depends on the type and strength of the interactions between asphaltenes and metal oxide surface (Nassar et al., 2011). Hematite Fe_2O_3 NPs have been reported for the efficient thiophene decomposition in the oxidative desulfurization process (Khalil et al., 2015) and photodegradation of some organic pollutants (Ali et al., 2017). The application of CNTs as a supporter for CoMo particles enhanced the hydrodesulfurization efficiency (Mohammed et al., 2017). The green synthesized bimetallic 2.12 nm Pt@ Cu NPs using the hot water extract of *Alchornea laxiflora* leaf, expressed higher oxidative desulfurization (≈99%) of dibenzothiophene in *n*-heptane model oil (1 g/L), relative to the conventional acetic acid catalyst (≈28%), in presence of H_2O_2, as an oxidant (Olajire et al., 2017).

The magnetic NPs have unique properties such as the extremely large surface area per volume substance, nonporous, highly stable, superparamagnetic behavior, and low degree of cytotoxicity (Malvindi et al., 2014; Saed et al., 2014). As mentioned above, the most important advantages of magnetic NPs its ease of separation, recycling, and reusability at least 5–10 times without losing their activity (Saed et al., 2014; Zakaria et al., 2015; Simonsen et al., 2018). Thus, it has different applications in the petroleum industry. For example, magnetic NPs have been used to improve the biodesulfurization and biodenitrogenation processes (Shan et al., 2005; Zakaria et al., 2015; Speight and El-Gendy, 2017; El-Gendy and Nassar, 2018). Magnetic NPs have been also reported to increase some other microbiological reaction rates, such as the biodegradation of different petroleum hydrocarbon pollutants (Saed et al., 2014). Magnetite NPs have been used as oil spill collector (Abdullah et al., 2018). Superparamagnetic Fe_3O_4 nano-scale aggregates naturally present in paraffin wax deposits are reported to control the physicochemical properties of crude oils (Lesin et al., 2010). Functionalized magnetic NPs are documented to be used for mapping reservoirs and as magnetic nanofluids for drilling, completion, and EOR (Kapusta et al., 2011; Cocuzza et al., 2012; Simonsen et al., 2018). Functionalized magnetic NPs can be applied in emulsion separation, which would solve many problems in upstream and downstream processes (Peng et al., 2012a,b). Considering that the smaller the scale of the NPs, the quicker the separation is (Ko et al., 2014, 2016). Functionalized Fe_3O_4 NPs can be applied for dewatering of heavy crude oil emulsions (Ali et al., 2015). Magnetic NPs have also been reported for efficient oil spill clean-up (Pavia-Sanders et al., 2013; Atta et al., 2015; Mirshahghassemi and Lead, 2015; Abdullah et al., 2016), produced water purification (Liang et al., 2015), and wastewater treatment from various petrochemical industries (Simonsen et al., 2018). Where, the higher the concentration of magnetic NPs is, the higher its efficiency is. A variety of compounds present in crude oils can contribute to deposit formation during production and processing operations. To prevent the formation of these deposits and thereby sidestep any flow assurance problems, magnetic NPs can be also used for the mitigation of undesired phase separation. For example, amino-functionalized magnetic NPs have been reported for selective extraction of naphthenic acids present in the crude oils which would cause deposit formation, corrosion, emulsion stabilization, and pollution of the produced water (Simonsen et al., 2019).

Indeed, there are many areas of the petroleum industry where nanotechnology can lead to more effective, less costly, and more environmentally friendly technologies than those that are readily available (Matteo et al., 2012; Zaman et al., 2012; El-Diasty and Ragab, 2013; Agista et al., 2018).

REFERENCES

Abdullah, M.M.S., Al-Lohedana, H.A., and Atta, A.M. 2016. Novel magnetic iron oxide nanoparticles coated with sulfonated asphaltene as crude oil spill collectors. RSC Adv. 6: 59242–59249.

Abdullah, M.M.S., Atta, A.M., Allohedan, H.A., Alkhathlan, H.Z., Khan, M., and Ezzat, A.O. 2018. Green synthesis of hydrophobic magnetite nanoparticles coated with plant extract and their application as petroleum oil spill collectors. Nanomaterials. 8: 855, doi:10.3390/nano8100855.

Adio, S.O., Rana, A., Chanabsha, B., BoAli, A.A.K., Essa, M., and Alsaadi, A. 2019. Silver nanoparticle-loaded activated carbon as an adsorbent for the removal of mercury from Arabian gas-condensate. Arab. J. Sci. Eng. 44: 6285–6293.

Agista, M.N., Guo, K., and Yu, Z. 2018. A state-of-the-art review of nanoparticles application in petroleum with a focus on enhanced oil recovery. Appl. Sci. 8: 871–900.

Ahamed, M., and Alhadlaq, H.A. 2014. Nickel nanoparticle-induced dose-dependent cytogenotoxicity in human breast carcinoma MCF-7 cells. Oncotargets Ther. 7: 269–280.

Ahmed, S., Annu, Ikram, S., and Salprima, Y.S. 2016. Biosynthesis of gold nanoparticles: a green approach. J. Phytochem. Phytobiol. B. 161: 141–153.

Alemán, J., Chadwick, A.V., He, J., Hess, M., Horie, K., Jones, R.G., Kratochvíl, P., Meisel, I., Mita, I., Moad, G., Penczek, S., and Stepto, R.F.T. 2007. Definitions of terms relating to the structure and processing of sols, gels, networks, and inorganic-organic hybrid materials (IUPAC Recommendations 2007). Pure Appl. Chem. 79(10): 1801–1829.

Ali, H.R., Nassar, H.N., and El-Gendy, N.Sh. 2017. Green synthesis of α-Fe_2O_3 using *Citrus reticulum* peels extract and water decontamination from different organic pollutants. Energy Sources A. 39(13): 1425–1434. https://doi.org/10.1080/15567036.2017.1336818.

Ali, N., Zhang, B., Zhang, H., Zaman, W., Li, X., Li, W., and Zhang, Q. 2015. Interfacially active and magnetically responsive composite nanoparticles with raspberry like structure; synthesis and its applications for heavy crude oil/water separation. Colloids Surf. A. 472: 38–49.

Alnarabiji, M.S., Yahya, N., Shafie, A., Soleman, H., Chandran, K., Abd Hamid, S.B., and Azizi, K. 2016. The influence of hydrophobic multiwall carbon nanotubes concentration on enhanced oil recovery. Procedia Eng. 148: 1137–1140.

Alonso, F., Riente, P., and Yus, M. 2011. Nickel nanoparticles in hydrogen transfer reactions. Acc. Chem. Res. 44(5): 379–391.

Alwan, A.M., Dheyab, A.B., and Allaa, A.J. 2017. Study of the influence of incorporation of gold nanoparticles on the modified porous silicon sensor for petroleum gas detection. Eng. Technol. J. 35 Part A.(8): 811–815.

Amani, H. 2017. Synergistic effect of biosurfactant and nanoparticle mixture on microbial enhanced oil recovery. J. Surfactants Deterg. 20: 589–597.

Annamalai, J., and Nallamuthu, T. (2016) Green synthesis of silver nanoparticles: characterization and determination of antibacterial potency. Appl. Nanosci. 6(2): 259–265.

Arikawa, M. 2006. Fullerenes—an attractive nano carbon material and its production technology. Nanotechnol. Percept. 2(3): 121–128.

Aromal, S.A., and Philip, D. 2012. Green synthesis of gold nanoparticles using *Trigonella foenum-graecum* and its size dependent catalytic activity. Spectrochim. Acta A. 97: 1–5.

ASTM E2456. 2018. Standard Terminology Relating to Nanotechnology. Annual Book of Standards, ASTM International, West Conshohocken, PA.

Atta, A.M., Al-Lohedan, H.A., and Al-Hussain, S.A. 2015. Functionalization of magnetite nanoparticles as oil spill collector. Int. J. Mol. Sci. 16: 6911–6931.

Baghbani-Arania, F., Movagharniaa, R., Sharifiana, A., Salehi, S., and Shandiz, S.A.S. 2017. Photo-catalytic, anti-bacterial, and anti-cancer properties of phyto-mediated synthesis of silver nanoparticles from *Artemisia tournefortiana* Rchb extract. J. Photochem. Photobiol. B: Biol. 173: 640–649.

Bakry, R., Vallant, R.M., Najam-ul-Haq, M., Rainer, M., Szabo, Z., Huck, C.W., and Bonn, G.K. 2007. Medicinal applications of fullerenes. Int. J. Nanomed. 2(4): 639–649.

Bakry, R., Vallant, R.M., Najam-ul-Haq, M., Rainer, M., Szabo, Z., Huck, C.W., and Bonn, G.K. 2007. Medicinal applications of fullerenes. Int. J. Nanomed. 2(4): 639–649.

Balasooriya, E.R., Jayasinghe, C.D., Jayawardena, U.A., Weerakkodige, U.A., Ruwanthika, R., Silva, D., De Udagama, R.M. and Vidya, P. 2017. Honey mediated green synthesis of nanoparticles: new era of safe nanotechnology. J Nanomater. 2017: 5919836. doi:10.1155/2017/5919836.

Barber, D.J., and Freestone, I.C. 1990. An investigation of the origin of the colour of the Lycurgus Cup by analytical transmission electron microscopy. Archaeometry. 32(1): 33–45.

Beck, M.T., and Mándi, G. 1997. Solubility of C60. Fullerenes Nanotubes Carbon Nanostruct. 5(2): 291–310.

Bello, B.A., Khan, S.A., Khan, J.A., Syed, F.Q., Anward, Y., Khan, S.B. 2017. Antiproliferation and antibacterial effect of biosynthesized AgNPs from leaves extract of Guiera senegalensis and its catalytic reduction on some persistent organic pollutants. J. Photochem. Photobiol. B. 175: 99–108.

Benelmekki, M. 2015. Chapter 1: Designing Hybrid Nanoparticles. Morgan and Claypool Publishers, San Rafael, CA.

Boca, S.C., Potara, M., Gabudean, A.M., Juhem, A., Baldeck, P.L., and Atilean, S. 2011. Chitosan-coated triangular nanoparticles as a novel class of biocompatible, highly effective photothermal transducers for in vitro cancer cell therapy. Cancer Lett. 311(2): 131–140.

Bogue, R. 2008. Biomimetic adhesives: a review of recent developments. Assembly Autom. 28(4): 282–288.

Boulenguez, J., Berthier, S., and Leroy, F. 2012. Multiple scaled disorder in the photonic structure of Morphorhetenor butterfly. Appl. Phys. A. 106: 1005–1011.

Bouville, F., Maire, E., Meille, S., Van de Moortèle, B., Stevenson, A.J., and Deville, S. 2014. Strong, tough and brittle bioinspired ceramics from brittle constituents. Nat. Mater. 13: 508–524.

Bratlie, K.M., Lee, H., Komvopoulos, K., Yang, P., Somorjai, G.A. 2007. Platinum nanoparticle shape effects on benzene hydrogenation selectivity. Nano Lett. 7: 3097–3101.

Brill, R.H. 1965. The chemistry of the Lycurgus Cup. Proc. 7th Int. Cong. Glass. 2(223): 1–13.

Brill, R.H., and Cahill, N.D. 1988. A red opaque glass from Sardis and some thoughts on red opaques in general. J. Glass Stud. 30: 16–27.

Cao, G., and Wang, Y. 2010. Nanostructures and Nanomaterials: Synthesis, Properties, and Applications, 2nd ed. World Scientific Publishing Ltd., London.

Capeness, M.J., Edmundson, M.C., and Horsfall, L.E. 2013. Nickel and platinum group metal nanoparticle production by Desulfovibrio alaskensis G20. New Biotechnol. 32(6): 727–731.

Chen, D., Li, X., Soule, T., Yorio, F., and Orr, L. 2016. Effects of solution chemistry on antimicrobial activities of silver nanoparticles against *Gordonia* sp. Sci. Total. Environ. 566: 360–367.

Chernousova, S., and Epple, M. 2013. Silver as antibacterial agent: ion, nanoparticle, and metal. Angew. Chem. Int. Ed. Engl. 52: 1636–1653.

Chithrani, B.D., Ghazani, A.A., and Chan, W.C.W. 2006. Determining the size and shape dependence of gold nanoparticle uptake into mammalian cells. Nano Lett. 6: 662–668.

Clark, D., and Pazdernik, N. 2015. Biotechnology: Applying the Genetic Revolution, 2nd ed. Elsevier BV, Amsterdam.

Cocuzza, M., Pirri, C., Rocca, V., and Verga, F. 2012. Current and future nanotech applications in the oil industry. Am. J. Appl. Sci. 9(6): 784–793.

Colomban, P. 2009. The use of metal nanoparticles to produce yellow, red and iridescent colour, from Bronze age to present times in lustre pottery any glass: solid state chemistry, spectroscopy and nanostructure. J. Nano Res. 8: 109–132.

Crundwell, F., Moats, M., Ramachandran, V., Robinson, T., and Davenport, W.G. 2011. Extractive Metallurgy of Nickel, Cobalt and Platinum-group Metals Overview. Elsevier Ltd., Oxford, 1–18.

Da Costa, M.V., Doughan, S., Han, Y., and Krull, U.J. 2014. Lanthanide upconversion nanoparticles and applications in bioassays and bioimaging: a review. Anal. Chim. Acta. 832: 1–33.

Dai, Y., Wang, Y., Liu, B., and Yang, Y. 2015. Metallic nano catalysis: an accelerating seamless integration with nanotechnology. Small. 11: 268–289.

Daisy, P., and Saipriya, K. 2012. Biochemical analysis of *Cassia fistula*aqueous extract and phytochemically synthesized gold nanoparticles as hypoglycemic treatment for diabetes mellitus. Int. J. Nanomed. 7: 1189–1202.

Das, S.K., Das, A.R., and Guha, A.K. 2010. Microbial synthesis of multishaped gold nanostructures. Small. 6(9): 1012–1021.

Delgado, J., Vilarigues, M., Ruivo, A., Corregidor, V., da Silva, R.C., Alves, L.C. 2011. Characterization of medieval yellow silver stained glass from Convento de Cristo in Tomar, Portugal. Nucl. Instrum. Methods B. 269: 2383–2388.

Douglas, K. 2016. DNA Nanoscience: From Prebiotic Origins to Emerging Nanotechnology. CRC Press, Taylor & Francis Group, Boca Raton, FL.

Dreaden, E.C., Alkilany, A.M., Huang, X., Murphy, C.J., and El-Sayed, M.A. 2012. The golden age: gold nanoparticles for biomedicine. Chem. Soc. Rev. 41: 2740–2779.

Duncan, T.V. 2011. Applications of nanotechnology in food packaging and food safety: barrier materials, antimicrobials and sensors. J. Coll. Interface. Sci. 363: 1–24.

Durán, N., Nakazato, G., and Seabra, A.B. 2016. Antimicrobial activity of biogenic silver nanoparticles, and silver chloride nanoparticles: an overview and comments. Appl. Microbiol. Biotechnol. 100(15): 6555–6570.

Elamawi, R.M., Al-Harbi, R.E., and Hendi, A.A. 2018. Biosynthesis and characterization of silver nanoparticles using *Trichoderma longibrachiatum* and their effect on phytopathogenic fungi. Egypt. J. Biol. Pest. CO. 28(28). DOI 10.1186/s41938-018-0028-1.

El-Diasty, A.I., and Ragab, A.M.S. 2013. Applications of nanotechnology in the oil & gas industry: latest trends worldwide & future challenges in Egypt. SPE 164716, The North Africa Technical Conference & Exhibition held in Cairo, Egypt.

El-Gendy, N.Sh., and Nassar, H.N. 2018. Biodesulfurization in Petroleum Refining. Wiley & Sons, Hoboken, NJ, and Scrivener Publishing LLC, MA, USA. https://doi.org/10.1002/9781119224075

El-Gendy, N.Sh., and Speight, J.G. 2016. Handbook of Refinery Desulfurization. CRC Press, Taylor & Francis Group, Boca Raton, FL. https://doi.org/10.1201/b19102

El-Gendy, N.Sh., El-Salamony, R.A., and Younis, S.A., 2016. Green synthesis of fluorapatite from waste animal bones and the photo-catalytic degradation activity of a new ZnO/green biocatalyst nano-composite for removal of chlorophenols. J. Water Process Eng. 12: 8–19. https://doi.org/10.1016/j.jwpe.2016.05.007

Elnashaie, S.S.E.H., Danafar, F., and Rafsanjani, H.H. 2018. Review about development from nanotechnology to nano-engineering. Res. Dev. Material. Sci. 6(3): RDMS.000636.2018. DOI: 10.31031/RDMS.2018.06.000636.

Eppler, A.S., Rupprechter, G., Anderson, E.A., and Somorjai, G.A. Thermal and chemical stability and adhesion strength of Pt nanoparticle arrays supported on silica studied by transmission electron microscopy and atomic force microscopy. J. Phys. Chem. B. 104(31): 7286–7292.

Esmaeilnezhad, E., Karimian, M., and Choi, H.J. 2019. Synthesis and thermal analysis of hydrophobic iron oxide nanoparticles for improving in-situ combustion efficiency of heavy oils. J. Ind. Eng. Chem. 71: 402–409.

Etim, U.J., Bai, P., and Yan, Z. 2018. Chapter 2, Nanotechnology applications in petroleum refining. In: Nanotechnology in Oil and Gas Industries, Topics in Mining, Metallurgy and Materials Engineering. T.A. Saleh (Editor). Springer International Publishing AG, Basel.

Fakoya, M.F., Patel, H., and Shah, S.N. 2018. Nanotechnology: innovative applications in the oil & gas industry. Int. J. Glob. Adv. Mater. Nanotechnol. 1(1): 16–30.

Fakruddin, M., Hossain, Z., and Afroz, H. 2012. Prospects and applications of nanobiotechnology: a medical perspective. J. Nanobiotechnol. 10: 31–37.

Falkiewicz-Dulik, M., and Macura, A.B. 2008. Nanosilver as substance biostabilising footwear materials in the foot mycosis prophylaxis. Mikologia Lekarska. 15: 145–150.

Faraday, M. 1857. Experimental relations of gold (and other metals) to light. Philos. Trans. R. Soc. Lond. 147: 145–181.

Felix, D.A., and Ahmed, W. (Editors). 2014. Nanobiotechnology. One Central Press Ltd., Manchester.

Feng, Q.L., Wu, J., Chen, G.Q., Cui, F.Z., Kim, T.N., and Kim, J.O. 2000. A mechanistic study of the antibacterial effect of silver ions on *Escherichia coli* and *Staphylococcus aureus*. J. Biomed. Mater. Res. 52: 662–668.

Ferrando, R., Jellinek, H., and Johnston, R.L. 2008. Nanoalloys: from theory to applications of alloy clusters and nanoparticles. Chem. Rev. 108(3): 845–910.

Feynman, R.P. 1960. There's plenty of room at the bottom. Eng. Sci. 23(5): 22–36.

Freestone, I., Meeks, N., Sax, M., and Higgitt, C. 2007. The Lycurgus Cup—a Roman nanotechnology. Gold Bull. 40(4): 270–277.

Galvis, H.M.T., Bitter, J.H., Khare, C.B., Ruitenbeek, M., Dugulan, A.I., and de Jong, K.P. 2012. Supported iron nanoparticles as catalysts for sustainable production of lower olefins. Science. 335: 835–838.

Gan, P.P., Ng, S.H., Huang, Y., and Li, S.F.Y. 2012. Green synthesis of gold nanoparticles using palm oil mill effluent (POME): a low-cost and eco-friendly viable approach. Bioresour. Technol. 113: 132–135.

Gazit, E. 2007. Plenty of Room for Biology at The Bottom: An Introduction to Bionanotechnology. Imperial College Press-World Scientific Publishing, London.

Gilaki, M. 2010. Biosynthesis of silver nanoparticles using plant extracts. J. Biol. Sci. 10: 465–467.

Gujrati, M., Malamas, A., Shin, T., Jin, E., Sun, Y., and Lu, Z.-R. 2014. Multifunctional cationic lipid-based nanoparticles facilitate endosomal escape and reduction-triggered cytosolic siRNA release. Mol. Pharm. 11, 2734–2744.

Gurunathan, S., Kyung-Jin, L., Kalishwaralal, K., Sheikpranbabu, S., Vaidyanathan, R., Eom, S.H., 2009. Antiangiogenic properties of silver nanoparticles. Biomaterials. 30: 6341–6350.

Guzmán, M., Jean, G.D., Stephan, G. 2009. Synthesis of silver nanoparticles by chemical reduction method and their antibacterial activity. Int. J. Chem. Biomol. Eng. 2: 104–111.

Han, X., Zheng, Y., Munro, C.J., Ji, Y., and Braunschweig, A.B. 2015. Carbohydrate nanotechnology: hierarchical assembly using nature's other information carrying biopolymers. Curr. Opin. Biotechnol. 2(34): 44–47.

Hasan, S.A. 2015. Review on nanoparticles: their synthesis and types. Res. J. Recent Sci. 4: 9–11.

He, L., Xu, J., and Bin, D. 2016. Application of nanotechnology in petroleum exploration and development. Petrol. Explor. Dev. 43(6): 1107–1115.

Heiligtag, F.J., and Niederberger, M. 2013. The fascinating world of nanoparticle research Mater. Today. 16(7/8): 262–271.

Hisatomi, T., Kubota, J., and Domen, K. 2014. Recent advances in semiconductors for photocatalytic and photoelectrochemical water splitting. Chem. Soc. Rev. 43: 7520–7535.

Huang, H., and Yang, X. 2004. Synthesis of polysaccharidestabilized gold and silver nanoparticles: a green method. Carbohydr. Res. 339: 2627–2631.

Huber, D.L. 2005. Synthesis, properties, and applications of iron nanoparticles. Small. 1(5): 482 –501.

Hunt, L.B. 1976. The true story of Purple of Cassius. Gold Bull. 9(4): 134–139.

Hurnaus, T., and Plank, J. 2015. Crosslinking of guar and HPG based fracturing fluids using ZrO_2 nanoparticles. SPE International Symposium on Oilfield Chemistry, 2015, The Woodlands, TX. https://doi.org/10.2118/173778-MS .

Husseiny, S.M., Salah, T.A., and Anter, H.A., 2015. Biosynthesis of size controlled silver nanoparticles by *Fusarium oxysporum*, their antibacterial and antitumor activities. Beni-Seuf Univ. J. Appl. Sci. 4: 225–231.

Iravani, S. 2011. Critical review—green synthesis of metal nanoparticles using plants. Green Chem. 13: 2638–2650.

Ivask, A., Kurvet, I., Kasemets, K., Blinova, I., Aruoja, V., Vija, H., Käkinen, A., Titma, T., Heinlaan, M., Visnapuu, M., Koller, D., Kisand, V., and Kahru, A. 2014. Size-dependent toxicity of silver nanoparticles to bacteria, yeast, algae, crustaceans and mammalian cells in vitro. PLoS ONE 9(7): e102108. DOI: 10.1371/journal.pone.0102108.

Jacob, S., Finub, J., and Narayanan, A., 2011. Synthesis of silver nanoparticles using Piper longum leaf extracts and its cytotoxic activity against Hep-2 cell line. Colloids Surf. B. Biointerfaces. 91: 212–214.

Jahagirdar, S.R. 2008. Oil-microbe detection tool using nano optical fibers. SPE Western Regional and Pacific Section AAPG Joint Meeting, Bakersfield, CA; 2008

Jena, J., Pradhan, N., Nayak, R.R., Dash, B.P., Sukla, L.B., Panda, P.K., and Mishra, B.K., 2014. Microalga *Scenedesmus* sp.: a potential low-cost green machine for silver nanoparticle synthesis. J. Microbiol. Biotechnol. 24: 522–533.

Jiang, J., Oberdorster, G., and Biswas, P. 2009. Characterization of size, surface charge and agglomeration state of nanoparticle dispersions for toxicological studies. J. Nanopart. Res. 11: 77–89.

Joerger, R., Klaus, T., and Granqvist, C.G., 2000. Biologically produced silver–carbon composite materials for optically functional thin-film coatings. Adv. Mater. 12: 407–409.

Johnson-McDaniel, D., Barrett, C.A., Sharafi, A., and Salguero, T.T. 2013. Nanoscience of an ancient pigment. J. Am. Chem. Soc. 135(5): 1677–1679.

José-Yacamán, M., Rendo´n, L., Arenas, J., and Serra Puche, M.C. 1996. Maya blue paint: an ancient nanostructured material. Science. 273: 223–225.

Jung, W.K., Koo, H.C., Kim, K.W., Shin, S., Kim, S.H., and Park, Y.H. 2008. Antibacterial activity and mechanism of action of the silver ion in *Staphylococcus aureus* and *Escherichia coli*. Appl. Environ. Microbiol. 74(7): 2171–2178.

Kakoulli, I., Radpour, R., Lin, Y., Svoboda, M., and Fischer, C. 2017. Application of forensic photography for the detection and mapping of Egyptian blue and madder lake in Hellenistic polychrome terracottas based on their photophysical properties. Dyes Pigments. 136: 104–115.

Kannan, N., Selvaraj, S., and Murty, R.V. 2010. Microbial production of silver nanoparticles. Dig. J. Nanomater. Bios. 5(1): 135–140.

Kapusta, S., Balzano, L., and te Riele, P.M. 2011. Nanotechnology applications in oil and gas exploration and production. International Petroleum Technology Conference. Society of Petroleum Engineers, Bangkok. https://doi.org/10.3997/2214-4609-pdb.280.iptc15152_noPW.

Karthik, L., Kumar, G., Kirthi, A.V., Rahuman, A., and Rao, K.B. 2014. *Streptomyces* sp. LK3 mediated synthesis of silver nanoparticles and its biomedical application. Bioprocess Biosyst. Eng. 37: 261–267.

Khademolhosseini, R., Jafaria, A., and Shabani, M.H. 2015. Micro scale investigation of enhanced oil recovery using nano/bio materials. Procedia Mater. Sci. 11: 171–175.

Khalil, M., Jan, B.M., Tong, C.W., and Berawi, M.A. 2017. Advanced nanomaterials in oil and gas industry: design, application and challenges. Appl. Energy. 191: 287–310.

Khalil, M., Lee, R.L., and Liu, N. 2015. Hematite nanoparticles in aquathermolysis: a desulfurization study of thiophene. Fuel. 145: 214–220.

Khandel, P., and Vishwavidyalaya, G.G. 2016. Microbes mediated synthesis of metal nanoparticles: current status and future prospects. Int. J. Nanomater. Bios. 6: 1–24.

Khodadadi, A., Torabiangajia, M., Talebizadeh Rafsanjania, A., and Yonesib, A. 2012. Adsorptive desulfurization of diesel fuel with nano copper oxide (CuO). Proceedings of the 4th International Conference on Nanostructures (ICNS4), 12–14.

Kim, K.-J., Sung, W.S., Moon, S.-K., Choi, J.-S., Kim, J.G., and Lee, D.G. 2008. Antifungal effect of silver nanoparticles on dermatophytes. J. Microbiol. Biotechnol. 18: 1482–1484.

Ko, S., Kim, E.S., Park, S., Daigle, H., Milner, T.E., Huh, C., Bennetzen, M.V., Geremia, G.A. 2016. Oil droplet removal from produced water using nanoparticles and their magnetic separation. SPE Annual Technical Conference and Exhibition. Society of Petroleum Engineers, Dubai. https://doi.org/10.2118/181893-MS.

Ko, S., Prigiobbe, V., Huh, C., Bryant, S.L., Bennetzen, M., and Mogensen, K. 2014. Accelerated oil droplet separation from produced water using magnetic nanoparticles. SPE Annual Technical Conference and Exhibition. Society of Petroleum Engineers, Amsterdam. https://doi.org/10.2118/170828-MS.

Krishnamoorti, R. 2006. Extracting the benefits of nanotechnology for the oil industry. J. Pet. Technol. 58(11): 24–26.

Krishnaraj, C., Jagan, E.G., Rajasekar, S., Selvakumar, P., Kalaichelvan, P.T., and Mohan, N. 2010. Synthesis of silver nanoparticles using *Acalypha indica* leaf extracts and its antibacterial activity against water borne pathogens. Colloid Surf. B. 76: 50–56.

Kröger, N., Deutzmann, R., and Sumper, M. 1999. Polycationic peptides from diatom biosilica that direct silica nanosphere formation. Science. 286: 1129–1132.

Kulkarni, R.R., Shaiwale, N.S., Deobagkar, D.N., and Deobagkar, D.D. 2015. Synthesis and extracellular accumulation of silver nanoparticles by employing radiation-resistant *Deinococcus radiodurans*, their characterization, and determination of bioactivity. Int. J. Nanomed. 10: 963–974.

Kumar, B., Smita, K., and Cumbal, L. 2016. One pot synthesis and characterization of gold nanocatalyst using *Sacha inchi* (*Plukenetia volubilis*) oil: green approach. IET Nanobiotechnol. 158: 55–60.

Kumar, C.G., and Mamidyala, S.K. 2011. Extracellular synthesis of silver nanoparticles using culture supernatant of *Pseudomonas aeruginosa*. Colloids Surf. B Biointerface. 84: 462–466.

Kumar, P.S., Balachandran, C., Duraipandiyan, V., Ramasamy, D., Ignacimuthu, S., and Al-Dhabi, N.A. 2015. Extracellular biosynthesis of silver nanoparticle using *Streptomyces* sp. 09 PBT 005 and its antibacterial and cytotoxic properties. Appl. Nanosci. 5: 169–180.

Kuppusamy, P., Mashitah, M.Y., Maniam, G.P, and Govindan, N. 2014. Biosynthesized gold nanoparticle developed as a tool for detection of HCG hormone in pregnant women urine sample. Asian Pac. J. Trop. Dis. 4(30): 237. DOI: 10.1016/S2222-1808(14)60538-7.

Kuppusamy, P., Yusoff, M.M., Maniam, G.P., Govindan, N. 2016. Biosynthesis of metallic nanoparticles using plant derivatives and their new avenues in pharmacological applications—an updated report. Saudi Pharm. J. 24: 473–484.

Leitão, I.M.V., and Seixas de Melo, J.S. 2013. Maya blue, an ancient guest–host pigment: synthesis and models. J. Chem. Educ. 90(11): 1493–1497.

Lesin, V.I., Koksharov, Y.A., and Khomutov, G.B. 2010. Magnetic nanoparticles in petroleum. Pet. Chem. 50(2): 102–105.

LewisOscar, F., MubarakAli, D., Nithya, C., Priyanka, R., Gopinath, V., Alharbi, N.S., and Thajuddin, N. 2015. One pot synthesis and anti-biofilm potential of copper nanoparticles (CuNPs) against clinical strains of *Pseudomonas aeruginosa*. Biofouling. 3(4): 379–391.

Li, Q., Wei, B., Xue, Y., Wen, Y., Li, J. 2016. Improving the physical properties of nano-cellulose through chemical grafting for potential use in enhancing oil recovery. J. Bioresour. Bioprod. 1(4): 186–191.

Li, Y., Leung, P., Yao, L., Song, Q.W., and Newton, E. 2006. Antimicrobial effect of surgical masks coated with nanoparticles. J. Hosp. Infect. 62: 58–63.

Liang, J., Du, N., Song, S., and Hou, W. 2015. Magnetic demulsification of diluted crude oilin-water nanoemulsions using oleic acid-coated magnetite nanoparticles. Colloids Surf. A. 466: 197–202.

Liang, M., Wei, S., Jian-Xin, L., Xiao-Xi, Z., Zhi, H., Wen, L., Zheng-Chun, L., Jian-Xin, T., 2017. Optimization for extracellular biosynthesis of silver nanoparticles by *Penicillium aculeatum* Su1 and their antimicrobial activity and cytotoxic effect compared with silver ions. Mater. Sci. Eng. C 77: 963–971.

Lima, E., Guerra, R., Lara, V., Guzmán, A. 2013. Gold nanoparticles as efficient antimicrobial agents for *Escherichia coli* and *Salmonella typhi*. Chem. Cent. J. 7(1): 11. DOI: 10.1186/1752-153X-7-11.

Liu, H., Jin, X., and Ding, B. 2016. Application of nanotechnology in petroleum exploration and development. Petrol. Explor. Develop. 43(6): 1107–1115.

Louis, C. 2012. Gold nanoparticles in the past: before the nanotechnology era. In: Gold Nanoparticles for Physics, Chemistry and Biology. Louis, C., Pluchery, O. (eds.). Imperial College Press, World Scientific Publishing, London.

Makhlouf, A.S.H., and Barhoum, A. (Editors). 2018. Fundamentals of Nanoparticles: Classifications, Synthesis Methods, Properties and Characterization. Elsevier BV, Amsterdam.

Malvindi, M.A., De Matteis, V., Galeone, A., Brunetti, V., Anyfantis, G.C., Athanassiou, A., Cingolani, R., and Pompa, P. 2014. Toxicity assessment of silica coated iron oxide nanoparticles and biocompatibility improvement by surface engineering. PLoS One 9(1): e85835. DOI: 10.1371/journal.pone.0085835.

Manikanth, S.B., Kalishwaralal, K., Sriram, M., Pandian, S.R.K., Hyung-seop, Y., Eom, S.H., and Gurunathan, S. 2010. Anti-oxidant effect of gold nanoparticles restrains hyperglycemic conditions in diabetic mice. J. Nanobiotechnol. 8: 77–81.

Mansha, M., Khan, I., Ullah, N., and Qurashi, A. 2017. Synthesis, characterization and visible-light-driven photoelectrochemical hydrogen evolution reaction of carbazole-containing conjugated polymers. Int. J. Hydrogen Energy. 42(16): 10952–1096.

Marambio-Jones, C., and Hoek, E.M.V. 2010. A review of the antibacterial effects of silver nanomaterials and potential implications for human health and the environment. J. Nanopart. Res. 12: 1531–1551.

Marcato, P.D., Duran, M., Huber, S.C., Rai, M., Melo, P.S., Alves, O.L., and Duran, N. 2012. Biogenic silver nanoparticles and its antifungal activity as a new topical transungual drug. J. Nano. Res. 20: 99–107.

Martinez-Castañón, G.A., Niño-Martınez, N., Martinez-Gutierrez, F., Martınez-Mendoza, J.R. and Ruiz, F. 2008. Synthesis and antibacterial activity of silver nanoparticles with different size, J. Nanopart. Res. 10: 1343–1348.

Mashaghi, S., Jadidi, T., Koenderink, G., and Mashaghi, A. 2013. Lipid nanotechnology. Int. J. Mol. Sci. 14, 4242–4282.

Matsunaga, T., Suzuki, T., Tanaka, M., and Arakaki, A. 2007. Molecular analysis of magnetotactic bacteria and development of functional bacterial magnetic particles for nano-biotechnology. Trends Biotechnol. 25: 182–188.

Matteo, C., Candido, P., Vera, R. R., and Francesca, V. 2012. Current and future nanotech applications in the oil industry. Am. J. Appl. Sci. 9(6): 784–793.

McFarland, A.D., and Van Duyne, R.P. 2003. Single silver nanoparticles as real-time optical sensors with zeptomole sensitivity. Nano Lett. 3: 1057–1062.

McNaught, A.D., and Wilkinson, A.R. (Editors). 1997. Compendium of Chemical Terminology: IUPAC Recommendations (2nd Edition). Blackwell Science, John Wiley & Sons Inc., Hoboken, NJ.
Mie, G. 1908. Beiträge zur Optik trüber Medien, speziell kolloidaler Metallösungen. Ann. Phys. Leipzig. 25: 377–445.
Mirshahghassemi, S., and Lead, J.R. 2015. Oil recovery from water under environmentally relevant conditions using magnetic nanoparticles. Environ. Sci. Technol. 49: 11729–11736.
Mo, R., Jiang, T., Di, J., Tai, W., and Gu, Z. 2014. Emerging micro and nanotechnology based synthetic approaches for insulin delivery. Chem. Soc. Rev. 43: 3595–3629.
Mobasser, S., and Firoozi, A.A. 2016. Review of nanotechnology applications in science and engineering. J. Civ. Eng. Urban. 6(4): 84–93.
Mohamedkhair, A.K., Drmosh, Q.A., and Yamani, Z.H. 2019. Silver nanoparticle-decorated tin oxide thin films: synthesis, characterization, and hydrogen gas sensing. Front. Mater. 6: 188. DOI: 10.3389/fmats.2019.00188.
Mohammed, M.I., Razak, A.A.A., and Shehab, M.A. 2017. Synthesis of nanocatalyst for hydrodesulfurization of gasoil using laboratory hydrothermal rig. Arab. J. Sci. Eng. 42(4): 1381–1387.
Montes, D., Cortés, F.B. and Franco, C.A. 2018. Reduction of heavy oil viscosity through ultrasound cavitation assisted by NiO nanocrystals-functionalized SiO_2 nanoparticles. DYNA. 85(207): 153–160.
Morais, M.G., Martins, V.G., Steffens, D., Pranke, P., and da Costa, J.A. 2014. Biological applications of nanobiotechnology. J. Nanosci. Nanotechnol. 14: 1007–1017.
Moret, R. 2006. Nanomonde: des nanosciences aux nanotechnologies. CNRS Editions, Paris.
Mubarak Ali, D., Thajuddin, N., Jeganathan, K., and Gunasekaran, M. 2011. Plant extract mediated synthesis of silver and gold nanoparticles and its antibacterial activity against clinically isolated pathogens. Colloid Surf. B. 85: 360–365.
Murugesan, S., Monteiro, O.R., and Khabashesku, V.N. 2016. Extending the lifetime of oil and gas equipment with corrosion and erosion resistant Ni-B-nanodiamond metal-matrix-nanocomposite coatings. Offshore Technology Conference, Houston, TX. https://doi.org/10.4043/26934-MS.
Muthuswamy, N., de la Fuente, J.L., Ochal, P., Giri, R., Raaen, S., Sunde, S., Rønninga, R., and Chen, D. 2013. Towards a highly-efficient fuel-cell catalyst: optimization of Pt particle size, supports and surface-oxygen group concentration. Phys. Chem. Chem. Phys. 15(11): 3803–3813.
Nagô, C., and Van de Voorde, M.H. 2014. Nanomaterials: doing more with less. In: Nanotechnology in a Nutshell. Atlantis Press, Paris, 55–70.
Nagy, A.J. and G. Mestl. 1999. High temperature partial oxidation reactions over silver catalysts. Appl. Catal. A-Gen. 188: 337–353.
Nassar, N.N., Hassan, A., and Pereira-Almao, P. 2011. Metal oxide nanoparticles for asphaltene adsorption and oxidation. Energy Fuels. 25: 1017–1023.
Nasr-El-Din, H.A., Gurluk, M.R., and Crews, J.B. 2013. Enhancing the performance of viscoelastic surfactant fluids using nanoparticles. EAGE Annual Conference and Exhibition Incorporating SPE Europec, London. https://doi.org/10.2118/164900-MS.
Needham, D., Arslanagic, A., Glud, K., Hervella, P., Karimi, L., Høeilund-Carlsen, P.-F., Kinoshita, K., Mollenhauer, J., Parra, E., Utoft, A., and Walke, P. 2016. Bottom up design of nanoparticles for anti-cancer diapeutics: "put the drug in the cancer's food". J. Drug Target. 1–21.

Niemeyer, C.M., and Mirkin, C.A. (Editors). 2005. Nanobiotechnology: Concepts, Applications and Perspectives. John Wiley & Sons Inc., Hoboken, New NJ.

Noorbakhsh, F. 2011. Antifungal effects of silver nanoparticle alone and with combination of antifungal drug on dermatophyte pathogen *Trichophyton rubrum*. Inter. Conf. Biosci. Biochem. Bioinf. 5: 364–367.

Nouailhat, A. 2008. An Introduction to Nanoscience and Nanotechnology. John Wiley & Sons Inc., Hoboken, NJ.

Olajire, A.A., Kareem, A., and Olaleke, A. 2017. Green synthesis of bimetallic Pt@Cu nanostructures for catalytic oxidative desulfurization of model oil. J. Nanostruct. Chem.7: 159–170.

Oliveira, M.M., Ugarte, D., Zanchet, D., and Zarbin, A.J.G. 2005. Influence of synthetic parameters on the size, structure, and stability of dodecanethiol-stabilized silver nanoparticles. J. Colloid Interface Sci. 292: 429–435.

Omran, B.A., Nassar, H.N., Fatthallah, N.A., Hamdy, A., El-Shatoury, E.H., and El-Gendy, N.Sh. 2018a. Waste upcycling of *Citrus sinensis* peels as a green route for the synthesis of silver nanoparticles. Energy Source. Part A. 40(2): 227–223. https://doi.org/10.1080/15567036.2017.1410597

Omran, B.A., Nassar, H.N., Fatthallah, N.A., Hamdy, A., El-Shatoury, E.H., and El-Gendy, N.Sh. 2018b. Characterization and antimicrobial activity of silver nanoparticles mycosynthesized by *Aspergillus brasiliensis*. J. Appl. Microbiol. 125: 370–382. https://doi.org/10.1111/jam.13776

Omran, B.A., Nassar, H.N., Younis, S.A., El-Salamony, R.A., Fatthallah, N.A., Hamdy, A., El-Shatoury, E.H., and El-Gendy, N.Sh. 2020. Novel mycosynthesis of cobalt oxide nanoparticles using *Aspergillus brasiliensis* ATCC 16404—optimization, characterization and antimicrobial activity. J. Appl. Microbiol. 128: 438–457. https://doi.org/10.1111/jam.14498

Omran, B.A., Nassar, H.N., Younis, S.A., Fatthallah, N.A., Hamdy, A., El-Shatoury, E.H., and El-Gendy, N.Sh. 2019. Physiochemical properties of *Trichoderma longibrachiatum* DSMZ 16517-synthesized silver nanoparticles for the mitigation of halotolerant sulphate-reducing bacteria. J. Appl. Microbiol. 126: 138–154. https://doi.org/10.1111/jam.14102

Oza, G., Pandey, S., Mewada, A., Kalita, G., and Sharon, M. 2012. Facile biosynthesis of gold nanoparticles exploiting optimum pH and temperature of freshwater alga *Chlorella pyrenoidusa*. Adv. Appl. Sci. Res. 3(3): 1405–1412.

Padovani, S., Sada, C., Mazzoldi, P., Brunetti, B., Borgia, I., Sgamellotti, A., Giulivi, A., D'Acapito, F., and Battaglin, G. 2003. Copper in glazes of renaissance luster pottery: nanoparticles, ions, and local environment. J. Appl. Phys. 93(12): 10058–10063.

Pal, S., Tak, Y.K., and Song, J.M. 2007. Does the antibacterial activity of silver nanoparticles depend on the shape of the nanoparticle? A study of the gram-negative bacterium Escherichia coli. Appl. Environ. Microbiol. 73(6): 1712–1720.

Palomo, J., and Filice, M. 2016. Biosynthesis of metal nanoparticles: novel efficient heterogeneous nanocatalysts. Nanomaterials. 6(5): 84–102.

Panigrahi, S., Kundu, S., Ghosh, S., Nath, S., and Pal, T. 2004. General method of synthesis for metal nanoparticles. J. Nanopart. Res. 6: 411–414.

Parashar, U.K., Saxena, P.S., and Srivastav, A. 2009. Bioinspired synthesis of silver nanoparticles. Dig. J. Nanomater. Bios. 4: 159–166.

Patel, V., Berthold, D., Puranik, P., and Gantar, M. 2015. Screening of cyanobacteria and microalgae for their ability to synthesize silver nanoparticles with antibacterial activity. Biotechnol. Rep. 5: 112–119.

Pavia-Sanders, A., Zhang, S., Flores, J.A., Sanders, J.E., Raymond, J.E., and Wooley, K.L. 2013. Robust magnetic/polymer hybrid nanoparticles designed for crude oil entrapment and recovery in aqueous environments. ACS Nano. 7(9): 7552–7561.

Peng, B., Tang., J., Luo, J., Wang, P., Ding, B., and Tam, K.C. 2018. Applications of nanotechnology in oil and gas industry: progress and perspective. Can. J. Chem. Eng. 96: 91–100.

Peng, J., Liu, Q., Xu, Z., and Masliyah, J. 2012a. Novel magnetic demulsifier for water removal from diluted bitumen emulsion. Energy Fuels. 26: 2705–2710.

Peng, J., Liu, Q., Xu, Z., and Masliyah, J. 2012b. Synthesis of interfacially active and magnetically responsive nanoparticles for multiphase separation applications. Adv. Funct. Mater. 22: 1732–1740.

Pereira, L., Dias, N., Carvalho, J., Fernandes, S., Santos, C., and Lima, N. 2014. Synthesis, characterization and antifungal activity of chemically and fungal-produced silver nanoparticles against *Trichophyton rubrum*. J. Appl. Microbiol. 117: 1601–1613.

Pileni, M.P. 1997. Nanosized particles made in colloidal assemblies. Langmuir 13, 3266–3276.

Prakasham, R.S., Buddana, S., Yannam, S., and Guntuku, G. 2012. Characterization of silver nanoparticles synthesized by using marine isolate *Streptomyces albidoflavus*. J. Microbiol. Biotechnol. 22: 614–621.

Premasudha, P., Venkataramana, M., Abirami, M., Vanathi, P., Krishna, K., and Rajendran, R. 2015. Biological synthesis and characterization of silver nanoparticles using *Eclipta alba* leaf extract and evaluation of its cytotoxic and antimicrobial potential. Bull. Mater. Sci. 38: 965–973.

Qj, F., and Lam, A. 2017. Biosynthesis of silver nanoparticles using peel extract of Raphanus sativus L. Biotechnol. Ind. J. 13: 1–10.

Qu, Y., Pei, X., Shen, W., Zhang, X., Wang, J., Zhang, Z., Li, S., You, S., Ma, F., and Zhou, J. 2017. Biosynthesis of gold nanoparticles by *Aspergillum* sp. WL-Au for degradation of aromatic pollutants. Physica E. 88: 133–141.

Quester, K., Avalos-Borja, M., and Castro-Longoria, E. 2013. Biosynthesis and microscopic study of metallic nanoparticles. Micron. 54–55: 1–27.

Rahimi, G., Alizadeh, F., and Khodavandi, A. 2016. Mycosynthesis of silver nanoparticles from *Candida albicans* and its antibacterial activity against *Escherichia coli* and *Staphylococcus aureus*. Trop. J. Pharm. Res. 15(2): 371–375.

Rai, M.K., Deshmukh, S. D., Ingle, A. P., Gade, A. K. 2012. Silver nanoparticles: the powerful nanoweapon against multidrug-resistant bacteria. Appl. Microbiol. 112(5): 841–852.

Rao, J.P., and Geckeler, K.E., 2011. Polymer nanoparticles: preparation techniques and size-control parameters. Prog. Polym. Sci. 36: 887–913.

Reibold, M., Paufler, P., Levin, A.A., Kochmann, W., Pätzke, N., and Meyer, D.C. 2006. Carbon nanotubes in an ancient Damascus sabre. Nature. 444: 286.

Reidy, B., Haase, A., Luch, A., Dawson, K.A., and Lynch, I. 2013. Mechanisms of silver nanoparticle release, transformation and toxicity: a critical review of current knowledge and recommendations for future studies and applications. Materials. 6: 2295–2350.

Rezk, M.Y., and Allam, N.K. 2019. Impact of nanotechnology on enhanced oil recovery: a mini review. Ind. Eng. Chem. Res. 58: 16287−16295.

Rezvani, M.A., Shojaie, A.F., and Loghmani, 2012. Synthesis and characterization of novel nanocomposite, anatase sandwich type polyoxometalate, as a reusable and green nano catalyst in oxidation desulfurization of simulated gas oil. Catal. Commun. 25: 36–40.

Roy, N., Gaur, A., Jain, A., Bhattacharya, S., and Rani, V. 2013. Green synthesis of silver nanoparticles: an approach to overcome toxicity. Environ. Toxicol. Pharmacol. 36(3): 807–812.

Rytwo, G. 2008. Clay minerals as an ancient nanotechnology: historical uses of clay organic interactions, and future possible perspectives. Macla. 9: 15–17.

Saed, D., Nassar, H.N., El-Gendy, N.Sh., Zaki, T., Moustafa, Y.M., and Badr, I.H.A. 2014. The enhancement of pyrene biodegradation by assembling MFe_3O_4 nano-sorbents on the surface of microbial cells. Energy Source Part A. 36(17): 1931–1937. https://doi.org/10.1080/15567036.2014.889782

Sanders, J.V. 1964. Color of precious opal. Nature. 204: 1151–1153.

Sathishkumar, M., Sneha, K., Won, S.W., Cho, C.W., Kim, S., and Yun, Y.S. 2009. Cinnamon zeylanicum bark extract and powder mediated green synthesis of nano-crystalline silver particles and its bactericidal activity. Colloid Surf. B. 73: 332–338.

Schaming, D., and Remita, H. 2015. Nanotechnology: from the ancient time to nowadays. Found. Chem. 17: 187–205.

Schröfel, A., Kratošová, G., Šafařík, I., Šafaříková, M., Raška, I., and Shor, L.M. 2014. Applications of biosynthesized metallic nanoparticles—a review. Acta Biomaterialia. 10: 4023–4042.

Sciau, P., Mirguet, C., Roucau, C., Chabanne, D., and Schvoerer, M. 2009. Double nanoparticle layer in a 12th century lustreware decoration: accident or technological mastery. J. Nano Res. 8: 133–139.

Semenov, K.N., Charykov, N.A., Keskinov, V.A., Piartman, A.K., Blokhin, A.A., and Kopyrin, A.A. 2010. Solubility of light fullerenes in organic solvents. J. Chem. Eng. Data. 55: 13–36.

ShamsiJazeyi, H., Miller, C.A., Wong, M.S., Tour, J.M., and Verduzco, R. 2014. Polymer-coated nanoparticles for enhanced oil recovery. J. Appl. Polym. Sci. 131: 40576. DOI: 10.1002/APP.40576.

Shan, G., Xing, J., Zhang, H., and Liu, H. 2005. Biodesulfurization of dibenzothiophene by microbial cells coated with magnetite nanoparticles. Appl. Environ. Microbiol. 71(8): 4497–4502.

Shankar, S.S., Rai, A., Ahmad, A., and Sastry, M. 2004. Rapid synthesis of Au, Ag and bimetallic Au core-Ag shell nanoparticles using Neem (Azadirachita indica) leaf broth. J. Colloid Interface Sci. 275: 496–502.

Shanthi, S., Jayaseelan, B.D., Velusamy, P., Vijayakumar, S., Chih, C.T., and Vaseeharan, B. 2016. Biosynthesis of silver nanoparticles using a probiotic *Bacillus licheniformis* Dahb1 and their antibiofilm activity and toxicity effects in *Ceriodaphnia cornuta*. Microb. Pathog. 93: 70–77.

Sharma, A.B, Sharma, M., and Pandey, R.K. 2009. Synthesis; properties and potential applications of semiconductor quantum particles. Asian J. Chem, 21(10): 5033–5038.

Sharma, G., Kumar, A., Sharma, S., Naushad, M., Dwivedi, R.P., ALOthman, Z.A., and Mola, G.T. 2017. Novel development of nanoparticles to bimetallic nanoparticles and their composites: a review. J. King Saud Univ. Sci. 31(2): 257–269.

Shi, D., Sadat, M.E., Dunn, A.W., and Mast, D.B. 2015. Photo-fluorescent and magnetic properties of iron oxide nanoparticles for biomedical applications. Nanoscale. 7: 8209–8232.

Shin, W.-K., Cho, J., Kannan, A.G., Lee, Y.-S., and Kim, D.-W. 2016. Cross-linked composite gel polymer electrolyte using mesoporous methacrylate-functionalized SiO_2 nanoparticles for lithium-ion polymer batteries. Sci. Rep. 6: 26332. http://dx.doi.org/10.1038/srep26332.

Shortland, A.J., Freestone, I.C., and Rohren, T. (Editors). 2009. From Mine to Microscope: Advances in the Study of Ancient Technology. Oxbow Books, Barnsley, South Yorkshire.

Simkiss, K., and Wilbur, K.M. 1989. Biomineralization, Academic Press, New York, NY, USA.

Simonsen, G., Strand, M., and Øye, G. 2018. Potential applications of magnetic nanoparticles within separation in the petroleum industry. J. Pet. Sci. Eng. 165: 488–495.
Simonsen, G., Strand, M., Norrman, J., and Øye, G. 2019. Amino-functionalized iron oxide nanoparticles designed for adsorption of naphthenic acids. Collids Surf. A. 568: 147–156.
Singh, T., Jyoti, K., Patnaik, A., Singh, A., Chauhan, R., and Chandel, S.S. 2017. Biosynthesis, characterization and antibacterial activity of silver nanoparticles using an endophytic fungal supernatant of *Raphanus sativus*. J. Gen. Eng. Biotechnol. 15: 31–39.
Sintubin, L., Windt, W.D., Dick, J., Mast, J., Ha, D., Verstraete, W., and Boon, N. 2009. Lactic acid bacteria as reducing and capping agent for the fast and efficient production of silver nanoparticles. Appl. Microbiol. Biotechnol. 84: 741–749.
Sleytr, U.B., Huber, C., Ilk, N., Pum, D., Schuster, B., and Egelseer, E.M. 2007. S-layers as a tool kit for nanobiotechnological applications. FEMS Microbiol. Lett. 267(2): 131–144.
Smitha, S.L., Nissamudeen, K.M., Philip, D., and Gopchandran, K.G. 2008. Studies on surface plasmon resonance and photoluminescence of silver nanoparticles. Spectrochim. Acta A. 71(1): 186–190.
Solga, A., Cerman, Z., Striffler, B.F., Spaeth, M., and Barthlott, W. 2007. The dream of staying clean: lotus and biomimetic surfaces. Bioinspir. Biomim. 2(4): 126–134.
Sosa, I.O., Noguez, C., and Barrera, R.G. 2003. Optical properties of metal nanoparticles with arbitrary shapes. J. Phys. Chem. B. 107(26): 6269–6275.
Speight, J.G. 2014. The Chemistry and Technology of Petroleum (5th Edition). CRC Press, Taylor & Francis Group, Boca Raton, FL, USA. https://doi.org/10.1016/C2015-0-02007-X
Speight, J.G. 2017. Handbook of Petroleum Refining. CRC Press, Taylor & Francis Group, Boca Raton, FL.
Speight, J.G., and El-Gendy, N.Sh. 2017. Introduction to Petroleum Biotechnology. Gulf Professional Publishing Company, Elsevier, Cambridge, MA, USA. https://doi.org/10.1016/C2015-0-02007-X
Sriramulu, M., and Sumathi, S. 2017. A mini review on fungal based synthesis of silver nanoparticles and their antimicrobial activity. Int. J. Chem. Tech. Res. 10: 367–377.
Strasser, P., Koh, S., Anniyev, T., Greeley, J., More, K., Yu, C., Liu, Z., Kaya, S., Nordlund, D., Ogasawara, H., Toney, M.F., and Nilsson, A. 2010. Lattice-strain control of the activity in dealloyed core–shell fuel cell catalysts. Nat. Chem. 2: 454–460
Sudha, S.S., Rajamanickam, K., and Rengaramanujam, J., 2013. Microalgae mediated synthesis of silver nanoparticles and their antimicrobial activity against pathogenic bacteria. Ind. J. Exp. Biol. 52: 393–399.
Suleimanov, B., Ismailov, F.S., and Veliyev, E.F. 2011. Nanofluid for enhanced oil recovery. J. Pet. Sci. Eng. 78(2): 431–437.
Sun, A., Zhang, Y., Chen, G., and Gai, Z. 2017. Application of nanoparticles in enhanced oil recovery: a critical review of recent progress. Energies. 10: 345–377.
Sun, S., 2000. Monodisperse FePt nanoparticles and ferromagnetic FePt nanocrystal superlattices. Science 80(287): 1989–1992.
Syafiuddin, A., Salmiati, Salim, M.R., Kueh, A.B.H., Hadibaratad, T., and Nur, H. 2017. A review of silver nanoparticles: research trends, global consumption, synthesis, properties, and future challenges. J. Chin. Chem. Soc. 64: 732–756.
Tamayo, L.A., Zapata, P.A., Vejar, N.D., Azócar, M.I., Gulppi, M.A., Zhou, X., Thompson, G.E., Rabagliati, F.M., and Páez, M.A. 2014. Release of silver and copper nanoparticles from polyethylene nanocomposites and their penetration into *Listeria monocytogenes*. Mater. Sci. Eng. C. 40: 24–31.

Taniguchi, N. 1974. On the basic concept of "nano-technology". Proceedings of International Conference on Production Engineering. Tokyo, Part II, Japan Society of Precision Engineering.

Thakkar, K.N., Mhatre, S.S., and Parikh, R.Y. 2010. Biological synthesis of metallic nanoparticles. Nanomedicine. 6: 257–262.

Tiwari, J.N., Tiwari, R.N., and Kim, K.S. 2012. Zero-dimensional, one dimensional, two-dimensional and three-dimensional nanostructured materials for advanced electrochemical energy devices. Prog. Mater Sci. 57: 724–803.

Vankar, P.S., and Bajpai, D. 2010. Preparation of gold nanoparticles from *Mirabilis jalapa* flowers. Indian J. Biochem. Biophys. 47: 157–160.

Vijayaraghavan, K., and Kamala Nalini, S.P. 2010. Biotemplates in the green synthesis of silver nanoparticles. Biotechnol. J. 5: 1098–1110.

Vivek, R., Thangam, R., Muthuchelian, K., Gunasekaran, P., Kaveri, K., and Kannan, S. 2012. Green biosynthesis of silver nanoparticles from *Annona squamosa* leaf extract and its in vitro cytotoxic effect on MCF-7 cells. Process Biochem. 47(12): 2405–2410.

Vollath, D. 2013. Nanomaterials: An Introduction to Synthesis, Properties and Applications (2nd Edition). John Wiley & Sons Inc., Hoboken, NJ.

Walter, P., Welcomme, E., Hallégot, P., Zaluzec, N.J., Deeb, C., Castaing, J., Veyssière, P., Bréniaux, R., Lévêque, J.L., and Tsoucaris, G. 2006. Early use of PbS nanotechnology for an ancient hair dyeing formula. Nano Lett. 6(10): 2215–2219.

Waszczuk, P., Barnard, T., Rice, M.C., Masel, R.I., and Wieckowsky, A. 2002. A nanoparticle catalyst with superior activity for electrooxidation of formic acid. Electrochem. Commun. 4: 599–603.

William, J.K.M., Ponmani, S., Samuel, R., Nagarajan, R., and Sangwai, J.S. 2014. Effect of CuO and ZnO nanofluids in xanthan gum on thermal, electrical and high pressure rheology of water-based drilling fluids. J. Pet. Sci. Eng. 117: 15–27.

Wu, D., Fan, W., Kishen, A., Gutmann, J.L., and Fan, B. 2014. Evaluation of the antibacterial efficiency of silver nanoparticles against *Enterococcus faecalis* biofim. J. Endod. 40(2): 285–290.

Yadav, K.K., Singh, J.K., Gupta, N., and Kumar, V. 2017. A Review of nanobioremediation technologies for environmental cleanup: a novel biological approach. J. Mater. Sci. Technol. 8: 740–757.

Yu, J., Berlin, J., Lu, W., Zhang, L., Kan, A., Zhang, P., Walsh, E.E., Work, S., Chen, W., Tour, J., Wong, M., and Tomson, M.B. 2010. Transport study of nanoparticles for oilfield application. SPE International Conference on Oilfield Scale, Aberdeen. https://doi.org/10.2118/131158-MS.

Zaheer, Z. 2018. Biogenic synthesis, optical, catalytic, and in vitro antimicrobial potential of Ag-nanoparticles prepared using Palm date fruit extract. J. Photochem. Photobiol. B: Biol. 178: 584–592.

Zakaria, B.S., Nassar, H.N., Saed, D., El-Gendy, N.Sh. 2015. Enhancement of carbazole denitrogenation rate using magnetically decorated *Bacillus clausii* BS1. Pet. Sci. Technol. 33(7): 802–811. https://doi.org/10.1080/10916466.2015.1014966

Zaman, M.S., Islam, M.R., and Mokhatab, S. 2012. Nanotechnology prospects in the petroleum industry. Pet. Sci. Technol. 30(10): 1053–1058.

Zhang, F., Nangreave, J., Liu, Y., and Yan, H. 2014. Structural DNA nanotechnology: state of the art and future perspective. J. Am. Chem. Soc. 136: 11198–11211.

CHAPTER 2

Phytosynthesis of Metal Nanoparticles

2.1 INTRODUCTION

Phytonanotechnology is referred to the science which depends on using different plant parts, such as stem, root, fruit, seed, callus, peel, leaves, and flower to synthesize metal/metal oxide nanoparticles (NPs) in various shapes and sizes (Figure 2.1) (Dubey et al., 2010a,b,c; Iravani, 2011; Mittal et al., 2013; Singh et al., 2016a; Yew et al., 2018). It has provided new avenues for the synthesis of NPs as it is eco-friendly, single-step technique, simple, rapid, stable, no need for high temperature, pressure, or toxic chemicals, and it is a cost-effective method (Firdhouse and Lalitha 2016a). Moreover, phytosynthesized NPs are produced mainly by using the universal solvent, that is, water, with the help of readily available plant materials' extract, as a reducing medium, which is nontoxic in nature. The biocompatibility of NPs, such as reduced metal cytotoxicity, is required for NPs in certain applications. Compared with physicochemically prepared NPs, biologically synthesized NPs are free from toxic by-products which could be attached to the NPs during the physiochemical synthesis and thus widen its applications (Ahmed and Ikram, 2015; Sabri et al., 2016). Not only this, but it has also no negative environmental impact compared with the physicochemical methods. It may solve some environmental issues, especially upon the utilization of agro-industrial wastes in the phytosynthesis of NPs. The most important advantage of phytosynthesis of NPs over microbial synthesis is easily scaled up for large-scale syntheses of NPs. Moreover, microbial synthesis is not of industrial feasibility due to the requirements of highly aseptic conditions and microbial cell maintenance. The rate of phytoreduction of metal ions has been found to be much faster compared to microbial ones, and the stability of the phytosynthesized NPs is much better (Zayed et al., 2012). Consequently, it is considered as the best platform for syntheses of NPs, as being free from toxic chemicals as well as providing natural capping agents for the stabilization of the NPs. Moreover, the use of plant extracts also reduces the cost of microorganisms' isolation, maintenance, and their culture media. All of this enhances the cost-competitive feasibility of phytosynthesis of NPs over microbial synthesis.

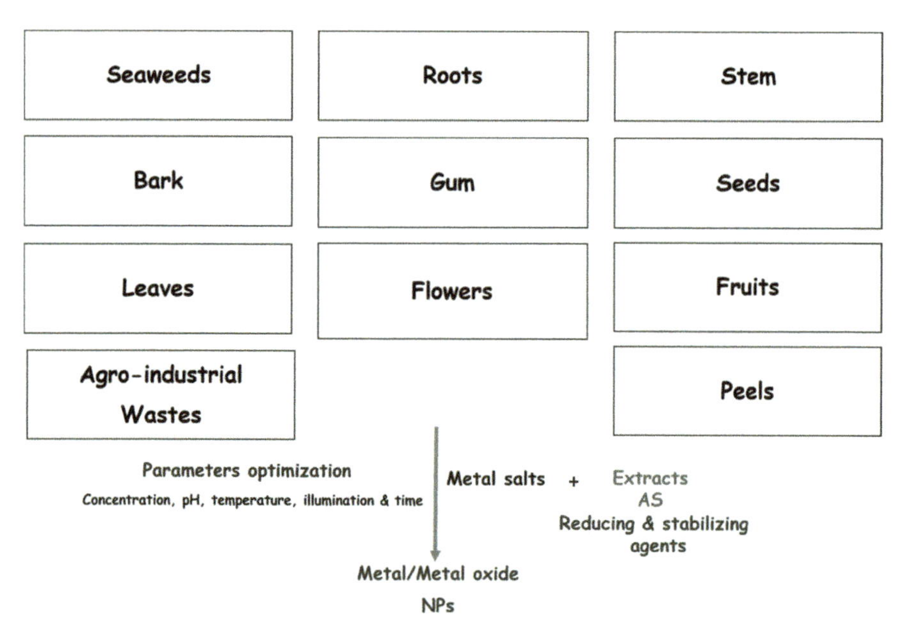

FIGURE 2.1 Materials for phytosynthesis of NPs.

The ability of plant extracts to reduce metal ions has been known since the early 1900s; however, the nature of the reducing agents involved was not well understood, until recently. Plants produce a vast and diverse assortment of organic compounds, the great majority of which do not appear to participate directly in growth and development. These substances traditionally are referred to as novel secondary metabolites, such as enzymes, vitamins, proteins, amino acids, phenolics, flavonoids, saponins, tannins, polysaccharides, alkaloids, and terpenoids (Figure 2.2). These compounds are mainly responsible for the reduction of ionic into bulk metallic NPs and act also as capping and stabilizing agents (Croteau et al., 2000; Vijayaraghavan and Nalini, 2010; Aromal and Philip, 2012; Tavakoli et al., 2015; Ahmed et al., 2016; Kuppusamy et al., 2016). Plant extracts act in one pot as reducing, capping, and stabilizing agents during the synthesis of NPs (Mittal et al., 2013). In the plant-based synthesis, several extracts (leaves, bark, stem, shoots, seeds, latex, secondary metabolites, roots, twigs, peel, fruit, seedlings, essential oils, tissue cultures, gum) were proved to generate particles (El-Gendy and Omran, 2019).

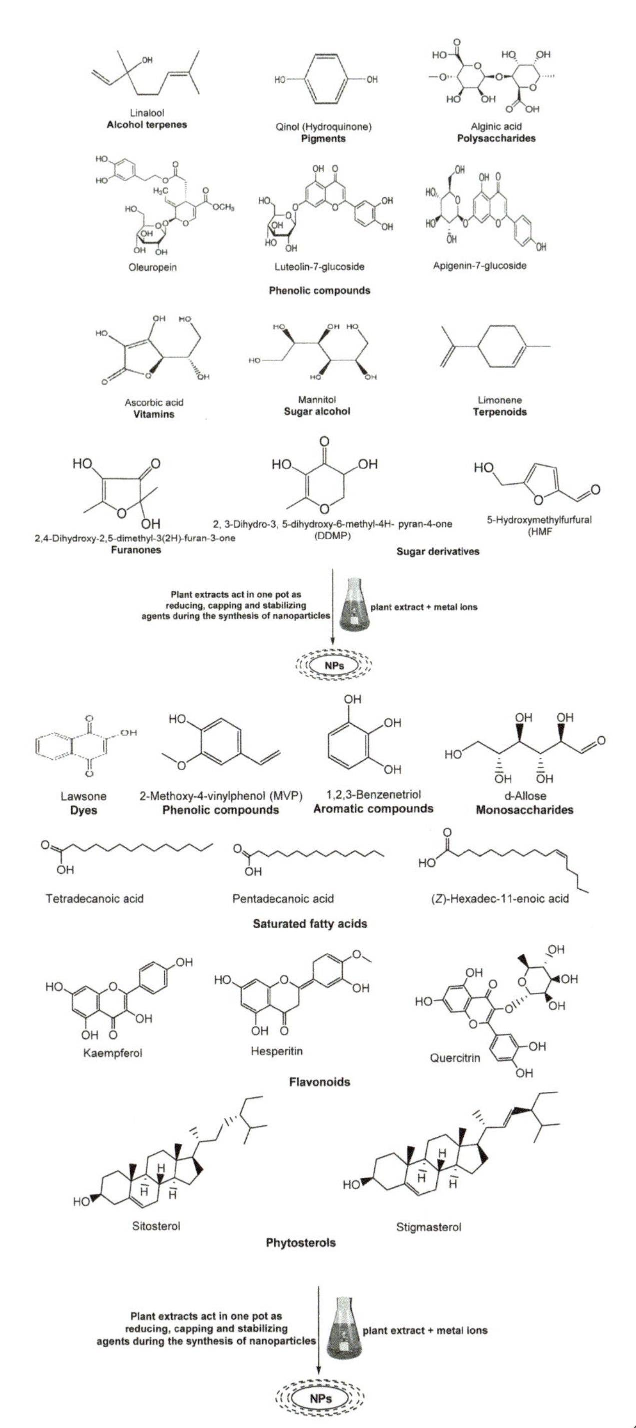
Linalool
Alcohol terpenes
Qinol (Hydroquinone)
Pigments
Alginic acid
Polysaccharides
Oleuropein
Luteolin-7-glucoside
Apigenin-7-glucoside
Phenolic compounds
Ascorbic acid
Vitamins
Mannitol
Sugar alcohol
Limonene
Terpenoids
2,4-Dihydroxy-2,5-dimethyl-3(2H)-furan-3-one
Furanones
2, 3-Dihydro-3, 5-dihydroxy-6-methyl-4H- pyran-4-one
(DDMP)
5-Hydroxymethylfurfural
(HMF
Sugar derivatives
Plant extracts act in one pot as reducing, capping and stabilizing agents during the synthesis of nanoparticles
plant extract + metal ions
NPs
Lawsone
Dyes
2-Methoxy-4-vinylphenol (MVP)
Phenolic compounds
1,2,3-Benzenetriol
Aromatic compounds
d-Allose
Monosaccharides
Tetradecanoic acid
Pentadecanoic acid
(Z)-Hexadec-11-enoic acid
Saturated fatty acids
Kaempferol
Hesperitin
Quercitrin
Flavonoids
Sitosterol
Stigmasterol
Phytosterols
Plant extracts act in one pot as reducing, capping and stabilizing agents during the synthesis of nanoparticles
plant extract + metal ions
NPs

(Continued)

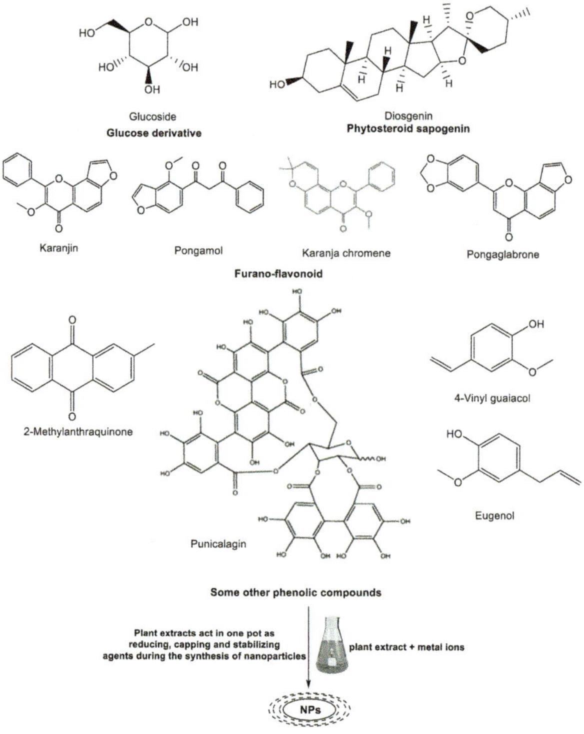

FIGURE 2.2 Some constituents of plant extracts involved in NPs phytosynthesis.

Terpenoids are a group of diverse organic polymers naturally synthesized in plants and composed of five-carbon isoprene units that display strong antioxidant activity (Makarov et al., 2014). Flavonoids are a large group of polyphenolic compounds that comprise several classes including anthocyanins, isoflavonoids, flavonols, chalcones, flavones, and flavanones, which can actively chelate and reduce metal ions to their NPs state (Makarov et al., 2014). Sugars are also exploited for the synthesis of metal NPs. It is known that monosaccharides such as glucose, due to their free aldehyde group act as reducing agents. Furthermore, the reducing ability of disaccharides and polysaccharides depends strongly on their type and concentration of individual monosaccharide components (Makarov et al., 2014). Proteins including different amino acids are capable of reducing several metal ions, resulting in the formation of NPs. Aqueous chitosan solution can act also as a stabilizing agent for the green synthesized NPs (Parida et al., 2011).

The constituents of the plant extract, its concentration, the concentration of the metal salt, the pH, temperature, contact time, mixing rate, and illumination intensities are known to affect the rate of production of the NPs, their quantity, shape, and other characteristics (Omran et al., 2018a).

Silver (Ag), gold (Au), and copper (Cu) NPs have been the particular focus of plant-based syntheses and usually the silver nitrate ($AgNO_3$), chloroauric acid (i.e., gold(III) chloride trihydrate $HAuCl_4 \cdot 3H_2O$), and copper(II) sulfate pentahydrate ($CuSO_4 \cdot 5H_2O$) are the main used precursors, respectively.

Several plants' parts have been reported for the phytosynthesis of metal NPs, such as Ag, Au, and Cu NPs. This chapter will summarize the most published reports concerning this and how the size and shape of metal NPs would differ with the changes in the physicochemical parameters of the phytogenesis process. The most important characterization instruments will be mentioned. The different extracts' components will be mentioned and a simplified discussion about the mechanisms involved in the phytogenesis of metal NPs will be also elucidated. The applications of some phytosynthesized metal NPs, for example, its antimicrobial efficiency, its application as a green reductants for some of the organic pollutants, and sensors for different pollutants will be illustrated. Since most of these aforementioned applications would be fruitful in the petroleum industry.

2.2 PHYTOSYNTHESIS OF SILVER NANOPARTICLES

Silver nanoparticles (AgNPs) are of great interest because of their unique properties, such as its size and shape depending on optical, electrical, and magnetic properties. Moreover, AgNPs are characterized by their stability in both alcoholic and aqueous suspensions without using any external capping agents (Singh et al., 2011). Also, because of its higher positive reduction potential, thus, its oxidation is thermodynamically unfavorable (Velmurugan et al., 2014). AgNPs are reported to have good electrical conductivity, photoelectrochemical activity, antimicrobial activity, and strong reduction power (Ping et al., 2018). Thus, AgNPs are intensively used in various industries and have wide applications. While AgNPs can be incorporated into antimicrobial applications, biosensor materials, composite fibers, cryogenic superconducting materials, cosmetic products, and electronic components. AgNPs have applications in water disinfection, textile industries, and agriculture (Ping et al., 2018). Moreover, AgNPs are reported to decrease cell proliferation and chemotaxis of the human mesenchymal stem cells, increase cytotoxicity, and oxidative stress of the human hepatoma HepG2 cells, and can have several other adverse effects (Syafiuddin et al., 2017). AgNPs can enhance the signal intensity of the surface-enhanced Raman scattering and can be used to improve the performance of solar cells (Syafiuddin et al.,

2017). The photocatalytic degradation capabilities of the green synthesized AgNPs on different dyes Methyl Violet, Safranin, Eosin methylene blue (MB), methyl orange (MO), and Rhodamine B (RhB) under sunlight illumination (Bhakya et al., 2015) and visible light irradiation Karthik et al. (2017) and also on 4-nitrophenol in presence of sunlight (Zaheer, 2018), has been reported. AgNPs can reduce several dyes and 4-nitophenol, in presence of $NaBH_4$ throughout the Langmuir–Hinshelwood model in which $NaBH_4$ acts as an electron and hydrogen donor changing the entire solution pH. Then, the surface of AgNPs becomes positively charged, consequently the BH_4^- and dyes become concomitantly adsorbed on the surface of AgNPs. Finally, the AgNPs then receive electrons from BH_4^- and transport them to the dye molecules or 4-nitrophenol. Moreover, the large volume of hydrogen supplied by $NaBH_4$ in the presence of AgNPs renders the hydrogenation of azo dyes and 4-nitophenol (Ahmed et al., 2015; Bonnia et al., 2016; Jyoti and Singh, 2016; Khan et al., 2016; Bello et al., 2017a,b; Saha et al., 2017; Ajitha et al., 2018; Ping et al., 2018). Furthermore, the large specific surface area of AgNPs helps in the desorption of the final product to colorless. AgNPs can be used in the synthesis of oil-based nanofluids, which have a lot of applications in the petroleum industry (Li et al., 2010). Furthermore, AgNPs are used as a catalyst in methanol oxidation to formaldehyde and ethylene oxide (Nagy and Mestl, 1999). One of the most important applications of AgNPs and CuNPs is their efficient application as antibiofilm agents (LewisOscar et al., 2015). AgNPs have been also reported as an efficient biocide for sulfate-reducing bacteria, the main cause of microbial corrosion in the oil industry (Omran et al., 2019). AgNPs-based adsorbate has been reported for the removal of mercury from gas condensate (Adio et al., 2019). AgNPs can be used for manufacturing high sensitivity sensors for low concentrations of gases, such as H_2, H_2S, CO, etc., (Mohamedkhair et al., 2019). However, it is reported in a review by Syafiuddin et al. (2017) that healthcare is the largest global consumer in AgNPs applications, with noticeable significant growth in its applications electronics and electrical fields.

In order to fulfill the requirement of AgNPs, several physical and chemical methods have been adopted for the synthesis and stabilization of AgNPs (Klaus-Joerger et al., 2001). Generally, conventional physical and chemical methods seem to be very expensive and hazardous (Gurunathan et al., 2015). However, the green chemistry approach for the synthesis of AgNPs shows much promise. Which seems to be simple, rapid, nontoxic, dependable, and eco-friendly, that can produce well-defined size and morphology under optimized conditions for translational research (Zhang et al., 2016). Interestingly, green synthesized AgNPs show high yield and high stability.

AgNPs have been usually synthesized using silver nitrate ($AgNO_3$). However, the applications of AgNPs are widely dependent on the size of these NPs.

2.2.1 THE ANTIMICROBIAL EFFECT OF AgNPs

One of the most important applications of AgNPs is their wide antimicrobial activity against different aerobic and anaerobic microorganisms which would form the corrosive biofilms that cause a drastic problem in the petroleum industry (Omran et al., 2018b, 2019).

There are several suggested mechanisms for the antimicrobial activity of AgNPs (Figure 2.3). It has been reported that silver is inert; however, on its contact with water, it is ionized to Ag^+, which can form complexes with nucleic acids and preferentially interact with the nucleosides rather than with the phosphate groups of nucleic acids. Moreover, electrostatic attraction between positively charged NPs and negatively charged bacterial cells can occur. It can accumulate inside the membrane and can subsequently penetrate the cells causing damage to the cell wall or cell membranes. Upon the entrance of Ag^+ ions in the cell, it intercalates between the purine and pyrimidine base pairs disrupting the hydrogen bonding between the two antiparallel strands and denaturing the DNA molecule. Furthermore, silver atoms bind to thiol groups (–SH) of enzymes forming stable S–Ag bonds with thiol-containing compounds, and then it causes the deactivation

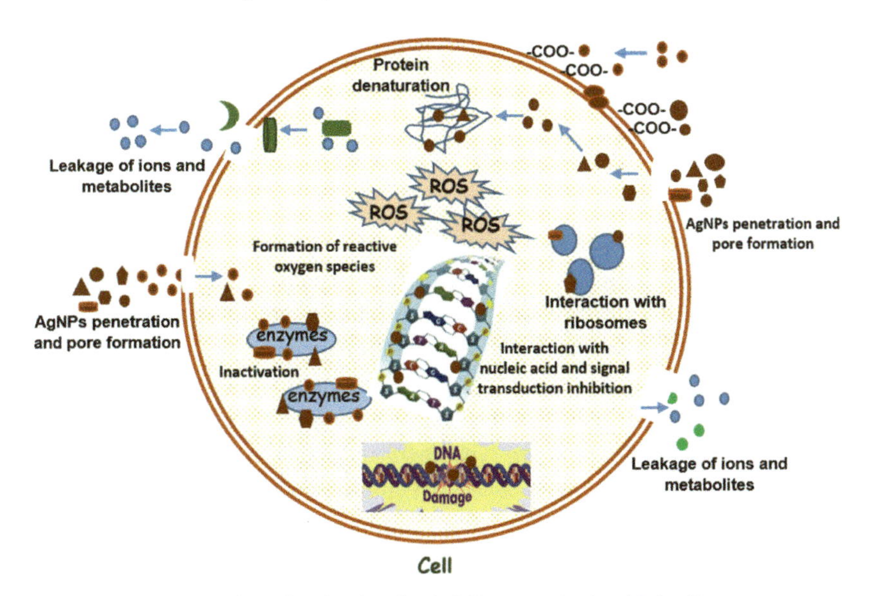

FIGURE 2.3 Suggested mode of action for AgNPs toward microbial cells.

of enzymes in the cell membrane that involves transmembrane energy generation and ion transport. Accumulation of AgNPs in bacterial membrane would cause an increase in permeability and death of the cell. Cell lysis is one of the reason for the antibacterial property of AgNPs. It also modulated phosphotyrosine profile of bacterial peptide that in turn affects signal transduction and inhibited the growth of microorganisms. Other mechanisms involve the interaction of silver molecules with biological macromolecules such as enzymes and DNA through an electron-release mechanism or free radical production. The inhibition of cell wall synthesis as well as protein synthesis was shown to be induced by silver NPs has been suggested by some literature with the proteomic data having evidence of accumulation of envelope protein precursor or destabilization of the outer membrane, which finally leads to adenosine triphosphate (ATP) leakage. Both silver NPs and silver ions can change the three-dimensional structure of proteins by interfering with disulfide bonds and block the functional operations of the microorganisms (Kuppusamy et al., 2016). It has been also reported that the phytogenic AgNPs cause membrane damage in *Candida* sp. and damage in fungal intercellular components and finally destroying cell function (Logeswari et al., 2012). Moreover, AgNPs have an efficient anti-fungal effect against spore-producing fungus and can effectively destroy fungal growth. The fungal cell wall is known to be made up of a high negatively charged polymer of fatty acid and protein which can interact efficiently with positively charged silver or gold ions, causing a significant change in fungal cell membrane structure (Gardea-Torresdey et al., 2002).

2.2.2 PHYTOSYNTHESIS OF AgNPs USING LEAVES' AND FLOWERS' EXTRACT

Geranium (*Pelargonium graveolens*) leaf extract has been reported to rapidly phytosynthesize a stable 16–40 nm AgNPs (Shankar et al., 2003). Moreover, Shankar et al. (2004a) proved that size of the phytosynthesized AgNPs (10–63 nm) differed according to the reaction pH and the concentration of neem (*Azadirachta indica*) leaf extract. Huang et al. (2007) reported that the phenolics, terpenoids, polysaccharides, and flavones compounds present in the leaf extract of *Cinnamomum camphora* reduced the Ag^+ ions into AgNPs with high bactericidal activity at a concentration of 45 μg/mL. Extracts of *Capsicum annuum* leaf were also used to phytosynthesize AgNPs (Li et al., 2007). While, flavonoid and terpenoid compounds present in the methanolic extract of *Eucalyptus hybrid* (safeda) leaves were claimed to be responsible for the synthesis of stabilization of AgNPs (Dubey et al., 2009). Leaf extract of

Datura metel that contains alkaloids, proteins, enzymes, amino acids, alcoholic compounds, and polysaccharides has been reported to phytosynthesize 16–40 nm AgNPs (Kesharwani et al., 2009). Moreover, the quinol and chlorophyll pigments present in the extract also contributed to the reduction of silver ions and stabilization of the NPs (Kesharwani et al., 2009). Monodispersed and spherical-shaped AgNPs with the average particle size of 9 nm and low surface plasmon peak of 446 nm, were phytosynthesized by *Hibiscus cannabinus* leaf extract. The ascorbic acid present in *H. cannabinus* leaf extract was suggested to be the main reducing agent. The phytosynthesized AgNPs expressed good antimicrobial activity against *E. coli*, *Proteus mirabilis*, and *Shigella flexneri* (Bindhu and Umadevi, 2013). The transmission electron microscopy (TEM), energy-dispersive spectroscopy (EDX), X-ray diffraction (XRD), and ultraviolet visible spectroscopy (UV/Vis) confirmed the phytoreduction of $AgNO_3$ to stable, spherical crystalline AgNPs with well-defined dimensions with an average size of 15–30 nm, using *Lippia citriodora* (Lemon Verbena) leaf aqueous extract (Cruz et al., 2010). The kinetic of phytoreduction was proportional to the reducing agent (i.e., the extract) concentration and was also enhanced by the increase of temperature from 25 to 95 °C. However, the time, temperature, and extract concentration did not influence significantly on the shape and size of the AgNPs. The high-performance liquid chromatography (HPLC) and mass spectrometry (MS) proved that the main compounds of the extract: verbascoside, isoverbascoside, chrysoeriol-7-O-diglucoronide, and luteonin-7-O-diglucoronide (Figure 2.4) were acting as reducing, capping, and stabilizing agents (Cruz et al., 2010).

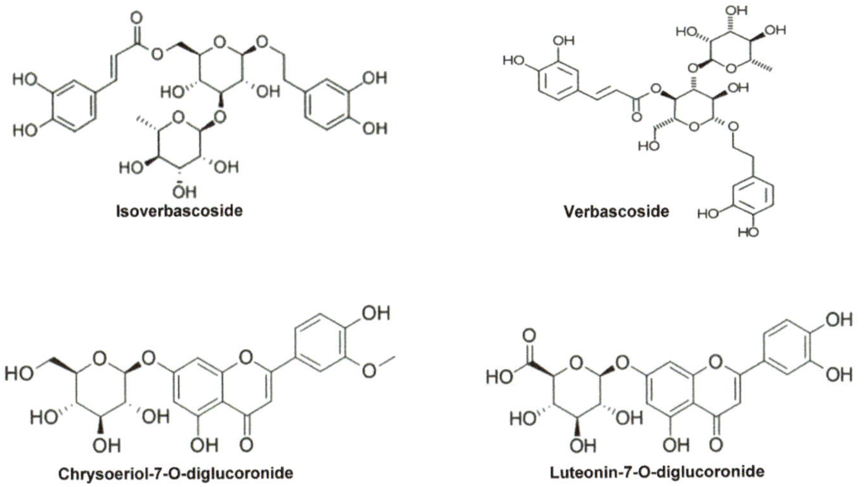

FIGURE 2.4 The main constituents of *Lippia citriodora* (Lemon Verbena) leaf aqueous extract.

Chenopodium album leaf extract was used to synthesize spherical-shaped AgNPs and gold nanoparticle (AuNPs) with the size range of 10–30 nm (Dwivedi and Gopal, 2010). Leaf extract of *Euphorbia hirta* phytosynthesized spherical-shaped 4–50 nm AgNPs (Elumalai et al., 2010). *Cycas* leaf was reported for the phytosynthesis of spherical-shaped 2–6 nm AgNPs (Jha and Prasad, 2010). Whereas the XRD-analysis revealed the face-centered cubic (fcc) unit cell structure of the prepared AgNPs having the sets of lattice planes (111), (200), (220), (311), and (222). The silver surface plasmon resonance (SPR) was observed at 449 nm, which steadily increased in intensity as a function of reaction time increment from 30 to 240 min. The nanotransformation (i.e., the reduction of Ag^+ ions to Ag^0 NPs) was attributed to the redox activities of ascorbic/dehydroascorbic acid and amento/hinoki flavones and the presence of ascorbates/glutathiones/metallothioneins (Scheme 2.1).

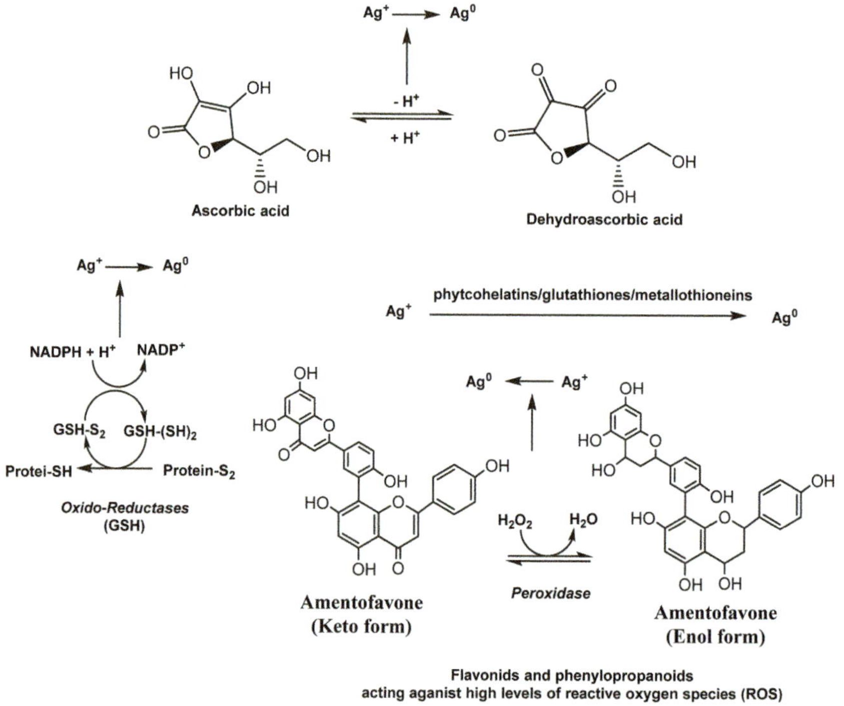

SCHEME 2.1 Proposed mechanism for phytosynthesis of AgNPs by *Cycas* leaf extract.

The rapid phytosynthesis of AgNPs was reported by Krishnaraj et al. (2010) using leaf extract of *Acalypha indica*, which was taken only 30 min.

The high-resolution transmission electron microscopy (HRTEM) analysis revealed spherical-shaped AgNPs with an average size of 20–30 nm. The phytosynthesized AgNPs were found to have effective antibacterial activity against water-borne pathogens; *E. coli* and *Vibrio cholera*, with the minimal inhibitory concentration of 10 μg/mL. That was attributed to the alteration in membrane permeability and respiration of the AgNPs treated bacterial cells (Krishnaraj et al., 2010).

AgNPs of triangular, decahedral, hexagonal, and spherical shapes were reported to be produced by varying the concentration of leaf extract of *Coleus amboinicu* (Narayanan and Sakthivel, 2010). Leaf extract of *Eucalyptus citriodora* (neelagiri) and *Ficus bengalensis* (marri) plants were reported to phytosynthesize AgNPs with an average size of 20 nm (Ravindra et al., 2010) in which cotton fibers loaded with these AgNPs expressed antibacterial toward *E. coli* (Ravindra et al., 2010). As reported by Raut et al. (2010) stable and crystalline AgNPs were formed by the reaction of an aqueous solution of 1 mM $AgNO_3$ with leaf extract of *Pongamia pinnata*. The UV/Vis spectroscopy studies were carried out to quantify the formation of AgNPs. TEM image disclosed that the phytosynthesized AgNPs were quite polydispersed and the size ranged from 20 to 50 nm with an average size of 38 nm. Water-soluble heterocyclic compounds such as flavones were mainly the responsible for the reduction and stabilization of the prepared NPs. The synthesized AgNPs were effective against *E. coli* (American type culture collection ATCC 8739), *S. aureus* (ATCC 6538p), *Pseudomonas aeruginosa* (ATCC 9027), and *K. pneumoniae* (clinical isolate). Kaviya et al. (2011a) reported the synthesis of silver nanoflakes using leaf extract of *Crossandra infundibuliform*. Prasad and Elumalai (2011) reported the phytosynthesis of 58 nm AgNPs by leaf extract of *Polyalthia longifolia*. The high concentration of ascorbic acid in *Ocimum sanctum* leaves extract was reported to be responsible for the rapid reduction of Ag^+ ions to 4–30 nm AgNPs, within only 8 min (Mallikarjun et al., 2011). Leaves extract of *A. indica* phytosynthesized 10 nm AgNPs within 30 min and when increasing the time to 4 h increased the size of the phytosynthesized AgNPs to 35 nm (Prathna et al., 2011a). AgNPs phytosynthesized by *O. sanctum* leaf extracts have been reported to express a high antimicrobial activity against both Gram-negative *E. coli* and Gram-positive *Streptococcus aureus* (Singhal et al., 2011). While Saxena et al. (2011) used an aqueous extract of *Ficus benghalensis* leaves to produce 16 nm AgNPs. The hot water extract of *Mollugo nudicaulis* leaves was used for the bioreduction of $AgNO_3$ to AgNPs with an average size of 9.3 nm (Anarkali et al., 2012). That showed good antimicrobial activity against *S. aureus*, *Vibrio cholerae*, *Micrococcus luteus*, and *Klebsiella pneumonia*. Neem (*A. indica*) leaf extract was used for the phytosynthesis of size-controlled AgNPs at room temperature (Balachandran et al., 2012), where the size ranged between 10 and 63 nm depending on the concentration of leaf extract and the pH of the reaction solution. Baskaralingam et al. (2012) reported antibacterial activity against *Vibrio alginolyticus* by AgNPs phytosynthesized by a leaf extract of *Calotropis gigantean*. Kouvaris et al. (2012) used a leaf extract of *Arbutus unedo* to produce AgNPs with a narrow size distribution. Spherical-shaped 4–15 nm AgNPs with fcc crystal structure were phytosynthesized by the hot water extract of *Paederia foetida* L. leaf extract and showed excellent antibacterial activity against different Gram classes of bacteria (Mollick et al., 2012). The highly stabilized 25–40 nm AgNPs, that were prepared using a leaf extract *Ocimum tenuiflorum*, were reported to express antibacterial toward the Gram-negative *E. coli* and Gram-positive *Cornyebacterium* and *B. subtilis* (Patil et al., 2012). Rapid phytosynthesis of crystalline fcc type AgNPs, within only 5 min, was reported using an aqueous extract of fresh leaves of *Prosopis juliflora* (Raja et al., 2012). A comparative study was performed by Zayed et al. (2012) to investigate the efficiency of five different plant leaf extracts: *Malva parviflora*, *Beta vulgaris* subsp. *Vulgaris*, *Anethum graveolens*, *Allium kurrat*, and *Capsicum frutescens*, for bioreduction of Ag+ ions to AgNPs. *M. parviflora* (Malvaceae) exhibited the best reducing and protecting action in terms of synthesis rate and monodispersity of the phytosynthesized AgNPs with an average size of 19–25 nm. The XRD analysis revealed a high degree of crystallinity and monophasic fcc structured AgNPs. The Fourier transform infrared spectroscopy (FTIR) analysis proved that the proteins secreted by the biomass acted as reducing, capping, and stabilizing agents. The single-pot process was used for phytosynthesis of spherical-shaped 5–40 nm AgNPs, within only 2 min, using carob (*Ceratonia siliqua*) leaf hot water extract, at room temperature. Atomic absorption spectroscopy analysis was used to follow up the conversion of Ag^+ ions into Ag^0, with time. The UV/Vis spectra showed the strong characteristic SPR band of AgNPs at 420 nm. The FTIR analysis showed that the carbonyl group of the protein amino acids present in carob leaf extract has a strong binding ability with metal, suggesting the formation of a layer covering AgNPs, and acting as a stabilizing agent to prevent agglomeration in the aqueous medium. Thus, the protein in carob extract acted as reducing, capping, and stabilizing agents. The phytosynthesized AgNPs expressed an effective antibacterial activity against the pathogenic *E. coli*, with minimum inhibitory concentration of 0.5 μg/L (Awwad et al., 2013). Gowri et al. (2013) reported the green synthesis of AgNPs using the hot water leaf extract of *P. santalinus*. UV/Vis spectra of the prepared

AgNPs showed an SPR peak at 418 nm. XRD results confirmed that the prepared AgNPs were fcc in structure. Energy dispersive X-ray (EDX) analysis showed peaks in the silver region at 3 keV indicating the presence of elemental silver. Aqueous leaf extract of *P. santalinus* contains steroids, saponins, tannins, phenols, tri-terpenoids, flavonoids, glycosides, and glycerides which were found to be responsible for bioreduction of $AgNO_3$ during the synthesis of AgNPs. Atomic force microscope (AFM) analysis demonstrated the size of the particles and was estimated to be 41 nm. The spherical shape of the prepared AgNPs was ascertained by scanning electron microscope (SEM). The synthesized AgNPs exhibited good antibacterial potential against Gram-positive and Gram-negative bacterial strains. Moreover, AgNPs with good antimicrobial activity were phytosynthesized by leaf extract of room dried leaves of *Catharanthus roseus* (Kotakadi et al., 2013). Whereas the average size range of 27 ± 2 and 30 ± 2 and Zeta potential of −63.1 mV, indicated the dispersion and stability of the prepared AgNPs. Koyyati et al. (2013) reported the utilization of *Raphanus sativus* var. longipinnatus leaf extract as a reducing agent in the synthesis of 5–22 nm AgNPs at 50 °C and within only 30 min. The prepared AgNPs expressed good antibacterial activity against *Pseudomonas putida, Klebsilla pneumonia, S. aureus*, and *B. subtilis*. It also showed good antioxidant capacity, thus can be used as a potential radical scavenger against deleterious damages caused by the free radicals. Logeswari et al. (2013) used the aqueous leaves extract of *Solanum tricobatum, Syzygium cumini, Centella asiatica*, and *Citrus sinensis* for the synthesis of 53, 41, 52, and 42 nm AgNPs, respectively. The FTIR spectroscopy confirmed the presence of protein acting as the stabilizing agent surrounding the AgNPs. The synthesized AgNPs expressed efficient antimicrobial activity against pathogenic bacteria. Prabakar et al. (2013) reported that the obvious color change to brown color and the SPR by UV/Visible spectroscopy of a well-observable peak at 440 nm confirming the phytosynthesis of AgNPs, using leaf extract of *Mukia scabrella*. Whereas the FTIR analysis indicated that the protein, in the leaf extract, acted as a possible capping agent. Energy dispersive X-ray (EDX) spectroscopy results showed a major signal for elemental silver. XRD analysis indicated the formation of metallic silver nanomaterials. Transmission electron microscopic study showed the NPs in the size range of 18–21 nm with the spherical shape. Zeta potential analysis showed −21.7 mV characteristic for stable AgNPs. The phytosynthesized AgNPs exhibited significant antimicrobial activity against the resistant Gram-negative bacteria MDR-GNB of the respiratory tract infections, for example, *Acinetobacter* sp., *K. pneumoniae*, and *P. aeruginosa*. The aqueous extract of *Lakshmi tulasi* (*O. sanctum*) leaf was used as a reducing and stabilizing agent, for the rapid phytosynthesis of AgNPs, within only 15 min (Rao et al., 2013). The XRD and the SEM coupled with X-ray energy dispersive spectroscopy (EDX) revealed that the prepared AgNPs were crystalline in nature and triangle shaped with an average size of 42 nm. The zeta potential of that prepared AgNPs was found to be −55 mV, where that large negative zeta potential value indicated the repulsion among AgNPs and thus, their dispersion stability. Singh et al. (2013a) reported the phytosynthesis of AgNPs using an aqueous extract of leaves of marine mangrove (*Rhizophora mucronata*). The extract acted as reducing, capping, and stabilizing agents. The alkaline pH favored the reduction of Ag^+ ions to Ag^0. Whereas small and highly dispersed AgNPs were formed at pH9.5 with the highest absorbance peak at 432 nm since large numbers of functional groups were available for silver binding. However, at pH 5.5 large NPs were formed with broadening of the SPR peak and redshift. Moreover, the size of NPs was also governed by the reaction temperature. The rate of reduction increased with increasing temperature, reaching its maximum with small AgNPs at 80 °C due to the reduction in an aggregation of the growing NPs. However, the yield decreased with higher temperature of 90 °C. However, a long duration of 96 h was required for the completion of the reaction. The FTIR analysis revealed that the AgNPs were surrounded by some proteins and secondary metabolites such as lignin, sterols, and alkaloids having functional groups of hydroxyl, amines, alcohols, phenol, and carboxylic acids. Moreover, the band shift in the hydroxyl groups and the increased band intensities for carbonyl groups in FTIR spectra confirmed the oxidation of these functional groups. Consequently, based on peak pattern, it can be inferred that both hydroxyl and carbonyl groups were involved in the synthesis of AgNPs. The XRD revealed the crystalline fcc structured AgNPs. The average size as calculated by Sherrer's equation was found to be 12 nm. The TEM proved that at higher temperatures most silver ions first formed nuclei, and the secondary growth of the particles stopped because the reaction rate was very high. The EDX analysis revealed an absorption peak, approximately at 3 keV due to the SPR of metallic silver. The antifungal activity of fluconazole and itraconazole was enhanced against some tested pathogenic fungi in the presence of those phytosynthesized Ag-NPs. A single-step green synthesis of AgNPs through the bioreduction of aqueous solution of $AgNO_3$ using *Psidium guajava* leaf water extract was reported by Gupta et al. (2014). A characteristic SPR peak appeared at around 487 nm. The XRD analysis proved the crystallinity nature and fcc geometry of the phytosynthesized AgNPs. TEM results revealed the presence of spherical-shaped particles with a mean diameter size of 60 nm. FTIR showed that the prepared NPs were capped with *P. guajava* bioactive molecules. The phytosynthesized AgNPs showed broad-spectrum antimicrobial activity against Gram-positive, Gram-negative human pathogenic bacteria, and fungi. Moreover, the phytosynthesized AgNPs showed prominent ability to inhibit the biofilms formed by *S. aureus*, *E. coli*, and *C. albicans* in a laboratory condition through crystal violet assay. Kumar et al. (2014), the green rapid syntheses of spherical-shaped 5–100 nm AgNPs using hot water extract of *Alternanthera dentate* leaves, within only 10 min. These AgNPs exhibited antibacterial activity against *P. aeruginosa, E. coli, K. pneumonia,* and *Enterococcus faecal*. Nakkala et al. (2014) reported the usage of the *Boerhaavia diffusa* plant extract as a reducing agent for green synthesis of 25 nm AgNPs. The XRD and TEM revealed fcc structure with the spherical shape, where the phytosynthesized AgNPs showed antibacterial activity against three fish bacterial pathogens; *Flavobacterium branchiophilum*, *Pseudomonas fluorescens*, and *Aeromonas hydrophila*. Rashidipour and Heydari (2014) reported a green synthetic approach using (*Olea europaea*) olive leaf water extract. The olive leaves' extract was prepared using an ultrasonic bath for 24 h. Parameters affecting the formation of AgNPs such as temperature, exposure time to extract, pH, concentration of silver nitrate ($AgNO_3$), and *O. europaea* extract were investigated and optimized. Optimum conditions for the synthesis of AgNPs were as follows: Ag^+ concentration 1 mM, extract concentration 8% (weight per volume) (w/v), pH 7, time 4 h and temperature 45 °C. SEM micrographs showed that the synthesized AgNPs were predominantly spherical in shape with an average size of 90 nm. Sivakamavalli et al. (2014) reported the role of acridone alkaloids, flavonoids, coumarines, terpenoids, and other volatile substances in the rapid phytosynthesis of AgNPs using *Ruta graveolens leaves aqueous* extract, which took only 30 min. The UV/Vis spectrum showed a characteristic sharp peak of AgNPs, at the wavelength of 440–560 nm. The XRD patterns confirmed that crystalline nature of AgNPs and FTIR results revealed that the phytochemical reaction of the constituents of the leaves' aqueous extract was responsible for the synthesis of AgNPs. The TEM showed spherical and triangular AgNPs with an average size around 30–50 nm. The phytosynthesized *R. graveolens* AgNPs showed an effective inhibitory activity against *S. aureus*, *P. aeruginosa*, and *C. albicans*. Ahlawat and Sehrawat (2015) reported the phytosynthesis of AgNPs using aqueous leaf extract of *Capparis decidua* (FORSK.). The stability of AgNPs after 2 months depicted almost no shift in the absorption intensity and the absorption maxima (452 nm) indicating the unchanged particle size. The FTIR analysis revealed that the stabilization was attributed to the presence of functional groups like amines, amides, alkynes, alkenes, bromoalkanes in the extract. However, the broadening of the UV-spectrum peak indicated that the particles are polydispersed. Moreover, the TEM analysis of AgNPs showed the formation of circular, triangular, rectangular, and oval-shaped NPs with average size ranged between 1.5 and 25 nm. Jafri et al. (2015) reported the phytosynthesis of spherical-shaped 50 nm AgNPs by marshmallow flower (*Althaea offinalis* L.), thyme (*Thmus vulgaris* L.), and pennyroyal (*Mentha pulegium* L.) leaf aqueous extracts with antimicrobial activity against fungi (*A. flavus* and P. *chrysogenum*) and bacteria (*E. coli* and *S. aureus*). According to Logeswari et al. (2015), *Ocimum tenuiflorum*, *S. tricobatum*, *S. cumini*, and *C. asiatica* leaves' hot water extracts were used to synthesize AgNPs. The formation and stability of the reduced AgNPs in the colloidal solution were monitored by UV/Vis spectrophotometer analysis. The mean particle diameter of AgNPs was calculated from the XRD pattern according to the line width of the plane and refraction peak using Scherer's equation. AFM showed the formation of AgNPs with an average size of 28, 65, 22.3, and 28.4 nm using leaf extract of *O. tenuiflorum*, *S. cumini*, *S. tricobatum*, and *C. asiatica*, respectively. Antimicrobial activity of the silver-green syhthesized NPs was performed by well diffusion method against *S. aureus*, *P. aeruginosa*, *E. coli*, and *K. pneumoniae*. The highest antimicrobial activity was expressed by AgNPs that were synthesized by *S. tricobatum* and *O. tenuiflorum* and was found to be 30 mm for *S. aureus* and *E. coli*. *Ziziphoratenuior* leaves were also used to prepare the AgNPs (Sadeghi and Gholamhoseinpoor, 2015). The TEM analysis revealed spherical and uniformly distributed AgNPs with average size from 8 to 40 nm. The FTIR analysis proved that the biomolecules that have primary amine groups, carbonyl groups, hydroxyl groups, and other stabilizing functional groups were responsible for the phytosynthesis of AgNPs. Ulug et al. (2015) reported the phytosynthesis of AgNPs by aqueous extract of *Ficuscarica* leaf and 5 mM silver nitrate, within 3 h and at 37 °C. Moreover, *Tinospora cordifolia* fresh leaves' hot water extract was used as natural reducing and capping agents (Selvam et al., 2016). The UV/Vis spectrum of AgNPs revealed a characteristic SPR peak at 430 nm. Different factors, which affect the production of AgNPs, were optimized and investigated using response surface methodology based Box Behnken design. Among the studied factors: $AgNO_3$ concentration, fresh weight of *T. cordifolia* leaf extract, incubation time, and pH. The optimum conditions were 1.25 mM of $AgNO_3$, 15 h for incubation time, 45 °C for temperature, and 4.5 for pH. The green synthesized AgNPs were characterized using XRD which revealed their crystalline nature and their average size was 30 nm as determined by using Scherer's equation. FTIR spectroscopy affirmed the role of *T. cordifolia* leaf extract as a reducing and capping agent. SEM–EDX showed the spherical shape of the prepared AgNPs and confirmed the presence of elemental silver. The antibacterial activity against *Staphylococcus* sp. (National center for biotechnology information NCBI-Accession: KC688883.1) and *Klebsiella* sp. (NCBI-Accession: KF649832.1) showed maximum inhibition zones of ≈ 13 and 12.3 mm, respectively, at 10 mg/mL of AgNPs. Ajitha et al. (2016) reported the phytosynthesis of AgNPs by hot water extract of *Sesbania grandiflora* leaves. Whereas the isotropic nature and spherical AgNPs were confirmed by the UV-spectrum with one symmetric SPR band centered at 416 nm. While the good crystalline fcc structure of the prepared AgNPs was confirmed by the XRD-analysis. The EDX ascertained the presence of metallic silver at ~3 keV. While the field emission scanning electron microscope (FESEM) and TEM proved dispersed, nonagglomerated, and spherical-shaped 16 nm AgNPs. The FTIR analysis proved the role of the polyphenolic compounds with hydroxyl and ketonic groups in the leaves' extract and their stronger ability to bind with Ag^+ ions, forming an intermediate complex with the free radical present in flavonoids of the extract. That subsequently undergoes oxidation to keto forms with consequent reduction of Ag^+ ions to AgNPs. Not only this, but the FTIR analysis also proved the role of the proteins in the leaves' extract in preventing the agglomeration of the AgNPs by forming a covering layer (i.e., capping of AgNPs) and thus, stabilizing the AgNPs in the medium. Whereas the main advantage of the phytosynthesized AgNPs was its high stability in an aqueous solution for 8 months at room temperature. Furthermore, the AgNPs expressed good antibacterial effect against pathogenic Gram-positive and negative bacteria; *Bacillus* spp., *Staphylococcus* spp., *E. coli*, and *Pseudomonas* spp., and antifungal effect against some pathogenic *A. niger* subsp., *A. flavus* subsp., and *Penicillium* spp. *Phlomis* species are reported to be rich sources of flavonoids, phenylpropanoids, and other phenolic compounds. Successful preparation of AgNPs was carried out using *Phlomis* hot water leaf extract (Allafchian et al., 2016). A strong broad peak around 440 nm of the prepared AgNPs was displayed using UV/Vis spectrophotometry. The XRD and FTIR confirmed the formation of AgNPs. While the SEM and TEM results indicated that AgNPs were spherical in shape with an average particle size of 25 nm. Moreover, the antibacterial activity of the phytosynthesized AgNPs against the Gram-positive bacteria *Staphyloccocus aureus* and *Bacillus cereus* and the Gram-negative bacteria *Salmonella typhimurium*, and *E. coli* was proved. The rapid phytoreduction of Ag^+ ions by aqueous leaf extract of *Polygonum hydropiper* was monitored using UV/Visible spectrophotometer and showed the formation of AgNPs within <30 min with a maximum absorption of AgNPs at 430 nm and average diameter of 60 nm (Bonnia et al., 2016). The phytosynthesized AgNPs acted as an effective photocatalyst in the reduction of MB dye in the presence of sodium borohydride $NaBH_4$ into photochemically inactive leucomethylene blue (LMB). In another study by Jyoti and Singh (2016), the aqueous leaf extract of *Zanthoxylum armatum* was used for phytoreduction of Ag^+ ions into spherical-shaped 15–50 nm AgNPs. In which, the green AgNPs demonstrated a catalytic degradation activity toward the hazardous dyes: Safranine O, Methyl Red, MO, and MB. Gorbe et al. (2016) reported the phytoreduction of $AgNO_3$ to functionalized AgNPs by hot water extract of pepino (*Solanum muricatum*) leaves, within 90 s, employing microwave oven irradiation. Although, microwave irradiation inactivates enzymes and proteins; however, the flavonoids and the phenolic compounds in the extract acted as reducing agents. Azizian-Shermeh et al. (2017) reported a rapid one-pot green synthesis of AgNPs using hot water extract of Osage orange (*Maclura pomifera*) leaf. The crystalline fcc structured and spherical-shaped AgNPs expressed SPR around 415 nm and an average size of 12 nm. The FTIR analysis confirmed the involvement of the carbonyl (C=O), hydroxyl (–OH), and amine (–NH) groups of leaf extracts in the fabrication of the Ag NPs. Those AgNPs expressed efficient antimicrobial activity against Gram-negative and positive bacteria and fungi. The aqueous extracts of *P. foetida* Linn. were used as reducing as well as a stabilizing agent during the synthesis of AgNPs and AuNPs (Bhuyan et al., 2017). Whereas the AgNPs with its small particle size (5–25 nm) showed good biocidal activity against the human pathogens; *B. cereus*, *E. coli*, *S. aureus*, and *A. niger*. As it easily penetrated the cell membrane, disturbed the metabolism, induced the production of reactive oxygen species destroyed the cell membrane, caused irretrievable damage, and finally led to the microbial cell death. However, the green synthesized AuNPs did not express any biocidal activity. However, both NPs showed good photocatalytic degradation activity, under solar irradiation, toward RhB in presence of $NaBH_4$. Bhuyan et al. (2017) illustrated a presumptive mechanism (Scheme 2.2). It is known that the BH_4^- ions are nucleophilic while dyes are electrophilic, and the nucleophilic ions can donate electrons to the metal particles, while the electrophilic ions can grab electrons from the metal particles. The BH_4^- ions and the dye molecules are firstly adsorbed on the surface of the NPs, whereas the AuNPs and Ag NPs are assumed to initiate shifting of an electron from BH_4^- ion (donor B_2H_4/BH_4^-) to (acceptor RhB) eventually leading to reduction of the dye.

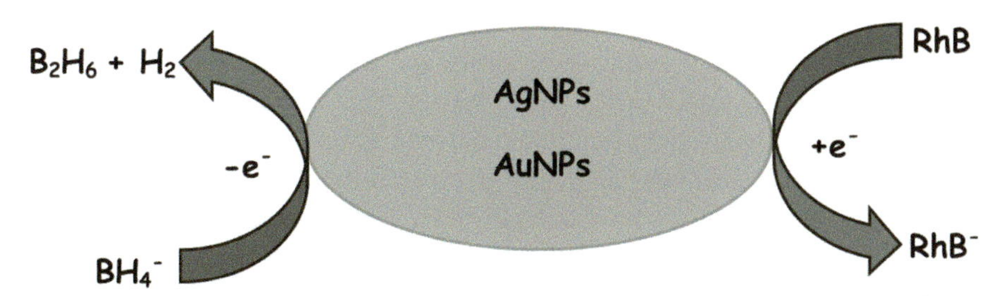

SCHEME 2.2 Presumptive mechanistic for photodegradation of RhB dye in presence of AgNPs and AuNPs.

The phytosynthesis of monodispersed spherical-shaped 50 nm AgNPs by hot water extract of *Guiera senegalensis* leaves within only 15 min has been reported (Bello et al., 2017a). Where only 3 mg of the AgNPs degraded 95% of both Congo red dye and the carcinogenic 4-nitrophenol in the presence of $NaBH_4$ within 22 and 36 min, respectively (Bello et al., 2017a). Divakar and Rao (2017) successfully prepared AgNPs by the aqueous extract of *Bombax cebia* leaf, under acidic condition, within 16 h. The prepared AgNPs were found to express antioxidant activity at the lowest concentration. Dogru et al. (2017) studied the effect of different concentrations of *Matricaria chamomilla* flower aqueous extract on the synthesis if different sizes of AgNPs. Whereas it was proved that increasing the extract concentration resulted in smaller sized AgNPs, since the number of extract molecules bound to the surface of Ag NPs increased, which may restrict and control the growth of Ag NPs and make them smaller compared to the usage of low extract concentration. It was also proved in that study, the decreasing in size of AgNPs causes blue shift (left shift), while the increase in size gives redshift (right shift) in the UV–Vis spectra. Not only this but it was also proved that, although some plant extract causes large and rapid aggregation of NPs after a few hours from synthesis; however, *M. chamomilla* flower aqueous extract acted as a strong capping agent to prevent the occurrence of large aggregation and enhance the stability of AgNPs. The interpreted results based on SEM images proved the instantaneous nucleation and growth initiation right after the addition of extract. It was also noticed that although longer reaction time allowed more growth of AgNPs; however, it might have triggered its aggregation. The higher concentrations of Ag^+ ions proved to enlarge the prepared AgNPs. Based on LaMer theory, Ag^+ concentration acts as a driving force to complete AgNPs formation via three consecutive steps: nucleation, growth, and completion. Thus, to properly produce uniform AgNPs, a homogeneous nucleation is needed. However, the monomer concentration (Ag^+)

must be reached to optimal value to exceed the energy barrier defined as Gibbs free energy of spherical seed formation, for the synthesis of Ag NPs. The different sized AgNPs, 70 ± 5 (Ag NP-1), 52 ± 5 (Ag NP-2), and 37 ± 4 nm (Ag NP-3) expressed antimicrobial activity against pathogens: *S. aureus*, *E. coli,* and *C. albicans* even at the low concentration of 6.25 ppm. Moreover, it was noticed that the small-sized AgNPs-3 expressed better inhibitory activity relative to the larger sized AgNPs-1 and AgNPs-2 and that was attributed to the smaller size and minimal aggregation of AgNPs-3 relative to the others. Smaller sized AgNPs have higher surface-to-volume ratio, which leads to the greater number of active atoms on the surface of AgNPs reacting with pathogenic cells. Thus, an increase in the number of active atoms contributes to the enhancement in the inhibitory effect of AgNPs-3. Additionally, when AgNPs are aggregated, the active surface area, which in contact with cells, is decreased or even lost; consequently, the antimicrobial activity of AgNPs is significantly decreased. Furthermore, the number of extract molecules and their availability on the surface of AgNPs contribute to their antimicrobial activity, and since the *M. chamomilla* extract itself has been used as a mild antimicrobial agent. The smaller size and minimal aggregation of AgNPs-3 increased the number of extract molecules bound on the AgNPs-3 surface and made the extract molecules more available. Thus, the *M. chamomilla* extract mediated AgNPs-3 exhibited drastically high antimicrobial property. Fafal et al. (2017) reported the reduction of Ag^+ ions by the aqueous extract of *Asphodelus aestivus* Brot. aerial par, where the flavonoids and phenols in the extract played a vital role in the reduction of Ag^+ ions to AgNPs. The XRD proved the monoclinic and cubic AgNPs. The SEM and TEM revealed spherical-shaped AgNPS with an average size of 23 nm. The EDX proved a high purity yield of AgNPs (85.9%) and a −22.2 mV Zeta potential indicated the high stability of the phytosynthesized AgNPs. Karthik et al. (2017) reported the production of crystalline fcc structured, spherical, and nonagglomerated AgNPs of SPR around 434 nm and size distribution of 12–25 nm, using the aqueous extract of *Camellia japonica* leaf within only 255 min. That was used for the synthesis of modified glassy carbon electrode (GCE) which exhibited good electrochemical detection ability to nitrobenzene with good selectivity, low detection limit, and wide linear response range at pH7. Moreover, those AgNPs expressed good photocatalytic degradation capabilities for Eosin yellowish (EY) dye, under visible light irradiation. Nicolás Gallucci et al. (2017) used the aqueous extract of Belgian endive (*Cichorium intybus* L. var. sativus) to produce quasispherical AgNPs with SPR of 420 nm and average size ranged between 19 and 64 nm. The

XRD pattern revealed biosynthetic fcc AgNPs, whereas the surface-enhanced Raman spectroscopy analysis revealed that the AgNPs were capped with bioactive molecules of the leaf extract, which were responsible for the bioreduction of Ag^+ ions. Those AgNPs showed high biocidal activity against *S. aureus*, *E. coli*, and *P. aeruginosa* in very low picomolar concentration levels. Anthocyanin in the red cabbage (*Brassica oleracea* L.) leaves aqueous extract proved to produce a high yield of monodispersed spherical-shaped 30 nm AgNPs (Ocsoy et al., 2017). The prepared AgNPs expressed enhanced inhibitory properties against Gram-positive *S. aureus*, Gram-negative *E. coli*, and fungus *C. albicans*, even at low concentration (9.37 ppm). Moreover, it was proved that the integration of hydrothermal approach to the green synthesis made the AgNPs more stable and retarded the aggregation for several weeks (Ocsoy et al., 2017). Raja et al. (2017) reported for the first time, the phytosynthesis of AgNPs using aqueous extract *Calliandra haematocephala* leaf, within 10 min and at 80 °C. The brown coloration was an indication for the reduction of Ag^+ ions to Ag^0 and the single characteristic peak of the spherical AgNPs at a wavelength of 414 nm was observed in the UV-Vis spectrum. The absorption band (λ_{max}) was constant for more than 30 days which substantiated the stability of the NPs. The SEM analysis revealed well-defined spherical SNPs without any agglomeration with an average size of 70 nm. The morphology of the silver NPs was characterized by SEM. The energy dispersive spectroscopy (EDS) showed the characteristic peak of elemental silver at 3 keV. The XRD ascertained the crystalline nature and purity of the AgNPs which implied the presence of (111) and (220) lattice planes of the fcc structure of metallic silver. Upon applying Debye–Scherrer equation, the average particle size was calculated to be 13.07 nm. The dynamic light scattering (DLS) measurements revealed the narrow distribution of the prepared AgNPs that ranged from 13.54 to 91.28 nm and had an average particle size of 104.3 nm. The zeta potential value of −17.2 mV proved the stability of the AgNPs and evaded the agglomeration of NPs. The negative potential value was attributed to the capping action of biomolecules present in the leaf extract of C. *haematocephala*. The FTIR spectroscopy proved the presence of various biomolecules in the leaf extract: L-cysteine amino acid, aldehydes, ketones and carboxylic acids, alcohols, phenols, proteins, enzymes, and polysaccharides, which played a major role in the reduction, capping, and stability of AgNPs. Moreover, the presence of gallic acid in the leaf extract was suggested to be responsible for the reduction of Ag^+ ions into Ag^0 (Scheme 2.3). Whereas the Ag^+ ions form intermediate complexes with phenolic groups present in gallic acid. These complexes consequently reduce Ag^+ into AgNPs with concomitant oxidation to quinone form. The prepared AgNPs expressed good antibacterial activity against the pathogenic bacteria, *E. coli*. Moreover, the prepared AgNPs, acted as a good sensor for hydrogen peroxide. Whereas the addition of AgNPs to H_2O_2 solution resulted in the formation of free radicals which initiated the degradation of the AgNPs. Subsequently, Ag^0 was oxidized to Ag^+ and a decrease in absorbance was observed. These findings suggested that the AgNPs can be successfully used to detect the concentration of H_2O_2 present in various samples. This is very important since hydrogen peroxide (H_2O_2) is a strong oxidizing agent and is widely used in the food, pharmaceutical, cosmetics, wood, pulp, water treatment, and petroleum sectors. However, the exposure and the presence of even a small amount of H_2O_2 in process streams result in various health and environmental hazards due to its toxicity (Tagad et al., 2013). Therefore, it is essential to develop accurate and fast methods to detect H_2O_2. Thus, the capacity of the phytosynthesized AgNPs to detect H_2O_2 would have different applications and in developing new antibacterial drugs and new biosensors for detecting H_2O_2.

The *Aegle marmelos* leaf extract mediated the synthesis of spherical prismatic AgNPs (Rao et al., 2017). That showed a good antimicrobial effect against the Gram-negative *E. coli* and *P. aeruginosa* and the phytopathogen *Fusarium solani*. Moreover, the prismatic Ag-nanostructures may be useful for surface-enhanced Raman spectroscopy application for the detection of low concentration organic molecules.

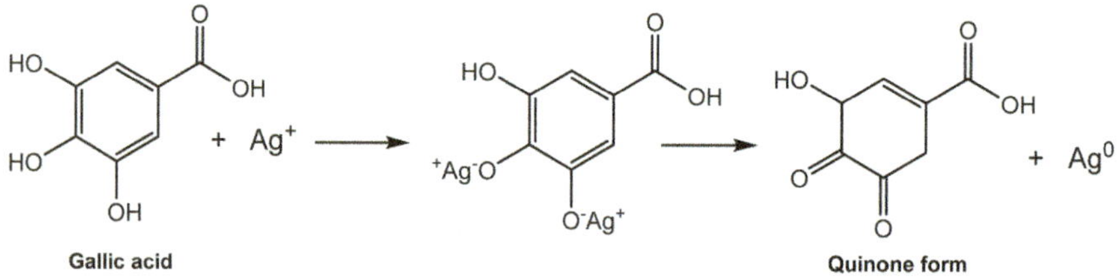

SCHEME 2.3 The possible mechanism for bioreduction of silver ions by gallic acid into AgNPs.

The reducing potential of *Artemisia vulgaris* leaves' methanolic extract (AVLE) was investigated for synthesizing AgNPs without the addition of any external reducing or capping agents (Rasheed et al., 2017). The appearance of blackish brown color evidenced the phytosynthesis of AgNPs. The synthesized AgNPs were characterized by UV/Vis spectroscopy which elucidated an absorption peak around 420 nm, which corresponds to the SPR of AgNPs. SEM image envisaged that the synthesized AVLE-AgNPs were spherical in shape with varying sizes ranged between 27 and 53 nm. AFM results indicated a considerable variability in the morphological features of AVLE-AgNPs. Height and width of the phytosynthesized AgNPs ranged from 7.7 to 29 nm and 0.049 to 2.27 nm, respectively. It was evident from the AFM images that the particles were uniform in size and polydispersed in nature. EDX revealed the presence of elemental silver peaks. TEM revealed that the average diameter of AgNPs was about 25 nm. The FTIR spectroscopy scrutinized the involvement of various functional groups during NPs synthesis. Sarkar and Paul (2017) used the hot water extract of *Mentha asiatica* (Mint) leaves for the synthesis of AgNPs. Those AgNPs expressed good bactericidal activity against the Gram-negative *E. coli, P. aeruginosa, Baccillus subtilis,* and the Gram-positive *S. aureus*. Tareq et al. (2017) used *Bryophyllum pinnatum* aqueous leaf extract to produce spherical-shaped 15–40 nm AgNPs within only 5 min. The FTIR analysis confirmed the presence of proteins residue; moreover, the UV/Vis absorption spectroscopy showed an absorption band of protein residues at 320 nm other than that of AgNPs at 447 nm. Although these protein residues acted as a capping agent to the formation and stabilization of AgNPs in the aqueous medium. The prepared AgNPs expressed a high antibacterial effect against the food pathogens Gram-negative *E. coli* and agricultural pathogens Gram-positive *Bacillus megaterium*. Ajitha et al. (2018) reported the synthesis of highly stable AgNPs with a zeta potential of −45 mV and flower-like structure with an average size of 36 nm, via an easy, rapid, and eco-friendly process using *Phyllanthus amarus* aqueous leaf extract. Those AgNPs expressed good bactericidal activity against the pathogenic *E. coli*, *Pseudomonas* spp., *Bacillus* spp., and *Staphylococcus* spp. and antifungal activity against the pathogenic *A. niger*, *A. flavus*, and *Penicillium* spp. Moreover, it expressed good catalytic degradation properties toward RhB dye, in the presence of $NaBH_4$. Kumar et al. (2018) reported the usage of leaf sap extract from Aloe *arborescens* as a reducing, stabilizing, and capping agent to produce a smooth surface, crystalline fcc structured AgNPs, with an average size of 45 nm, whereas the direct sunlight acted as inducer, during the green synthesis process. Those AgNPs expressed excellent bactericidal effects against the human pathogens *P. aeruginosa* and *S. aureus*. Diterpenes extracted from *Iboza Riparia* were used for the synthesis of highly stable fcc crystal structured AgNPs in presence of trisodium citrate, $NaBH_4$, and ascorbic acid (Sabela et al., 2018). The prepared AgNPs of SPR around 405.9 nm were of an average size of 156 nm and conductivity of 0.494 mS that signified the poor reduction and capping of AgNPs by diterpene.

2.2.3 PHYTOSYNTHESIS OF AgNPs USING SEEDS' EXTRACT

Kumar et al. (2010) reported a comparative study for the phytosynthesis of AgNPs using an aqueous extract of *S. cumini* leaf and seed. The extract of seed was found to contain more polar soluble alkaloids, polyphenols, saponins, and flavonoids than of leaves. Thus, the seed extract showed a greater ability for phytosynthesis of AgNPs than the leaf extract, producing 73 nm and 29 nm AgNPs, respectively. The size and shape of the AgNPs were found to be depending on the polyphenolic content in the extract. The UV–Vis showed the characteristic peak of AgNPs at 450 nm. The peak and color intensity of the solution increased with time, indicating higher production of AgNPs. However, the smaller the size of AgNPs, the greater the color shift toward red. Moreover, the position of the peak changed with increasing incubation time and shifted to a longer wavelength, with a larger particles size. This reduction in AgNPs size was well correlated with the reduction in active species like polyphenolic content in the reaction mixture, which consequently, reduces the synthesis rate and nucleation process. An increase in nucleation rate provided less time for surfactant/capping agents to bind the nucleated AgNPs. This prevented further nucleation. The faster reaction rate was found to increase the number and size of AgNPs. Thus, if the reacting solution contains more reducing molecules and less surfactants, the size of synthesized AgNPs would be larger. This might be one of the reasons for the synthesis of larger-sized AgNPs by seed extract, whereas reducing molecules (i.e., polyphenolic compounds) were higher but surfactants were lower, relative to those in the leaf extract. Banerjee and Narendhirakannan (2011) used extracts of *S. cumini* (jambul) seeds to phytosynthesize AgNPs. The NPs produced using the seed extract were found to have strong antioxidant properties. That was attributed to the high concentration of the adsorbed polyphenolic antioxidants on the surface of the particles. Green synthesis of AgNPs from $AgNO_3$ was carried out using *Sinapis arvensis* seed water extract (Khatami et al., 2015). Different concentrations of $AgNO_3$ (1, 2.5, 3, 4, and 5 mM) were prepared then added to 5 mL of *S. arvensis* seed exudates. The mixtures were kept at 25 °C. The phytosynthesis of AgNPs was indicated by the change of the color from light yellow to brown. Maximum absorption appeared at 412 nm. TEM analysis showed that the prepared AgNPs were spherical-shaped-particles with an average size of 14 nm. The antifungal activity of the synthesized AgNPs was proved. Paul and Yadav (2015) used the aqueous extract of seeds of *Pisum sativum* (Black pea) for the reduction of $AgNO_3$ to spherical-shaped AgNPs with SPR of 423 nm and an average size of 3–36 nm. Such phytosynthesized AgNPs expressed good antimicrobial efficiency against *E. coli* and yeast *C. albicans*. Khan et al. (2016) developed a simple and environmental friendly method for the synthesis of AgNPs using *Dimocarpus longan* seed water extract. A characteristic SPR peak of AgNPs appeared at 432 nm which confirmed the phytosynthesis of AgNPs using UV/Vis spectroscopy. TEM revealed that the phytosynthesized AgNPs were approximately 40 nm in size. Maximum production of AgNPs was obtained under the following conditions: 10 mL of plant extract and 2 mM $AgNO_3$ within 180 min of incubation. FTIR showed that the polyphenolic compounds were majorly involved in the reduction of Ag^+ into Ag^0. The catalytic activities of AgNPs were assessed against the catalytic degradation of MB and chemo catalytic reduction of 4-nitrophenol to 4-aminophenol. Results revealed that the prepared AgNPs had strong chemo catalytic activity as it aided in the complete reduction of 4-nitrophenol (4-NP) to 4-aminophenol (4-AP) within 10 min. Besides, the AgNPs exhibited a promising antioxidant activity in scavenging 2,2-diphenyl-1-picrylhydrazyl (DPPH) radicals. The findings of this study concluded that the phytosynthesized AgNPs are promising agents possessing strong catalytic and reducing properties (Khan et al., 2016).

2.2.4 PHYTOSYNTHESIS OF AgNPs USING FRUIT EXTRACT

Fruit extract of *Tanacetum vulgare* was used to phytosynthesize Ag and AuNPs (Dubey et al., 2010c). The FTIR analysis revealed that the carbonyl groups were involved in the reduction of metal ions into NPs. The zeta potential of the AgNPs was shown to vary with pH: a low zeta potential obtained at the strongly acidic condition. Larger particle size was obtained under acidic conditions (Dubey et al., 2010c). The hot water extract from *Allium Cepa* (Onion) was used for reducing $AgNO_3$ to spherical-shaped AgNPs with an average size of 33.67 nm (Saxena et al., 2010). That expressed antibacterial activity against *E. coli* and *S. typhimurium* at a concentration of 50 μg/mL. The hot water extract of *A. Cepa* (Onion) was also used in another study, as a reducing and capping agent in the synthesis of AgNPs with an average size of 40 nm, with good antibacterial activity against *E. coli* (Benjamin and Bharathwaj, 2011). Prathna et al. (2011b) phytosynthesized also spherical-shaped 50 nm AgNPs nm using 4:1 v/v juice of *Citrus limon* (lemon; 20 g/L citric acid, 5 g/L ascorbic acids) and 10–20 M silver nitrate within 4 h. It was found that citric acid was responsible for reducing Ag^+ ions into AgNPs. Rad et al. (2013) reported the green synthesis of 40 nm AgNPs and 30 nm AuNPs by the hot water extract of *Nitraria schoberi* fruits. Whereas the alkaloids, tannins, flavonoids, diosgenin and phenolic compounds, and enzymes in the extract acted as reducing, stabilizing, and capping agents. Mittal et al. (2014) prepared highly stable monodispersed, spherical, crystalline fcc structured AgNPs, with an average size of 12.5 nm, using the aqueous methanol *S. cumini* fruit extract at room temperature. Balamanikandan et al. (2015) used the aqueous extract of Onion (*A. cepa*) for the synthesis of AgNPs with a characteristic maximum SPR of 440 nm. That showed antimicrobial activity against *Bacillus streptococcus*, *B. subtilis, E. coli, K. pneumoniae, Streptococcus* sp., *Aspergillus terres, A. flavus, Aspergillus ochraceus*, and *A. niger*. The phytosynthesized 20 nm AgNPs by aqueous extract of *Hyphaene thebaica* fruit, reduced the 4-nitrophenol into 4-aminophenol in the presence of $NaBH_4$. However, the Ag complex formation of 4-nitophenol with the AgNPs was observed (Bello et al., 2017b). Demirbas et al. (2017) proved that the anthocyanins-rich berries (*Rubus fruticosus* L. (blackberry); *Fragaria vesca* L. (strawberry) and *Rubus idaeus* L. (raspberry) aqueous extracts produced AgNPs with relatively higher antioxidant activity toward 1-Diphenyl-2-picryl-hydrazyl (DPPH) and higher antimicrobial activities against Gram-negative *E. coli* ATCC 25922 and *Salmonella enterica* subsp. *enterica* NCTC 8394, Gram-positive *S. aureus* ATCC 29213 and *B. cereus* ATCC 11778 and fungus *C. albicans* ATCC 10231 compared to the berry extracts itself. Kumar et al. (2017) proved that the Andean blackberry fruit extracts as both a reducing and capping agent, in the green synthesis of AgNPs. Notwithstanding, the characteristic SPR of spherical and aggregated AgNPs appeared as a broad absorption peak with λ_{max} 435 nm. The TEM and selected area electron diffraction (SAED) study revealed spherical-shaped AgNPs with size ranged between 12–50 nm and the diffraction rings of the AgNPs have been indexed as (111), (200), and (220) consistent with the fcc structure of Ag which are typical of the polycrystalline AgNPs structure. Moreover, the polydispersity index (PDI) value >0.1 proved the polydispersity of AgNPs. The XRD patterns revealed the large crystalline domain sizes of that AgNPs powders, which corresponded to pure Ag metal with fcc symmetry, and consisted of those obtained from SAED pattern. The FTIR-analysis suggested that the adsorbed O–H and C=O groups on the surface of AgNPs were involved in the reduction process. The green synthesized AgNPs expressed good antioxidant efficiency against 1,1-diphenyl-2-picrylhydrazyl. Naraginti et al. (2017) reported the green synthesis of AgNPs and AuNPs using kiwi fruit extract at 80 °C. The broadening as well as the shift of SPR band from 425 to 442 nm for AgNPs and 538 to 549 nm for AuNPs have been noticed and were attributed to the increase in particle size and was also confirmed by the TEM micrographs. Furthermore, the increase in temperature from 30 to 80 °C increased the rate of formation of NPs. It is worth to know that the reduction potential of Au^{3+} ions is higher than the Ag^{+} ions. The FTIR analysis proved that the proteins present in the kiwi fruit extract might have acted as a capping agents for the synthesized NPs, via the free amine groups or cysteine residue. Those NPs expressed good catalytic reduction of two organic pollutants; 4-NP and MB to 4-aminophenol (4-AP) and LMB in presence of $NaBH_4$, respectively. It is important to know that the reduction of 4-NP to 4-AP by aqueous $NaBH_4$ is thermodynamically favored but the kinetic barrier decreases the feasibility of the reaction due to the large potential difference between donor and acceptor molecules. The metal NPs catalyze this reaction by promoting electron relay from the donor BH_4^- to acceptor 4-NP to overcome the kinetic barrier. Same as in the case of the reduction of MB, the redox potential of NPs should be located between the redox potential of the donor ($NaBH_4$) and acceptor (MB) system for efficient catalysis that related to the electron relay effect, whereas the metal NPs act as electron transfer mediators between MB and $NaBH_4$ during their action as a redox catalyst. Moreover, they performed good bactericidal activity against pathogenic *P. aeruginosa* and *S. aureus*. It is very important to know that as the particle size decreased, the reduction potential and bactericidal activity increased due to their higher surface area. Ndikau et al. (2017) reported the green synthesis of spherical-shaped 17.96 nm AgNPs using hot water extract of watermelon (*Citrullus lanatus*) fruit rind. In another study by Saha et al. (2017), the hot water extract of *Gmelina arborea* was used to phytosynthesize spherical-shaped AgNPs with an average size of 17 nm, with SPR of 418 nm, within 4 h. The phytosynthesized AgNPs expressed 100% catalytic reduction to 10 mM MB dye in presence of $NaBH_4$, within only 30 min, using 1.5 mL AgNPs and 10 min upon the usage of 3 mL AgNPs, at ambient conditions. The aqueous extract of palm date fruit pericarp (*Phoenix dactylifera* L.) was used for the green synthesis of AgNPs (Zaheer, 2018). Those have the characteristic SPR band around 425–440 nm, whereas the shape of the spectra, intensity of the peak, and position of the SRP band depended very much on the concentration of the Ag^{+} ions. However, redshift did not occur confirming the no or little aggregation of the AgNPs has occurred. However, the shape of the spectra, absorbance, and position of the SRP band depended very much on the concentration of the extract. Although red shift occurred with the increase of the extract concentration (>2 cm^3). However, that was attributed to the possible adsorption of sugars moiety of the extract onto the surface of AgNPs. The prepared AgNPs were highly stable spherical shaped with size ranged between 3 and 30 nm. The FTIR analysis proved that the major organic constituents of date extract (monosaccharides) prevented the AgNPs from aggregation. The prepared AgNPs expressed good antibacterial and antifungal activities against the *S. aureus*, *E. coli*, and *C. albicans* human pathogens. It performed also catalytic and photocatalytic degradation of 4-NP by AgNPs in the absence and presence of sunlight (Zaheer, 2018).

2.2.5 PHYTOSYNTHESIS OF AgNPs USING PLANT STEM EXTRACT

Extract of rhizome of *Dioscorea batatas* was used to phytosynthesize AgNPs with good antimicrobial against the yeasts *C. albicans* and *Saccharomyces cerevisiae* (Nagajyothi and Lee, 2011). Vanaja et al. (2013) reported the green synthesis of AgNPs using *Cissus quadrangularis* stem extract. Optimization of the parameters affecting the phytosynthesis of AgNPs was performed. The maximum synthesis of AgNPs was attained within 1 h, at pH 8, 1 mM $AgNO_3$, and 70 °C. The SEM analysis revealed a mixture of spherical, rod, and triangular shapes with sizes ranging from 37 to 44 nm. The FTIR showed that the functional groups, carboxyl, amine, and phenolic compounds of stem extract, were involved in the reduction of Ag^+. The synthesized AgNPs showed antibacterial activity against *Klebsiella planticola* and *B. subtilis.* Shalaby et al. (2015) used the hot water extract of fresh ginger rhizome (*Zingiber officinale*) for the synthesis of spherical-shaped stable and crystalline fcc structured AgNPs with SPR that occurred around 430 nm and an average size of 3.1 nm. That performed good antibacterial activity against the pathogenic *S. aureus* and *E. coli* strains. The ascorbic and oxalic acids were the main components of the extract and played the main role in the reduction of Ag^+ ions into AgNPs. While the main possible steps were nucleation, condensation, surface reduction, and stabilization. The reduction is accomplished by electron transfer that covers Ag^+ into Ag^0 followed by the nucleation of particles. Then, the formed AgNPs condensed into bigger particles. Those were then immediately bounded by the layer(s) of oxalic acid and/or ascorbic acid through electrostatic forces of the ginger extract. Moreover, other extract components like zingerone and phenylpropanoids could have been aggregated along with other compounds, around the layers of reducing acids, thus, consequently acted as a protective material for the stabilization of AgNPs. The FTIR analysis proved the presence of various functional groups that were mainly derived from heterocyclic compounds, alkanoids, and flavonoids in the extract that functioned as the capping ligands for the synthesis of AgNPs. Phytosynthesis of AgNPs was also carried out using *Caesalpinia pulcherrima* hot water stem extract (Moteriya and Chanda, 2018). The colorless solution of *C. pulcherrima* stem extract changed to a brown color indicating the preliminary formation of AgNPs. The synthesized AgNPs showed characteristic absorption maxima in the visible region of 350–750 nm. Optimum conditions for green synthesis of AgNPs by *C. pulcherrima* stem extract was 10 min boiling time for stem extract preparation, 12 mL stem extract addition to the reaction medium, 1 mM $AgNO_3$ concentration, pH 10 of the reaction medium, and the reaction time for synthesis of AgNPs was 24 h. The thermal stability and capping action of the biomolecules surrounding the surface of AgNPs was confirmed by the thermogravimetric analysis. The initial weight loss of about 6% at the temperature of 100 °C was due to the loss of water molecules from AgNPs. The second weight loss observed in the temperature range of 300–400 °C was found to be around 49%. There was a steady weight loss when the temperature was increased up to 800 °C. This weight loss was due to the degradation of bioorganic molecules present on the surface of AgNPs. The crystalline nature of the prepared AgNPs was confirmed via XRD. TEM revealed that the synthesized AgNPs were spherical in shape with an average size of 8 nm.

2.2.6 PHYTOSYNTHESIS OF AgNPs USING BARK EXTRACT

Cinnamomum zeylanicum bark is rich in linalool, methyl chavicol, and eugenol which were reported for being responsible for the bioreduction of silver and palladium ions to the corresponding NPs in separate experiments (Sathishkumar et al., 2009a, b). Moreover, the protein from *C. zeylanicum* bark helped to stabilize the synthesized AgNPs (Sathishkumar et al., 2009b). *Afzelia quanzensis* bark hot water extract was tested for the green synthesis of AgNPs (Moyo et al., 2015). Based on UV/Vis spectrum analysis, the characteristic absorption band of the phytosynthesized AgNPs was observed at 427 nm. Furthermore, the SEM analysis revealed spherical-shaped AgNPs with a size range of 10–80 nm. In addition, the XRD analysis showed that the AgNPs were crystalline in nature and had a fcc structure. Furthermore, based on the performed FTIR analysis, the presence of phytochemical functional groups such as carboxyl (–C=O) and amine (N–H) in *A. quanzensis* seeds' extract, confirmed to be the reducing agents responsible for NPs' formation. Nayak et al. (2016) reported the phytosynthesis potentials of *F. benghalensis* and *A. indica* bark hot water extract for the production of AgNPs without the usage of any external reducing or capping agents. The appearance of dark brown color indicated the synthesis of AgNPs which was further confirmed via UV/Vis spectrophotometer. The morphology of the synthesized particles was characterized by FESEM and AFM. The XRD patterns clearly illustrated the crystalline phase of the synthesized AgNPs. The synthesized AgNPs showed promising antimicrobial activity against Gram-negative *E. coli*, *P.*

aeruginosa, and *V. cholerae* and Gram-positive *B. subtilis* bacteria. Saponin from the root bark of *Ilex Mitis* was extracted and used for the synthesis of highly stable fcc crystal structured AgNPs in presence of trisodium citrate, $NaBH_4$, and ascorbic acid (Sabela et al., 2018). The prepared AgNPs of SPR around 413.2 nm were of an average size of 50 nm and conductivity of 0.39 mS that small size and low conductivity were attributed to the structure of saponin itself, which have hydrophilic glycoside moieties combined with a lipophilic triterpene derivative.

2.2.7 PHYTOSYNTHESIS OF AgNPs USING GUM EXTRACT

A natural biopolymer, gum kondagogu (GK) (*Cochlospermum gossypium*) acted as a reducing and stabilizing agent in the phytosynthesis of AgNPs, upon autoclaving $AgNO_3$ with the gum aqueous solution at 121 °C and 15 psi (Kora et al., 2010). The AgNPs were suggested to be probably capped and stabilized by the polysaccharides along with the proteins present in the gum. As that, carbohydrate polymer is very complex and it is categorized as rhamnogalacturonans due to the abundance of rhamnose, galactose, and uronic acids, thus, it is most likely that more than one mechanism is involved in the complexation and subsequent reduction of silver ions by GK during autoclaving. It has an abundance of hydroxyl, carbonyl, and carboxylic functional groups, and uronic acid content and shows a pH of 4.9–5.0. Moreover, the presence of negatively charged groups is also confirmed from the negative zeta potential value of −23.4 mV for the gum. Consequently, the large number of hydroxyl and carboxylic groups of this biopolymer facilitated the complexation of silver ions. Subsequently, these silver ions oxidized the hydroxyl groups to carbonyl groups, during which the silver ions are reduced to elemental silver. Moreover, the dissolved air also caused oxidation of the existing hydroxyl groups to carbonyl groups, such as aldehydes and carboxylates. In turn, those powerful reducing aldehyde groups along with the other existing abundant carbonyl groups reduced more and more silver ions to elemental silver. To optimize the NP synthesis, the influence of different parameters such as gum particle size, concentration of gum, concentration of silver nitrate, and reaction time was studied, and the optimum conditions were reported to be 38 μm, 0.5%, 5 mM, and 60 min, respectively. Those produced monodispersed and size-controlled spherical NPs of around 3 nm. The formed AgNPs were highly stable and had significant antibacterial action on both the Gram classes of bacteria (Kora et al., 2010).

In another study, Vinod et al. (2011) reported the phytosynthesis of Ag, Au, and Pt using an aqueous medium containing GK and $AgNO_3$, $HAuCl_4$, and H_2PtCl_6, respectively. The UV/Vis studies showed a distinct SPR at 412 and 525 nm due to the formation of Au and Ag NPs, respectively, within the gum network. XRD studies indicated that the NPs were crystalline in nature with fcc geometry. The crystalline phases obtained with the formation of Ag, Au, and Pt NPs within the GK matrix could be indexed to (111), (200), (220), and (311) planes. The TEM analysis revealed that the noble metal NPs prepared in the present study appears to be homogeneous with the particle size ranging between 2 and 10 nm. The PtNPs, AgNPs, and AuNPs showed average size in increasing order of 2.4 ± 0.7 nm of 5.5 ± 2.5 nm and 7.8 ± 2.3 nm, respectively. Aqueous extract of gum olibanum (*Boswellia serrata*), as a renewable natural plant biopolymer, was used for the phytosynthesis of spherical-shaped 7.5 ± 3.8 nm AgNPs by autoclaving at 121 °C and 103 kPa (Kora et al., 2012). Whereas the water-soluble compounds in the gum served as dual-functional reducing and stabilizing agents. The optimum conditions were found to be 5 mM $AgNO_3$, 0.5% gum concentration and 60 min. The FTIR spectroscopy and Raman spectroscopy proved that both amino and carboxylate groups of the gum were involved in the capping and stabilizing of the AgNPs. However, the FTIR proved that both hydroxyl and carbonyl groups of the gum are involved in the reduction of silver ions to AgNPs. The gum olibanum has been categorized under glucuronoarabino-galactan type and its hydrolysis by autoclaving yields polyhydroxylated reducing aldose sugars like arabinose, galactose, and xylose (Sen et al., 1992; Shukla et al., 2005). Thus, a large number of hydroxyl along with the available glucuronic acid groups of the gum were suggested to complex the silver ions. Then, these silver ions would have been reduced to elemental silver possibly by in-situ oxidation of hydroxyl groups; and by the inherent aldehyde groups as well as those produced by the air oxidation. Furthermore, the produced AgNPs exhibited substantial antibacterial activity on both the Gram classes of bacteria (Kora et al., 2012). A natural biopolymer, gum acacia was also reported to act as both a reducing and stabilizing agent in the phytosynthesis of AgNPs (Venkatesham et al., 2012). This reaction was carried out in an autoclave at a pressure of 15 Psi and a temperature of 120 °C, for 2 min. The efficiency of synthesis increased with the increase of $AgNO_3$ and gum concentrations recording its maximum at 0.5% of both. The UV-Vis, spectra showed the characteristic peak of AgNPs at 410–430 nm. The FTIR analysis revealed that the –OH and –COOH groups were involved in the stabilization of the AgNPs. The XRD-analysis revealed the crystalline fcc structured

AgNPs with an average crystalline size of 8 nm and the TEM analysis revealed that 70% of the particles were in the size range from 1 to 10 nm and few particles were also observed above the 20 nm range. Furthermore, the prepared AgNPs were found to be highly stable and expressed significant antibacterial effect on *E. coli* and *M. luteus* (Venkatesham et al., 2012). Rastog et al. (2014) reported the green synthesis of AgNPs by GK which acted as both reducing and stabilizing agent. The average size of those stable AgNPs was 5 nm and those AgNPs proved to be a direct colorimetric sensor for the sensitive detection of Hg^{2+} in various groundwater samples. Kora and Sashidhar (2015) performed a comparative study for the green synthesis of AgNPs using gum ghatti and gum olibanum via autoclaving at 121 °C and 15 psi for 30 min. Antibacterial activity and cytotoxicity of the AgNPs prepared by gum ghatti (Ag NP-GT) were greater than those prepared by gum olibanum (Ag NP-OB). This could be attributed to the smaller size (5.7 nm), higher monodispersity, and zeta potential (−22.4 mV) of the Ag NP-GT. The study suggests that Ag NP-GT can be employed as a cytotoxic bactericidal agent, whereas Ag NP-OB (7.5 nm and zeta potential of −14.9 mV) as a biocompatible bactericidal agent. In another study by Velusamy et al. (2015), AgNPs were prepared via the reduction of $AgNO_3$ upon mixing with the gum extract of neem (*A. indica*) and autoclaving at 121 °C and 15 psi. The prepared crystalline and spherical AgNPs had the characteristic SPR around 418 nm and average diameter ranged between 12.09 and 29.65 nm. The EDX revealed the strong signal of Ag peak at approximately 3 keV. The prepared AgNPs proved to have a greater affinity to the sulfur-containing proteins in the bacterial cell membrane, inside the cells, and the phosphorus-containing elements like DNA. Whereas the amount of DNA damage within the bacterial cells increased with the increasing concentration of the AgNPs, and the damage to DNA of the Gram-negative *S. enteritidis* was higher than that of the Gram-positive *B. cereus*, implying that *S. enteritidis* might have suffered greater membrane damage than *B. cereus*. Kora and Sashidhar (2018) used rhamnogalacturonan gum for the green synthesis of highly stable crystalline fcc structured AgNPs with an average particle size of 4.5 nm and zeta potential of −32.7 mV. The minim inhibitory concentrations of those AgNPs for the Gram-negative *S. aureus* 25923, *P. aeruginosa* 27853, *E. coli* 25922, and *E. coli* 35218 were 10.0, 5.0, 2.0, and 2.0 μ/mL, respectively. It expressed also antibiofilm activity to the studied strains at 2 μg/mL, which can have implications in the treatment of biocide resistant bacterial caused by biofilms and biofouling control.

2.2.8 PHYTOSYNTHESIS OF AgNPs USING ROOTS' EXTRACT

Aqueous extract of Ginger (*Zingiber officinale*) roots was used for phytosynthesis of fcc structured AgNPs with an average size of 2.89 nm (Priyaa and Satyan, 2014). The brown color due to the SPR of AgNPs appeared within only 30 min. The maximum absorbance peak of AgNPs shifted from 410 to 450 mm with the increase in temperature from room temperature to 60 °C. The formation of quasispherical AgNPs and spherical icosahedral AuNPs by the roots of Chinese herbal *Angelica pubescens* was confirmed from the SPR intensified at ~414 and ~540 nm, and the EDX spectra of highest optical absorption peaks at 3 and 2. 3 keV, respectively (Markus et al., 2017). The TEM analysis revealed an average particles size of 12.48 and 7.44 nm, respectively. Whereas the primary and secondary metabolites of *A. pubescens* Maxim, such as coumarins, terpenes, sesquiterpenes, phenols, and flavonoids, and phenols acted as strong reducing agents, reduced silver and gold salts into their respective NPs, and attached on the NP surface. Thus, provided stabilization through electrostatic interaction and formed a protective capping layer around the metallic NPs, and prevented the agglomeration of the NPs (Markus et al., 2017). The 2,2-diphenyl-1-picrylhydrzyl (DPPH) is known to consist of stable free radical molecules and is readily reduced by accepting hydrogen or electron from NPs. Those green-synthesized AgNPs and AuNPs expressed antioxidant activity against radicals with IC_{50} of 1.01 and 1.23 mg/mL, respectively. That was attributed to the antioxidant property of *Angelicae Pubescentis* Radix via the protective capping layer of AgNPs and AuNPs, whereas the flavonoids, sesquiterpenes, and phenols seemed to be the major contributors to the free radical scavenging activity. However, only the AgNPs exhibited *antimicrobial activity against E. coli, S. aureus, P. aeruginosa*, and *S. enterica*. The 1 min microwave, hot water extract of dried root of *Z. officinale* (dry ginger) was used to synthesize AgNPs using $AgNO_3$ solution in a microwave oven (1200 W, 50 Hz), within only 60 s (Vijaya et al., 2017). The shagaols, gingerol, paradol, zingerone, and starch present in the extract act as reducing and stabilizing agents. Whereas the alkanoids, flavoinds, and alkaloids, the active components of dried root of *Z. officinale* extract, act as the capping agent. The XRD-analysis and the SAED pattern proved the fcc structure and nanocrystallinity of the phytosynthesized AgNPs. The HR-SEM and HR-TEM proved homogeneous, uniform, and spherical-shaped 10 nm AgNPs. The phytosynthesized AgNPs expressed good antibacterial effects against the Gram-positive: *B. cereus*, *B. subtilis*, *S. aureus*, *M. luteus*, and *Enterococcus* sp. and Gram-negative: *P. aeruginosa*, *Salmonella typhi*, *E. coli*, and *K. pneumonia*.

2.2.9 PHYTOSYNTHESIS OF AgNPs USING AGRO-INDUSTRIAL WASTES

Extract of banana (*Musa paradisiaca*) peels has been reported for phytosynthesis of AgNPs that expressed antifungal activity against the yeasts *Candida lipolytica* and *C. albicans*, and antibacterial activity against *E. coli*, *Shigella* sp., *Klebsiella* sp., and *Enterobacter aerogenes* (Bankar et al., 2010). Kaviya et al. (2011b) reported the effect of temperature on the size of the phytosynthesized AgNPs. Whereas the AgNPs phytosynthesized by peel extract of *C. sinensis* which have a broad-spectrum antibacterial activity found to have an average size of 35 and 10 nm at 25 °C and 60 °C, respectively. Satsuma mandarin (*Citrus unshiu*) peel hot water extract was tested for its ability to phytosynthesize AgNPs (Basavegowda and Lee, 2013). The formed AgNPs were in the size range of 5–20 nm. The SPR peak which indicated the synthesis of AgNPs was observed within 10 min of reaction time at 440 nm and then steadily increased till it reached the maximum absorption at 72 h. The presence of elemental silver was proven by EDX analysis as a typical absorption peak appeared approximately at 3 keV. Most of the formed NPs were spherical in shape as was revealed by TEM with various particle sizes ranged between 5 and 20 nm. The hydroxyl and aldehydic sugars of the agro-industrial bagasse were reported to bioreduce Ag^+ ions into spherical-shaped 50–150 nm AgNPs within only 4 min in a microwave assist method (Mishra and Sardar, 2013). Yang and Li (2013) reported a simple route using mango peel hot water extract as a reducing as well as a capping agent for the green synthesis of AgNPs. Mango (*Mangifera indica*) is consumed all over the world. After consumption of the pulp, the peel is generally discarded. Different factors which affect the AgNPs green synthesis using mango peels' extract were optimized. The optimum conditions were temperature 80 °C, $AgNO_3$ concentration of 0.5 mM, pH 11, reaction time of 90 min, and 0.1 ml of the *M. indica* peels' extract. The prepared AgNPs were stable for about 3 months. The TEM analysis revealed that the green synthesized AgNPs were in the size range of 7 and 27 nm. Orange peel (*C. sinensis*) has been reported to prepare well-dispersed spherical articles of 3–12 nm AgNPs (Veeraputhiran, 2013).

A comparative study for one-step microwave-assisted green synthesis of AgNPs using aqueous extracts from the peels of citrus fruits (orange, grapefruit, tangelo, lemon, and lime) (Kahrilas et al., 2014). Box–Benhken design for three factors (time, temperature, and pressure) was applied to optimize the only successful one using the orange peel extract. The TEM analysis revealed nanospheres 7.36 ± 8.06 nm AgNPs were successfully synthesized within only 15 min. The orange peel extract acted as reducing and capping agents. The gas chromatography–mass spectroscopy (GC/MS) proved that the putative compounds responsible for successful AgNP synthesis with orange extract were aldehydes (Kahrilas et al., 2014). Ibrahim (2015) reported an eco-friendly, cost-efficient, rapid and easy method for the synthesis of AgNPs using banana peel hot water extract (BPE) which served as a reducing and capping agent. The different factors affecting silver reduction were investigated. The optimum conditions were $AgNO_3$ (1.75 mM), BPE (20.4 mg dry weight), pH (4.5), and incubation time (72 h). BPE was capable of reducing Ag^+ into AgNPs within 5 min after heating the reaction mixture (40–100 °C) as indicated by the appearance of the reddish-brown color. The UV/Vis spectrum of AgNPs revealed a characteristic SPR peak at 433 nm. The phytosynthesized AgNPs were characterized using different spectroscopic and microscopic techniques. XRD revealed the crystalline nature of the prepared AgNPs. FESEM showed spherical-shaped and monodispersed NPs. TEM confirmed the spherical nature with a size average of 23.7 nm. EDX analysis showed peaks in the silver region confirming the presence of elemental silver. FTIR affirmed the role of BPE as a reducing and capping agent of silver ions. *Solanum melongena* waste peel water extract was used for the green synthesis of AgNPs as assumed by Sharma et al. (2016). UV/Vis spectra showed maximum absorbance at 430 nm which confirmed AgNPs synthesis. TEM analysis revealed spherical-shaped AgNPs with a size of 20 nm. FTIR results suggested that the phytochemicals and certain proteins present in the vegetable peels' extract might have been involved in the synthesis, capping, and stabilizing of the produced AgNPs. Furthermore, the AgNPs showed a strong antibacterial potential against the two most pathogenic Gram-negative bacteria, *E. coli,* and *K. pneumoniae*. Li et al. (2017) reported the usage of the aqueous extract of rice straw biomass as a reducing agent in the preparation of AgNPs from $AgNO_3$, within 140 min, at room temperature, and under optimal light illumination of 60,000 lx. The prepared AgNPs expressed a good antibacterial effect even at 8 μg/mL. The green synthesis of AgNPs has been also successfully accomplished using the hot water extract of the leafy outer waste part of *Zea mays*, which is a waste material produced from the corn industry (Patra and Baek, 2017). The green synthesized AgNPs were characterized by UV/Vis spectrophotometry which revealed a peak with SPR at 450 nm. The synthesized AgNPs were evaluated for their antibacterial activity against foodborne pathogenic bacteria *B. cereus* ATCC 13061, *Listeria monocytogenes* ATCC 19115, *S. aureus* ATCC 49444, *E. coli* ATCC 43890, and *S. typhimurium* ATCC 43174. The anticandidal activity of AgNPs was also recorded against *Candida* species (*C. albicans* KACC 30003 and

KACC 30062, *C. glabrata* KBNO6P00368, *C. geochares* KACC 30061, and *C. saitoana* KACC 41238). The AgNPs displayed the moderate antimicrobial activity of ≈9.26–11.57 mm inhibition zone. The valorization of the domestic waste, hull of black gram (*Vigna mungo*), into 100 nm AgNPs with antioxidant and anticoagulant activities has been reported (Varadavenkatesan et al., 2017). Whereas the FT-IR analysis revealed the role of the functional groups of the ketones, alcohols, amines, carboxylic acids, and polyphenols present in the seed coat of *V. mungo* in the reduction mechanism and further AgNPs stabilization. Omran et al. (2018) reported that the major components of hot water extract of *C. sinensis* (sweet orange) peel (SOPE); 4-vinyl guaiacol, eugenol, aromatics, terpenes, sugar derivatives, and saturated fatty acids are involved in the bioreduction, stabilization, and capping of the highly stable and monodispersed AgNPs with PDI and zeta potential values of 0.276 and −20.3 mV, respectively. Whereas one-factor-at-a-time technique was applied for the optimization of the phytosynthesis process and revealed optimum conditions of 4000 ppm of SOPE, 3 mM $AgNO_3$, pH 9, 40 °C, 100 rpm, 24 h, and under illumination using fluorescent light 36W/6400K. The FTIR spectrum of SOPE (Figure 2.5a) revealed the characteristic absorbance of the O–H groups at 3365 cm^{-1}. The bands between 1419 and 1616 cm^{-1} indicated the presence of phenolic compounds. The strong band at 1056 cm^{-1} was attributed to the ether linkages. The band at 921 cm^{-1} was assigned to C–H bending in alkenes. The band observed at 634 cm^{-1} was assigned to C–H stretching in alkenes. While the two weak bands at 818 cm^{-1} and 634 cm^{-1} were assigned to the out-of-plane bending vibrations of the –O–H and C–H groups, respectively. Briefly, the FTIR results confirmed the presence of –C=O, –OH, C–O stretch acidic, C=C aromatics, C–N stretch of aliphatic amines, and C–H stretching alkane. The FTIR spectrum of the green synthesized AgNPs (Figure 2.5b) showed a shift of the aforementioned absorption bands. The observed four intense XRD peaks (Figure 2.6), corresponding to the (111), (200), (220), and (311) reflections of metallic silver, that plane at 2θ angles of 38.01°, 46.06°, 64.645°, and 77.41°, respectively, and confirmed the crystallinity of the phytosynthesized AgNPs with a predicted size of 15 nm. The HRTEM (Figure 2.7a) and FESEM (Figure 2.7b) showed spherical-shaped 3–12 nm AgNPs, which were relatively smaller than that of DLS (23.81 nm). In the DLS technique, the hydrodynamic diameter (particles and capping agents) is taken into account.

Ping et al. (2018) correlated the effect of temperature on the size of the phytosynthesized AgNPs by hot water extract of grape seed, as waste from grape processing, and the efficiency of its catalytic degradations properties of hazardous dyes, in the presence of $NaBH_4$. Whereas the elevated temperature

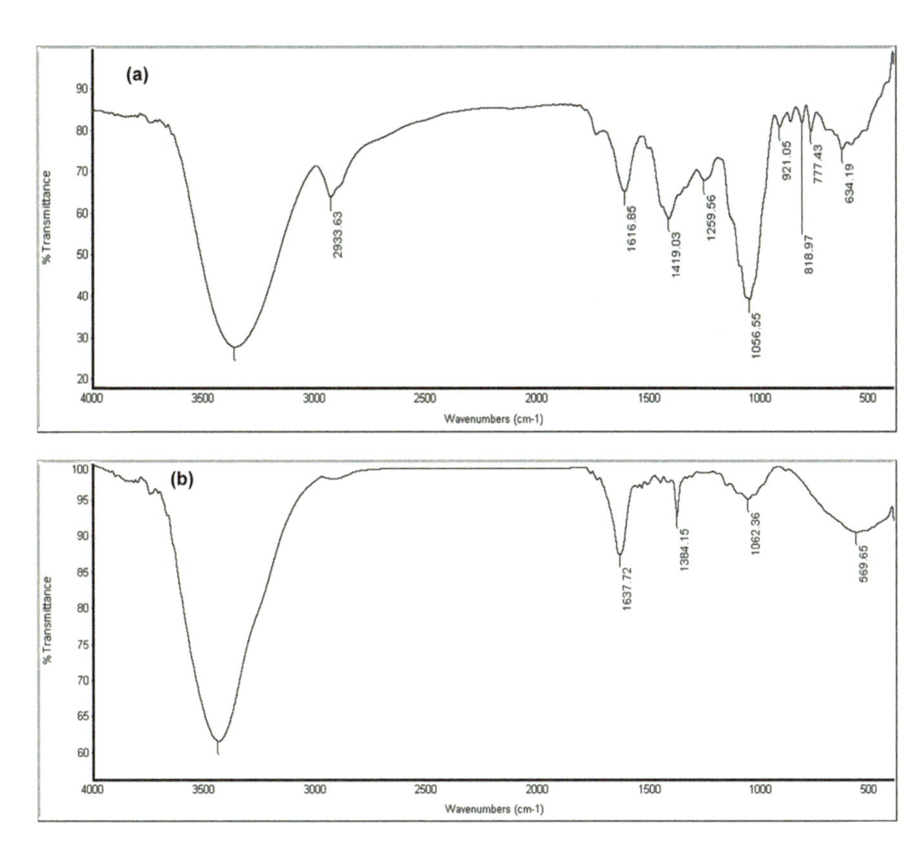

FIGURE 2.5 The FTIR spectra of sweet orange peel extract (SOPE) (a) and (b) phytosynthesized.

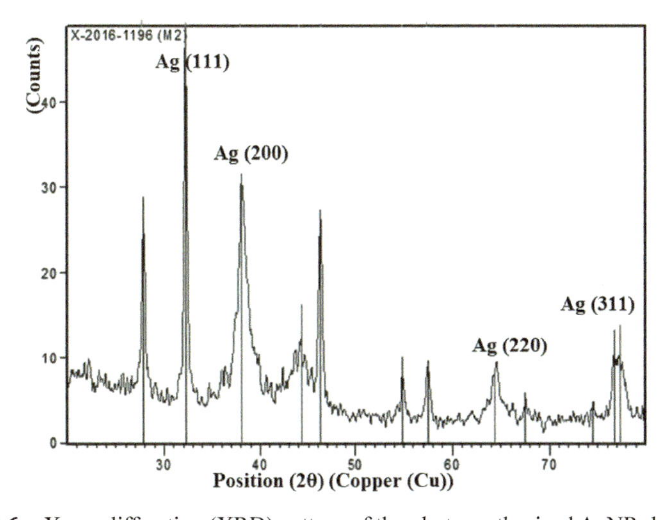

FIGURE 2.6 X-ray diffraction (XRD) pattern of the phytosynthesized AgNPs by SOPE.

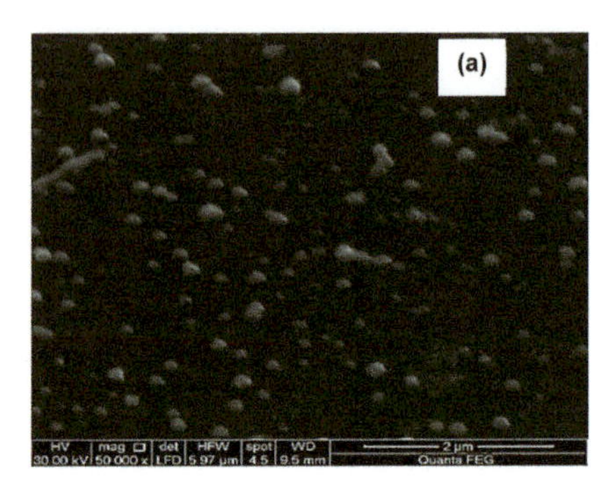

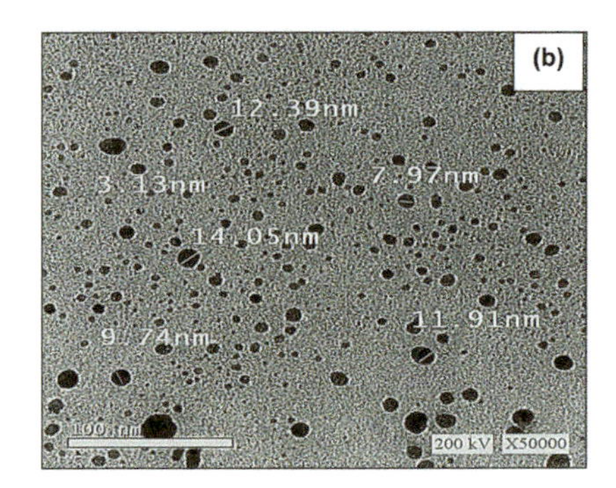

FIGURE 2.7 Field emission scanning electron microscope (FESEM) micrograph (a), and high-resolution transmission electron microscope (HRTEM) image (b) of the phytosynthesized AgNPs by SOPE.

(>40 °C) made silver particles grow bigger with lower specific surface area, due to the faster nucleation and so reduced the effectiveness of surface catalysis. The proanthocyanidins present in the seed extract played dual roles in the green synthesis of AgNPs, reducing silver ions and capping AgNPs. First, by the formation of Ag^{+}-procyanidin complex than, the conversion of Ag+ ions into Ag^{0} atoms followed by coalescence, cluster formation, and finally, their eventual growth yielding AgNPs.

2.2.10 PHYTOSYNTHESIS OF AgNPs USING SEAWEEDS

Seaweeds or marine macroalgae are called "bionanofactories" since the live and dead dried biomass was used for the synthesis of metallic NPs. They are low-cost and environmentally effective, macroscopic structured material, and have the distinct advantage of its high metal uptake capacity (Davis et al., 2003). Several seaweeds, for example, *Fucus vesiculosus* (Mata et al., 2009), *Chaetomorpha linum* (Kannan et al., 2013), *Spirogyra varians* (Salari et al., 2014), *Enteromorpha flexuosa* (Yousefzadi et al., 2014), *T. conoides* (Rajeshkumar et al., 2012), *Ulva faciata* (El-Rafie et al., 2013), *Sargassum muticum* (Lodeiro and Sillanpaa 2013), *T. conoides* (Rajeshkumar et al., 2013a), *Padina gymnospora* (Singh et al., 2013b), and *Ulva reticulata* (Sudha et al., 2013) have been used for the phytosynthesis of different sizes and shapes AgNPs.

Marimuthu et al. (2011) reported the phytosynthesis of AgNPs with antifungal property against *Humicola insolens* (MTCC 4520), *Fusarium dimerum* (MTCC 6583), *Mucor indicus* (MTCC 3318), and *Tricoderma reesei* (MTCC 3929), using the aqueous extract of red seaweed *Gelidiella acerosa* as the reducing agent. While Sahayaraj et al. (2012) used the marine algae *Padina pavonica* (Linn.) thallus broth in the extracellular synthesis of highly stable spherical polydispersed AgNPs of average size 46.8 nm. Whereas the terpenoids were mainly responsible for the reduction of metal ions by oxidation of aldehydic groups in the molecules to carboxylic acids. Moreover, the metal particles were observed to be stable in the solution even 2 years after synthesis. Rajeshkumar et al. (2013b) used the hot water extract of brown marine seaweed *Sargassum longifolium* for the green synthesis of AgNPs. The FTIR analysis revealed that the soluble organic compounds and proteins in the extract might have bound with silver ions then reduced to AgNPs NPs. While the carboxylic groups in the extract might have enhanced the stability of the produced AgNPs. Those AgNPs expressed excellent antifungal activity against pathogenic: *A. fumigatus*, *C. albicans*, and *Fusarium* sp.

Moreover, Passos de Aragã et al. (2016) reported the green synthesis of AgNPs by a naturally occurring polysaccharide extracted from red marine algae (*Gracilaria birdiae*), abundant in the Northeastern Brazilian coast. Those polysaccharides acted a triple role: the first, forming complexes with the silver ions and thereby controlling the process of reduction; the second, the reducing groups of the polysaccharide (i.e., reducing agent), and finally the third, stabilized and protected the particles from aggregation (i.e., stabilizing agent). Passos de Aragã et al. (2016) proved that the size and dispersion of silver NPs were affected by the concentration of polysaccharide and the pH. The FTIR analysis documented that the reduction of the silver ions was coupled by the oxidation of the hydroxyl and carbonyl groups in the extract. The synthesized AgNPs showed antibacterial activity with greater efficacy against Gram-negative bacteria than the Gram-positive bacteria, which was attributed to the differences in the cell wall composition. Whereas the presence of a thicker peptidoglycan layer in Gram-positive bacteria prevents the entry of AgNPs. Generally, Gram-positive bacteria are composed of three-dimensional thick peptidoglycan (~20–80 nm) layer compared to that of Gram-negative bacteria (~7–8 nm). The peptidoglycan layer possessing linear polysaccharide chain is crosslinked by more short peptides thus forming a complex structure leading to difficult penetration of AgNPs into Gram-positive bacteria compared to that of Gram-negative bacteria (Feng et al., 2000). Whereas big changes in the membrane structure of Gram-negative bacteria would occur as a result of the interaction with silver cations and would lead to the increased membrane permeability of the bacteria (Dibrov et al., 2002).

2.3 PHYTOSYNTHESIS OF AuNPs

The properties of AuNPs are very different from that of bulk, as the AuNPs are wine red solution while the bulk gold is yellow solid (Ahmed and Ikram, 2015). The AuNPs can be manufactured into a variety of shapes including nanorods, nanospheres, nanocages, nanostars, nanobelts, and nanoprisms (Thakor et al., 2013). The size and shape of AuNPs strongly affect its properties. For example, the triangular-shaped NPs show attractive optical properties in comparison to the spherical ones. Moreover, the spherical AuNPs exhibit a range of colors (e.g., brown, orange, red, and purple) in an aqueous solution (Ahmed and Ikram, 2015). These unique properties of AuNPs can potentially be exploited in a diverse range of industrial applications using their optical and electronic properties in optics, medical diagnostics, treatment, sensors, and coatings (Cortie, 2004). AuNPs have been mainly synthesized from gold chloride ($HAuCl_4 \cdot 3H_2O$). Because of the increased demand for gold in many industrial applications, there is a growing need for cost effectiveness processes to yield AuNPs by the implementation of green chemistry in the development of new NPs (Corti and Holliday, 2004). This has motivated the researchers to synthesize AuNPs using green synthetic procedures/routes which appeal to be possible environmentally friendly nanofactories (Iravani, 2011). Ahmad et al. (2017) reported several mechanisms for the phytosynthesis of AuNPs. Polyphenols like ellagic acid is a gallic acid dimer. Both ellagic and gallic acid possess hydroxyl groups that can act as reducing agents by the charge transfer mechanism. Vitamins such as ascorbic acid have the potential to convert atmospheric oxygen to water. This reaction speeds up in the presence of light radiation and metallic ions. The oxidation of ascorbic acid to dehydroascorbic acid enhances its reducing potential and results in the reduction of other oxidized chemical units thus forming AuNPs. Moreover, plant metabolites that are soluble in water serve as reducing units. Some of the common examples include phenolic compounds like alkaloids, terpenoids, and some coenzymes. Vitamin C in the water extract of *A. cepa* is reported to reduce Au^{3+} ions to AuNPs (Parida et al., 2011). Furthermore, coenzymes, such as nicotinamide adenine dinucleotide (NAD), are found in living cells, in two forms within the cell: NAD^+ and NADH. NAD^+ is an oxidizing unit and acts as an electron acceptor from other molecules and gets itself reduced. While the formed NADH in this reaction acts as an electron donor (Scheme 2.4) and helps in the transport of electrons in redox reactions. Au^{3+} ions can be reduced by the oxidation of hemiacetal and hydroxyl groups of polysaccharides to carbonyl and acetate groups (Scheme 2.4). The glycolysis reaction involves ten transformation reactions forming ten intermediates and results in the release of energy as ATP. Along with ATP, hydrogen ions are also produced in excess, which also acts as reducing agents. Reducing agents can act as stabilizing agents too. Proteins carry free carboxylate ions of amino acid or amine group that can bind to AuNPs, acting as capping and stabilizing agents. Moreover, the reduction sites of the polysaccharide can give away amino groups that might stabilize the formed metal NPs. Thus, both amino groups and carbohydrates steadfastly bind to the hydrophilic surface provided by the AuNPs. Not only this but the morphology and the size of NPs can be controlled to some extent by changing pH or time of the reaction (Klekotko et al., 2015).

$$2HAuCl_4 \longrightarrow 2Au^+ + 4HCl$$

$$NAD^+ + e^- \longrightarrow NAD$$

$$NAD + H^+ \longrightarrow NADH + e^-$$

$$e^- + Au^+ \longrightarrow Au^0$$

$$AuCl_4 + 3R\text{-}OH \longrightarrow Au^0 + 3R{=}O + 3H^+ + 4Cl$$

SCHEME 2.4 Possible mechanism for bioreduction of gold ions into AuNPs.

2.3.1 PHYTOSYNTHESIS OF AuNPs USING LEAVES' AND FLOWERS' EXTRACT

The synthesis of AuNPs by leaf extracts of *P. graveolens* and *A. indica* was reported by Shankar et al. (2003, 2004a), where the terpenoid and flvanone components of the leaf broth kept the NPs stable (Shankar et al., 2004a). Although Shankar et al. (2004b) reported the green synthesis of triangular gold nanoprisms by a single-step, room-temperature reduction of aqueous chloroaurate ions ($AuCl_4^-$) by the extract of the plant lemongrass (*Cymbopogon flexuosus*). Whereas it was proved that the nanotriangles are formed by assemblies of spherical NPs that seem to be "liquid-like" and fluidity occurred due to the NP surface complexation of aldehydes/ketones present in the extract. Narayanan and Sakthivel (2008) used leaf extract of *Coriandrum sativum* (coriander) to phytosynthesize AuNPs that had diverse shapes (spherical, triangular, decahedral), with average particles size of 7–58 nm. The secondary hydroxyl and carbonyl groups of apiin, which are found in the leaf extract of henna, were reported to be responsible for the reduction of silver and gold ions, where the size and shape of the NPs could be controlled by changing the appin concentration (Kasthuri et al., 2009). The alteration of the shapes and size of the phytosynthesized AuNPs with reaction temperature was also reported by Song et al. (2009). Whereas a mixture of the plate (triangles, pentagons, and hexagons) and spherical structures (size, 5–300 nm) AuNPs were obtained at lower temperatures and concentrations of leaf extracts of *Magnolia kobus* and *Diopyros kaki*, while smaller spherical shapes were obtained at higher temperatures and leaf extracts. The phytosynthesis of various shapes and sizes of Au and Ag NPs was reported by the leaf extract of *Hibiscus rosa sinensis* (Philip, 2010) that depended on the ratio of the metal salt and extract in the reaction medium and the variation of pH of the reaction medium for Au and AgNPs. The FTIR spectra showed that the AuNPs and AgNPs were bound to amine groups and carboxylate ion groups. Dwivedi and Gopal (2010, 2011) reported the phytosynthesis of quasispherical-shaped 10–30 nm AuNPs, using leaves extract of *C. album*. Das et al. (2011) reported the green synthesis of spherical-shaped AuNPs with the average size of 20 nm, using *Nyctanthes arbortristis* flower extract. The phytosynthesis of spherical-shaped 4–20 nm AuNPs using Bael (*A. marmelos*) leaves was also reported (Jha and Prasad, 2011). A leaf extract of *Cassia auriculata* was also reported to phytosynthesize spherical and triangular AuNPs of average size ranged between 15 and 25 nm, at room temperature and within only 10 min at room temperature (Kumar et al., 2011a). FTIR spectroscopy analysis revealed that carbohydrate, polyphenols, and protein molecules found in leaf extract of *Lonicera japonica* plant were involved in the synthesis and capping of Ag and AuNPs of different sizes and shapes. That varied between 36 and 72 nm spherical to a few plate-like polyshaped AgNPs and polyshaped nanoplates of 40–92 nm AuNPs (Kumar and Yadav, 2011). *Murraya Koenigii* leaf extract was proved as a reducing and stabilizing agent upon producing well dispersed, 10 nm AgNPs at room temperature and 20 nm AuNPs at around 100 °C, within a short time of approximately 5 min. The UV/Vis showed a symmetric SPR band centered at 411 nm and 532 nm, for AgNPs and AuNPs, respectively. The crystallinity of the phytosynthesized NPs was confirmed from the high-resolution TEM images, SAED, and XRD patterns (Philip et al., 2011). Jha and Prasad (2012) performed a comparative study, at room temperature, for the phytosynthesis of AuNPs by three different leaves extract of *Mentha*, *Ocimum*, and *Eucalyptus*. The UV/Vis analysis revealed the SPR of AuNPs, at 539, 536, and 537 nm using the aforementioned plant extracts, respectively. The XRD analysis confirmed the production of AuNPs. The TEM analysis showed spherical-shaped NPs, with an average size of 3–16 nm, whereas *Ocimum* leaf provided the finer particles relative to the others. The phytosynthesis of various shapes of AuNPs, triangular, hexagonal, and spherical, was reported by Khalil et al. (2012) using olive (*O. europaea*) leaf hot water extract. A very short period of time, only 20 min, was required for the bioconversion and bioreduction to AuNPs at room temperature. The formed AuNPs were 50 nm in size. The high phenolic contents (i.e., Oleuropein, apigenin-7-glucoside, and luteolin-7-glucoside) of the hot water extract of olive leaves aided in the reduction of gold cations to metallic gold (Au^0). Arunachalam et al. (2013) reported one-pot phytosynthesis and characterization of AuNPs from *Memecylon umbellatum* leaf water extract. The plant is known to comprise saponins, phenols, phytosterols, and quinones which help in bioreduction of gold ions. The transmission electron microscope images showed the formation of spherical, hexagonal, and triangular-shaped AuNPs. They were monodispersed with an average size of 17 nm. EDX analysis showed an elemental gold signal at 2.3 keV. The *M. umbellatum* leave extract contain triterpenoids, alkaloids, and tannins, hence these biomolecules functioned as reducing agents for the synthesis of AuNPs. Ahmed et al. (2014) synthesized AuNPs from *Salicornia brachiata* leaf aqueous extract and studied their antibacterial activity against *E. coli*, *Pseudomonas areuoginosa*, *Salmonella typhi*, and *S. aureus*. The produced AuNPs exhibited a purple color with a characteristic SPR at 532 nm. SEM and TEM confirmed the polydispersed nature of the synthesized AuNPs with a size average of 22–35 nm size. The XRD confirmed the purity and crystallinity of the prepared AuNPs. It was surprising that the AuNPs were not formed when a mixture of $HAuCl_4 \cdot 3H_2O$ and the plant extract was heated to 60 °C under the sunlight. However, the reaction was catalyzed by the addition of trace amounts of 10 μM sodium borohydride, and shear change in color was observed within 5 min which did not diminish even after 30 days. These AuNPs were investigated for their catalytic activity for the reduction of nitrophenol and MB. The reduction of 4-NP into 4-aminophenol (4-AP) was followed by UV/Vis spectrophotometer that was accompanied by a change in color from yellowish-green to colorless. Reduction of MB into LMB by AuNPs in presence of sodium borohydride ($NaBH_4$) was also indicated by diminishing the characterized MB color (Suvith and Philip, 2014). Initially, the aqueous MB showed two absorption peaks at 664 and 614 nm which disappeared after reduction. The antibacterial activity of AuNPs, ofloxacin, and a combination of NPs with ofloxacin were investigated. It was important to note that the combined activity of AuNPs with ofloxacin was much higher than either the NPs or antibiotic individually. Consequently, a more effective biocompatible mixture of a suitable antibiotic and AuNPs is recommended. Klekotko et al. (2015) reported the green synthesis of AuNPs using the hot water extract of mint (*Mentha piperita*). Whereas the UV/Vis spectrum showed two characteristics peaks: a narrow and broad one at 540 nm and 900–1000 nm indicating the presence of anisotropic shapes of NPs. The transmission electron microscope proved the presence of polydispersed NPs (90% nanospheres, 8% nanoprisms, and 2% hexagonal nanoplates), with sizes ranged between 10 and 300 nm. Sadeghi et al. (2015) reported the fabrication of AuNPs from *Stevia rebaudiana* leaf water extract. TEM and SEM images showed average sizes of the AuNPs in the range of 5–20 nm. FTIR spectra showed that they were functionalized with biomolecules containing a primary amine group (NH_2), carbonyl group, OH groups, and other stabilizing functional groups. It was found from the XRD study that AuNPs were 17 nm and had fcc structure. The zeta potential measurement of the prepared AuNPs was stable in a wide range of pH (6–12). AuNPs were also fabricated from a mixture of *Carica papaya* and *C. roseus* leaf water extract (Muthukumar et al., 2016). A change in color from yellow to ruby red which is a characteristic feature of AuNPs was observed. A characteristic SPR band appeared between 500 and 600 nm regions. The electron microscopy results showed that AuNPs were mostly spherical; however, triangle and hexagonal-shaped morphologies were also noticed. The infrared spectroscopic studies revealed that alkaloids, flavonoids, and proteins were present along with AuNPs that were responsible for their stabilization and nonagglomeration. Ahmad et al. (2016) reported the phytosynthesis of AuNPs using aqueous *Elaise guineensis* (oil palm) leaf extract without the addition of any external stabilizing agents. TEM image showed the formation of predominant spherical AuNPs with a mean particle diameter of 27.89 ± 14.59 nm. DLS data showed that AuNPs were 55.22 ± 42.86 nm in diameter, indicating the formation of multilayer coatings derived from *E. Guineensis* leaf extract biomolecules surrounding the prepared AuNPs. Spherical, triangular, pentagonal, and hexagonal AuNPs were observed. The FTIR analysis revealed that carboxylic and phenolic compounds present in leaf extract played dual roles as reducing and stabilizing agents during the phytosynthesis of AuNPs. *Panax ginseng* leaves extract have been reported to phytosynthesize Au and Ag NPs via the bioreduction of auric chloride and silver nitrate, respectively (Singh et al., 2016b). These heavy metals have been found to be highly resistant toward the microbial biofilm (Singh et al., 2016b). Balalakshmi et al. (2017) reported the effective fabrication of AuNPs using *Sphaeranthus indicus* leaf hot water extract. The UV/Vis spectrum of AuNPs showed a characteristic SPR peak at 531 nm. TEM revealed the spherical shape of AuNPs with a mean particle size of 25 nm. It is worth mentioning that the green synthesized *S. indicus* extract-synthesized AuNPs treatment did not show any mortality against the nontarget microcrustacean *Artemia nauplii*. These results were in agreement with the ones observed by Kesarla et al. (2014), who previously reported that the green synthesized AuNPs using *Terminalia bellirica* aqueous extract were nontoxic to the nontarget organism *Artemia salina*. Recently, Murugan et al. (2015) proved that the *Cymbopogon citratus* synthesized AuNPs had no toxicity against the nontarget copepod *Mesocyclops aspericornis*. According to Dutta et al. (2017), the average size of the AuNPs synthesized by the *Syzygium jambos* (*Myrtaceae* family) leaf hot water extract was lower (5 nm) than the AuNPs synthesized using the bark (8–10 nm). FTIR spectra indicated that the saccharides and phenolics present in the *S. jambos* extracts were the major contributors responsible for the synthesis and stabilization of NPs. Haq Khan et al. (2017) reported the green synthesis of dispersed and spherical-shaped AuNPs of average size 2–12 nm, using hot water extract of *Sueda fruciotosa* (family *Amaranthaceae*). The prepared AuNPs expressed good phtoctalytic degradation potential on MB. Moreover, an AuNPs modified GCE was applied for the electrochemical decomposition of phenolic Azo dyes. Seetharaman et al. (2017) reported the successful phytosynthesis of 32.89 nm AuNPs using *Crescentia cujete* leaf water extract. The aqueous extract of *C. cujete* was employed as a reducing agent to reduce Au ions into AuNPs. XRD study depicted fcc crystalline structure and the zeta potential revealed that synthesized AuNPs were highly stable and negatively charged. The prepared AuNPs performed an effective bactericidal activity against both Gram-positive and Gram-negative bacterial pathogens.

2.3.2 PHYTOSYNTHESIS OF AuNPs USING SEEDS' AND BUDS' EXTRACT

Raghunandan et al. (2010) reported that the flavonoids present in the extract of dried clove (*Syzygium aromaticum*) buds are responsible for the reduction of Au^{3+} ions producing and stabilizing irregularly shaped AuNPs. Krishnamurthy et al. (2011) reported the phytosynthesis of spherical-shaped $\leq$10 nm AuNPs using the seed extract of *Cuminum cyminum*. Moreover, *Trigonella foenum-graecum* (Aromal and Philip, 2012) and *Sesbania drummondii* (Nellore et al., 2012) seeds extracts used as bioreductants, stabilizing, and capping agent of spherical-shaped 15–25 nm and 6–20 nm AuNPs, respectively. Jayaseelan et al. (2013) reported the utilization of hot water extract of *Abelmoschus esculentus* seeds for the green synthesis of spherical-shaped AuNPs with sizes ranged between 45 and 75 nm and with efficient antifungal effect against *Puccinia graminis tritci*, *A. flavus*, *A. niger*, and *C. albicans*. Seetharaman et al. (2017) reported the benign synthesis of AuNPs using *C. cujete seed* water extract. In a short period of time, the bioreduction reaction was completed when the aqueous extract of *C. cujete* was employed as a reducing agent to reduce Au^{3+} ions to their NP state. Generated AuNPs mean size was 32.89 nm. *Vitis vinifera* seed extract was used for the synthesis of AuNPs (Ismail et al., 2014). The UV/Vis spectra characteristic of AuNPs appeared at 565 nm. The FTIR assured the existence of aromatic amines and secondary alcohols as capping and reducing agents of the phytosynthesized AuNPs.

2.3.3 PHYTOSYNTHESIS OF AuNPs USING FRUITS AND NUTS

The phenolic compounds in the fruit extract of Elephant apple (*Dillenia indica*) were found to be involved in the bioreduction of Au^{3+} ions into AuNPs (Nadeem et al., 2017). Hot water extract of *Areca catechu* nuts has been applied to the phytosynthesis of AuNPs (Rajan et al., 2015). Although broad- and red-shifted bands of UV/Vis spectrum indicate the presence of anisotropic AuNPs with increased particle size. While highly symmetrical, sharp, intense, and blue-shifted absorption bands correspond to the enhanced formation of monodispersed AuNPs. The microwave irradiation of 2450 MHz and reaction temperature of 100 °C produced AuNPs within only 1 min with average particles size of 13.7 nm and 18.1 nm, respectively. However, higher yield (3-fold increase), with predominance of spherical-shaped AuNPs, was obtained applying the microwave irradiation, due to the enhancement in the kinetics of the reaction. While the reaction took 5 h at room temperature with the formation of large anisotropic AuNPs with average particles size of 22.2 nm. The FTIR-analysis revealed the involvement of proteins and flavonoids present in the nuts' extract in the reduction of chloroaurate ions and stabilization of AuNPs. The lower the particles size the higher its specific surface area and consequently, the higher its catalytic activity for the reduction of organic pollutants; MB, MO, EY, and 4-NP. A huge difference between the redox potential of the acceptor and donor restricts the efficient transfer of electrons. An enhanced reduction rate with the addition of the catalyst is owing to the intermediate redox potential value between that of the donor and acceptor. Taking into consideration, the catalytic efficiency is surface-dependent and the organic pollutants could be degraded only if they adsorb on the surface of the catalyst. Moreover, the antioxidant potential of the green synthesized AuNPs has been determined through nitric oxide (NO) and 2,2-diphenyl-1-picrylhydrazyl radical (DPPH) scavenging activities. Whereas it was found to be also dependent on the size and concentration of AuNPs. The lower the size, the higher is its antioxidant potential. The higher the AuNPs concentration, the higher is its antioxidant potential. The extract had a high percentage of tannins and polyphenols. Thus, the synergic action of these bioactive compounds as reducing and stabilizing agents had been assumed to be the reason for the antioxidant potential of the green synthesized AuNPs. Furthermore, the efficient antibacterial activity against the human pathogens, *E. coli*, *K. pneumonia*, *P. aeruginosa*, *Enterobacter* sp., and *S. aureus,* was found to be also dependent on the size of AuNPs. Nevertheless, AuNPs bactericidal effect was summarized based on the following mechanism, the attachment of AuNPs to the surface of the cell membrane, which disturbed its power functions such as permeability and respiration, led to the depletion of antioxidants and production of ROS. Furthermore, the AuNPs might have released metal ions that penetrated the cell wall causing DNA damage and protein dysfunctioning. Yuan et al. (2017) used the aqueous pulp extract of *C. annuum* var. *grossum* as a reducing agent for Au^{3+} ions into AuNPs with various shapes (triangle, hexagonal, and quasispherical shapes) and size range of 6–37 nm, within only 10 min, at room temperature. Whereas the effect of pH was very obvious, at low pH < 2, the reduction stopped and no AuNPs were obtained and at high pH ≥ 14, the wine red color of AuNPs appeared immediately after adding $AuCl_4$, but turned deep blue within 20 min and was not stable and particles deposited within an hour. It was also noticed that the reaction rate was faster with weak alkaline treatment than that with acidic treatment that was explained as excessive hydrogen ions at low pH kept chloroauric acid in molecular forms and inhibit the reduction. While with pH increasing, a shift from $[HAuCl_4]$ to [AuCl4]− occurred and the gold ions tended to form $[Au(OH)_4]^-$. Whereas Yuan et al. (2017) suggested the possible reduction mechanism (Scheme 2.5).

$$HAuCl_4 \longrightarrow AuCl_4^- + H^+$$

$$AuCl_4^- \longrightarrow AuCl \longrightarrow Au^0$$

SCHEME 2.5 Possible phytoreduction of gold ions in basic medium.

Thus, it was obvious that the change in pH highly affected the shape and size of AuNPs. The dissociation or hydrolysis rate and equilibrium state of $AuCl_4^-$ in solution could be strongly influenced by pH. The FTIR analysis proved that the chemical groups including amino, sulfhydryl, and carboxylic groups in the biomolecules were responsible for the formation of AuNPs. It is well known that the chemical species of the biomolecules present in the extract are mediated by pH. Thus, both the substrates and reactants were affected by pH, thus, the nucleation and aging process could be affected by pH. Consequently, the morphology, in turn, could be influenced by pH conditions. The steps of the synthesis of the AuNPs were first summarized as the Au^{3+} ions were first reduced as a gold atom by the extract biomolecules or reducing agents. Then, the reduced atom collided and aggregated together to assemble a AuNP. They formed small NP grew by aggregating more small particles with the increment of incubation time. The effect of initial Au^{3+} ions concentration was well discussed. As the initial ions concentration increased, the chance of nucleation was higher and the particles grew into large particles by aggregation. Since excessive Au^{3+} ions in the solution would have consumed most of the biomolecules (i.e., the reducing agents), and the remained biomolecules would not have been enough to act as capping and stabilizing agents, thus, could not prevent the particles from aggregation and consequently, the precipitation of large AuNPs from the bulk solution was performed. The same occurred at the high reaction temperature. Although high temperature increased the rate of reaction and decreased the formation time of AuNPs but decreased the stability of the formed NPs. Higher temperature benefits the aggregation and similarly accelerated the precipitation to occur. The contrary occurred at low initial Au^{3+} ions concentration. Whereas excessive biomolecules would be found, acting as both reducing and capping agents. Thus, the small AuNPs could have been prevented from aggregation and kept the as-formed NPs dispersed at a small size distribution. The produced AuNPs expressed good catalytic activity for reduction of the pale yellow 4-nitrophenol in $NaBH_4$ solution, to the colorless 4-aminophenol via the formation of the golden yellow 4-nitrophenolate ion as an intermediate.

2.3.4 PHYTOSYNTHESIS OF AuNPs USING SHOOT/STEM EXTRACT

AuNPs were phytosynthesized from *Cucurbita pepo* shoot water extract (Gonnelli et al., 2015). In this case, the bioreduction of $HAuCl_4.3H_2O$ was carried out at 40 °C for 30 min and exhibited an absorption peak at about

570 nm. Mishra et al. (2016) fabricated water well dispersed AuNPs from *Hibiscus sabdariffa* leaf and stem extracts. High amounts of AuNPs were obtained at pH 4–6 (i.e., acidic conditions) at 100 °C within a short time of approximately 6 min. To know the chemical state of the prepared AuNPs, X-ray photoelectron spectroscopy (XPS) analysis was performed. The Au 4f7/2 spectrum had two peaks at 81.0 and 84.5 eV, corresponding to the binding energies of Au^0 and gold ions (Au^{n+}). Firdhouse and Lalitha (2016b) reported the successful biogenic synthesis of AuNPs using the water extracts of *Kedrostis foetidissima* fresh leaves and stem. The phytosynthesized AuNPs were spherical in shape and uniformly distributed as confirmed by SEM analysis. The size was below 50 nm and they were flower shaped as confirmed by TEM. The phytoconstituents of the aqueous extracts of *K. foetidissima* fresh leaves and stem served as excellent reducing and capping agents as revealed from the FTIR spectra.

2.3.5 PHYTOSYNTHESIS OF AuNPs USING BARK EXTRACT

Daisy and Saipriya (2012) phytosynthesized 55–98 nm AuNPs (55–98 nm) using the hypoglycemic aqueous extract of *Cassia fistula* bark. That was attributed to the action of the particles on concentrating the hypoglycemic agent from the extract on their surfaces. The bark of *Hypericum perforatum* and *Hamamelis virginiana* have been reported for phytosynthesis of spherical and polyhedral shaped 4–6 nm AuNPs (Pasca et al., 2014). Spherical and triangular-shaped 20–50 nm AuNPs have been phytosynthesized by *Terminalia arjuna* bark extract (Suganthy et al., 2018). In an earlier study by Rakhi and Gopal (2012), *T. arjuna* bark phytosynthesized a mixed triangular, tetragonal, pentagonal, hexagonal, rod-like, and spherical shape of 15–20 nm AuNPs. The triterpenoids, saponines, tannins, flavonoids present in the bark extract of *T. arjuna* are reported to act as bioreductants, stabilizing, and capping agents for AuNPs (Rakhi and Gopal, 2012). Moreover, such polyshaped AuNPs expressed potential catalytic reduction of 4-nitrophenol into 4-aminophenol in presence of sodium borohydride (Rakhi and Gopal, 2012).

2.3.6 PHYTOSYNTHESIS OF AuNPs USING ROOTS' EXTRACT

Kumar et al. (2011b) reported the phytosynthesis of 5–15 nm spherical-shaped AuNPs using root extract of *Z. officinale*. Pasca et al. (2014) reported the phytosynthesis of mixed shapes (spherical, ovals, and polyhedral) of 3–4 nm AuNPs using root extract of *Angelica archangelica*. Suman et al. (2014)

reported the phytosynthesis of cubic-shaped 12.17–38.26 nm AuNPs by root extract of *Morinda citrifolia*. Salunke et al. (2014) reported the phytosynthesis of 20–30 nm gold nanospheres from *Plumbago zeylanica* root hot water extract. It was observed that the AuNPs formation was accelerated with increasing temperature and the metal salt concentration with a maximum of 50 °C temperature and 0.7 mM concentration of $HAuCl_4.3H_2O$, respectively. Characterization was performed using UV/Vis spectrophotometer, FTIR, XRD, EDX, TEM, high-performance thin layer liquid chromatography, and GC/MS. The TEM images showed some noteworthy features of AuNPs shapes, they were a mixture of spherical, triangular, and hexagonal-shaped particles. They also exhibited antimicrobial and antibiofilm activities against *E. coli*, *Acinetobacter baumannii*, and *S. aureus*. The phytosynthesized AuNPs showed significantly higher antimicrobial activity relative to those synthesized chemically. It appeared that the diffusion and organic contents of the root extract enhanced the AuNPs activity. Ahmed et al. (2017) summarized the antimicrobial mechanism as follows: AuNPs reduce the amount of ATP-synthase, thus, the bacterial cells become deficient of energy for processing their activities, consequently limiting the metabolic processes. It destructs the ribosomal subunit that is normally required to bind the ribosome to tRNA, collapsing the overall biological mechanism required for a healthy bacterial cell. Moreover, small-sized AuNPs with large specific surface area bound strongly to proteins present in the cytoplasm and cell wall of bacteria disturbing the normal functioning of cells leading to cell death. Furthermore, phosphorus-containing proteins and sulfur-containing DNA molecules are chosen targets for AuNPs attack. Moreover, binding of AuNPs to coenzymes such as NADH affects the respiratory chains and releases oxygen to creating an oxidative stress. All of these cause severe damage to the structure and function of the cell, leading to cell death. It is worth to know that the antifungal effect of AuNPs is lower than the antibacterial effect. Ahmed et al. (2017) attributed this to the mechanism of action of AuNPs with microbial cells. Where, in bacteria, respiration occurs through the cell membrane. Thus, the surface adsorption of AuNPs on the bacterial surface hampers the dehydrogenation process during respiration. While in the case of fungi, respiration takes place via mitochondrial membrane, a thus larger dose of AuNPs is needed to enter the fungal cells and target the mitochondrial membrane.

M. citrifolia root hot water extract was used for the phytosynthesis of AuNPs as reported by Suman et al. (2014). A change in color to pink ruby red color took place within 24 h and the maximum absorption peak was obtained at 540 nm. The FTIR bands at 1318 and 1089 cm^{-1} were due to C–N stretching vibration of aromatic and aliphatic amines. Synthesis and characterization of AuNPs from *Panicum maximum* root extract was also reported by Agarwal and Srivastava (2014). After reaction, the color solution changed to pinkish-red within 5 min and the characteristic AuNPs SPR occurred at 540 nm. According to Abbasi et al. (2015), mostly the polysaccharides and proteins played a key role in the bioreduction of gold ions to their NP state during the phytosynthesis of AuNPs using *Ipomoea carnea* root extract. The reduction of gold ions was completed within 6 h. *P. ginseng* root extract has been reported to phytosynthesize Au and Ag NPs with anticoagulant properties, via the bioreduction of auric chloride ($AuCl_3$) and silver nitrate ($AgNO_3$), respectively (Singh et al., 2016c). Ahn et al. (2018) reported the rapid phytosynthesis of AuNPs using *Acanthopanax sessiliflorus* root water extract. Bioreduction of $HAuCl_4.3H_2O$ by the studied extract resulted in the formation of gold nanoflowers within 8 s at room temperature which was an astonishing result. Umamaheswari et al. (2018) were the first to report the green synthesis of monodispersed and spherical-shaped AuNPs with an average size of 10.5 nm, using 5,7-dihydroxy-6-metoxy-3′,4′methylenedioxyisoflavone (Dalspinin), extracted from the roots of *Dalbergia coromandeliana*. It was highly stable up to 5 months, without the formation of any aggregation, and showed good catalytic degradation potentials for Congo red and MO, in the presence of $NaBH_4$.

2.3.7 PHYTOSYNTHESIS OF AuNPs USING AGRO-INDUSTRIAL WASTES

Waste management represents an important challenge in the agro-food-based industries and demands an integrated approach in the context of recycling, upcycling, reuse, and recovery of such wastes. According to Bankar et al. (2010), AuNPs were synthesized using banana (*M. paradisiaca*) peel hot water extract (BPE) as a simple, nontoxic, and eco-friendly "green material." The boiled, crushed, acetone precipitated, air-dried peel powder was used to reduce $HAuCl_4.3H_2O$. The maximum production of AuNPs was obtained at pH 5, 1 mM of $HAuCl_4.3H_2O$, and 10 mg ml^{-1} of BPE at 80 °C. The sugar beet pulp was reported as an effective reductant for making gold nanowires at room temperature. The NPs formed initially and then joined to form chains and nanowires. Whereas the formation of nanowires and nanorods depended on the conditions of the reaction, particularly the pH (Castro et al.,

2011). Ahmad et al. (2012) presented a simple and eco-friendly method for the biosynthesis of AuNPs using pomegranate (*Punica granatum*) peel hot water extract. Peel extract of pomegranate was subjected to $HAuCl_4.3H_2O$ at room temperature. The TEM studies showed that the average particle size of AuNPs was found to be 10 ±1.5 nm. The XRD studies confirmed the fcc lattice of the synthesized AuNPs. The reduction of gold ions and the formation of stable NPs occurred rapidly within an hour of reaction, making it one of the fastest bioreducing methods. Patra et al. (2016) reported a suitable, eco-friendly, nontoxic, one-step synthesis procedure of AuNPs via the utilization of the food waste material, that is, (*A. cepa*) onion peels (OPs) water extract. SPR spectra characteristic to AuNPs were obtained at 535 nm with an average particle size of 45.42 nm. The green synthesized OP-AuNPs were surface capped with a number of phenolic bioactive compounds, including cysteine derivatives. OP-AuNPs displayed a strong synergistic antibacterial and anticandidal potential along with antioxidant.

2.3.8 PHYTOSYNTHESIS OF AuNPs USING SEAWEEDS

Tannis and flavonoids are defined as naturally occurring seaweed polyphenolic compounds and are found in marine algae. Whereas marine red algae are rich sources of phenolic compounds especially bromophenols (Li et al., 2010). Singaravelu et al. (2007) and Oza et al. (2012) reported the action of nitrate reductase in *Sargassum wightii* extract as reducing and stabilizing the gold ions into gold NPs under alkaline pH and at room temperature. The brown algae *Stoechospermum marginatum* has been reported for the synthesis of AuNPs with antibacterial effect on pathogenic bacteria (Rajathi et al., 2012). Whereas the hydroxyl groups present in the diterpenoids of the brown alga were responsible for the bioreduction of Au^{3+} ions. The chlorophycean algae *Ulva intestinalis* and *Rhizoclonium fontinale* were reported as bionanofactories for AuNPs (Parial et al., 2012). Castro et al. (2013) reported the change of the size and shape of gold NP with the initial pH value using the red seaweed *Chondrus crispus*. *Galaxaura elongate* is rich in R-, C-phycoerythrin, fucoxanthins, sulfated polysaccharides, polypeptides, proteins, tannins, flavonoids, halogenated terpenes, and polyphenols. However, El-Kassas and El-Sheekh (2014) attributed the formation and stabilization of the green synthesized AuNPs to the hydroxyl functional group from polyphenols and carbonyl group from algal proteins used extract. Whereas the aqueous extract of the red seaweed *Corallina officinalis* biosynthesized AuNPs with average particle size of 14.6 ± 1 nm. Furthermore, Naveena and Prakash (2013) reported the green synthesis of AuNPs with average size of 45–57 nm, using the water extract of the red marine alga *Gracilaria corticata*. The prepared AuNPs proved a good bactericidal effect against the Gram-positive *S. aureus* and *Enterococcus faecalis* and Gram-negative *E. coli* and *E. aerogenes*. It also expressed a good antioxidant activity, as the synthesized AuNPs showed a good capacity for scavenging the DPPH free radical and ferric-ion reducing ability antioxidant power (FRAP). Rajeshkumar et al. (2013a) used also the hot water extract of marine brown algae *T. conoides* for the green synthesis of spherical-shaped polydispersed AuNPs with an average size 6–10 nm. Moreover, Vijayan et al. (2014) reported the green synthesis of silver and gold NPs with average particles size of 2–17 nm and 2–19 nm, respectively, by the hot water extract of the seaweed *T. conoides*. But, although, the synthesized AgNPs were efficient in controlling the bacterial biofilm formation; however, AuNPs did not show any remarkable antibiofilm activity. Varun et al. (2014) reported that the carboxylic, amine and polyphenolic groups studied in the aqueous extract of the brown seaweed *Dictyota bartayresiana* acted as the reducing agent during the green synthesis of AuNPs.

2.4 PHYTOSYNTHESIS OF COPPER NANOPARTICLES (CuNPs)

Copper is an essential micronutrient that plays an important role in the health of all living organisms. Copper is incorporated into an array of proteins and metalloenzymes required for performing various metabolic functions by the plant cells/organisms. Copper has an outstanding electrical conductivity, good catalytic behavior, and surface-enhanced Raman scattering activity, and hence CuNPs have attracted the attention of researchers for using it as an essential component in the future nanodevices (Chandra et al., 2014). CuNPs have also gained importance due to their widespread applications as antimicrobials, gas sensors, electronics and coating on textiles, batteries, solar energy conversion tools, and high-temperature superconductors (Chen et al., 2012). CuNPs effectively inhibit the growth of many pathogenic bacteria like *S. aureus, B. subtilis, E. coli, Shigella dysenteriae, Salmonella typhi, K. pneumonia* (Ahamed et al., 2014; Sutradha et al., 2014; Naika et al., 2015) and possess good antifungal activity against various pathogenic fungi (Kanhed et al., 2014; Shende et al., 2015). The green synthesized CuNPs under solar irradiation is a very promising eco-friendly catalyst for the degradation of pollutants of environmental concern (Nazar et al., 2018). Upon solar irradiation, CuNPs can be excited, forming electrons (e^-) and holes (h^+). Such active holes are responsible for the breaking of water molecules into H^+ and $^{\bullet}OH$. But the free electrons are responsible for the formation of superoxide radicals $HO_2^{\bullet}$, which consequently produces hydrogen peroxide (H_2O_2) and $O_2^{\bullet}$. The strong oxidizing agents $^{\bullet}OH$, $O_2^{\bullet}$, and H_2O_2 degraded the pollutants into low molecular weight compounds and finally, into carbon dioxide and water (Nazar et al., 2018).

CuNPs have been synthesized from various salts of copper like copper acetate, copper chloride, copper sulfate, and copper nitrate using extracts made from various plants belonging to different groups. Although CuNPs have been synthesized by physical and chemical methods for quite some time now, the biological synthesis of CuNPs has started recently.

The aqueous extract of *Impatiens balsamina* L. contains lawsone, lawsone methyl ether, and a few other strong reducing organic compounds (e.g., kaempferol, bilawsone, etc.) (Yang et al., 2001). The powerful reducing ability of these organic compounds was utilized to create a synthesis platform for CuNPs. To verify the viability of this presumption, the leaf extract reacted with copper sulfate ($CuSO_4$) solution. Expectedly, CuNPs were produced as the components of the leaf extract (mostly lawsone) targeted Cu^{2+} ions present in the medium and reduced them to Cu^0. The absorbance reached its maximum value near 550 nm after 72 h. The particles were mostly spherical shaped with a mean diameter 5–10 nm as was revealed using TEM. From the XRD pattern, three peaks at $2\theta = 43.42°$, $50.14°$, and $74.31°$ correlating to, respectively, (111), (200), and (220) crystal planes of copper, were detected. This further confirmed the crystalline nature and purity of the green synthesized CuNPs. The FTIR study supported the role of aromatic compounds like lawsone along with amides in the production of the CuNPs and their surface stabilization mechanism. CuNPs afforded efficient photocatalytic property to degrade the toxic organic dyes under solar irradiation (Yang et al., 2001). The phytosynthesized CuNPs using hot water extract of *M. kobus* leaves expressed efficient biocidal activity on *E. coli* (Lee et al., 2011). Aqueous hot water extract of *M. kobus* leaf extract was used for phytosynthesis of 37–110 nm CuNPs from $CuSO_4 \cdot 5H_2O$ (Lee et al., 2013). Whereas the conversion after 24 h was about 70% at 25 °C and 80%–100% at 60 and 95 °C. The average particle size decreased from 110 nm at 25 °C to 37 nm at 95 °C. That was explained as, with the increase in reaction temperature, the reaction rate increases and thus most copper ions are consumed in the formation of nuclei, stopping the secondary reduction process on the surface of the preformed nuclei. Although the reaction rate was highest at 20% leaf broth concentration, however, with increasing leaf broth concentration, the average particle size decreased up to 15% leaf broth concentration and then increased at 20% leaf broth concentration. That was attributed to the presence of too many capping materials from the plant leaf broth, which caused some aggregation of copper particles at a high leaf broth concentration of 20%, probably due to the interaction between NPs, which are surrounded by proteins and metabolites such as terpenoids and reducing sugars. Due to the difficulty in initially forming copper nuclei, the phytosynthesis of CuNPs by *M. kobus* leaf extract was slower than that of silver (Song and Kim, 2009), gold (Song et al., 2009), and platinum (Song et al., 2010). The UV-Vis, spectroscopy showed the characteristic peak of CuNPs at 560 nm. Moreover, the XPS confirmed the characteristic peaks of CuNPs. The energy-dispersive X-ray spectroscopy (EDS) profile showed copper signals along with oxygen and carbon peak, which were attributed to the biomolecules that might have been bounded to the surface of the CuNPs. The FTIR analysis revealed that the reduction of copper ions and stabilization of synthesized copper NPs were attributed to some proteins and metabolites such as terpenoids and reducing sugars having the functional groups of amines, alcohols, ketones, aldehydes, and carboxylic acids. The antibacterial tests were carried out by counting viable *E. coli* cells after 24 h growth in shake flasks containing latex foams coated CuNPs, which showed higher antibacterial activity compared with untreated foams and foams treated with chemically synthesized CuNPs using sodium borohydride and Tween20. Moreover, the antibacterial activities were found to be inversely proportional to the average NPs-sizes. Thus, the smaller NPs synthesized at higher temperature (95 °C) and extract concentration of (15%) showed higher antibacterial activity due to the larger specific surface area and higher content of CuNPs (Lee et al., 2013). Hot water extract of *Artabotrys odoratissimus* leaves has been also reported for phytosynthesis of CuNPs (Kathad and Gajera, 2014). Nasrollahzadeh et al. (2014) reported that the formation of CuNPs was possibly facilitated by flavonoids and phenolics acids present in the *Euphorbia esula* leaf hot water extract. The formation of CuNPs with flavonoid and phenolics acids would take place via the following steps: (1) complexation with copper metal salts, (2) simultaneous reduction of copper ions, and (3) capping with oxidized polyphenols. UV/Vis spectrum of CuNPs using *E. esula* leaf extract showed an absorbance maximum peak at around 580 nm which is characteristic for CuNPs. The size of the phytosynthesized CuNPs was almost in the range of 20–110 nm in diameter as revealed by TEM. The prepared CuNPs showed a green route for the catalytic degradation of 4-NP. The *Ginkgo biloba* belongs to the family of Ginkgoaceae. The leaves of this species are extensively used as a source of herbal medicine due to its medicinal phytochemicals. Green synthesis of CuNPs using *G. biloba* hot water leaf extract as a reducing and stabilizing agent has been successfully established by (Nasrollahzadeh and Sajadi, 2015). The synthesized CuNPs showed a maximum absorption peak at 560 nm. The synthesized CuNPs by this method was quite stable and no obvious variance in the shape, position, and symmetry of the absorption peak is observed even after one month indicating the stability of the phytosynthesized CuNPs. The presence of flavonoid and other phenolics within the extract as has been confirmed by FTIR could be responsible for the reduction of copper chloride dihydrate and formation of the corresponding CuNPs. The TEM observations showed that the diameter of the CuNPs was in the range of 15–20 nm. Nagaonkar et al. (2015) reported the *Citrus medica* (L.) fruit extract mediated phytosynthesis of CuNPs. The CuNPs were also reported to be green synthesized via *Plantago asiatica* leaf hot water extract as natural source and reaction biomedia (Nasrollahzadeh et al., 2017). It was noticed that the usage of *P. asiatica* leaf extract made a simple, eco-friendly, and cost-effective method for the preparation of CuNPs which had the ability to reduce copper ions into metallic copper (Cu^0) within 5 min of reaction time without using any external reducing or stabilizing agents. The progress of the reaction was monitored using UV/Vis spectroscopy. A characteristic absorption peak of CuNPs appeared at 555 nm after 5 min of reaction. The TEM analysis demonstrated that the CuNPs were uniform with spherical morphology and a particle size in the range of 7–35 nm. Polyphenolic compounds and flavonoids could be adsorbed on the surface of CuNPs, possibly by interaction through π-electrons interaction which might aid in the reduction, stabilization, and capping of the prepared NPs. The catalytic activity of the CuNPs was evaluated by cyanation of aldehydes in the extract. This method provided several advantages such as an easy, green synthetic process, shorter reaction time, and higher yield. CuNPs were also prepared using *S. aromaticum* (clove) bud hot water extract (Rajesh et al., 2017). The addition of *S. aromaticum* bud extract to copper acetate solution resulted in a color change of the solution from blue to pale bluish-green due to the formation of CuNPs. The UV/Vis absorption spectrum showed the characteristic absorption peak of CuNPs at ~580 nm. The zeta potential of the CuNPs verified the stability of the NPs as it recorded −21.3 mV. The high crystalline nature of CuNPs with a fcc phase was evidenced from the XRD pattern. The EDX demonstrated high intense metallic peak of copper (Cu) and low intense peaks of carbon (C), oxygen (O), chlorine (Cl), and phosphorus (P) elements due to the capping action of biomolecules of bud extract during CuNPs formation. The FESEM images revealed the presence of relatively spherical-shaped CuNPs with an average size of 20 nm. The TEM micrographs showed that the particles were monodispersed, almost spherical in nature, and well segregated without any agglomeration with a diameter of ≈15 nm. The selected area diffraction (SAED) pattern depicted the circular fringes corresponding to (111), (200), and (220) planes of the fcc structure of Cu which was in accordance with XRD results. The antimicrobial activity was investigated against selected pathogens using the phytosynthesized CuNPs. Zone of inhibitions of 8 mm and 6 mm was attained against *Bacillus sp.* and *Penicillium sp.*, respectively. Roy et al. (2017) also reported the green synthesis of CuNPs using *I. balsamina* leaf water extract. *I. balsamina* or garden balsam is an annual plant found in southern Asia. Its leaf extract is used as a traditional medicine for treating snakebites, unwanted moles, and many other skin ailments. The hot water extract of *P. granatum* (Pomegranate) seeds has been reported for phytosynthesis of semispherical-shaped 40–80 nm CuNPs with SPR of 553 nm (Nazar et al., 2018). The green synthesized CuNPs expressed efficient photocatalytic degradation of methylene blue under solar light (Nazar et al., 2018). Nagar and Devra (2018) proved that the chemical constituents of the hot water extract of *A. indica* leaves terpenoids, nimbaflavone, and polyphenols reduced Cu^{2+} ions into CuNPs and acted also as stabilizing and capping agents. The size of the phytosynthesized CuNPs was found to be dependent on the concentrations of the precursors. Moreover, the high temperature enhanced the nucleation over the agglomeration. But pH affected the capping and stabilizing efficiency. The optimum conditions were found to be 20% leaf extract, 7.5×10^{-3} M $CuCl_2$, pH6.6, and 85 °C. The highly crystalline and cubical CuNPs with SPR of 560 nm and an average size of 48 nm were stable for 2 months at 4 °C, with a zeta potential value of −17.5 mV (Nagar and Devra, 2018). In another study by Khani et al. (2018), the fruit extract of *Ziziphus spina-christi* (L.) Willd was used for the preparation of spherical shaped and well dispersed 5–20 nm CuNPs. That expressed efficient adsorption of crystal violet dye and biocidal effect on the Gram-positive *S. aureus* and the Gram-negative *E. coli* (Khani et al., 2018). Rajesh et al. (2018) used the bud hot water extract of *S. aromaticum* (clove) which contained proteins, tannins, flavonoids, alkaloids, and carotenoids as reducing, stabilizing, and capping agents in the phytosynthesis of monodispersed spherical-shaped 15 nm CuNPs. Zeta potential value of −21.3 mV proved the high stability of such phytosynthesized CuNPs. Moreover, the phytosynthesized CuNPs proved an efficient bactericidal effect on pathogenic Gram-positive and Gram-negative bacteria *Staphylococcus* spp., *E. coli*, *Pseudomonas* spp., and *Bacillus* spp. and fungicidal effect on *Penicillium* spp. Hot water extract of *Rhus coriaria* L. (sumaq) fruits was used as a stabilizing agent in the synthesis of CuNPs using hydrazine hydrate as a reducing agent and sodium hydroxide as a catalyst (Ismail, 2020). The produced spherical-shaped 7–10 nm expressed efficient biocidal activity against the Gram-positive bacteria *B. subtilis* and *S. aureus* and the Gram-negative bacteria *E. coli.*

REFERENCES

Abbasi, T., Anuradha, J., Ganaie, S.U., and Abbasi, S.A. 2015. Gainful utilization of the highly intransigent weed ipomoea in the synthesis of gold nanoparticles. J. King Saud Univ. Sci. 27: 15–22.

Agarwal, K., and Srivastava, M. M. 2014. Chemistry synthesis and characterization of gold nanoparticles embedded with extract of the plant *Panicum maximum* with enhanced antioxidant behavior. Int. J. Sci. Res. 63: 2–4.

Ahamed, M., Alhadlaq, H.A., Khan, M.A.M., Karuppiah, P., and Al-Dhabi, N.A. 2014. Synthesis, characterization, and antimicrobial activity of copper oxide nanoparticles. J. Nanomater. 2014, Article ID 637858, 1–4.

Ahlawat, J., and Sehrawat, A.R. 2015. Biological synthesis of silver nanoparticles using aqueous leaf extract of *Capparis decidua* (FORSK.) EDGEW: a better alternative. J. Pharm. Res. 9(4): 244–249.

Ahmad, B., Hafeez, N., Bashir, S., Rauf, A., and Mujeeb-ur-Rehman. 2017. Phytofabricated gold nanoparticles and their biomedical applications. Biomed. Pharmacother. 89: 414–425.

Ahmad, T., Irfana, M., and Bhattacharjee, S. 2016. Parametric study on gold nanoparticle synthesis using aqueous *Elaise Guineensis* (Oil palm) leaf extract: effect of precursor concentration. Proc. Eng. 148: 1396–1401.

Ahmad, N., Sharma, S., and Rai, R. 2012. Rapid green synthesis of silver and gold nanoparticles using peels of *Punica granatum*. Adv. Mater. Lett. 3(5): 376–380.

Ahmed, S., Ahmad, M., Swami, B.L., and Ikram, S. 2016. A review on plants extract mediated synthesis of silver nanoparticles for antimicrobial applications: a green expertise. J. Adv. Res. 7: 17–28.

Ahmed, S., and Ikram, S. 2015. Synthesis of gold nanoparticles using plant extract: an overview. Nano Res. Appl. 1: 1. http://nanotechnology.imedpub.com/archive.php.

Ahmed, K.B., Senthilnathan, R., Megarajan, S., and Anbazhagan, V. 2015. Sunlight mediated synthesis of silver nanoparticles using redox phytoprotein and their application in catalysis and colorimetric mercury sensing. J. Photochem. Photobiol. B, Biol. 151: 39–45.

Ahmed, K.B.A., Subramanian, S., Sivasubramanian, A., Veerappan, G., and Veerappan, A. 2014. Preparation of gold nanoparticles using *Salicornia brachiate* plant extract and evaluation of catalytic and antibacterial activity. Spectrochim. Acta A. 130: 54–58.

Ahn, S., Singh, P., Jang, M., Kim, Y.J., Castro-Aceituno, V., Simu, S.Y., Kim, Y.J., and Yang, D.C. 2018. Gold nanoflowers synthesized using Acanthopanacis cortex extract inhibit inflammatory mediators in LPS-induced RAW264.7 macrophages via NF-B and AP-1 pathways. Colloids Surf. B Biointerfaces. 162: 398–404.

Ajitha, B., Reddy, Y.A.K., Jeon, H.-J., and Ahn, C.W. 2018. Synthesis of silver nanoparticles in an eco-friendly way using *Phyllanthus amarus* leaf extract: antimicrobial and catalytic activity. Adv. Powder Technol. 29: 86–93.

Ajitha, B., Reddy, Y.A.K., Rajesh, K.M., and Reddy, P.S. 2016. *Sesbania grandiflora* leaf extract assisted green synthesis of silver nanoparticles: antimicrobial activity. Mater. Today Proc. 3: 1977–1984.

Allafchian, A.R., Mirahmadi-Zare, S.Z., Jalali, S.A.H., Hashemi, S.S., and Vahabi, M.R. 2016. Green synthesis of silver nanoparticles using *Phlomis* leaf extract and investigation of their antibacterial activity. J. Nanostruct. Chem. 6: 129–135.

Anarkali, J., Vijaya Raj, D., Rajathi, K., and Sridhar, S. 2012. Biological synthesis of silver nanoparticles by using *Mollugo nudicaulis* extract and their antibacterial activity. Arch. Appl. Sci. Res. 4(3): 1436–1441.

Aromal, S.A., and Philip, D. 2012. Green synthesis of gold nanoparticles using *Trigonella foenumgraecum* and its size dependent catalytic activity. Spectrochim. Acta A. 97: 1–5.

Arunachalam, K.D., Annamalai, S.K., and Hari, S. 2013. One–step green synthesis and characterization of leaf extract–mediated biocompatible silver and gold nanoparticles from *Memecylon umbellatum*. Int. J. Nanomed. 8: 1307–1315.

Awwad, A.M., Salem, N.M., and Abdeen. A.O. 2013. Green synthesis of silver nanoparticles using carob leaf extract and its antibacterial activity. Int. J. Ind. Chem. 4: 29. http://www.industchem.com/content/4/1/29.

Azizian-Shermeh, O., Einali, A., and Ghasemi, A. 2017. Rapid biologically one-step synthesis of stable bioactive silver nanoparticles using Osage orange (*Maclura pomifera*) leaf extract and their antimicrobial activities. Adv. Powder Technol. 28: 3164–3171.

Balachandran, Y.L., Peranantham, P., Selvakumar, R., Gutleb, A.C., and Girija, S. 2012. Size-controlled green synthesis of silver nanoparticles using dual functional plant leaf extract at room temperature. Int. J. Green Nanotechnol. 4(3): 310–325.

Balalakshmi, C., Gopinath, K., Govindarajan, M., Lokeshd, R., Arumugam, A., Alharbi, N.S., Kadaikunnane, S., Khalede, J.M., and Benelli, G. 2017. Green synthesis of gold nanoparticles using a cheap *Sphaeranthus indicus* extract: impact on plant cells and the aquatic crustacean *Artemia nauplii*. Photochem. Photobiol. B: Biol. 173: 598–605.

Balamanikandan, T., Balaji, S., and Pandiarajan, J. 2015. Biological synthesis of silver nanoparticles by using onion (*Allium cepa*) extract and their antibacterial and antifungal activity. World Appl. Sci. J. 33(6): 939–943.

Banerjee, J., and Narendhirakannan, R. 2011. Biosynthesis of silver nanoparticles from *Syzygium cumini* (L.) seed extract and evaluation of their in vitro antioxidant activities. Dig. J. Nanomater. Biostruct. 6: 961–968.

Bankar, A., Joshi, B., Kumar, A.R., and Zinjarde, S. 2010. Banana peel extract mediated novel route for the synthesis of silver nanoparticles. Colloids Surf. A. Physicochem. Eng. Asp. 368: 58–63.

Basavegowda, N., and Lee, Y.B. 2013. Synthesis of silver nanoparticles using Satsuma mandarin (*Citrus unshiu*) peel extract: a novel approach towards waste utilization. Mater. Lett. 109: 31–33.

Baskaralingam, V., Sargunar, C.G., Lin, Y.C., and Chen, J.C. 2012. Green synthesis of silver nanoparticles through *Calotropis gigantea* leaf extracts and evaluation of antibacterial activity against *Vibrio alginolyticus*. Nanotechnol. Dev. 2: e3. http://dx.doi.org/10.4081/nd.2012.e3.

Bello, B.A., Khan, S.A., Khan, J.A., Syed, F.Q., Anward, Y., Khan, S.B. 2017a. Antiproliferation and antibacterial effect of biosynthesized AgNPs from leaves extract of *Guiera senegalensis* and its catalytic reduction on some persistent organic pollutants. J. Photochem. Photobiol. B. 175: 99–108.

Bello, B.A., Khan, S.A., Khan, J.A., Syed, F.Q., Mirza, M.B., Shah, L., and Khan, S.B. 2017b. Anticancer, antibacterial and pollutant degradation potential of silver nanoparticles from *Hyphaene thebaica*. Biochem. Biophys. Res. Commun. 490: 889–894.

Benjamin, G., and Bharathwaj, S. 2011. Biological synthesis of silver nanoparticles from *Allium cepa* (onion) & estimating its antibacterial activity. Int. Conf. Biosci., Biochem. Bioinf. IPCBEE. IACSIT Press, Singapore, 5: 35–38.

Bhakya, S., Muthukrishnan, S., Sukumaran, M., Muthukumar, M., Senthil Kumar, T., and Rao, M.V. 2015. Catalytic degradation of organic dyes using synthesized silver nanoparticles: a green approach. J. Bioremed. Biodeg. 6: 5. http://dx.doi.org/10.4172/2155–6199.1000312.

Bhuyan, B., Paula, A., Paula, B., Dhara, S.S., Dutta, P. 2017. *Paederia foetida* Linn. promoted biogenic gold and silver nanoparticles: synthesis, characterization, photocatalytic and in vitro efficacy against clinically isolated pathogens. J. Photochem. Photobiol. B: Biol. 173: 210–215.

Bindhu, M.R., and Umadevi, M. 2013. Synthesis of monodispersed silver nanoparticles using *Hibiscus cannabinus* leaf extract and its antimicrobial activity. Spectrochim. Acta A. Mol. Biomol. Spectrosc. 101: 184–190.

Bonnia, N.N., Kamaruddin, M.S., Nawawi, M.H., Ratimd, S., Azlinae, H.N., and Ali, E.S. 2016. Green biosynthesis of silver nanoparticles using '*Polygonum hydropiper*' and study its catalytic degradation of Methylene Blue. Procedia Chem. 19: 594–602.

Castro, L., Blázquez, M.L., Muñoz, J.A., González, F., and Ballester, A. 2013. Biological synthesis of metallic nanoparticles using algae. IET Nanobiotechnol. doi: 10.1049/iet-nbt.2012.0041

Castro, L., Blázquez, M.L., Muñoz, J.A., González, F., García-Balboa, C., and Ballester, A. 2011. Biosynthesis of gold nanowires using sugar beet pulp. Process Biochem. 46: 1076–1082.

Chandra, S., Kumar, A., and Tomar, P.K. 2014. Synthesis and characterization of copper nanoparticles by reducing agent. *J. Saudi Chem. Soc.*, 18: 149–15.

Chen, Y., Wang, D., Zhu, X., Zheng, X., and Feng, L. 2012. Long-term effects of copper nanoparticles on wastewater biological nutrient removal and N_2O generation in the activated sludge process. *Environ. Sci. Technol.*, 46: 12452–12458.

Corti, C.W., and Holliday, R.J. 2004. Commercial aspects for gold applications. Gold Bull. 37: 20–27.

Cortie, M.B. 2004. The weird world of nanoscale gold. Gold Bull. 37(1–2): 12–19.

Croteau, R., Kutchan, T.M., and Lewis, N.G. 2000. Natural products (secondary metabolites). Biochem. Mol. Biol. Plant. 24: 1250–1319.

Cruz, D., Fale, P.L., Mourato, A., Vaz, P.D., Serralheiro, M.L., and Lino, A.R.L. 2010. Preparation and physicochemical characterization of Ag nanoparticles biosynthesized by *Lippia citriodora* (Lemon Verbena). Colloids Surf. B. Biointerfaces. 81(1): 67–73.

Daisy, P., and Saipriya, K. 2012. Biochemical analysis of *Cassia fistula* aqueous extract and phytochemically synthesized gold nanoparticles as hypoglycemic treatment for diabetes mellitus. Int. J. Nanomed. 7: 1189–1202.

Das, R.K., Gogoi, N., and Bora, U. 2011. Green synthesis of gold nanoparticles using *Nyctanthes arbortristis* flower extract. Bioprocess Biosys. Eng. 34(5): 615–619.

Davis, T.A., Volesky, B., and Mucci, A. 2003. A review of the biochemistry of heavy metal biosorption by brown algae. Water Res. 37(18): 4311–4330.

Demirbas, A., Yilmaz, V., Ildiz, N., Baldemir, A., and Ocsoy, I. 2017. Anthocyanins-rich berry extracts directed formation of AgNPs with the investigation of their antioxidant and antimicrobial activities. J. Mol. Liq. 248: 1044–1049.

Dibrov, P., Dzioba, J., Gosink, K., and Hase, C. 2002. Chemiosmotic mechanism of antimicrobial activity of Ag^+ in *Vibrio cholerae*. Antimicrob. Agents Chemother. 46: 2668–2670.

Divakar, T.E., and Rao, Y.H. 2017. Biological synthesis of silver nano particles by using *Bombax ceiba* plant. J. Chem. Pharm. Sci. 10(1): 574–576.

Dogru, E., Demirbas, A., Altinsoy, B., Duman, F., and Ocsoy, I. 2017. Formation of *Matricaria chamomilla* extract-incorporated Ag nanoparticles and size-dependent enhanced antimicrobial property. J. Photochem. Photobiol. B: Biol. 174: 78–83.

Dubey, M., Bhadauria, S., and Kushwah, B.S. 2009. Green synthesis of nanosilver particles from extract of *Eucalyptus hybrid* (safeda) leaf. Dig. J. Nanomater. Biostruct. 4(3): 537–543.

Dubey, S.P., Lahtinen, M., and Sillanpaa, M. 2010a. Green synthesis and characterizations of silver and gold nanoparticles using leaf extract of *Rosa rugosa*. Colloids Surf. A. Physicochem. Eng. Asp. 364: 34–41.

Dubey, S.P., Lahtinen, M., and Sillanpää, M. 2010b. Tansy fruit mediated greener synthesis of silver and gold nanoparticles. Process Biochem. 45: 1065–1071.

Dubey, S.P., Lahtinen, M., Särkkä, H., and Sillanpää, M. 2010c. Bioprospective of *Sorbus aucuparia* leaf extract in development of silver and gold nanocolloids. Colloid Surf. B. 80: 26–33.

Dutta, P.P., Bordoloi, M., Gogoi, K., Roy, S., Narzary, B., Bhattacharyya, D.R., Mohapatra, P.K., and Mazumder, B. 2017. Antimalarial silver and gold nanoparticles: green synthesis, characterization and in vitro study. *Biomed. Pharmacother.* 91: 567–580.

Dwivedi, A.D., and Gopal, K. 2010. Biosynthesis of silver and gold nanoparticles using *Chenopodium album* leaf extract. Physicochem. Eng. Asp. 369: 27–33.

Dwivedi, A.D., and Gopal, K. 2011. Plant-mediated biosynthesis of silver and gold nanoparticles. J. Biomed. Nanotechnol. 7: 163–164.

El-Gendy, N.Sh., and Omran, B.A. 2019. Green synthesis of nanoparticles for water treatment. In: Nano and Bio-Based Technologies for Wastewater Treatment: Prediction and Control Tools for the Dispersion of Pollutants in the Environment. John Wiley & Sons, Inc., Hoboken, NJ and Scrivener Publishing LLC, Beverly, MA. https://doi.org/10.1002/9781119577119.ch7

El-Kassas, H.Y., and El-Sheekh, M. 2014. Cytotoxic activity of biosynthesized gold nanoparticles with an extract of the red seaweed *Corallina officinalis* on the MCF-7 human breast cancer cell line. Asian Pac. J. Cancer Prev. 15: 4311–4317.

El-Rafie, H.M., El-Rafie, M.H., and Zahran, M.K. 2013. Green synthesis of silver nanoparticles using polysaccharides extracted from marine macro algae. Carbohydr. Polym. 96(2): 403–410.

Elumalai, E., Prasad, T., Hemachandran, J., Therasa, S.V., Thirumalai, T., and David, E. 2010. Extracellular synthesis of silver nanoparticles using leaves of *Euphorbia hirta* and their antibacterial activities. J. Pharm. Sci. Res. 2: 549–554.

Fafal, T., Taştan, P., Tüzün, B.S., Ozyazici, M., and Kivcak, B. 2017. Synthesis, characterization and studies on antioxidant activity of silver nanoparticles using *Asphodelus aestivus* Brot. aerial part extract. S. Afr. J. Bot. 112: 346–353.

Feng, Q.L., Wu, J., Chen, G.Q., Cui, F.Z., Kim, T.N., and Kim, J.O. 2000. A mechanistic study of the antibacterial effect of silver ions on *Escherichia coli* and *Staphylococcus aureus*. J. Biomed. Mater. Res. 52: 662–668.

Firdhouse, M.J., and Lalitha, P. 2016a. Biogenic silver nanoparticles—synthesis, characterization and its potential against cancer inducing bacteria. J. Mol. Liquids 222: 1041–1050.

Firdhouse, M. J., and Lalitha, P. 2016b. Flower-shaped gold nanoparticles synthesized using *Kedrostis foetidissima* and their anti-proliferative activity against bone cancer cell lines. Int. J. Ind. Chem. 7: 347–358.

Gardea-Torresdey, J.L., Parsons, J.G., Gomez, E., Peralta-Videa, J., Toiani, H.E., Santiago, P., and Jose Yacaman, M. 2002. Formation and growth of Au nanoparticles inside live alfalfa plants. Nano Lett. 2: 397–401.

Gonnelli, C., Cacioppo, F., Giordano, C., Capozzoli, L., Salvatici, C., Salvatici, M.C., Colzi, I., Del Bubba, M., Ancillotti, C., and Ristori, S. 2015. *Cucurbita pepo* L. extracts as a versatile hydrotropic source for the synthesis of gold nanoparticles with different shapes. Green Chem. Lett. Rev. 8: 39–47.

Gorbe, M., Bhat, R., Aznar, E., Sancenón, F., Marcos, M.D., Herraiz, F.J., Prohens, J., Venkataraman, A., and Martínez-Máñez, R. 2016. Rapid biosynthesis of silver nanoparticles using pepino (*Solanum muricatum*) leaf extract and their cytotoxicity on HeLa cells. Materials. 9: 325. doi: 10.3390/ma9050325.

Gowri, K.G.S., and Arumugam, A. 2013. Phytosynthesis of silver nanoparticles using *Pterocarpus santalinus* leaf extract and their antibacterial properties. J. Nanostruct. Chem. 3: 68–75.

Gupta, K., Hazarika, S.N., Saikia, D., Namsa, N.D., and Mandal, M. 2014. One step green synthesis and anti-microbial and anti-biofilm properties of *Psidium guajava* L. leaf extract-mediated silver nanoparticles. Mater. Lett. 125: 67–70.

Gurunathan, S., Park, J.H., Han, J.W., and Kim, J.H. 2015. Comparative assessment of the apoptotic potential of silver nanoparticles synthesized by *Bacillus tequilensis* and *Calocybe indica* in MDA-MB-231 human breast cancer cells: Targeting p53 for anticancer therapy. Int. J. Nanomed. 10: 4203–4222.

Haq Khan, Z.U., Khan, A., Chen, Y., Khan, A., Shah, N.S., Muhammad, N., Murtaza, B., Tahir, K., Khan, F., and Wan, P. 2017. Photocatalytic applications of gold nanoparticles synthesized by green route and electrochemical degradation of phenolic Azo dyes using AuNPs/GC as modified paste electrode. J. Alloys Compd. 725: 869–876.

Huang, J.L., Li, Q.B., Sun, D.H., Lu, Y.H., Su, Y.B., Yang, X., Wang, H., Wang, Y., Shao, W., He, N., Hong, J., and Chen, C. 2007. Biosynthesis of silver and gold nanoparticles by novel sundried *Cinnamomum camphora* leaf. Nanotechnology. 18(10): 105104. doi: 10.1088/0957–4484/18/10/105104.

Ibrahim, Z.H.H. 2015. Green synthesis and characterization of silvernanoparticles using banana peel extract and their antimicrobial activity against representative microorganisms. J. Rad. Res. Appl. Sci. 8: 265–275.

Iravani, S. 2011. Green synthesis of metal nanoparticles using plants. Green Chem. 13: 2638–2650.

Ismail, M.I.M. 2020. Green synthesis and characterizations of copper nanoparticles. Mater. Chem. Phys. 240: 122283.

Ismail, E.H., Khalil, M.M.H., Al Seif, F.A., El-magdoub, F., Bent, A.N., and Rahman, A. 2014. Biosynthesis of gold nanoparticles using extract of grape (*Vitis vinifera*) leaves and seeds. Progress Nanotechnol. Nanomater. 3: 1–12.

Jayaseelan, C., Ramkumar, R., Abdul Rahuman, A., and Perumal, P. 2013. Green synthesis of gold nanoparticles using seed aqueous extract of *Abelmoschus esculentus* and its antifungal activity. Ind. Crops Prod. 45: 423–429.

Jha, A.K., and Prasad, K. 2010. Green synthesis of silver nanoparticles using *Cycas* leaf. Int. J. Green Nanotechnol. Phys. Chem. 1(2): P110–P117.

Jha, A.K., and Prasad, K. 2011. Biosynthesis of gold nanoparticles using bael (*Aegle marmelos*) leaf: mythology met technology. J. Green Nanotechnol. 3(2): 92–97.

Jha, A.K. and Prasad, K. 2012. Biosynthesis of gold nanoparticles using common aromatic plants. Int. J. Green Nanotechnol. 4(3): 219–224.

Jyoti, K., and Singh, A. 2016. Green synthesis of nanostructured silver particles and their catalytic application in dye degradation. J. Genet. Eng. Biotechnol. 14: 311–317.

Kahrilas, G.A., Wally, L.M., Fredrick, S.J., Hiskey, M., Prieto, A.L., and Owens, J.E. 2014. Microwave-assisted green synthesis of silver nanoparticles using orange peel extract. ACS Sustain. Chem. Eng. 2(3): 367–376.

Kanhed, P., Birla, S., Gaikwad, S., Gade, A., Seabra, A.B., Rubilar, O. Duran, N., and Rai, M. 2014. In vitro antifungal efficacy of copper nanoparticles against selected crop pathogenic fungi. Mater. Lett. 115: 13–17.

Kannan, R.R., Arumugam, R., Ramya, D., Manivannan, K., and Anantharaman, P. 2013. Green synthesis of silver nanoparticles using marine macroalgae *Chaetomorpha linum*. Appl. Nanosci. 3: 229–233.

Karthik, R., Govindasamy, M., Chen, S.-M., Cheng, Y.-H., Muthukrishnan, P., Padmavathy, S., and Elangovan, A. 2017. Biosynthesis of silver nanoparticles by using *Camellia japonica* leaf extract for the electrocatalytic reduction of nitrobenzene and photocatalytic degradation of Eosin-Y. J. Photochem. Photobiol., B: Biol. 170: 164 –172.

Kasthuri, J., Veerapandian, S., and Rajendiran, N. 2009. Biological synthesis of silver and gold nanoparticles using apiin as reducing agent. Colloids Surf. B Biointerfaces. 68: 55–60.

Kathad, U., and Gajera, H.P. 2014. Synthesis of copper nanoparticles by two different methods and size comparision. Int. J. Pharm. Biosci. 5(3): 533–540.

Kaviya, S., Santhanalakshmi, J., and Viswanathan, B. 2011a. Biosynthesis of silver nanoflakes by *Crossandra infundibuliformis* leaf extract. Mater. Lett. 67: 64–66.

Kaviya, S., Santhanalakshmi, J., Viswanathan, B., Muthumary, J., and Srinivasan, K. 2011b. Biosynthesis of silver nanoparticles using *Citrus sinensis* peel extract and its antibacterial activity. Spectrochim. Acta. A Mol. Biomol. Spectrosc. 79: 594–598.

Kesarla, M.K., Mandal, B.K., and Bandapalli, P.R. 2014. Gold nanoparticles by *Terminalia bellirica* aqueous extract—a rapid green method. J. Exp. Nanosci. 9: 825–830.

Kesharwani, J., Yoon, K.Y., Hwang, J., and Rai, M. 2009. Phytofabrication of silver nanoparticles by leaf extract of *Datura metel*: hypothetical mechanism involved in synthesis. J. Bionanosci. 3: 39–44.

Khalil, M.M.H., Ismail, E.H. and El-Magdoub, F. 2012. Biosynthesis of Au nanoparticles using olive leaf extract. Arab. J. Chem. 5: 431–437.

Khan, F.U., Chen, Y., Khan, N.U., Khan, Z.U., Khan, A.U., Ahmad, A., Tahir, K., Wang, L., Khan, M.R., and Wan, P. 2016. Antioxidant and catalytic applications of silver nanoparticles using *Dimocarpus longan* seed extract as a reducing and stabilizing agent. J. Photochem. Photobiol. B: Biol. 164: 344–351.

Khani, R., Roostaei, B., Bagherzade, G., and Moud, M. 2018. Green synthesis of copper nanoparticles by fruit extract of *Ziziphus spina-christi* (L.) Willd.: application for adsorption of triphenylmethane dye and antibacterial assay. J. Mol. Liq. 225: 541–549.

Khatami, M., Pourseyedi, S., Khatami, M., Hamidi, H., Zaeifi, M., and Soltani, L. 2015. Synthesis of silver nanoparticles using seed exudates of *Sinapis arvensis* as a novel bioresource, and evaluation of their antifungal activity. Bioresour. Bioprocess. 2: 19–26.

Klaus-Joerger, T., Joerger, R., Olsson, E., and Granqvist, C.G. 2001. Bacteria as workers in the living factory: metal-accumulating bacteria and their potential for materials science. Trends Biotechnol. 19: 15–20.

Klekotko, M., Matczyszyn, K., Siednienko, J., Olesiak-Banska, J., Pawlik, K., and Samoc, M. 2015. Bio-mediated synthesis, characterization and cytotoxicity of gold nanoparticles. Phys. Chem. Chem. Phys. 17: 29014–29019

Kora, A.J., and Sashidhar, R.B. 2015. Antibacterial activity of biogenic silver nanoparticles synthesized with gum ghatti and gum olibanum: a comparative study. J. Antibiot. 68: 88–97.

Kora, A.J., and Sashidhar, R.B. 2018. Biogenic silver nanoparticles synthesized with rhamnogalacturonan gum: antibacterial activity, cytotoxicity and its mode of action. Arab. J. Chem. 11: 313–323.

Kora, A.J., Sashidhar, R.B., and Arunachalam, J. 2010. Gum kondagogu (*Cochlospermum gossypium*): a template for the green synthesis and stabilization of silver nanoparticles with antibacterial application. Carbohydr. Polym. 82(3): 670–679.

Kora, A.J., Sashidhar, R.B., and Arunachalam, J. 2012. Aqueous extract of gum olibanum (*Boswellia serrata*): a reductant and stabilizer for the biosynthesis of antibacterial silver nanoparticles. Process Biochem. 47(10): 1516– 1520.

Kotakadi, V.S., Rao, Y.S., Gaddam, S.A., Prasad, T.N.V.K.V., Reddy, A.V., and Gopal, D.V.R.S. 2013. Simple and rapid biosynthesis of stable silver nanoparticles using dried leaves of *Catharanthus roseus*. Linn. G. Donn and its anti microbial activity. Colloids Surf. B. Biointerfaces. 105: 194–198.

Kouvaris, P., Delimitis, A., Zaspalis, V., Papadopoulos, D., Tsipas, S., and Michailidis, N. 2012. Green synthesis and characterization of silver nanoparticles produced using *Arbiutusunedo* leaf extract. Mater. Lett. 76: 18–20.

Koyyati, R., Nagati, V., Merugu, R., and Manthurpadigya, P. 2013. Biological synthesis of silver nanoparticles using *Raphanus sativus var. longipinnatus* leaf extract and evaluation of their antioxidant and antibacterial activity. Int. J. Med. Pharm. Sci. 3(4): 89–100.

Krishnamurthy, S., Sathishkumar, M., Lee, S.Y., Bae, M.A., and Yun, Y.S. 2011. Biosynthesis of Au nanoparticles using cumin seed powder extract. J. Nanosci. Nanotechnol. 11: 811–1814.

Krishnaraj, C., Jagan, E.G., Rajasekar, S., Selvakumar, P., Kalaichelvan, P.T., and Mohan, N. 2010. Synthesis of silver nanoparticles using *Acalypha indica* leaf extracts and its antibacterial activity against water borne pathogens. Colloids Surf. B. Biointerfaces. 76(1): 50–56.

Kumar, S.S.D., Houreld, N.N., Kroukamp, E.M., and Abrahamse, H. 2018. Cellular imaging and bactericidal mechanism of green-synthesized silver nanoparticles against human pathogenic bacteria. J. Photochem. Photobiol., B: Biol. 178: 259–269.

Kumar, D.A., Palanichamy, V., and Roopan, S.M. 2014. Green synthesis of silver nanoparticles using *Alternanthera dentata* leaf extract at room temperature and their antimicrobial activity. Spectrochim. Acta Part A: Mol. Biomol. Spectrosc. 127: 168–171.

Kumar, K.P., Paul, W., and Sharma, C.P. 2011b. Green synthesis of gold nanoparticles with *Zingiber officinale* extract: characterization and blood compatibility. Process Biochem. 46: 2007–2013.

Kumar, B., Smita, K., Cumbal, L., and Debut, A. 2017. Green synthesis of silver nanoparticles using Andean blackberry fruit extract. Saudi J. Biol. Sci. 24: 45–50.

Kumar, V., and Yadav, S.K. 2011. Synthesis of stable, polyshaped silver, and gold nanoparticles using leaf extract of *Lonicera japonica* L. Int. J. Green Nanotechnol. 3(4): 281–291.

Kumar, V., Gokavarapu, S., Rajeswari, A., Dhas, T., Karthick, V., and Kapadia, Z. 2011a. Facile green synthesis of gold nanoparticles using leaf extract of antidiabetic potent *Cassia auriculata*. Colloids Surf B Biointerfaces. 87: 159–163.

Kumar, V., Yadav, S.C., and Yadav, S.K. 2010. *Syzygium cumini* leaf and seed extract mediated biosynthesis of silver nanoparticles and their characterization. J. Chem. Technol. Biotechnol. 85(10): 1301–1309.

Kuppusamy, P., Yusoff, M.M., Maniam, G.P., and Govindan, N. 2016. Biosynthesis of metallic nanoparticles using plant derivatives and their new avenues in pharmacological applications—an updated report. Saudi Pharm. J. 24: 473–484.

Lee, H.-J., Lee, G., Jang, N.R., Yan, J.H., Song, J.Y., and Kim, B.S. 2011. Biological synthesis of copper nanoparticles using plant extract. NSTI-Nanotech. 1: 371–374.

Lee, H., Song, J.Y., Kim, B.S. 2013. Biological synthesis of copper nanoparticles using *Magnolia kobus* leaf extract and their antibacterial activity. J. Chem. Tech. Biotechnol. 88(11): 1971–1977.

LewisOscar, F., MubarakAli, D., Nithya, C., Priyanka, R., Gopinath, V., Alharbi, N.S., and Thajuddin, N. 2015. One pot synthesis and anti-biofilm potential of copper nanoparticles (CuNPs) against clinical strains of *Pseudomonas aeruginosa*. Biofouling. 3(4): 379–391.

Li, J., Ma, Q., Shao, H., Zhou, X., Xia, H., and Xie, J. 2017. Biosynthesis, characterization, and antibacterial activity of silver nanoparticles produced from rice straw biomass. BioResources. 12(13): 4897–4911.

Li, S., Shen, Y., Xie, A., Yu, X., Qui, L., Zhang, L., and Zhang, Q. 2007. Green synthesis of silver nanoparticles using *Capsicum annum* L. extract. Green Chem. 9: 852–858.

Li, W., Xie, X.B., Shi, Q.S., Zeng, H.Y., Yng, Y.O.U., and Chen, Y.B. 2010. Antibacterial activity and mechanism of silver nanoparticles on *Escherichia coli*. Appl. Microbiol. Biotechnol. 85: 1115–1122.

Lodeiro, P., and Sillanpaa, M. 2013. Gold recovery from artificial seawater using synthetic materials and seaweed biomass to induce gold nanoparticles formation in batch and column experiments. Mar. Chem. 152: 11–19.

Logeswari, P., Silambarasan, S., and Abraham, J. 2012. Synthesis of silver nanoparticles using plant extracts and analysis of their antimicrobial activity. J. Saudi Chem. Soc. 4: 23–45.

Logeswari, P., Silambarasan, S., and Abraham, J. 2013. Ecofriendly synthesis of silver nanoparticles from commercially available plant powders and their antibacterial properties. Sci. Iran. F. 20(3): 1049–1054.

Logeswari, P., Silambarasan, S., and Abraham, J. 2015. Synthesis of silver nanoparticles using plants extract and analysis of their antimicrobial property. J. Saudi Chem. Soc. 19: 311–317.

Makarov, V., Love, A., Sinitsyna, O., Yaminsky, S.M.I., Taliansky, M., and Kalinina, N. 2014. Green nanotechnologies: synthesis of metal nanoparticles using plants. Acta Nature 6(1): 35–44.

Mallikarjun, K., Narsimha, G., Dillip, G., Praveen, B., Shreedhar, B., and Lakshmi S. 2011. Green synthesis of silver nanoparticles using *Ocimum leaf* extract and their characterization. Dig. J. Nanomater. Biostruct. 6: 181–186.

Marimuthu, V., Palanisamy, S.K., Sesurajan, S., and Sellappa, S. 2011. Biogenic silver nanoparticles by *Gelidiella acerosa* extract and their antifungal effects. Avicenna J. Med. Biotechnol. 3: 143–148.

Markus, J., Wang, D., Kim, Y.-J., Ahn, S., Mathiyalagan, R., Wang, C., and Yang, D.C. 2017. Biosynthesis, characterization, and bioactivities evaluation of silver and gold nanoparticles mediated by the roots of Chinese herbal *Angelica pubescens* Maxim. Nanoscale Res. Lett. 12:46. doi: 10.1186/s11671–017-1833–2.

Mata, Y.N., Torres, E., Blazquez, M.L., Ballester, A., Gonzalez, F., and Munoz, J.A. 2009. Gold (111) biosorption and bioreduction with the brown alga *Fucus vesiculosus*. J. Hazard. Mater. 166(2–3): 612–618.

Mishra, P., Ray, S., Sinha, S., Das, B., Khan, M.I., Behera, S.K., Yun, S.I., Tripathy, S.K., and Mishra, A. 2016. Facile bio–synthesis of gold nanoparticles by using extract of *Hibiscus sabdariffa* and evaluation of its cytotoxicity against U87 glioblastoma cells under hyperglycemic condition. Biochem. Eng. J. 105: 264–272.

Mishra, A., and Sardar, M. 2013. Rapid biosynthesis of silver nanoparticles using sugarcane bagasse—an industrial waste. J. Nanoeng. Nanomanufact. 3(3): 217–219.

Mittal, A.K., Bhaumik, J., Kumar, S., and Banerjee, U.C. 2014. Biosynthesis of silver nanoparticles: elucidation of prospective mechanism and therapeutic potential. J. Colloid Interface. Sci. 415: 39–47.

Mittal, A.K., Chisti, Y., and Banerjee, U.C. 2013. Synthesis of metallic nanoparticles using plant extracts. Biotechnol. Adv. 31: 346–356.

Mohamedkhair, A.K., Drmosh, Q.A., and Yamani, Z.H. 2019. Silver nanoparticle-decorated tin oxide thin films: synthesis, characterization, and hydrogen gas sensing. Front. Mater. 6: 188. doi: 10.3389/fmats.2019.00188.

Mollick, M.M.R., Bhowmick, B., Maity, D., Mondal, D., Bain, M.K., Bankura, K., Sarkar, J., Rana, D., Acharya, K., and Chattopadhyay, D. 2012. Green synthesis of silver nanoparticles using *Paederia foetida* L. leaf extract and assessment of their antimicrobial activities. Int. J. Green Nanotechnol. 4(3): 230–239.

Moteriya, P., and Chanda, S. 2018. Biosynthesis of silver nanoparticles formation from *Caesalpinia pulcherrima* stem metabolites and their broad spectrum biological activities. J. Genet. Eng. Biotechnol, 16(1): 105–113.

Moyo, M., Gomba, M., and Nharingo, T. 2015. *Afzelia quanzensis* bark extract for green synthesis of silver nanoparticles and study of their antibacterial activity. Int. J. Indian Chem. 6: 329–338.

Murugan, K., Benelli, G., Panneerselvam, C., Subramaniam, J., Jeyalalitha, T., Dinesh, D., Nicoletti, M., Hwang, J.S., Suresh, U., and Madhiyazhagan, P. 2015. *Cymbopogon citratus*-synthesized gold nanoparticles boost the predation efficiency of copepod *Mesocyclops aspericornis* against malaria and dengue mosquitoes. Exp. Parasitol. 153: 129–138.

Muthukumar, T., Sudhakumari, B., Sambandam, A., Aravinthan, T., Sastry, P., and Kim, J. H. 2016. Green synthesis of gold nanoparticles and their enhanced synergistic antitumor activity using HepG2 and MCF7 cells and its antibacterial effects. Process Biochem. 51: 384–391.

Nadeem, M., Abbasi, B.H., Younas, M., Ahmad, W., and Khan, T. 2017. A review of the green syntheses and anti-microbial applications of gold nanoparticles. Green Chem. Lett. Rev. 10(4): 216–227.

Nagajyothi, P., and Lee, K. 2011. Synthesis of plant-mediated silver nanoparticles using *Dioscorea batatas* rhizome extract and evaluation of their antimicrobial activities. J. Nanomater. 2011(49). http://dx.doi.org/10.1155/2011/573429.

Nagaonkar, D., Shende, S., and Rai, M. 2015. Biosynthesis of copper nanoparticles and its effect on actively dividing cells of mitosis in *Allium cepa*. Biotechnol. Prog. 31(2): 557–565.

Nagar, N., and Devra, V. 2018. Green synthesis and characterization of copper nanoparticles using *Azadirachta indica* leaves. Mater. Chem. Phys. 213: 44–51.

Nagy, A.J., and Mestl, G. 1999. High temperature partial oxidation reactions over silver catalysts. Appl. Catal. A-Gen. 188: 337–353.

Naika, H.R., Lingarajua, K., Manjunath, K., Kumar, D., Nagaraju, G., Suresh, D., and Nagabhushana, H. 2015. Green synthesis of CuO nanoparticles using *Gloriosa superba* L. extract and their antibacterial activity. J. Taibah Uni. Sci. 9: 7–12.

Nakkala, J.R., Mata, R., Gupta, A.K., and Sadras, S.R. 2014. Green synthesis and characterization of silver nanoparticles using *Boerhaavia diffusa* plant extract and their antibacterial activity. Ind. Crop Prod. 52: 562–266.

Naraginti, S., Tiwari, N., and Sivakumar, A. 2017. Green synthesis of silver and gold nanoparticles for enhanced catalytic and bactericidal activity. IOP Conf. Ser.: Mater. Sci. Eng. 263: 022009. doi:10.1088/1757–899X/263/2/022009.

Narayanan, K.B., and Sakthivel, N. 2008. Coriander leaf mediated biosynthesis of gold nanoparticles. Mater. Lett. 62: 4588–4590.

Narayanan, K.B., and Sakthivel, N. 2010. Phytosynthesis of gold nanoparticles using leaf extract of *Coleus amboinicus Lour*. Mater. Charact. 61: 1232–1238.

Nasrollahzadeh, M., Momeni, S.S., and Sajadi, M. 2017. Green synthesis of copper nanoparticles using *Plantago asiatica* leaf extract and their application for the cyanation of aldehydes using $K_4Fe(CN)_6$. J. Colloid Interface Sci. 506: 471–177.

Nasrollahzadeh, M., and Sajadi, S.M. 2015. Green synthesis of copper nanoparticles using *Ginkgo biloba* L. leaf extract and their catalytic activity for the Huisgen [3 + 2] cycloaddition of azides and alkynes at room temperature. J. Colloid Interface Sci. 457: 141–147.

Nasrollahzadeh, M., Sajadib, S.M., and Khalaj, M. 2014. Green synthesis of copper nanoparticles using aqueous extract of the leaves of *Euphorbia esula* L and their catalytic activity for ligand-free Ullmann coupling reaction and reduction of 4-nitrophenol. RSC Adv. 4: 47313–47319.

Naveena, B.E., and Prakash, S. 2013. Biological synthesis of gold nanoparticles using marine algae *gracilaria corticata* and its application as a potent antimicrobial and antioxidant agent. Asian J. Pharm. Clin. Res. 6: 179–182.

Nayak, D., Ashe, S., Rauta, P.R., Kumari, M., and Nayak, B. 2016. Bark extract mediated green synthesis of silver nanoparticles: evaluation of antimicrobial activity and antiproliferative response against osteosarcoma. Mater. Sci. Eng. C. 58: 44–52.

Nazar, N., Bibi, I., Kamal, S., Iqbal, M., Nouren, S., Jilani, K., Umair, M., and Ata, S. 2018. Cu nanoparticles synthesis using biological molecule of *P. granatum* seeds extract as reducing and capping agent: growth mechanism and photo-catalytic activity. Int. J. Biol. Macromol. 106: 1203–1210.

Ndikau, M., Noah, N.M., Andala, D.M., and Masika, E. 2017. Green synthesis and characterization of silver nanoparticles using *Cirullus lanatus* fruit rind extract. Int. J. Anal. Chem. Volume 2017, Article ID 8108504. https://doi.org/10.1155/2017/8108504

Nellore, J., Pauline, P.C., and Amarnath, K. 2012. Biogenic synthesis of *Sphearanthus amaranthoids* towards the efficient production of the biocompatible gold nanoparticles. Dig. J. Nanomater. Biostruct. 7: 123–133.

Nicolás Gallucci, M., Fraire, J.C., Ferreyra Maillard, A.P.V., Páez, P. L., Aiassa Martínez, I.M., Pannunzio Miner, E.V., Coronado, E.A., and Dalmasso, P.R. 2017. Silver nanoparticles from leafy green extract of Belgian endive (*Cichorium intybus* L. var. sativus): biosynthesis, characterization, and antibacterial activity. Mater. Lett. 197: 98–101.

Ocsoy, I., Demirbas, A., McLamor, E.S., Altinsoy, B., Ildiz, N., and Baldemir, A. 2017. Green synthesis with incorporated hydrothermal approaches for silver nanoparticles formation and enhanced antimicrobial activity against bacterial and fungal pathogens. J. Mol. Liq. 238: 263–269.

Omran, B.A., Nassar, H.N., Fatthallah, N.A., Hamdy, A., El-Shatoury, E.H., and El-Gendy, N.Sh. 2018. Waste upcycling of *Citrus sinensis* peels as a green route for the synthesis of silver nanoparticles. Energy Sources A. 40(2): 227–236. https://doi.org/10.1080/15567036.2017.1410597

Omran, B.A., Nassar, H.N., Younis, S.A., Fatthallah, N.A., Hamdy, A., El-Shatoury, E.H., and El-Gendy, N.Sh. 2019. Physiochemical properties of *Trichoderma longibrachiatum* DSMZ 16517-synthesized silver nanoparticles for the mitigation of halotolerant sulphate-reducing bacteria. J. Appl. Microbiol. 126: 138–154. https://doi.org/10.1111/jam.14102

Oza, G., Pandey, S., Mewada, A., Kalita, G., and Sharon, M. 2012. Facile biosynthesis of gold nanoparticles exploiting optimum pH and temperature of freshwater alga *Chlorella pyrenoidusa*. Adv. Appl. Sci. Res. 3(3): 1405–1412.

Parial, D., Patra, H.K., Dasgupta, A.K., Pal, R. 2012. Screening of different algae for green synthesis of gold nanoparticles. Eur. J. Phycol. 47(1): 22–29.

Parida, U.K., Bindhani, B.K., and Nayak, P. 2011. Green synthesis and characterization of gold nanoparticles using onion (*Allium cepa*) extract. World J. Nano Sci. Eng. 1: 93–98.

Pasca, R.D., Mocanu, A., Cobzac, S.C., Petean, I., Horovitz, O., and Tomoaia–Cotisel, M. 2014. Biogenic syntheses of gold nanoparticles using plant extracts. Part. Sci. Technol. 32: 131–137.

Passos de Aragã, A., Maria de Oliveira, T., Quelemes, P.V., Perfeito, M.L.G., Araújo, M.C., Santiago, J.A.S., Cardoso, V.S., Quaresma, P., Leite, J.R.S.A., and Alves da Silva, D. 2016. Green synthesis of silver nanoparticles using seaweed *Gracilaria birdiae* and their antibacterial activity. Arab. J. Chem. 12(8): 4182–4188.

Patil, R., Kokate, M., and Kolekar, S. 2012. Bioinspired synthesis of highly stabilized silver nanoparticles using *Ocimum tenuiflorum* leaf extract and their antibacterial activity. Spectrochim. Acta A Mol. Biomol. Spectrosc. 91: 234–238.

Patra, J.K., and Baek, K.H. 2017. Antibacterial activity and synergistic antibacterial potential of biosynthesized silver nanoparticles against foodborne pathogenic bacteria along with its anticandidal and antioxidant effects. Front. Microbiol. 8: 167–181.

Patra, J.K., Kwon, Y., and Baek, K.H. 2016. Green biosynthesis of gold nanoparticles by onion peel extract: synthesis, characterization and biological activities. Adv. Powder Technol. 27: 2204–2213.

Paul, N.S., and Yadav, R.P. 2015. Biosynthesis of silver nanoparticles using plant seeds and their antimicrobial activity. Asian J. Biomed. Pharm. Sci. 5(45): 26–28.

Philip, D. 2010. Green synthesis of gold and silver nanoparticles using *Hibsicus rosa sinensis* Physica E Low Dimens. Syst. Nanostruct. 42(5): 1417–1424.

Philip, D., Unni, C., Aromal, S.A., and Vidhu, V.K. 2011. *Murraya Koenigii* leaf-assisted rapid green synthesis of silver and gold nanoparticles. Spectrochim. Acta A Mol. Biomol. Spectrosc. 78(2): 899–904.

Ping, Y., Jun, Z., Tieling, X., Guoqiang, C., Ran, T., and Kwang-Ho, C. 2018. Green synthesis of silver nanoparticles using grape seed extract and their application for reductive catalysis of Direct Orange 26. J. Ind. Eng. Chem. 58: 74–79.

Prabakar, K., Sivalingam, P., Rabeek, S.I.M., Muthuselvam, M., Devarajan, N., Arjunan, A., Karthick, R., Suresh, M.M., and Wembonyama, J.P. 2013. Evaluation of antibacterial efficacy of phyto fabricated silver nanoparticles using *Mukia scabrella* (Musumusukkai) against drug resistance nosocomial Gram-negative bacterial pathogens. Colloids Surf. B Biointerfaces. 104: 282–288.

Prasad, T.N.V.K.V., and Elumalai, E. 2011. Biofabrication of Ag nanoparticles using *Moringa oleifera* leaf extract and their antimicrobial activity. Asian Pac. J. Trop. Biomed. 1: 439–442.

Prathna, T.C., Chandrasekaran, N., Raichur, A.M., and Mukherjee, A. 2011a. Kinetic evolution studies of silver nanoparticles in a bio-based green synthesis process. *Colloid. Surface* A. 377: 212–216.

Prathna, T.C., Raichur, A.M., Chandrasekaran, N., and Mukherjee, A. 2011b. Biomimetic synthesis of silver nanoparticles by *Citrus limon* (lemon) aqueous extract and theoretical prediction of particle size. Colloids Surf B Biointerfaces. 82: 152–159.

Priyaa, G.H., and Satyan, K.B. 2014. Biological synthesis of silver nanoparticles using ginger (*Zingiber officinale*) extract. J. Environ. Nanotechnol. 3(4): 32–40.

Rad, M.S., Rad, J.S., Heshmati, G.A., Miri, A., and Sen, D.J. 2013. Biological synthesis of gold and silver nanoparticles by *Nitraria schoberi* fruits. Am. Adv. Drug Del. 1(2): 174–179.

Raghunandan, D., Bedre, M.D., Basavaraja, S., Sawle, B., Manjunath, S., and Venkataraman, A. 2010. Rapid biosynthesis of irregular shaped gold nanoparticles from macerated aqueous extracellular dried clove buds (*Syzygium aromaticum*) solution. Colloids Surf. B Biointerfaces. 79: 235–240.

Raja, S., Ramesh, V., and Thivaharan, V. 2017. Green biosynthesis of silver nanoparticles using *Calliandra haematocephala* leaf extract, their antibacterial activity and hydrogen peroxide sensing capability. Arab. J. Chem. 10: 253–261.

Raja, K., Saravanakumar, A., and Vijayakumar, R. 2012. Efficient synthesis of silver nanoparticles from *Prosopis juliflora* leaf extract and its antimicrobial activity using sewage. Spectrochim. Acta A. Mol. Biomol. Spectrosc. 97: 490–494.

Rajan, A., Vilas, V., and Philip, D. 2015. Studies on catalytic, antioxidant, antibacterial and anticancer activities of biogenic gold nanoparticles. J. Mol. Liq. 212: 331–339.

Rajathi, A.A.F., Parthiban, C., Kumar, G.V., and Anantharaman, P. 2012. Biosynthesis of antibacterial gold nanoparticles using brown alga, *Stoechospermum marginatum* (kutzing). Spectrochim. Acta A Mol. Biomol. Spectrosc. 99, 166–73.

Rajesh, K.M., Ajitha, B., Reddy, Y.A.K., Suneetha, Y., and Reddy, P.S. 2017. Assisted green synthesis of copper nanoparticles using *Syzygium aromaticum* bud extract: physical, optical and antimicrobial properties. Optik, 154: 593–600.

Rajesh, K.M., Ajitha, B., Reddy, Y.A.K., Suneetha, Y., and Reddy, P.S. 2018. Assisted green synthesis of copper nanoparticles using *Syzygium aromaticum* bud extract: physical, optical and antimicrobial properties. Optik. 154: 593–600.

Rajeshkumar, S., Kannan, C., and Annadurai, G. 2012. Green Synthesis of silver nanoparticles using marine brownalgae *Turbinaria conoides* and its antibacterial activity. Int. J. Pharma Bio. Sci. 3(4): 502–510.

Rajeshkumar, S., Malarkodi, C., Gnanajobitha, G., Paulkumar, K., Vanaja, M., Kannan, C., and Annadurai, G. 2013a. Seaweed-mediated synthesis of gold nanoparticles using *Turbinaria conoides* and its characterization. J. Nanostruct. Chem. 3(4): 1–7.

Rajeshkumar, S., Malarkodi, C., Paulkumar, K., Vanaja, M., Gnanajobitha, G., and Annadurai, G. 2013b. Algae mediated green fabrication of silver nanoparticles and examination of its antifungal activity against clinical pathogens. Int. J. Metals. Volume 2014, Article ID 692643. http://dx.doi.org/10.1155/2014/692643.

Rajeshkumar, S., Malarkodi, C., Vanaja, M., and Annadurai, G. 2016. Anticancer and enhanced antimicrobial activity of biosynthesizd silver nanoparticles against clinical pathogens. J. Mol. Struct. 1116: 165–173.

Rakhi, M., and Gopal, B. 2012. *Terminalia arjuna* bark extract mediated size controlled synthesis of polyshaped gold nanoparticles and its application in catalysis. Int. J. Res. Chem. Environ. 2(4): 338–344.

Rao, Y.S., Kotakadi, V.S., Prasad, T.N.V.K.V., Reddy, A.V., and Gopal, D.V.R.S. 2013. Green synthesis and spectral characterization of silver nanoparticles from *Lakshmi tulasi* (*Ocimum sanctum*) leaf extract. Spectrochim. Acta A. Mol. Biomol. Spectrosc. 103: 156–159.

Rao, K.J., Praneeth, N.V.S., and Paria, S. 2017. A promising technique of *Aegle Marmelos* leaf extract mediated self-assembly for silver nanoprism formation. AIChE J. 63(9): 3670–3680.

Rasheed, T., Bilal, M., Iqbal, H.M.N., and Li, C. 2017. Green biosynthesis of silver nanoparticles using leaves extract of *Artemisia vulgaris* and their potential biomedical applications. Colloids Surf. B Biointerfaces. 158: 408–415.

Rashidipour, M., and Heydari, R. 2014. Biosynthesis of silver nanoparticles using extract of olive leaf: synthesis and in vitro cytotoxic effect on MCF-7 cells. J. Nanostruct. Chem. 4: 112–118.

Raut, R.W., Kolekar, N.S., Lakkakula, J.R., Mendhulkar, V.D., and Kashid, S.B. 2010. Extracellular synthesis of silver nanoparticles using dried leaves of *pongamia pinnata* (L) pierre. Nano-micro Lett. 2(2): 106–113.

Ravindra, S., Mohan, Y.M., Reddy, N.N., and Raju, KM. 2010. Fabrication of antibacterial cotton fibres loaded with silver nanoparticles via "green approach". Colloids Surface. A. 367: 31–40.

Roy, K., Ghosh, C.K., and Sarkar, A.K. 2017. Degradation of toxic textile dyes and detection of hazardous Hg^{2+} by low-cost bioengineered copper nanoparticles synthesized using *Impatiens balsamina* leaf extract. Mater. Res. Bull. 94: 257–262.

Sabela, M.I., Makhanya, T., Kanchi, S., Shahbaaz, M., Idress, D., and Bisetty, K. 2018. One-pot biosynthesis of silver nanoparticles using *Iboza Riparia* and *Ilex Mitis* for cytotoxicity on human embryonic kidney cells. J. Photochem. Photobiol., B: Biol. 178: 560–567.

Sabri, M.A., Umer, A.A., Awan, G.H., Hassan, M.H., and Hasnain, A. 2016. Selection of suitable biological method for the synthesis of silver nanoparticles. Nanomater. Nanotechnol. 6: 29. doi: 10.5772/62644.

Sadeghi, B., and Gholamhoseinpoor, F. 2015. A study on the stability and green synthesis of silver nanoparticles using *Ziziphora tenuior* (Zt) extract at room temperature. Spectrochim. Acta Part A: Mol. Biomol. Spectrosc. 134: 310–315.

Sadeghi, B., Mohammadzadeh, M., and Babakhani, B. 2015. Green synthesis of gold nanoparticles using *Stevia rebaudiana* leaf extracts: characterization and their stability. J. Photochem. Photobiol. B: Biol. 148: 101–106.

Saha, J., Begum, A., Mukherjee, A., and Kumar, S. 2017. A novel green synthesis of silver nanoparticles and their catalytic action in reduction of Methylene Blue dye. Sustain. Environ. Res. 27: 245–250.

Sahayaraj, K., Rajesh, S., and Rathi, J.M. 2012. Silver nanoparticles biosynthesis using marine alga *padina pavonica* (linn.) and its microbicidal activity. Dig. J. Nanomater. Bios. 7(4): 1557–1567.

Salunke, G.R., Ghosh, S., Kumar, R.J.S., Khade, S., Vashisth, P., Kale, T., Chopade, S., Pruthi, V., Kundu, G., Bellare, J.R., and Chopade, B.A. 2014. Rapid efficient synthesis and characterization of silver, gold, and bimetallic nanoparticles from the medicinal plant *Plumbago zeylanica* and their application in biofilm control. Int. J. Nanomed. 9: 2635–2653.

Sarkar, D., and Paul, G. 2017. Green synthesis of silver nanoparticles using *Mentha asiatica* (mint) extract and evaluation of their antimicrobial potential. Int. J. Curr. Res. Biosci. Plant Biol. 4(1): 77–82.

Sathishkumar, M., Sneha, K., Kwak, I.S., Mao, J., Tripathy, S., and Yun, Y.S. 2009a. Phyto-crystallization of palladium through reduction process using *Cinnamom zeylanicum* bark extract. J. Hazard. Mater. 171: 400–404.

Sathishkumar, M., Sneha, K., Won, S., Cho, C.W., Kim, S., and Yun, Y.S. 2009b. *Cinnamon zeylanicum* bark extract and powder mediated green synthesis of nano-crystalline silver particles and its bactericidal activity. Colloids Surf B Biointerfaces. 73: 332–338.

Saxena, A., Tripathi, R.M., and Singh, R.P. 2010. Biological synthesis of silver nanoparticles by using onion (*Allium cepa*) extract and their antibacterial activity. Dig. J. Nanomater. Biostruct. 5(2): 427–432.

Saxena, A., Tripathi, R., Zafar, F., and Singh, P. 2011. Green synthesis of silver nanoparticles using aqueous solution of *Ficus benghalensis* leaf extract and characterization of their antibacterial activity. Mater. Lett. 67: 91–94.

Seetharaman, P., Chandrasekaran, R., Gnanasekar, S., Mani, I., and Sivaperumal, S. 2017. Biogenic gold nanoparticles synthesized using *Crescentia cujete* L. and evaluation of their different biological activities. Biocatal. Agric. Biotechnol. 11: 75–82.

Selvam, K., Sudhakar, C., Govarthanan, M., Thiyagarajan, P., Sengottaiyan, A., Senthilkumar, B., and Selvankumar, T. 2016. Eco-friendly biosynthesis and characterization of silver nanoparticles using *Tinospora cordifolia* (Thunb.) Miers and evaluate its antibacterial, antioxidant potential. J. Radiat. Res. Appl. Sci. 10: 6–12.

Sen, A.K., Das, A.K., Banerji, N., and Vignon, M.R. 1992 Isolation and structure of a 4-O-methyl-glucuronoarabino-galactan from *Boswellia serrata*. Carbohydr. Res. 223: 321–327.

Shalaby, T.I., Mahmoud, O.A., El Batouti, G.A., and Ibrahim, E.E. 2015. Green synthesis of silver nanoparticles: synthesis, characterization and antibacterial activity. J. Nanosci. Nanotechnol. 5(2): 23–29.

Shankar, S.S., Ahmad, A., and Sastry, M. 2003. Geranium leaf assisted biosynthesis of silver nanoparticles. Biotechnol. Prog. 19: 1627–1631.

Shankar, S.S., Rai, A., Ahmad, A., and Sastry, M. 2004a. Rapid synthesis of Au, Ag, and bimetallic Au core–Ag shell nanoparticles using Neem (*Azadirachta indica*) leaf broth. J. Colloid. Interface Sci. 275: 496–502.

Shankar, S.S., Rai, A., Ankamwar, B., Singh, A., Ahmad, A., and Sastry, M. 2004b. Biological synthesis of triangular gold nanoprisms. Nat. Mater. 3: 482–488.

Sharma, K., Kaushik, S., and Jyoti, A. 2016. Green synthesis of silver nanoparticles by using waste vegetable peel and its antibacterial activities. J. Pharm. Sci. Res. 8(5): 313–316.

Shende, S., Ingle, A.P., Gade, A., and Rai, M. 2015. Green synthesis of copper nanoparticles by *Citrus medica* Linn. (Idilimbu) juice and its antimicrobial activity. World J. Microbiol. Biotechnol. 31(6): 865–873.

Shukla, P.K., Bhatnagar, P., and Yadav, R. 2005. *Boswellia serrata*: a gum-oleoresin yielding tree. Vaniki Sandesh. 29: 23–26.

Singaravelu, G., Arokiamary, J.S., Kumar, V.G., and Govindaraju, K. 2007. A novel extracellular synthesis of mondisperse gold nanoparticles using marine alga, *Sargassum wightti Greville*. Colloids Surf. B Biointerfaces. 57(1): 97–101.

Singh, M., Kalaivani, R., Manikandan, S., Sangeetha, N., and Kumaraguru, A.K. 2013b. Facile green synthesis of variable metallic gold nanoparticle using *Padina gymnospora*, a brown marine macroalga. Appl. Nanosci. 3(2): 145–151.

Singh, P., Kim, Y.J., and Yang, D.C. 2016b. A strategic approach for rapid synthesis of gold and silver nanoparticles by Panax ginseng leaves. Artif. Cells Nanomed. Biotechnol. 44: 1949–1957.

Singh, P., Kim, Y.J., Wang, C., Mathiyalagan, R., and Yang, D.C. 2016c. The development of a green approach for the biosynthesis of silver and gold nanoparticles by using *Panax ginseng* root extract, and their biological applications. Artif. Cells Nanomed. Biotechnol. 44(4): 1150–1157.

Singh, P., Kim, Y.J., Zhang, D., and Yang, D.C. 2016a. Biological synthesis of nanoparticles from plants and microorganisms. Trends Biotechnol. 34(7): 588–600.

Singh, M., Kumar, M., Kalaivani, R., Manikandan, S., and Kumaraguru, A.K. 2013a. Metallic silver nanoparticle: a therapeutic agent in combination with antifungal drug against human fungal pathogen. Bioprocess Biosyst. Eng. 36: 407–415.

Singh, C., Sharma, V., Naik, P.K., Khandelwal, V., and Singh, H. 2011. A green biogenic approach for synthesis of gold and silver nanoparticles using *Zingiber officinale*. Dig. J. Nanomater. Biostruct. 6: 535 –542.

Singhal, G., Bhavesh, R., Kasariya, K., Sharma, A.R., and Singh, R.P. 2011. Biosynthesis of silver nanoparticles using *Ocimum sanctum* (Tulsi) leaf extract and screening its antimicrobial activity. J. Nanopart. Res. 13: 2981 –2988.

Sivakamavalli, J., Deepa, O., and Vaseeharan, B. 2014. Discrete nanoparticles of *Ruta graveolens* induces the bacterial and fungal biofilm inhibition. Cell Commun. Adhes. 21(4): 229–238.

Song, J.Y., Jang, H.K., and Kim, B.S. 2009. Biological synthesis of gold nanoparticles using *Magnolia kobus* and *Diopyros kaki* leaf extracts. Process Biochem. 44(10): 1133–1138.

Song, J.Y., and Kim, B.S. 2009. Rapid biological synthesis of silver nanoparticles using plant leaf extracts. Bioprocess Biosyst. Eng. 32: 79–84.

Song, J.Y., Kwon, E-Y., and Kim, B.S. 2010. Biological synthesis of platinum nanoparticles using *Diopyros kaki* leaf extract. Bioprocess Biosyst. Eng. 33: 159–164.

Sudha, S.S., Rajamanikam, K., and Rengaramanujam, J. 2013. Microalgae mediated synthesis of silver nanoparticles and their antibacterial activity against pathogenic bacteria. Indian J. Exp. Biol. 52: 393–399.

Suganthy, N., Sri Ramkumar, V., Pugazhendhi, A., Benelli, G., and Archunan, G. 2018. Biogenic synthesis of gold nanoparticles from Terminalia arjuna bark extract: assessment of safety aspects and neuroprotective potential via antioxidant, anticholinesterase, and antiamyloidogenic effects. Environ. Sci. Pollut. Res. Int. 25(11): 10418–10433.

Suman, T.Y., Rajasree, S.R.R., Ramkumar, R., Rajthilak, C., and Perumal, P. 2014. The green synthesis of gold nanoparticles using an aqueous root extract of *Morinda citrifolia* L. Spectrochim. Acta A. 118: 11–16.

Sutradhar, P., Saha, M., and Maiti, D. 2014. Microwave synthesis of copper oxide nanoparticles using tea leaf and coffee powder extracts and its antibacterial activity. J. Nanostruct. Chem. 4: 86–92.

Suvith, V.S., and Philip, D. 2014. Catalytic degradation of methylene blue using biosynthesized gold and silver nanoparticles. Spectrochim. Acta Part A. 118: 526–532.

Syafiuddin, A., Salmiati, S., Salim, M.R., Kueh, A.B.H., Hadibaratad, T., and Nur, H. 2017. A review of silver nanoparticles: research trends, global consumption, synthesis, properties, and future challenges. J. Chin. Chem. Soc. 64: 732–756.

Tagad, C.K., Kim, H.U., Aiyer, R.C., More, P., Kim, T., Moh, S.H., Kulkarni, A., and Sabharwal, S.G. 2013. A sensitive hydrogen peroxide optical sensor based on polysaccharide stabilized silver nanoparticles. RSC Adv. 45: 22940–22943.

Tareq, F.K., Fayzunnesa, M., and Kabir, M.S. 2017. Antimicrobial activity of plant-median synthesized silver nanoparticles against food and agricultural pathogens. Microb. Pathog. 109: 228–232.

Tavakoli, F., Salavati-Niasari, M., and Mohandes, F., 2015. Green synthesis and characterization of graphene nanosheets. Mater. Res. Bull. 63: 51–57.

Thakkar, K.N., Mhatre, S.S., and Parikh, R.Y. 2010. Biological synthesis of metallic nanoparticles. Nanomed. Nanotechnol. Biol. Med. 6: 257–262.

Thakor, A.S., Jokerst, J., Zavaleta, C., Massoud, T.F., and Gambhir, S.S. 2011. Gold nanoparticles: a revival in precious metal administration to patents. Nano Lett. 11: 4029–4036.

Ulug, B., HalukTurkdemir, M., Cicek, A., and Mete, A. 2015. Role of irradiation in the green synthesis of silver nanoparticles mediated by fig (*Ficus carica*) leaf extract. Spectrochim. Part A: Mol. Biomol. Spectrosc. 135: 153–161.

Umamaheswari, C., Lakshmanan, A., and Nagarajan, N.S. 2018. Green synthesis, characterization and catalytic degradation studies of gold nanoparticles against congo red and methyl orange. J. Photochem., Photobiol. B: Biol. 178: 33–39.

Vanaja, M., Gnanajobitha, G., Paulkumar, K., Rajeshkumar, S., Malarkodi, C., and Annadurai, G. 2013. Phytosynthesis of silver nanoparticles by *Cissus quadrangularis*: influence of physicochemical factors. J. Nanostruct. Chem. 3: 17–25.

Varadavenkatesan, T., Vinayagam, R., and Selvaraj, R. 2017. Structural characterization of silver nanoparticles phyto-mediated by a plant waste, seed hull of *Vigna mungo* and their biological applications. J. Mol. Struct. 1147: 629–635.

Varun, S., Sudha S., and Senthil Kumar, P. 2014. Biosynthesis of gold nanoparticles from aqueous extract of *Dictyota bartayresiana* and their antifungal activity. Indian J. Adv. Chem. Sci. 2(3): 190–193.

Veeraputhiran, V. 2013. Bio-catalytic synthesis of silver nanoparticles. Int. J. Chem. Tech. Res. 5(5): 255–2562.

Velmurugan, P., Anbalagan, K., Manosathyadevan, M., Lee, K.J., Cho, M., Lee, S.M., Park, J.H., Oh, S.G., Bang, K.S., and Oh, B.T. 2014. Green synthesis of silver and gold nanoparticles using *Zingiber officinale* root extract and antibacterial activity of silver nanoparticles against food pathogens. Bioprocess Biosyst. Eng. 37: 1935–1943.

Velusamy, P., Das, J., Pachaiappan, R., Vaseeharan, B., and Pandian, K. 2015. Greener approach for synthesis of antibacterial silver nanoparticles using aqueous solution of neem gum (*Azadirachta indica* L.). Ind. Crops Prod. 66: 103–109.

Venkatesham, M., Ayodhya, D., Madhusudhan, A., and Veerabhadram, G. 2012. Synthesis of stable silver nanoparticles using gum acacia as reducing and stabilizing agent and study of its microbial properties: a novel green approach. Int. J. Green Nanotechnol. 4(3): 199–206.

Vijaya, J.J., Jayaprakasha, N., Kombaiaha, K., Kaviyarasuc, K., Kennedye, L.J., Ramalingam, R.J., Al-Lohedan, H.A., Mansoor-Ali, V.M., and Maaza, M. 2017. Bioreduction potentials of dried root of *Zingiber officinale* for a simple synthesis of silver nanoparticles: Antibacterial studies. J. Photochem. Photobiol. B. 177: 62–68.

Vijayan, S.R., Santhiyagu, P., Singamuthu, M., Ahila, N.K., Jayaraman, R., and Ethiraj, K. 2014. Synthesis and characterization of silver and gold nanoparticles using aqueous extract of seaweed, *Turbinaria conoides*, and their antimicrofouling activity. Sci. World J. Volume 2014, Article ID 938272. http://dx.doi.org/10.1155/2014/938272.

Vijayaraghavan, K., and Nalini, S.P.K. 2010. Biotemplates in the green synthesis of silver nanoparticles. Biotechnol. J. 5: 1098–1110.

Vinod, V.T.P., Saravanan, P., Sreedhar, B., Devi, D.K., and Sashidhar, R.B. 2011. A facile synthesis and characterization of Ag, Au and Pt nanoparticles using a natural hydrocolloid gum kondagogu (*Cochlospermum gossypium*). Colloid Surface B. 83(2): 291–298.

Yang, N., and Li, W. 2013. Mango peel extract mediated novel route for synthesis of silver nanoparticles and antibacterial application of silver nanoparticles loaded onto non-woven fabrics. Ind. Crops Prod. 48: 81–88.

Yang, X., Summerhurst, D.K., Koval, S.F., Ficker, C., Smith, M.L., and Bernards, M.A. 2001. Isolation of an antimicrobial compound from *Impatiens balsamina* L. using bioassay-guided fractionation. Phytother. Res. 15(8): 676–680.

Yew, Y.P., Shameli, K., Miyake, M., Bahiyah Bt Ahmed Khairudin, N., Eva Bt Mohamad, S., Naiki, T., and Lee, K.X. 2018. Green biosynthesis of superparamagnetic magnetite Fe_3O_4 nanoparticles and biomedical applications in targeted anticancer drug delivery system: a review. Arab. J. Chem. 13(1): 2287–2308.

Yousefzadi, M., Rahimi, Z., and Ghafori, V. 2014. The green synthesis, characterization and antimicrobial activities of silver nanoparticles synthesized from green alga *Enteromorpha flexuosa* (wulfen). J. Agardh. Mater. Lett. 137: 1–4.

Yuan, C.-G., Huo, C., Yu, Sh., and Gui, B. 2017. Biosynthesis of gold nanoparticles using *Capsicum annuum* var. *grossum* pulp extract and its catalytic activity. Physica E 85: 19–26.

Zaheer, Z. 2018. Biogenic synthesis, optical, catalytic, and in vitro antimicrobial potential of Ag-nanoparticles prepared using Palm date fruit extract. J. Photochem. Photobiol., B: Biol. 178: 584–592.

Zayed, M.F., Eisa, W.H., and Shabaka, A.A. 2012. *Malva parviflora* extract assisted green synthesis of silver nanoparticles. Spectrochim. Acta A. Mol. Biomol. Spectrosc. 98: 423–428.

Zhang, X.F., Liu, Z.G., Shen, W., and Gurunathan, S. 2016. Silver nanoparticles: synthesis, characterization, properties, applications, and therapeutic approaches. Int. J. Mol. Sci. 17: 1534–156.

CHAPTER 3

Microbial Synthesis of Metal Nanoparticles

3.1 INTRODUCTION

Metals are known to be essential for the life processes of microorganisms. For example, calcium, cobalt, iron, potassium, sodium, and so on, are required as nutrients and they act as the catalysts for biochemical reactions, stabilizers of protein structures and bacterial cell walls, serve in maintaining osmotic balance, involve in redox processes, or stabilize various enzymes and DNA through electrostatic forces (Bruins et al., 2000). However, in order to survive in environments containing high levels of metals, microorganisms have adapted themselves by evolving mechanisms to cope with metals. Where, metal nanoparticle (NP) formation is thought to be a way of microbial self-protection; in another word, NPs formation is the "by-product" of a resistance mechanism against a specific metal. These mechanisms may involve altering the chemical nature of the toxic metal so that it no longer causes toxicity, resulting in the formation of NPs of the metal concerned. Thus, these NPs formation is the "by-product" of a resistance mechanism against a specific metal, and this can be used as an alternative way of producing them; whereas, the cells convert toxic metals from their ionic form into atomic one, thereby decreasing their solubility and preventing their penetration into the cell (Hallmann et al., 1997; Sabry et al., 1997). Since marine environment usually contains a lot of metal salt, whatever, naturally occurring or as contaminants. Thus, high-metal tolerance, among marine bacteria, is a common phenomenon which makes them ideal model candidates for exploring metal/metal oxide NP synthesis in the laboratory environment (Sabry et al., 1997). Not only this, but also in nature there are also a variety of nanomaterials that can be synthesized by biological processes. For example, the magnetotactic bacteria synthesize intracellular magnetite or greigite nanocrystallites, the diatomas synthesize siliceous materials and S-layer bacteria produce gypsum and calcium carbonate layers. Liu et al. (2012) reported that *Gluconacetobacter xylinum* secretes chloride ions from the cytoplasm and generates reductases to overcome metal stress, via the bioreduction of the silver ions to form Ag/AgCl NPs as by-product.

Although phytosynthesis is less time-consuming but it produces polydispersed metal NPs due to the multiple components; flavonoids, terpenoids, and polyphenols participate in the reduction of metal ions (Ghosh et al., 2012; Salunkhe et al., 2014). Geographical and seasonal variations may also affect phytosynthesis (Singh et al., 2013). While, microbial synthesis of NPs overcomes such drawbacks, but regular maintenance of culture and sterile conditions for NP synthesis is required (Salunkhe et al., 2014). Dameron et al. (1989) reported the first mycosynthesis of NPs, producing CdS NPs by *Candida albicans*.

A variety of diverse groups of biological agents such as bacteria, fungi, yeast, algae, and actinomycetes can be used for biological synthesis of NPs. This can be occurred via two methods: (1) bioreduction, in which metal ions are biologically reduced into more stable forms. Whereas, many organisms have the ability to utilize dissimilatory metal reduction, in which the reduction of a metal ion is coupled with the oxidation of an enzyme, with the production of stable and inert metallic NPs. (2) biosorption throughout the binding of metal ions from an aqueous or soil sample onto the organism itself (i.e., on the cell wall) and does not require the input of energy. Whereas, the microorganism produces peptides or has a modified cell wall which binds to metal ions, and these are able to form stable complexes in the form of NPs.

Microbial synthesis of NPs can be categorized according to its performance via intracellular and extracellular techniques. However, the extracellular mode of synthesis is preferred over the intracellular mode owing to easy recovery of NPs. Extracellular production of metal NPs using both living or dried biomass occurs via two routes: (1) biomolecules released into the external medium reduces the metal ions to metal NPs and/or (2) metal NPs formed inside the cell and then secreted outside. Extracellular metal NPs may remain attached to the microbial cell wall and would require mild sonication for recovery. Intracellular method of synthesis requires additional steps to recover the accumulated NPs from cells and therefore, it is less preferred. Ultrasonication of bacterial cells is the most common technique to recover metal NPs (Kalishwaralal et al., 2010). Besides this, heat treatment like autoclaving and the use of detergents and salts can also be employed to lyse the cells (Fesharaki et al., 2010; Sneha and Yun, 2013). The intracellular synthesis of NPs is carried out via the reduction of the metal ions using α-NADH-dependent nitrate reductase. While, during the extracellular synthesis of NPs, the extracellular polysaccharides (EPS) produced by biomass are used for the reduction of noble metal ions. Moreover, the ability of microorganisms to trap metal ions and reduce them into NPs is due to the presence of sulfur-containing extra cellular proteins and other biomolecules produced by the microorganisms known as secondary metabolites, for example organic acids, fatty acids, quinines, and so on (Abirami and Kannabiran, 2016). Zhang et al. (2005) reported that under anaerobic conditions, many microorganisms are able to conserve energy through the bioreduction of metals. However, various enzymes are believed to take an active part in the bioreduction process of transporting electrons from certain electron donors to metal electron acceptors. Not only this but some microorganisms also have a strong biosorptive capacity to metal ions, reduce and precipitate them in their metallic form (Sintubin et al., 2009). This is suggested to be nonenzymatic reduction mechanism, where some organic functional groups of microbial cell walls could be responsible for the bioreduction process under certain conditions (Lin et al., 2001). Gericke and Pinches (2006) reported that the biosynthesis of NPs is governed by the microbial growth phase and its metabolic status.

These diverse microbial groups are reported to have many advantages over physical and chemical methods such as easy and simple scale-up, easy downstream processing, simpler biomass handling and recovery, and economic viability (Rai et al., 2009; Thakkar et al., 2011; Renugadevi and Aswini, 2012). These different biological agents therefore demonstrate immense biodiversity in the synthesis of NPs and lead to green nanotechnology (Vaseeharan et al., 2010; Singh et al., 2013; Thakkar et al., 2011). Among all biological agents, fungi present higher tolerance and metal bioaccumulation abilities. Not only this but also fungi are considered as efficient secretors of extracellular enzymes, thus ease in the scale-up, which makes the entire process more cost-effective. Moreover, one of the biggest advantages of biological synthesis via fungal enzymes is the big opportunity of evolving a balanced approach for the biosynthesis of NPs over a wide range of chemical compositions, which would not be possible by other microbe-based methods. All of these add to the advantageous characteristics for the bioproduction of NPs using fungi (van den Hondel et al. 1992; Mandal et al., 2006; Rahimi et al., 2016; Syafiuddin et al., 2017; Omran et al., 2018a; 2019). Nevertheless, microalgae attract many researchers, because of their capabilities to remediate toxic metals to nontoxic forms. That occurs via accumulation of metals by chelation, chemical transformation, and production of bio-mineral structures and metal NPs (Mahdieh et al., 2012). Algae have the advantages of growing rapidly with a minimum growth requirement of sunlight, atmospheric CO_2, and a few common mineral salts to reproduce their biomass very rapidly. Nevertheless, biosynthesis of NPs using algal extracts usually takes shorter time than other biosynthesizer. Where, there are a variety of biomolecules responsible for the reduction of metals, including polysaccharides, peptides, and pigments. Besides, proteins through aminogroups or cysteine residues and sulphated polysaccharides act as stabilizing and capping the metal NPs in aqueous solutions.

AgNPs have gained a lot of attentiveness over other metal NPs. Because the surface plasmon resonance (SPR) energy of AgNPs is located away from the interband transition energy, and also due to its high thermal stability and its low toxicity to human cells. Moreover, remarkable applications were reported for the AgNPs in the fields of catalysis, optoelectronics, water treatment, biosensors, antimicrobials, and therapeutics. Nevertheless, AuNPs have different applications in optoelectronic devices, ultrasensitive chemical and biological sensors, and as catalysts. AuNPs can be detected by numerous technique, such as optic absorption fluorescence and electric conductivity.

It is well-known that metals such as Au, Ag, and Pt can catalyze the decomposition of H_2O_2 to oxygen. Moreover, Ag can be used as a catalyst for the oxidation of ethylene to ethylene oxide and methanol to formaldehyde (Nagy and Mestl, 1999). AgNPs can be used as optical sensors for the formation of small molecule adsorbates (McFarland and Van Duyne, 2003). AgNPs, AuNPs, and CuNPs have been reported for the reduction of nitro-aromatics for their corresponding amino-aromatics, in the presence of sodium borohydride ($NaBH_4$) (Zaheer, 2018). One of the most important applications of AgNPs and CuNPs is their efficient application as antibiofilm agents (LewisOscar et al., 2015). AgNPs have been used also for preparation of H_2 gas sensors (Mohamedkhair et al., 2019). This would have applications in H_2-fueled engines and hydrocracking process, decreasing the threat of fire and explosions. AgNPs can be applied for the preparation of Hg adsorbents in gas condensate (Adio et al., 2019). Nevertheless, very active catalysts can be produced when AuNPs <5 nm is supported on base metal oxide or carbon. Moreover, gas sensors based on AuNPs have been developed for detecting a number of gases, including CO and NOx. The most catalytical active material has an Au core (submonolayer Pd shell) nanostructure. Moreover, Pd-coated silver NPs are very effective catalyst for remediation of trichloroethene and common organic pollutant in ground water. From the economic point of view, one of the main advantage of Au catalysts is that it offers compared with other precious metal catalysts a lower cost and a greater price stability because gold is substantially cheaper and is considerable more plentiful than platinum. The polymer–gold NPs composites possess interesting electrical properties and the nanocomposites composed of gold and biopolymer are employed as a novel biosensor. AgNPs incorporated on textile fabric and polymeric medical devices (such as surgical masks) express significant antimicrobial efficiency. Moreover, AuNPs can be used for labeling applications as they are a very attractive contrast agent. Thus, because of the interaction of AuNPs with light they can be used for the visualization of particles. Moreover, AuNPs absorb light and have been explored as a method of heating. Not only this but AuNPs also provide nontoxic routes to drug and gene delivery application (Hainfeld et al., 2014).

This chapter deals with the diversity of microbes involved in the synthesis of silver, gold, and copper NPs and their expected direct and indirect applications in petroleum industry. The possible mechanisms for the microbial synthesis of metal NPs are also discussed. The alteration of size and shapes of the biosynthesized NPs with the physicochemical variations in the microbial process and the characterization instruments will be also covered in this chapter.

3.2 MICROBIAL SYNTHESIS OF SILVER NPS

Silver NPs have gained more attention due to its distinctive properties: electrical conductivity, chemical stability, and catalytic and antibacterial activity (Bhui and Misra, 2012). It has been reported as the most commercialized nanomaterials, with more than 200 consumer products (Pulit-Prociak and Banach, 2016). Although, several silver salts are used as precursors for the synthesis of AgNPs, such as $AgSO_4$, $AgBF_4$, AgF_3SO_3, $AgSbF_6$, Ag_2O, $Ag(NH_3)_2NO_3$, $AgClO_4$, $AgC_2H_3O_2$, and $AgNO_3$ (Vijayaraghavan and Kamala Nalini, 2010). However, $AgNO_3$ is the predominant precursor due to its availability, high chemical stability, and low cost. AgNPs can be produced via an extracellular pathway, whereas the bioreduction of Ag^+ ions can occur by reductase enzymes and electron shuttle quinones (Durán et al., 2005) and/or an intracellular pathway, whereas the ions are bioreduced by electrons produced by the organisms to overcome toxicity in the presence of enzymes such as NADH-dependent reductases (Kumar et al., 2007). Yin et al. (2012) found that sunlight could also induce the synthesis of AgNPs with dissolved organic matter in environmental waters.

When bioreduction of AgNPs occurred, a change in color is usually observed in the $AgNO_3$ solution, where it turned from yellow into brown, which is an indicator of the SPR of AgNPs (Chan and Don, 2013b). The position of the plasmonic band detected on the solutions of metallic NPs is dependent on several parameters such as size, shape, and polydispersity of particles. The more the narrow is the band, the bigger is the uniformity index of AgNPs size distribution (Becaro et al., 2015). Various studies have established that the UV/Vis broad spectrum within the range of 410–450 nm is assigned for the SPR of nano-sized AgNPs within a range of 2–100 nm (Shanthi et al., 2016; Gupta et al., 2017; Omran et al., 2018b). However, the absorbance band between 380 and 420 nm is reported also to indicate the formation of spherical or roughly spherical Bio-AgNPs (Pal et al., 2007; Pereira et al., 2014).

3.2.1 ALGAL PREPARATION OF AgNPs

Several microalgal strains have been reported for the preparation of AgNPs. An extract of the unicellular green alga *Chlorella vulgaris* is reported to synthesize single crystalline Ag nanoplates of rod-like particles with a mean length of 44 nm and a width of 16–24 nm at room temperature (Xie et al., 2007). Proteins in the extract were involved in the biological synthesis, providing the dual function of Ag ion reduction and shape-controlled synthesis of AgNPs. Hydroxyl groups in tyrosine residues and carboxyl groups in aspartic acid and/or glutamic acid residues were identified as the most active functional groups for Ag ion reduction and for directing the anisotropic growth of Ag nanoplates. However, Brayner et al. (2007) reported the intracellular biosynthesis of Au, Ag, Pd, and Pt by *Anabaena*, *Calothrix*, and *Leptolyngbya* cyanobacteria and then naturally released in the culture medium, where they are stabilized by algal polysaccharides, allowing their easy recovery. Whereas, the intracellular nitrogenase enzyme is responsible for the metal reduction and the cellular environment is involved in the colloid growth process. The size of the recovered particles as well as the reaction yield is shown to depend on the cyanobacteria genus. Brayner et al. (2007) reported that the nitrogenase and hydrogenase which are class of reducing enzymes, present in cyanobacteria, can reduce Ag^+ ions to form Ag^0 NPs. Not only this but the concentration of the cellular nitrogenase also governs the size of AgNPs, whereas a higher concentration in heterocysts leads to rapid formation of larger shaped AgNPs near the cell wall. While the intermediate concentration forms small and un-aggregated NPs colloids.

Furthermore, Lengke et al. (2007) reported the intra- and extracellular biosynthesis of spherical-shaped AgNPs at 100 °C (<10 nm and 1–200 nm, respectively) by the cyanobacterium; *Plectonemaboryanum* UTEX 485 with 28 d. Whereas, UTEX 485 utilized nitrate as the major source for the bioreduction of silver ions into AgNPs, generation of metabolic energy, and redox balancing. The presence of AgNPs in the cells indicated entry of Ag^+

and NO_3^- into the cyanobacteria cells and their dissemination. Where, the nitrate reduced by cyanobacterial metabolic processes first to nitrite and then to ammonium (Scheme 3.1). The latter is then fixed as the amide group of glutamine.

$$AgNO_3 \longrightarrow Ag^+ + NO_3^-$$
$$NO_3^- + 2H^+ + 2e^- \longrightarrow NO_2^- + H_2O$$
$$NO_2^- + 8H^+ + 6e^- \longrightarrow NH_4^+ + 2H_2O$$

SCHEME 3.1 Intracellular bioreduction of silver ions.

Devina Merin et al. (2010) reported the biosynthesis of 53–72 nm AgNPs by the cell filtrate of marine microalgae; *C. calcitrans*, *C. salina*, *I. galbana*, and *T. gracilis*, within 72 h. Whereas the metabolites excreted by the culture bioreduced the silver ions via electron shuttle or reducing agents released into the solution under light condition and not in the darkness. The stability of the prepared AgNPs was measured over a period of 30 days and no change in the intensity of the Ag^0 characteristic SPR absorption peak at 420 nm was observed. The appearance of the absorption peak at 234 nm of the amide bond confirmed the involvement of secreted proteins in the bioreduction, capping, binding to the NPs, and enhancement of stability.

Mubarak Ali et al. (2011) reported that marine cyanobacterium, *Oscillatoria willei* ntdm01 intracellularly synthesizes AgNPs. Whereas, the Ag^+ ions reduced into neutral silver by the protein molecules present inside the cell, and then the protein-capped AgNPs released outside the dead cell. However, harvested *Spirulina platensis* biomass from an exponential growth phase was reported for bioreduction of $AgNO_3$ to AgNPs (Mahdieh et al., 2012). That was suggested to occur as follows; trapping of metal ions on the surface of algal cells via electrostatic interaction between the ions and negatively charged carboxylate groups present in the algal cell surface. Thereafter, the ions are enzymatically reduced, possibly by the cellular reductases released by *S. platensis* into the solution forming the nuclei, which subsequently grow through the further bioreduction of metal ions and accumulation of these nuclei. Thus, the secreted cofactor NADH plays an important role (Senapati et al., 2005). Furthermore, cyanobacteria localized reducing conditions may be produced by a bacterial electron transport chain via energy-generating reactions within the cells (Lengke and Southam, 2006).

Barwal et al. (2011) reported that the protein concentration is directly proportional to the rate of particle formation and inversely proportional to the size of the particles. Whereas, the protein-depleted fraction of *Chlamydomonas* cell extract treated with $AgNO_3$ solution resulted in the formation of large-sized AgNPs in comparison with a whole-cell extract. Jena et al. (2012) reported the intracellular and extracellular biosynthesis of AgNPs by microalga *Chlorococcum humicola*. While, Sudha et al. (2013) reported the bioreduction of $AgNO_3$ to polydispersed spherical-shaped 44–79 nm AgNPs by cyanobacterial *Microcoleus* sp. biomass and filtrate. Whereas the UV/Vis shoulder appeared at 370 nm was denoted to the transverse plasmon vibration in silver NPs, while the peak at 440 nm was attributed to the excitation of longitudinal plasmon vibrations of AgNPs. In later study, Jena et al. (2104) explored the intracellular and extracellular biogenic syntheses of AgNPs using the unicellular green microalga *Scenedesmus* sp. The intracellular NP biosynthesis was initiated by a high rate of Ag^+ ion accumulation in the microalgal biomass, biochemical reduction of Ag^+ ions, and subsequent formation of spherical crystalline 15–20 nm AgNPs. However, the extracellular synthesis using boiled extract showed the formation of well scattered, highly stable, spherical smaller 5–10 nm AgNPs. But, the intracellular synthesis of AgNPs using whole cells was faster than the extracellular AgNPs synthesis using raw and boiled algal extract. That was attributed to the presence of active biomolecules in the living cell. The mechanism of the intracellular biosynthesis of AgNPs was elucidated as follows; first, metal ions adsorption (metabolism independent) on the cell surface and absorption to organelles/cytoplasmic ligands (metabolism dependent). The algae cells were found to possess pores 3–5 nm wide, which permit the passage of low molecular weight substances such as water, inorganic ions, gases, and other small nutrient substances required for growth and metabolism. Moreover, the TEM analysis of the exposed biomass revealed that the synthesized AgNPs were 4–35 nm in size, which is quite a bit larger than pores size. Moreover, they were widely distributed throughout the cytoplasm and more NPs were localized toward the compact region, which constitutes different organelles. Consequently, the metal ions (Ag^+) might have been trapped on the surface of the algal cell, possibly via electrostatic interaction between the negatively charged functional groups present in the cell surface, followed by the reduction of metal ions by several enzymes, leading to the formation of nuclei in the cytoplasm. Simultaneously, growth and accumulation of these nuclei occur by the subsequent reduction of metal ions to metal particles. The synthesis and accumulation of a larger number of AgNPs toward the compact region of the cell was also explained as upon metal stress, the photosynthetic machinery (chloroplast) responds first to the metal by the overexpression of proteins: ATP synthase, RuBP carboxylase, and oxygen-evolving enhancer protein (Barwal et al., 2011). Thus, the enhanced activity of these enzymes may have led to the cell-mediated biosynthesis of AgNPs. The FTIR analysis confirmed the that amino acid residues of the protein moiety have a strong binding ability with metals, suggesting the formation of a layer surrounding the AgNPs and acting as a capping agent to prevent agglomeration, thereby providing their stability. This study confirmed the technical feasibility of applying microalgae for green synthesis of AgNPs owing to its economic viability and the possibility of scale-up at a low cost. In view of the high rate of silver accumulation and its ability to form a large amount of intracellular AgNPs and it may be also used as a potential source for phytomining of silver from industrial wastes in large-scale systems.

The cyanobacterium *S. platensis* and the two green algae (*C. vulgaris* and *Scendesmus* obliquus) were reported for their capability to biosynthesize AgNPs with average particles size of 20, 8.0, and 8.8 nm, respectively (El-Sheekh and El-Kassas, 2014a). Patel et al. (2015) performed a comparative study of bioreduction of Ag^+ ions into AgNPs using the biomass and the cell-free culture liquid of eight cyanobacterial and eight green algae strains, in the presence or absence of light. Moreover, the C-phycocyanin was isolated and purified from the cyanobacterial strain *Limnothrix* sp. 37-2-1 and EPS of *Scenedesmus* sp. 145-3 were tested also for biosynthesis of AgNPs. Only three cyanobacterial strains *Anabaena* sp. 66-2, *Limnothrix* sp. 37-2-1, and *Synechocystis* sp. 48-3 and one green algal strain; *Coelastrum* sp. 46-4 were able to synthesize Ag-NPs under both dark and light conditions. Furthermore, all cell free cultures produced AgNPs only under light conditions. The same occurred in case of EPS of *Scenedesmus* sp. 145-3 and the protein-based pigment C-phycocyanin from *Limnothrix* sp. 37-2-1. However, the latter was denatured by $AgNO_3$ within 12 h. From the SEM analysis, it was revealed that the cell walls of *Limnothrix* sp. 37-2-1 might have served as nucleation sites at which Ag^+ ions get deposited and transformed into AgNPs, where the AgNPs were present and evenly distributed throughout the biomass. The TEM analysis confirmed that the shape and size of the biosynthesized AgNPs differ according to the applied microorganisms (Table 3.1) and also differ with the type of the applied microalgae. The C-phycocyanin from *Limnothix* sp. 37-2-1 has been reported to produce a mixture of spherical and elongated 25.65 ± 2 AgNPs, while that of *Spirulina* sp. produced spherical 13.85 ± 2 AgNPs. The C-phycocyanin is a blue-colored photosynthetic accessory pigment consisting of two polypeptide chains that carry covalently attached

TABLE 3.1 Various Algal Strains Involved for Biosynthesis of AgNPs

Algae	**UV–Vis (λ_{nm})**	**Size (nm)**	**Shape**	**Activity**	**References**
Spirulina platensis	430	12	Spherical	–	Mahdieh et al. (2012)
Microcoleus sp	440	55	Spherical	Antibacterial activity	Sudha et al. (2013)
Scenedesmus sp.	420	5–10	Spherical	Antibacterial activity	Jena et al. (2014)
Limnothrix sp. 37-2-1	400–450	31.86 ± 1	Elongated	−ve antibacterial activity	Patel et al. (2015)
Anabaena sp. 66-2		24.13 ± 2	Irregular	+ve antibacterial activity	
Synechocystis sp. 48-3		14.64 ± 2	Irregular	+ve antibacterial activity	
Botryococcus braunii		15.67 ± 1	Spherical	+ve antibacterial activity	
Coelastrum sp. 143-1		19.28 ± 1	Spherical	+ve antibacterial activity	

linear tetrapyrrole–phycocyanobilin (Glazer, 1994) and can bind to heavy metals (Gelagutashvili, 2013). C-phycocyanin is part of phycobilisomes, structures attached to thylakoids involved in light harvesting and transferring electrons toward photosystem II reaction centers (Chen and Berns, 1979; MacColl, 1998). Bekasova et al. (2008) reported that the red pigment R-phycoerythrin reduced Ag+ ions into AgNPs without the need for a reductant. Thus, Patel et al. (2015) assumed that C-phycocyanin might have utilized the same mechanism, since they are similar in structure and function. Moreover, Sharma et al. (2015) reported the extracellular biosynthesis of well-scattered, highly stable, spherical AgNPs of average size 30–50 nm by aqueous extract of *S. platensis*.

Sedlakova-Kadukova et al. (2017) proved that the size distribution and mean diameter of the AgNPs, which are crucial for various applications can be controlled by the organism selection (Thakkar et al., 2011). Living cells of *Parachlorella kessleri*, *Dictyosphaerium chlorelloides*, and *Desmodesmus quadricauda* were successfully used for the preparation of AgNPs. Where, the UV/Vis spectroscopy and transmission electron microscope (TEM) analyses revealed that the slowest NPs formation with narrow NPs size and shape distribution occurred with *P. kessleri*, recording spherical AgNPs of 7.6 nm. While, the average size of the produced NPs by *D. chlorelloides* and *D. quadricauda* was 23.4 and 24 nm with a broad size distribution of 10–43 and 7–50 nm, respectively (Sedlakova-Kadukova et al., 2017).

3.2.2 FUNGAL PREPARATION OF AgNPs

Although the rates of bioreduction by using cell extract or purified enzyme are faster, these two methods need coenzymes (such as NADH, NADPH, FAD, etc.) for the continuation of the reaction. As they are expensive, the use of whole cells is preferred because the coenzymes will be recycled during the pathways in intact living cells (Korbekandi et al., 2013). Serval fungal strains have been reported for biosynthesis of AgNPs (Table 3.2). Whereas, the mycosynthesis of AgNPs can be performed throughout the following steps; first, fungi capture metal ions by electrostatic interaction of cellular surface charged negatively with silver ions charged positively, and then ions bind by sticky polysaccharide compounds secreted by the cell. Further, the reduction of ions into their atomic metal form occurs by cellular enzymes. Consequently, the metal clusters are formed, adsorbing new ions, and then these metal clusters grow, forming NPs. The proteins form a protective cover on the surface of the particles, preventing the repeated solution of metal and

TABLE 3.2 Various Fungal Strains Involved for Biosynthesis of AgNPs

Fungi	UV–Vis (λ_{nm})	Size (nm)	Activity	References
Fusarium oxysporum	413	5–15	–	Ahmad et al. (2003)
Aspergillus fumigatus	420	5–25	–	Bhainsa and D'Souza (2006)
Aspergillus flavus	420	8.92 ± 1.61	–	Vigneshwaran et al. (2007)
T. asperellum	410	13–18	–	Mukherjee et al. (2008)
Aspergillus flavus	425	7	–	Moharrer et al. (2012)
Cryphonectria sp.	440	30–70	Antimicrobial activity	Dar et al. (2013)
Fusarium oxysporum	430	25–50	–	Korbekandi et al. (2013)
Humicola sp.	415	5–25	–	Syed et al. (2013)
Aspergillus oryzae MUM 97.19	400–550	14–76	Antifungal activity	Pereira et al. (2014)
Aspergillus terreus MALEX	420	15–29	Antimicrobial activity	Abdel-Hadi et al. (2014)
Penicillium chrysogenum MUM 03.22	430	6–100	Antifungal activity	Pereira et al. (2014)
Penicillium sp. J3	425	60	–	Lima et al. (2014)
A. niger PFR6	430	8.7 ± 6	–	Devi and Joshi (2015)
Aspergillus tamarii PFL2	419	3.5 ± 3	–	Devi and Joshi (2015)
F. oxysporum f. sp. lycopersici	420	5–13	Antibacterial and cytotoxicity activities	Husseiny et al. (2015)
Penicillium ochrochloron PFR8	430	7.7 ± 4.3	–	Devi and Joshi (2015)
Arthroderma fulvum	420	15.5±2.5	Antifungal activity	Xue et al. (2016)
Cladosporium sphaerospermum F16 (KU199685)	435	15.1 ± 1		Abdel-Hafez et al. (2016)
T. viride	620	1–50	Antibacterial	Elgorban et al. (2016)

TABLE 3.2 *(Continued)*

Fungi	UV–Vis (λ_{nm})	Size (nm)	Activity	References
Alternaria sp.	426	10–30	Antibacterial activity	Singh et al. (2017)
Penicillium aculeatum Su1	422	4–55	Antimicrobial activity and cytotoxicity activity toward lung adenocarcinoma cells	Liang et al. (2017)
Rhizopus stolonifer	420	2.86 ± 0.3	–	Abdel Rahim et al. (2017)
Trichoderma harzianum	420		–	Elamawi et al. (2018)
T. viride	420		–	
T. longibrachiatum	385	10	Antifungal	
Cladosporium cladosporioides	440	30–60	Antimicrobial activity	Manjunath and Joshi (2019)

NPs aggregation. Thus, stability of mycoprepared NPs for several months occurred (Tyupa et al., 2016).

Basavaraja et al. (2008) reported that the exposure of fungal cells to Au and Ag ions resulted in the release of nitrate reductase which is an enzyme in the nitrogen cycle and responsible for the conversion of nitrate to nitrite. Nitrate reductase is essential for ferric iron reduction and the subsequent formation of highly stable AuNPs and AgNPs in solution (Figure 3.1).

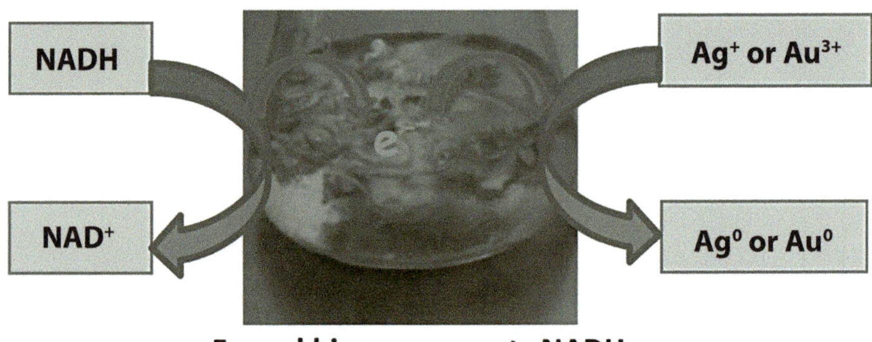

FIGURE 3.1 Extracellular enzymatic reduction of metal ions by fungi.

Maliszewska et al. (2009) reported that the absorption peak at 210 nm is assigned to the strong absorption of peptide bonds in the fungal cell filtrate, while the absorption at 280 nm indicates the presence of tryptophan, tyrosine, and/or phenylalanine residues in the protein and indicates the release of proteins into the filtrate.

Information on the location of the AgNPs relative to the fungal cells is very important in understanding the mechanism of their formation. The deposition of AgNPs on the surface of fungal mycelium incubated with $AgNO_3$ solution suggests that the AgNPs are formed on the surface of the mycelium not in solution. Where, the first step involves trapping of the Ag ions on the surface of the fungal cells via electrostatic interaction between the silver ion and negatively charged carboxylate groups in protein present on the cell wall of the mycelia, the reduction of the Ag^+ ions may occur due to reductases released by the fungus into the solution. The NADH and NADH-dependent nitrate reductase enzyme are important factors in the biosynthesis of metal NPs. Where, the reduction may occur by means of the electron from NADH where the NADH-dependent reductase can act as a carrier. The appearance of the organic matrix (probably protein) in the TEM and/or SEM micrographs present among the NPs and adhered to their surface elucidates the stability of the NPs in the solution may be due to the stabilization of proteins containing cysteine. Usually, the dynamic light scattering (DLS) measurements, reveals higher particle size than the TEM analysis, since the particle size obtained is augmented substantially by contributions from the hydrated capping agents (probably protein) and also from solvation effects (the hydrodynamic diameter could be as high as 1.3 times the original diameter of the capped particles) (Mukherjee et al., 2008).

A novel biological method for the synthesis of AgNPs using the fungus intracellular reduction of Ag^{+} ions and formation of 25 ± 12 nm AgNPs by *Verticillium* was reported (Mukherjee et al., 2001a). The electron microscopy analysis of thin sections of the fungal cells indicated that the AgNPs were formed below the cell wall surface, possibly due to reduction of the metal ions by enzymes present in the cell wall membrane. Not only this but also the metal ions were not toxic to the fungal cells and the cells continued to multiply after biosynthesis of the AgNPs. Kowshik et al. (2002) reported the extracellular synthesis of 2–5 nm AgNPs by silver tolerant yeast strain MKY3 within its logarithmic growth phase. The extracellular synthesis of pyramidal-shaped 50–200 nm AgNPs by a white rot fungus, *Phaenerochaete chrysosporium* was reported (Vigneshwaran et al., 2006). However, extracellular and intracellular synthesis of AgNPs was demonstrated using growing and whole cell systems of *Fusarium oxysporum* (Nair, 2009). Where, the metal ions were reduced by the nitrate-dependent reductase that is conjugated with electron donor (quinone shuttle) (Scheme 3.2; Durán et al., 2005).

It was also reported that proteins can bind to NPs either through free amine groups or cysteine residues in the proteins (Vigneshwaran et al., 2007). However, Pereira et al. (2014) reported that the presence of absorption band at 270 nm in bio-AgNPs UV/Vis spectra is attributed to the electronic excitation of both aromatic tyrosine and tryptophan residues in proteins (Lakowicz, 2006) confirming the presence of extracellular proteins in the filtrate solution and the possible involvement of the amino acids in the reduction and stability of metallic ions (Mandal et al., 2006). Moreover, the presence of the signature peaks of amino acids in the FTIR spectra of the bio-AgNPs supports the presence of proteins in fungal cell-free filtrate and its involvement in the bioreduction of Ag ions, capping and stabilization of the AgNPs. Nevertheless, Devi and Joshi (2015) mentioned that the UV/Vis absorption peaks at around λ_{280nm} is assigned to the strong absorption of peptide bonds in fungal filtrate and indicates the presence of aromatic acid such as tryptophan and tyrosine residues in the protein. This suggests possible mechanism for the reduction of silver ions present in the solution via the reductases released by the fungus into the solution. In addition, Basavaraja et al. (2008) reported that the proteins could most possibly play a role in forming a coat covering the AgNPs, extracellularly biosynthesized by *Fusarium semitectum*, that is, capping it to prevent its agglomeration and consequently stabilizing it in the medium.

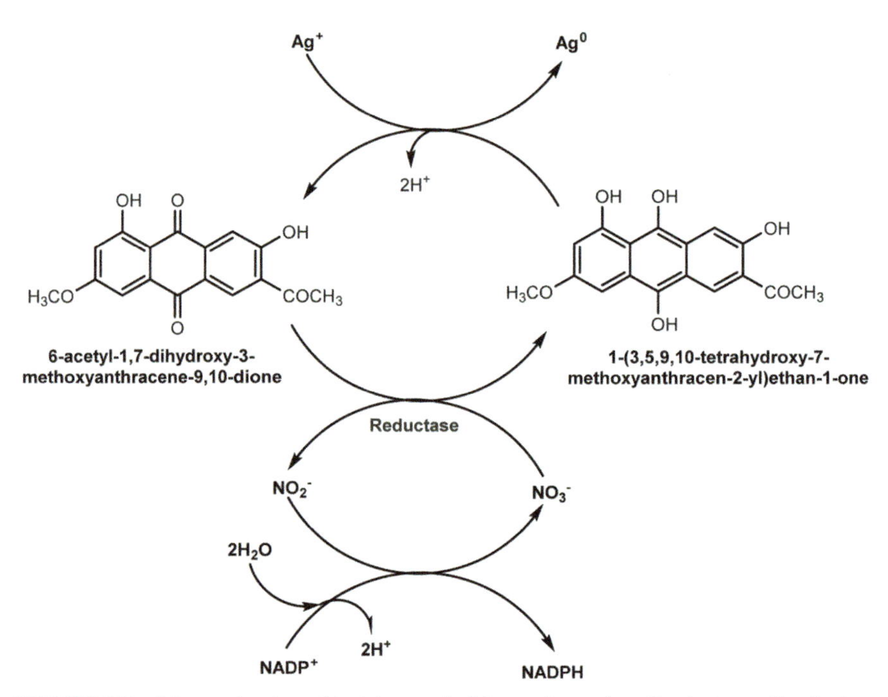

SCHEME 3.2 Mycoreduction of Ag^+-ions to Ag^0 due to the conjugation between the electron shuttle and the nitrate-dependent reductase.

Sanghi and Verma (2009) reported that under alkaline conditions, the intracellular and extracellular biosynthesis of spherical-shaped 25–75 nm AgNPs by the white rot fungi, *Coriolus versicolor* (MUCL) was much faster (1 h) and could easily proceed at room temperature even without stirring. Whereby, other than the fungal proteins, glucose and the surface S–H groups of the fungus played a major role in the bioreduction of Ag-ions and stabilization of the formed AgNPs. One factor at a time (OFAT) technique was applied to optimize the fungal synthesis of AgNPs using *Pycnoporus sanguineus* as the source of reducing agent and the anion sodium dedocyl sulfate was used as the stabilizer. Whereas, three parameters were found to significantly influence the size of AgNPs; the $AgNO_3$ concentration, the incubation temperature, and the agitation speed. Response surface methodology (RSM) based on Box-Behnken design was applied to optimize these three factors. The statistical analysis showed that the optimum operating conditions were 0.001 M of $AgNO_3$, 38 °C, and 200 rpm that yielded the smallest size of AgNPs recording approximately 14.86 nm (Chan and Don, 2013a). Abdel-Hafez et al. (2016) suggested that a variety of biomolecules are involved in biological NP synthesis, such biomolecules are likely to be inactivated like polysaccharides and proteins under the extremely acidic conditions (pH 3.0) and start to work effectively in neutral and slightly alkaline conditions.

AgNPs have been also produced from silver ions placed into contact with the cell filtrate of different *Aspergillus* strains; *A. fumigatus* (Bhainsa and D'Souza, 2006), *A. flavus* (Vigneshwaran et al., 2006), and *A. niger* (Gade et al., 2008). In vitro, silver NPs were formed by *F. oxysporum* by reducing the silver ions in the presence of nitrate reductase (NADPH-dependent) and stabilizing the AgNPs by the capping peptide; phytochelatin (Anil Kumar et al., 2007). *F. oxysporum* PTCC 5115 synthesized silver NPs due to the reduction of silver nitrate by nitrate reductase enzyme (Karbasian et al., 2008). *F. acuminatum* Ell. and Ev. (USM-3793) isolated from infected ginger (*Zingiber officinale*) have been reported for extracellular synthesis of polydispersed, spherical-shaped 13 nm AgNPs, via the reduction of silver nitrate by NADH-dependent nitrate reductase (Ingle et al., 2008). *Fusarium semitectum* has been also reported for the extracellular synthesis of the spherical-shaped AgNPs (Basavaraja et al., 2008). The pathogenic *Fusarium solani* (USM-3799) that causes a disease in onion is reported to synthesize extracellular and polydispersed, spherical-shaped 5–35 nm AgNPs (Ingle et al., 2009). Extracellular biosynthesis of AgNPs has been also reported using different *Trichoderma* strains; *T. asperellum* (Mukherjee et al., 2008), *T. asperellum* (Mukherjee et al., 2008), and *T. viride* (Fayaz et al., 2009; Amanulla et al., 2010). Extracellular reduction of silver nitrate into silver NPs has also been investigated using different *Penicillium* strains; *P. brevicompactum* WA 2315 (Shaligram et al., 2009) and *P. fellutanum* (Kathiresan et al., 2009). Whereas, *P. fellutanum* was able to successfully reduce silver ions into silver NPs under dark conditions, which added to the advantageous of biological synthesis of NPs as it lowers the consumption of energy. The extracellular synthesis of AgNPs has been also reported by *Alternaria alternate* (Gajbhiye et al., 2009), *Verticillium* sp. (Priyabrata et al., 2001), *Aspergillus fumigatus* (Bhainsa and D' Souza, 2006), *Cladosporium cladosporioides* (Balaji et al., 2009), mycelium of fungus *Phoma* sp.3.2883

(Chen et al., 2003), and yeast strain MKY3 (Kowshik et al., 2003). However, Korbekand et al. (2013) reported that the intracellular biosynthesis of AgNPs by *Fusarium oxysporum*. On the other hand, the extracellular cell-free filtrate of *Penicillium chrysogenum* MUM 03.22 and *Aspergillus oryzae* MUM 97.19 were reported for the mycosynthesis of AgNPs (Pereira et al., 2014). Ortega et al. (2015) reported the synthesis of AgNPs by the culture supernatant of the yeast *Cryptococcus laurentii* (BNM 0525). Moreover, Gudikandula and Maringanti (2016) reported the mycosynthesis of AgNPs by the extracellular cell-free filtrate of *Pycnoporus* sp (HE792771) and biosynthesized AgNPs expressed better antimicrobial activity against the pathogenic bacteria comparatively with the chemical synthesized AgNPs. The biosynthesis of spherical shaped 2–5 nm AgNPs by baker's yeast (*Saccharomyces cerevisiae*) has been also reported (Prasad et al., 2010). Whereas, the bioreduction occurred due to the high carbon source-dependent partial pressure of gaseous hydrogen (i.e. high reduction potential r-H_2) and the pH-sensitive membrane bound oxidoreductases, under high pH values (Figure 3.2).

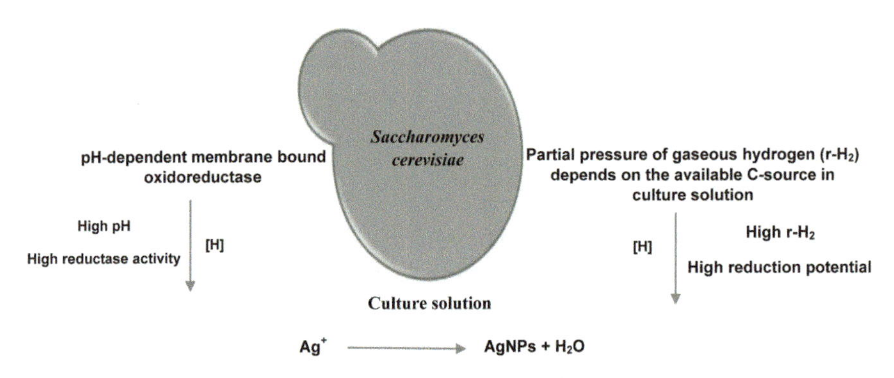

FIGURE 3.2 Possible mechanism for the biosynthesis of AgNPs.

Apte et al. (2013) reported that the 3,4-dihydroxy-L-phenylalanine (L-DOPA) and chloroauric acid facilitated the extracellular bioreduction of Ag^+ and Au^{3+} ions to their corresponding metal NPs by yeast *Yarrowia lipolytica* NCIM 3590. Whereas, the melanin present in the yeast isolate acted as an electron exchanger either in oxidizing or reducing metals and the conversion of the hydroxyl groups to the quinone groups produces reducing agents for the conversion of metal ions into elemental nanostructures (Scheme 3.3). The optimum conditions for maximum productivity of Ag and Au NPs were 100 °C and alkaline pH 10–12. That was attributed to the heat stability of melanin. Moreover, its solubility is high under alkaline conditions and these favor the optimal synthesis of nanostructures.

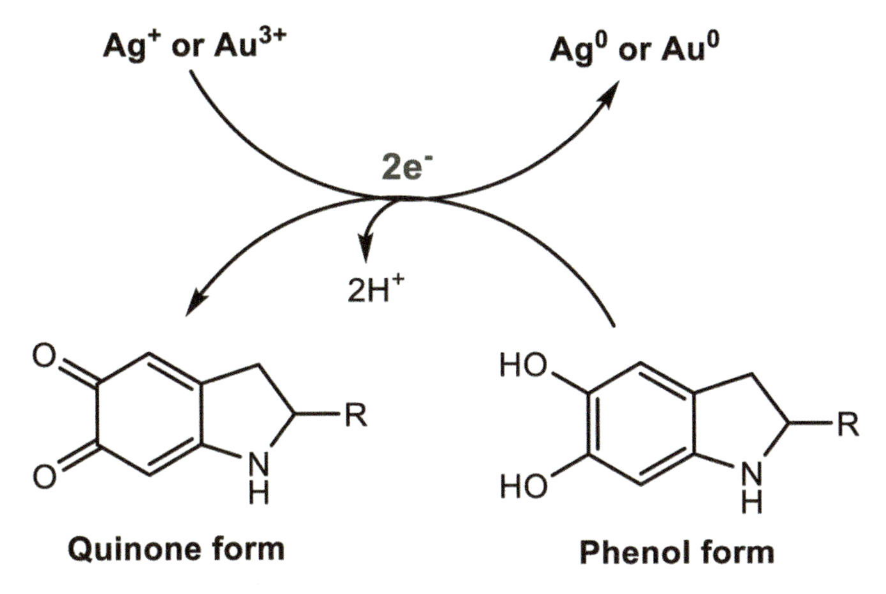

SCHEME 3.3 The formation of silver and gold NPs by L-DOPA-induced melanin derived from yeasts.

The extracellular biosynthesized AgNPs by *Humicola* sp. were reported to be stabilized by the secreted protein in the reaction mixture which appeared at around 270 nm in the UV spectrum (Syed et al., 2013). *Fusarium oxysporum* DSM 841 has been reported for intracellular mycosynthesis of AgNPs, where the presence of $AgNO_3$ (0.1 mM) in the culture as enzyme inducer and glucose (560 mM) as electron donor had positive effects on NP production (Korbekandi et al., 2013). The SEM micrographs revealed spherical AgNPs attached to the surface of biomass, single (25–50 nm) or in aggregates (100 nm). The appearance of the dark brown color in the culture is an indication for colloidal suspension (hydrosol) of AgNPs, which is due to the excitation of surface longitudinal plasmon resonance (vibration) in the AgNPs. Korbekandi et al. (2013) proved that the $AgNO_3$ as the inducer activates the transcription of genes involved in the synthesis of reducing enzymes (i.e., gene expression) and can increase the productivity of AgNPs, which is important for scaling up this process. Moreover, the presence of glucose as electron donor provides electrons for reduction of cofactors, and makes the reduction of silver ions (Ag^+) to silver metal (Ag^0) faster. In the absence of electron donor, the electron reservoirs in reaction mixture for the recovery and revival of cofactors, after a short time are finished and the reaction declines. Korbekandi et al. (2013) mentioned also the importance of knowing the exact concentration of $AgNO_3$ to avoid its toxic effect on the cell biomass and maximize the AgNPs yield. The TEM micrographs revealed that the AgNPs were produced inside the cytoplasm, aggregated in vesicles, which were then secreted through cell membrane by exocytosis.

A comparative study for the extracellular biosynthesis of AgNPs via four fungal strains; *Rhizopus nigricans*, *Fusarium semitectum*, *Colletotrichum gloeosporioides*, and *Aspergillus nidulans* has been conducted. Where, 13–74 nm AgNPs were prepared within 60–300 s with the appearance of the characteristic peak of AgNPs $\lambda_{420\text{–}430\,nm}$ (Ravindra and Rajasab, 2014). Another comparative study has been performed for the biosynthesis of AgNPs by the filtrate of three endophytic fungi *Aspergillus tamarii* PFL2, *Aspergillus niger* PFR6, and *Penicllium ochrochloron* PFR8 isolated from an ethno-medicinal plant *Potentilla fulgens* L. That proved the effect of microbial strain on the size of the prepared AgNPs. Where, the electron microscopy study revealed the formation of spherical-shaped AgNPs with different sizes; 3.5 ± 3 nm, 8.7 ± 6 nm, and 7.7 ± 4.3 nm, respectively (Devi and Joshi, 2015). Tyupa et al. (2016) performed a comparative study for the mycosynthesis of AgNPs using three fungal strains; *Penicillium glabrum*, *Fusarium nivale*, and Fusarium *oxysporum*. Whereas, their isolation from an aerobic activated sludge of wastewater treatment facilitated their enhanced tolerance to silver ion toxicity. The mycosynthesis was examined in cell filtrate and in cultural liquid with biomass, under both light and dark conditions. Using the culture liquid with biomass was not so promising, since the presence of the anions of Cl^-, SO_4^{2-}, and HPO_4^{2-} in the cultural liquid resulted in binding of Ag^+ cations; thus, complicating their enzymatic reduction into Ag^0, consequently, decreasing the NPs yield. The negative effect of the growth medium components can be decreased by the separation of biomass from the cultural liquid and its transfer into the distilled water containing silver ions. However, the metal silver formation proceeded actively but only on the surface of fungal granules and biosynthesis of AgNPs in the solution was impossible because of lack of enzymes, electron donor molecules, proteins stabilizing particles, and other biomolecules, which are typical for cultural medium and necessary for reduction of Ag^+ ions with the formation of NPs. Thus, it was concluded that to successfully perform mycosynthesis of AgNPs in the microbial growth medium it is necessary to optimize the medium composition by eliminating adverse ions such as Cl^-, SO_4^{2-}, and HPO_4^{2-} or decreasing their contents to the level required. Whereas, a nine-fold increase in the AgNPs yield was achieved using *F. oxysporum* in the medium containing; KNO_3 1 g/l and starch 4 g/L of starch. It was noticed that at the low Ag^+ ions concentration (<70 mg/L for *F. nivale* and *F. oxysporum*, and <50 mg/L for *Penicillium glabrum*), the mycosynthesis was slow, and at their high concentration (>200 mg/L) the yield of NPs also decreased as a result of the inhibiting effect of the metal ions on the cells and aggregations of particles. That was explained as follows; one of the main factors preventing aggregation of NPs is their protective covering with a biopolymer layer. The surface of metal NP carried a slight positive charge insufficient for emergence of forces of electrostatic repulsion between the particles. Proteins secreted by the cells had a negative potential. When proteins are adsorbed on the particles, the surface of the latter received a strong negative charge that increased their resistance to aggregation. Thus, at high concentration of Ag^+ ions the resources of the proteins stabilizing particles are exhausted, changing the kinetics of forming NPs, where proteins cannot be adsorbed on a surface of fast-growing particles, and the latter aggregated and precipitated. Tyupa et al. (2016) reported the strong effect of changes in pH on the mycosynthesis of AgNPs and attributed this to the changes of the superficial charges of particles, cells, and biopolymers at varying the pH of the medium, as well as because of changes in the properties of enzymes and other protein molecules near to their isoelectric point either in strongly acid or alkaline solutions. Also, it affects the ζ-potential of NPs and the protein stabilizers covering them, changing force of adsorption of the latter to surfaces of the NPs. In the alkaline medium, the ζ-potential of particles decreased and changed its sign from positive to negative; the forces of electrostatic repulsion between NPs and negatively charged proteins emerged, the particles lost their stabilizing biopolymer cover. On the contrary, at low pH values protein molecules are actively adsorbed on the surface of positively charged particles, preventing them from aggregation and sedimentation on the surface of the cells. The slight acidic conditions (pH 5) favored the mycosynthesis of AgNPs by *F. nivale* and *F. oxysporum*. The mixing conditions (i.e., the shaking) is also important for NP bioformation as diffusion processes are intensified at high values of the rate of shaking, accelerating a number of chemical reactions; and also aeration could significantly increase the rate of metabolism for aerobic microorganisms. The optimum shaking speed was 200 rpm. The intensification of mixing the medium decreased the degree of silver particle adsorption on the cellular surface, promoting the NPs to escape into the culture liquid, and this effect was enhanced in the following series of the cultures: *P. glabrum* < *F. nivale* < *F. oxysporum*. Whereas, the increment of shaking speed from 50 up to 250 rpm, the AgNPs yield increased nearly one and a half times in case of *F. oxysporum*, and that effect was less manifested for *F. nivale*, and it was practically absent in case of *P. glabrum*. Not only had this but the stage of growth of microorganisms also played a decisive role in the formation of AgNPs. Where, culturing *F. nivale* and *F. oxysporum* up to 80–96 h and the cultivation time of *P. glabrum* up to 160–168 h provided the maximum yield of NPs. Furthermore, Tyupa et al. (2016) reported that in the absence of light the rate of NP bioformation decreased, and it increased with lighting intensity growth, reaching a maximum at 750–800 lx. The results of TEM showed that *F. nivale*, *F. oxysporum*, and *P. glabrum* formed mainly NPs of more than 100 nm, 30–50 nm, and 20–30 nm in size, respectively, present both mainly in the cultural liquid and on the surface of the fungal cells. Finally, a continuous process for mycosynthesis of AgNPs by *F. oxysporum* has been performed and to maximize the amount of NPs synthesized for the minimum time, the optimum conditions were found to be; the flow rate being 200 mL/h, and the concentration of Ag^+ ions being 20 mg/L. That yielded a content of AgNPs in the reactor of nearly 16 mg/L. Thus, this is very promising as higher yield with shorter time and lower $AgNO_3$ concentration was obtained in the continuous process than in the batch one (80% and 65% yield, respectively).

The cell filtrate of *Trichoderma viride* was reported to mycosynthesize AgNPs within 24 h in darkness and at 25 °C (Elgorban et al., 2016). Whereas, the SEM and TEM revealed a spherical shaped, stabilized, and polydispersed globular AgNPs with a wide range of particles size ranged between 1 and 50 nm. The AgNPs were not in direct contact even in the aggregations and were enclosed by thin layer of organic material, which acted as a capping agent and provided the stability for the AgNPs. The carbon and oxygen peaks in the energy dispersive spectroscopy spectrum were attributed to the X-ray emissions from the existing proteins in the fungus and the Ag peak appeared at 3 KeV. Xue et al. (2016) reported also the mycosynthesis of spherical-shaped AgNPs by cell filtrate of *Arthroderma fulvum* HT77 (accession number AB193716.1) at optimum operating conditions of 1.5 mM, alkaline pH, reaction temperature of 55 °C, and reaction time of 10 h. Ottoni et al. (2017) have performed a comparative study on 20 different fungal strains isolated from sugarcane plantation soil for extracellular biosynthesis of AgNPs using their cell free filtrate. Where, four fungal strains *Rhizopus arrhizus* IPT1011, *Rhizopus arrhizus* IPT1013, *Trichoderma gamsii* IPT853, and *Aspergillus niger* IPT856 expressed the highest efficiency for biosynthesis of AgNPs, expressing the characteristic peak of spherical-shaped AgNPs at 418 nm, 420 nm, 426 nm, and 430 nm, respectively, with average size of 50–70 nm. The appearance of a peak between 279 and 285 nm, with nitrate reductase activity of the isolates supported the hypothesis of enzymatic reduction of $AgNO_3$ into AgNPs. Abdel Rahim et al. (2017) reported that the mycelia extract of *Rhizopus stolonifer* that was isolated from naturally infected tomato fruits bioreduced $AgNO_3$ to monodispersed spherical-shaped AgNPs. Abdel Rahim et al. (2017) reported also that the presence of Plasmon band at 420 nm due to dipole plasmon resonance shows that the AgNPs have a spherical shape (Sathishkumar et al., 2009; Kannan et al., 2011). From the XRD analysis, the full-width at half-maximum (FWHM) provides useful tool for determination size of NPs and their distribution in the medium (Mock et al., 2002). Thus, Abdel Rahim et al. (2017) concluded that a FWHM of 79.42 nm is mostly an indication of a small size distribution. The Fourier transform infrared (FTIR) analysis confirmed that the carbonyl group resulted from amino acid residue and peptide protein strongly bounded to metal, and the protein acts as capping for the biosynthesized AgNPs preventing its agglomeration and stabilized the AgNPs particles within the medium. Abdel Rahim et al. (2017) also mentioned the importance of temperature effect on the fungal biosynthesis of AgNPs. Where, 10 °C and 80 °C no AgNPs were prepared due to the denaturation or inactivation of enzymes and active molecules which are involved in biogenesis of AgNPs. However, small monodispersed AgNPs with average size of 2.86 ± 0.3 nm were produced at 40 °C. Large AgNPs with average particle size of 25.89 ± 3.8 and 48.43 ± 5.2 nm were produced at 20 °C and 60 °C, respectively. That was attributed to the low activity of the enzymes involved in AgNPs biogenesis at unsuitable temperature (Birla et al., 2013; Sherif et al., 2015). At an increased temperature, the kinetic energy of the AgNPs in the solution also increases and collision frequency between the particles also rises, resulting in a higher rate of agglomeration (Sarkar et al., 2007). However, Abdel-Hafez et al. (2016) reported the extracellular biogenic synthesis of AgNPs by filtrate of *Cladosporium sphaerospermum* F16 (KU199685) pronounced dark reddish brown color and more intense absorbance peaks of AgNPs at 50–70 °C. Where, a sharp narrow UV spectra peak at lower wavelength region (412 nm at 70 °C) was developed, which indicated the formation of smaller NPs. Sintubin et al. (2012) and Othman et al. (2017) stated that at higher temperatures silver ion reduction was favorable and proceed at higher rate. But, at lower reaction temperature, the peaks observed at higher wavelength regions (440 nm at 30 °C) with an increase in AgNPs size. That was explained as, at higher temperature, the reactants are consumed rapidly leading to the formation of smaller NPs. The effect of $AgNO_3$ concentration was also studied by Abdel Rahim et al. (2017), and proved that it controlled the size of the biosynthesized AgNPs. Where, the smallest AgNPs size (2.86 ± 0.3 nm) was obtained at 10^{-2} M

AgNPs and 40 °C. While, large AgNPs with average particle size 54.67 ± 4.1 and 14.23 ± 1.3 nm were prepared at concentration 10^{-1} and 10^{-3} M of $AgNO_3$, respectively.

The culture filtrate of the endophytic *C. cladosporioides* (GenBank accession no. KT384175) that was isolated from brown algae, *Sargassum wighti* proved for the capability for the mycosynthesis of spherical-shaped AgNPs within 1 h at room temperature (Manjunath and Joshi, 2019). The FTIR analysis proved the involvement of the proteins, enzymes, and polyphenols of *C. cladosporioides* extract in capping as well as bioreduction of AgNPs. Taking into consideration that the NADPH alone did not reduce Ag^+ ions to Ag^0, while the dialyzed cell filtrate mixed with $AgNO_3$ in presence of NADPH produced the AgNPs. Thus, confirmed the possible role of NADPH-dependent reductase in bioreduction (Manjunath and Joshi, 2019). Singh et al. (2017) reported also the mycosynthesis of spherical-shaped 10–30 nm AgNPs within only 20 min, using the supernatant of endophytic fungus *Alternaria* sp. isolated from the healthy leaves of *Raphanus sativus*.

RSM was applied to optimize the mycosynthesis of AgNPs. Whereas, round to oval 4–16 nm AgNPs were prepared using the culture filtrate of *Trichoderma viride* ATCC36838 at predicted optimum conditions of; 2 mM silver nitrate and 28% (v/v) of culture filtrate at pH 7.0, 70 °C within 34 h (Othman et al., 2017). Liang et al. (2017) reported the biosynthesis of AgNPs using the cell free filtrate of the fungus strain *Penicillium aculeatum* Su1 (GenBank accession number: KJ554994) which was isolated from heavy metal-contaminated soil. The stability of the prepared AgNPs was attributed to the protein coating the surface of the AgNPs, which prevented the agglomeration and increment in particle size caused by particle collisions, which maintained the high stability of the colloidal AgNPs. The decrease in the amount of biosynthesized AgNPs with longer incubation period (>72 h) was attributed to the release of more bioactive substances, whereas high concentrations of bioactive substances in the reaction mixture limited the production of AgNPs. The sterilized fungal biomass at (121 °C, 0.1 MPa) did not produce AgNPs due to the inactivation of bioactive substances (especially proteins). Agglomeration of AgNPs occurred at high temperature (>37 °C) due to the too rapid biosynthesis. Moreover, excessive concentrations of silver ions (>2 mM) accelerated the aggregation of NPs and generated larger AgNPs, which consequently led to a red-shift effect of the absorbance peak (i.e., shifted to longer wavelength). The binding vibrations of the amide I and amide II bands of proteins and the –C–N stretching vibrations, belonging to aromatic and aliphatic amines, were observed in the FTIR-spectrum of the prepared AgNPs. That confirmed the involvement of proteins in the reduction of silver ions and its binding to AgNPs to prevent their agglomeration. Liang et al. (2017) suggested that proteins can bind with NPs via free amine groups, cysteine residues or the electrostatic attraction of the negatively charged carboxylate groups in cell free filtrate from fungal mycelia. The enzymes of Su1 possessed higher catalytic activity at alkaline pH8. Elamawi et al. (2018) reported the mycosynthesis of monodispersed spherical shaped 10 nm AgNPs by the filtrate of *Trichoderma longibrachiatum*, under optimum conditions of 10 g fungal biomass, 28 °C, 72 h and without shaking. Moreover, there was no recorded significant difference between nonfiltered and filtered samples. Omran et al. (2018a) applied the OFAT to study and optimize the effect of different physicochemical parameters (the reaction time, pH, temperature, different stirring rates, illumination, and finally, the different concentrations of silver nitrate and fungal biomass) to maximize the mycosynthesis of AgNPs by the mycelial cell free filtrate (MCFF) of *Aspergillus brasiliensis* ATCC 16404. The preliminary visual observation of characteristic brown color indicated the bioreduction of Ag^+ ions to Ag^0. The UV/Vis spectrophotometric technique displayed the characteristic sharp peak of small-sized AgNPs at λ_{440nm}. Moreover, the UV/Vis examination of the MCFF revealed the characteristic peaks of proteins 220 and 280 nm. The DLS analysis revealed good dispersion and average particle size 35.8 nm. The narrow single peak of polydispersed index (PDI) value of 0.366 with an intercept of 0.879, indicated the good quality of the mycosynthesized AgNPs. Moreover, the PDI value was <0.5, indicated the monodispersity of the mycosynthesized AgNPs. The negatively charged zeta potential of the dispersed AgNPs of −16.7 mV signified the presence of repulsion between the AgNPs and increased the formulation stability. Roy et al. (2013) reported that a large negative or positive values of NPs-zeta potential, increases the repulsion between the prepared NPs and lowers the tendency of their assembling and increases the long-term stability in the solution (Kotakadi et al., 2016). However, the vice versa occurs with low values of zeta potential, as there is no force preventing the NPs from aggregation, assembling, or flocculation. The energy dispersive X-ray spectroscopy displayed a maximum elemental distribution of silver elements. The X-ray diffraction spectroscopy demonstrated the crystallinity of the mycosynthesized AgNPs. The field emission scanning electron microscope and high-resolution transmission electron microscope revealed monodispersed spherical shaped AgNPs with average particle size of 6–21 nm. The FTIR analysis showed the major peaks of proteins providing the possible role of MCFF in the synthesis and stabilization of the AgNPs. The mycosynthesized AgNPs expressed good biocidal activity against different pathogenic microorganisms causing some water-related diseases and health problems to local residents (Omran et al., 2018a).

Different mushroom strains have been reported for the extracellular preparation of AgNPs, such as *Phaenerochaete chrysosporium* (Vigneshwaran et al., 2006) *Pleurotus sajor caju* (Nithya and Ragunathan, 2009), and *Coriolus versicolor* (Rashmi and Preeti, 2009).

3.2.3 BACTERIAL PREPARATION OF AgNPs

As a biomanufacturing unit, bacteria have the added advantage of ease of handling compared to fungi. Silver-resistant genes, c-type cytochromes, peptides, cellular enzymes like nitrate reductase, and reducing cofactors play significant roles in bacteriagenic synthesis of AgNPs. Organic materials released by bacteria act as natural capping and stabilizing agents for AgNPs, thereby preventing their aggregation and providing stability for a longer time. Moreover, the controlling of reaction conditions has been suggested to control the morphology, dispersion, and yield of NPs.

Serval bacterial strains have been reported for the biosynthesis of AgNPs (Table 3.3). *Pseudomonas stutzeri* AG259 isolated from a silver mine reduced $AgNO_3$, yielding AgNPs of 200 nm (Klaus et al., 1999; Joerger et al., 2000). That was the first report on bacteriagenic synthesis of AgNPs. AG259 can survive in an extreme silver-rich environment and accumulate AgNPs inside the cells.

Binoj and Pradeep (2002) reported the biosynthesis of 10–25 nm AgNPs by *Lactobacillus* sp. *Corynebacterium* strain SH09 bioreduced diamine silver complexes, resulting in AgNPs ranging in size between 10 and 15 nm (Haoran et al., 2005). The biomass of *Aeromonas* SH10 synthesized AgNPs via the reduction of $[Ag(NH_3)_2]^+$ to Ag° in solution (Mouxing et al., 2006). Shahverdi et al. (2007a) reported the extracellular biosynthesis of AgNPs within only 5 min, using the culture supernatant of *Enterobacteria*. This was attributed to the presence of hydroquinones with excellent redox properties that could act as electron shuttle in the metal reduction and also due to the presence of other reducing agents released in the supernatant of *Enterobacteria*. However, upon the addition of piperitone, an inhibitor of nitroreductases, the production of NPs was prevented, pointing out that the oxygen-insensitive nitroreductase NfsA is a key player in the reduction of silver ions. Some bacteria have been reported for both extra- and intracellular biosynthesis

TABLE 3.3 Various Bacterial Strains Involved for Biosynthesis of AgNPs

Bacteria	UV-Vis (λ_{nm})	Size (nm)	Activity	References
Bacillus licheniformis	440	50	–	Kalimuthu et al. (2008)
Bacillus licheniformis	420–430	50	–	Kalishwaralal et al. (2008)
Bacillus licheniformis Dahb1	422	18.69–63.42	Antibiofilm activity	Shanthi et al. (2016)
Bacillus subtillus IA751	450	50–80	Antimicrobial activity	Kannan and Subbalaxmi (2011)
Brevibacterium casei	420	10–50	–	Kalishwaralal et al. (2010)
Corynebacterium strain SH09	430–440	10–15	–	Zhang et al. (2005)
Deinococcus radiodurans R1 ATCC BAA-816	426	16.82	Antibacterial, anti-biofouling and cytotoxicity activities	Kulkarni et al. (2015)
Lactobacillus casei subsp. (DSM 20 011)	436	25–100	–	Korbekandi et al. (2012)
Lactobacillus sp.	425	11.2	Antimicrobial activity	Sintubin et al. (2009)
Pseudomonas aeruginosa strain BS-161R	430	8–24	Antimicrobial activity	Kumar and Mamidyala (2011)
Rhodococcus NCIM 2891	420	10	–	Otari et al. (2012)
Staphylococcus aureus		8–14		Senapati et al. (2005)
Streptomyces albidoflavu	410	10–40	Antimicrobial activity	Prakasham et al. (2012)
Streptomyces albogriseolus	409	16.25 ± 1.6	Antimicrobial activity	Samundeeswari et al. (2012)
Streptomyces ghanaensis VITHM1	420	30–50	Antibacterial activity	Abirami and Kannabiran (2016)
Streptomyces hygroscopicus	420	20–30	Antimicrobial activity	Sadhasivam et al. (2010)
Streptomyces olivaceous	420	500	Antimicrobial activity	Rita Evelyne and Subbiayh (2014)
Streptomyces sp JAR1	420	68.13	Antimicrobial activity	Chauhan et al. (2013)
Streptomyces sp. BHUMBU-80	450	21 ± 1	Antimicrobial activity	Gupta et al. (2017)

TABLE 3.3 *(Continued)*

Bacteria	UV-Vis (λ_{nm})	Size (nm)	Activity	References
Streptomyces sp. II	450	5–40	Antimicrobial activity	Sukanya et al. (2013)
Streptomyces sp. JF714876	425	90	Antimicrobial activity	Vidyasagar et al. (2012)
Streptomyces sp. LK3	420	5	Antiparasitic activity	Karthik et al. (2014)
Streptomyces sp. SS2	420	67.95 ± 18.52	Antimicrobial activity	Mohanta and Behera (2014)
Streptomyces sp. VDP-5	425	30–40	Antimicrobial activity	Singh et al. (2014)
Streptomyces sp. VITSJK10	420	20–70	Antimicrobial, cytotoxicity	Subashini et al. (2014)
Streptomyces sp.09 PBT 005	440	98–595	Antimicrobial, cytotoxicity	Kumar et al. (2015)
Streptomyces viirdochromogene	400	2.15–7.27	Antimicrobial activity	El-Naggar and Abdelwahed (2014)
Ureibacillus thermosphaericus	420	10–100	–	Motamedi Juibari et al. (2011)

of AgNPs; *Lactobacillus* sp. produced 500 nm AgNPs (Nair and Pradeep, 2002), *Aeromonas* sp. SH10 produced 6.4 nm AgNPs (Mouxing et al., 2006), *Proteus mirabilis* produced also spherical-shaped 10–20 nm AgNPs (Samadi et al., 2009), and *Vibrio alginolyticus* produced also spherical-shaped 50–100 nm AgNPs (Rajeshkumar et al., 2013).

The thermophilic, Gram-positive *Bacillus licheniformis* synthesized AgNPs of 50 nm (Kalimuthu et al., 2008). Where, the reduction of $AgNO_3$ was suggested to occur by means of the electrons from NADH where the NADH-dependent reductase can act as a carrier. Since *B. licheniformis* is known to secrete the cofactor NADH and NADH-dependent enzymes, especially nitrate reductase that might be responsible for the bioreduction of Ag^+ to Ag^0 and the subsequent formation of AgNPs. He et al. (2007) reported that the NADH and NADH-dependent nitrate reductase enzyme is induced by nitrate ions and reduces metal ions to metallic NPs, through electron shuttle enzymatic metal reduction process (Figure 3.3).

Mokhtari et al. (2009) proved the involvement of piperitone (3 methyl-6-1 methylethyl)-2 cyclohexan-1-one) and the nitroreductase enzymes in the cells of the Gram-positive *Enterobacteria* into the bioreduction of silver ions to silver. Moreover, the light role was also illustrated in the bioreduction of $AgCl_2$ to AgNPs by *K. pneumonia* (Figure 3.4) (Mokhtari et al., 2009).

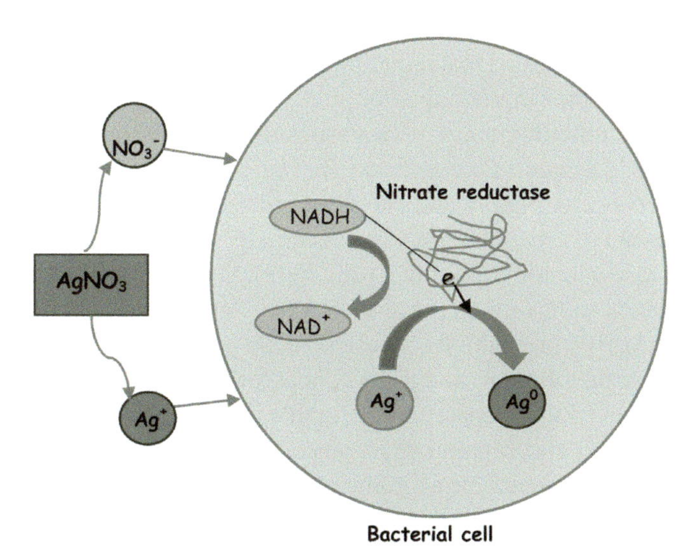

FIGURE 3.3 The possible mechanism for the biosynthesis of silver NP involving NADH-dependent nitrate reductase enzyme that may convert Ag^+ to Ag^0 through electron shuttle enzymatic metal reduction process.

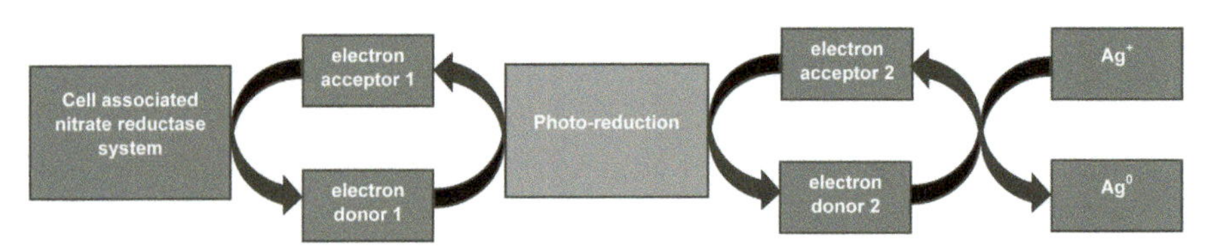

FIGURE 3.4 Light role in bioreduction of silver ions into AgNPs.

The *Bacillus* sp. which was exposed to an aqueous solution of $AgNO_3$ under atmospheric conditions proved its ability to synthesize AgNPs (Pugazhenthiran et al., 2009). Syntheses using microwave-irradiated culture supernatant of *Bacillus subtilis* yielded AgNPs in the range 5–60 nm (Saifuddin et al., 2009). Sintubin et al. (2009) reported different Gram-positive and Gram-negative bacteria with the capability of the biological synthesis of AgNPs, for example, *Lactobacillus* spp., *Pediococcus pentosaceus*, *Enterococcus faecium*, and *Lactococcus garvieae*. Extracellular synthesis of AgNPs was also achieved using culture supernatant of *Klebsiella pneumonia* (Shahverdi et al., 2007b), *Enterobacter cloacae* (Ahmad et al., 2007), and *Escherichia coli* (Ahmad et al., 2007; Gurunathan et al., 2009). However, the nonenzymatic reduction of silver ions to AgNPs via their interaction with cell wall functional groups has been reported for *Lactobacillus* A09, *Corynebacterium* sp. SH09, and *Aeromonas* SH10 (Fu et al., 2006; Lin et al., 2005; Zhang et al., 2005, 2007). Lin et al. (2005) suggested that reducing sugars of bacterial cells may function as the electron donor in the reduction of Ag^+ into its metallic form Ag^0. The capacity of alkalified biomass of Gram-positive bacteria *Lactobacillus* spp., *Enterococcus faecium*, *Lactococcus garvieae*, *Pediococcus pentosaceus*, and *Staphylococcus aureus* and the Gram-negative bacteria *Pseudomonas aeruginosa* and *Escherichia coli* to precipitate silver from diamine silver complex was evaluated (Sintubin et al., 2009). Only the lactic acid bacteria were confirmed to have the ability to produce Ag^0. *L. fermentum* expressed the highest AgNPs productions, smallest mean particle size of 11.2 nm, the narrowest size distribution, and most NPs associated with the outside of the cells. Thus, proved the preparation of AgNPs using the biomatrix of bacteria both as a reducing and capping agent since the cell wall controls the growth of the NPs and prevents their agglomeration. Furthermore, the most alkaline pH tested, pH 11.5, gave the greatest recovery of silver within shortest time of 1 min. Moreover, the association of the NPs with the bacterial biomass facilitates the concentration of the NPs by centrifugation. Not only this but it was also applied in a very concentrated Ag^+ solution (1 g/L), which is much higher compared to enzymatic processes. Enzyme-catalyzed silver reduction is restricted to lower silver concentrations ranging between 10 and 100 mg/L Ag^+. Sintubin et al. (2009) attributed the good efficiency of the Gram-positive bacteria for production of AgNPs to the structure and constituents of their cell wall. Since, the Gram-positive bacteria possess a cell wall that consists of a thick layer of peptidoglycan, teichoic acids, lipoteichoic acids, proteins, and polysaccharides. Thus, these bacteria have many anionic surface groups which can act as sites for the biosorption and subsequent reduction of silver cations (van Hullebusch et al. 2003). Furthermore, Lactic acid bacteria are known to produce exopolysaccharides (EPS) that can protect the bacterial cells against toxic compounds such as metal ions (van Hullebusch et al., 2003). Taking into consideration that the EPS consists of repeating sugar units, mainly; glucose, galactose, and rhamnose in different ratios, which can provide additional sites for biosorption of silver ions (Jolly et al., 2002). Sintubin et al. (2009) suggested the following reduction mechanism for production of AgNPs (Scheme 3.4), based on the increase of pH, would enhance the bioreduction. Since, high pH catalyzes the ring opening of monosaccharides such as glucose to their open chain aldehyde which delivers the reducing power. When metal ions are present, the aldehyde will be oxidized to the corresponding carboxylic acid, and at the same time, the metal ions will be reduced. Prasad et al. (2010) reported the biosynthesis of spherical shaped 2–6 nm AgNPs by *Lactobacillus*. The biosynthesis depended mainly on the partial pressure of gaseous hydrogen (r-H_2) as it decided the reduction potential. Whereas, under high r-H_2, the reduction of metal ions and the production of AgNPs occur. Moreover, the pH-dependent oxidoreductase played a vital role. Whereas, at high pH, the reductase is activated, and the reduction and production of AgNPs occur.

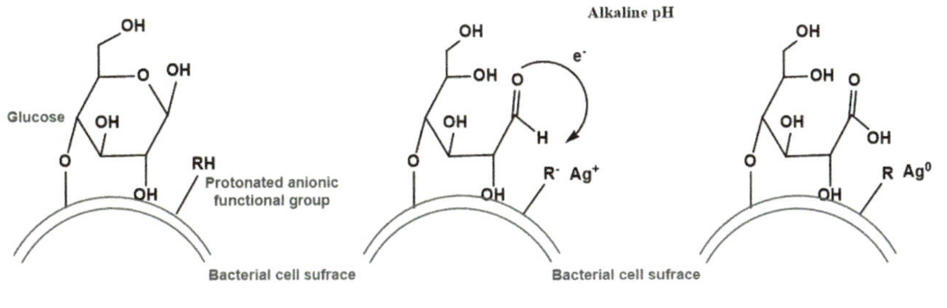

SCHEME 3.4 Suggested bacterial reduction of Ag^+ to Ag^0 at the cell surface (Sintubin et al., 2009).

Kannan and Subbalaxmi (2011) proved the extracellular biogenic synthesis of AgNPs by *Bacillus subtillus* IA751 with the appearance of broad peak at about 450 nm. The laser diffraction particle size analyzer revealed polydispersed mixture of NPs with average size ranged between 50 and 80 nm. Motamedi Juibari et al. (2011) reported the extracellular production of spherical-shaped 10–100 nm AgNPs by the thermophilic *Ureibacillus thermosphaericus* at 80 °C and high $AgNO_3$ concentration (0.01 M). It is worth to mention the important observation noticed by Motamedi Juibari et al. (2011); the average size of Ag-NPs at the silver ion concentration of 0.001 M and different temperatures of 60, 70, and 80 °C was 57, 29, and 13 nm, respectively. Whereas, recorded 51, 67, and 75 nm, respectively, when the silver ion concentration of 0.01 M was used. The increasing trend observed in the average particle size at the silver ion concentration of 0.01 M $AgNO_3$ by increasing the temperature was attributed to the possible increment in the secondary reduction process which is a common phenomenon in high ion concentrations. However, the decreasing trend observed in lower ion concentration (0.001 M $AgNO_3$) by increasing the temperature was attributed to the possible increase in the nuclei formation and consequently stopping the secondary reduction process because of the lower available free silver ions in the reaction solution. But, generally it was concluded that; these two phenomena are accelerated by increasing the reaction temperature (Motamedi Juibari et al., 2011).

Korbekandi et al. (2012) reported the optimization of the bioreductive synthesis of AgNPs by *Lactobacillus casei* subsp. (DSM 20 011). Where, they proved that the presence of silver nitrate as an enzyme activity inducer seems to be important for scaling up this process, as triggers the transcription of genes involved in synthesis of reducing enzymes (gene expression) and increases the productivity of NP synthesis. Nevertheless, the presences of electron donors provide electrons for reduction of cofactors, and thus, prolong the reduction of silver ions to silver nanometals. But in the absence of the electron donor, electron reservoirs in the reaction mixture for the recovery and revival of cofactors are depleted after short time and consequently, the reaction would decrease. Thus, the presence of $AgNO_3$ (0.1 mmol/L) in the culture as the enzyme inducer, and glucose (56 mmol/L) as the electron donor in the reaction mixture had positive effects on NP production. They have attributed the bioreduction of $AgNO_3$ to AgNPs to several aspects; the nitrate reductase which is an enzyme in the nitrogen cycle responsible for the conversion of nitrate to nitrite. The reduction, mediated by the presence of the enzyme in organisms, has been suggested to be responsible for the synthesis. Moreover, the Gram-positive bacteria have a cell wall which consists of a thick layer of peptidoglycan, teichoic acid, lipoteichoic acid, proteins, and polysaccharides. *L. casei* has many anionic surface groups, which may act as sites for the biosorption and subsequent bioreduction of silver cations. Furthermore, lactic acid bacteria are able to produce EPS, which may serve in the protection of the cells against toxic compounds such as metal ions (i.e., high concentrations of Ag^+). Thus, EPS was suggested to provide additional sites for biosorption of Ag^+ ions and the cell wall may control the growth of the NPs and prevent their agglomeration. The TEM micrographs showed that individual spherical-shaped AgNPs as well as a number of aggregates of about 25–100 nm were located inside, on the surface of the cells, and outside the cells. That proved, the Ag^+ ions first were trapped by *L. casei* cells, and then were reduced by enzymes present within the cytoplasm and/or AgNPs were formed on the surface of the cytoplasmic cell membrane, inside the cytoplasm and outside the cells, possibly due to the bioreduction of the metal ions by enzymes present on the cytoplasmic membrane and within the cytoplasm. Otari et al. (2012) reported the bioreduction of $AgNO_3$ to 10 nm spherical-shaped AgNPs in living culture of *Rhodococcus* NCIM 2891. That might have been carried out intracellularly by the enzyme systems of the *Rhodococcus* sp. present inside the cell, for example; α-NADH dependent nitrate reductase. Extracellular biosynthesis of AgNPs by a radiation-resistant *Deinococcus radiodurans* R1 ATCC BAA-816 has been also reported (Kulkarni et al., 2015). Extracellular synthesis of spherical, monodispersed and nonagglomerated AgNPs using *Streptomyces ghanaensis* VITHM1 strain has been reported by Abirami and Kannabiran (2016). Where, the FTIR analysis proved the presence of different functional groups such as amide linkages and –CO– which are responsible for capping and stabilization the AgNPs. Not only this but also the negative values of zeta potential also confirmed the repulsion among the particles due to the adsorption of –OH, which increased the stability of the formulation. The GC/MS of the ethyl acetate extract of *Streptomyces ghanaensis* VITHM1 proved the presence of; 4-phenylbutanal, N-(2-(1hindol-3-yl) ethyl)-3-chloropropanamide, and 3-benzylhexahydropyrrolo [1,2-a] pyrazine-1,4-dione, which might have acted as reducing and stabilizing agents for the synthesis of AgNPs. Abirami and Kannabiran (2016) elucidated the exact mechanism for the biosynthesis of AgNPs; the hydroxyl and carbonyl groups might have responsible for the reduction of metal ions to NPs and the stabilizations of formed NPs, respectively (Scheme 3.5).

The probiotic *Bacillus licheniformis* Dahb1 cell free extract was used for biosynthesis of spherical-shaped 18.69–63.42 nm AgNPs (Shanthi et al., 2016). Gupta et al. (2017) reported the extracellular synthesis of spherical-shaped AgNPs by *Streptomyces* sp. BHUMBU-80 within 80 min at 7.0% supernatant dose and 4 mM $AgNO_3$. The FTIR spectra confirmed the involvement of various functional groups present in culture supernatant responsible for the reduction of Ag^+ ions into Ag^0. Which were OH of the glucose, phenolic compounds, NH of the proteins, and C=O of the amides. The SEM and TEM analyses confirmed the presence of spherical AgNPs with an average size of 21 ± 1 nm. The XRD analysis revealed AgNPs with average crystallite size of 6.7 nm. The XPS analysis confirmed the presence of two individual peaks which attributed to the Ag $3d_{3/2}$ and Ag $3d_{5/2}$ binding energies corresponding to the presence of metallic silver.

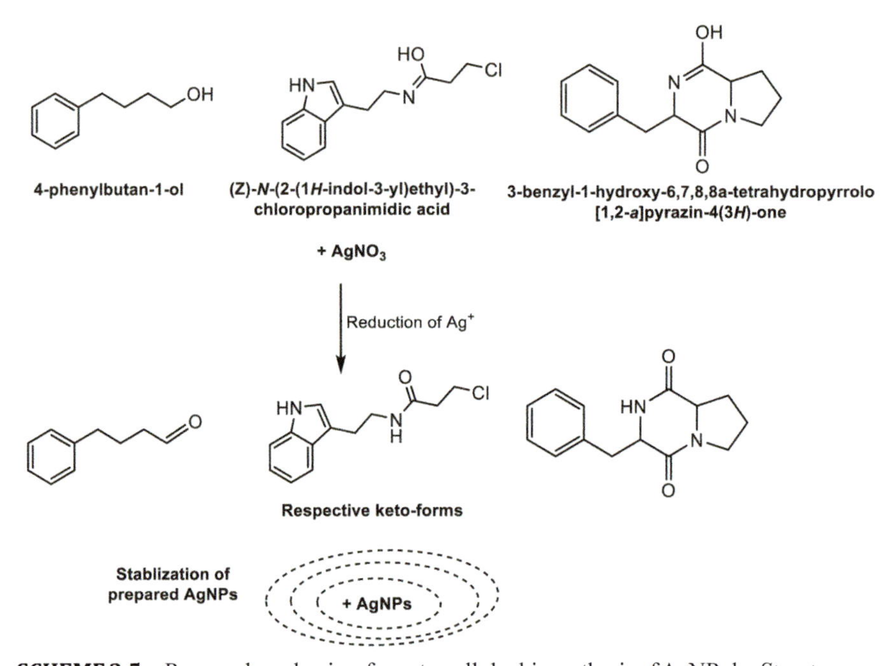

SCHEME 3.5 Proposed mechanism for extracellular biosynthesis of AgNPs by *Streptomyces ghanaensis* VITHM1.

Singh et al. (2015) proposed a mechanism for bacteriagenic synthesis of AgNPs (Figure 3.5). Accumulation of silver ions may occur in two stages: (1) nonspecific and energy-independent attachment to the cell surface and (2) intracellular accumulation (Shakibai et al. 2003). Wang et al. (2012) The important role played by the cell wall in the biogenesis of NPs via its and enzymes (Figure 3.5). This involves the nucleation of clusters of silver ions during the initial phase of AgNPs synthesis which causes an electrostatic interaction between the ions and negatively charged carboxylate groups of the cell wall. Then the trapped silver ions on the bacterial surface are thereby reduced by the cellular reductases and other redox proteins released by the cells to AgNPs (Mahdieh et al., 2012; Debabov et al., 2013). Nair and Pradeep (2002) observed the diffusion of these nanoclusters through the cell wall. Prakash et al. (2011) reported the involvement of the S-layer of bacteria, which physically masks the negatively charged peptidoglycan sheet of the cell wall and, thus, included in the bacteria–metal surface interaction. Prakash et al. (2011) pointed also for another potential mechanism across the cell wall, which conferred in bacteria with electro-kinetic potential via the generation of transmembrane proton gradient. This gradient can indirectly drive active symport of sodium along with silver ions from the surroundings. ATP binding employs special silver-binding proteins attached to membrane lipids on the external surface of bacteria, which attract the silver ions readily initiating AgNPs synthesis.

Jha and Prasad (2010) pointed out that the silver reduction machinery involves electron shuttle enzymatic silver reduction process. NADH generated during bacterial glycolysis and electron transport chain via energy-generating reactions creates a cellular reducing environment, due to hydrogen atoms, conducive for the synthesis of AgNPs. Where, the nitrate ions of $AgNO_3$ salt induce nitrate reductase. The enzyme gains electron from NADH and oxidizes it to NAD^+ and then reduces the Ag^+ ions to Ag^0 NPs (Figure 3.5). Further, nitrate ions (NO_3^-) converted to nitrogen dioxide (NO_2), then nitrogen oxide (NO) and nitrous oxide (N_2O), and finally to gaseous nitrogen (N_2) (Karthik and Radha 2012). Jha and Prasad (2010) glutathione and thioredoxin systems are significant to indirectly maintain the reducing conditions and regulate the activity of enzymes. Prasad et al. (2010) reported that high pH and partial pressure of gaseous H_2 are also important factors in bacteria-mediated AgNPs production. Since, high pH also activates reductases of oxidoreductase enzymes. Moreover, the reducing cofactors generated by the activity of various spore-associated enzymes, like glucose oxidase, alkaline phosphatase, laccase, and catalase, can stimulate the biogenesis of AgNPs (Hosseini-Abari et al., 2013). Peptides which are involved in biosynthesis and stabilization of AgNPs are also affected by solution pH. Peptides, containing amino acids like arginine, cysteine, lysine, methionine, glutamic acid, and aspartic acid are known for recognition and interaction with silver ions in solution and generate a reducing environment around them leading to their reduction and the formation of polydispersed AgNPs (Naik et al., 2002; Nam et al., 2008). Physiological conditions govern the metal–peptide interfacial interactions. For example, tyrosine undergoes conversion to a semiquinone structure under alkaline conditions through ionization at the phenol group, which reduces silver ions (Selvakannan et al., 2004). Tryptophan is converted to transient tryptophyl radical at high pH, which donates electron to reduce silver ions (Si and Mandal, 2007). Moreover, peptides with disulfide linkage can also be used for peptide-coated AgNP synthesis (Graf et al. 2009).

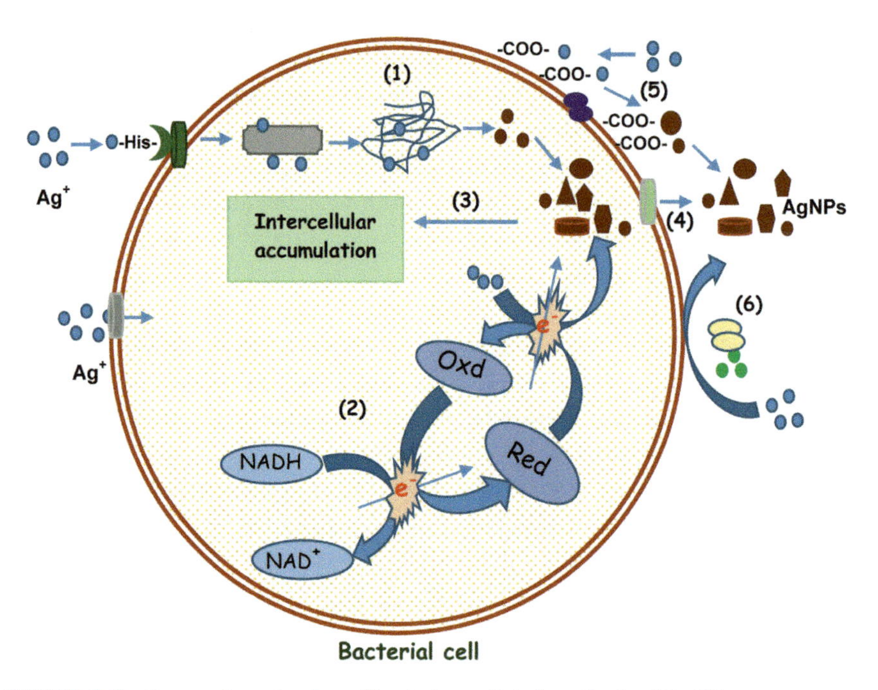

FIGURE 3.5 Proposed mechanism of bacteria-mediated synthesis of AgNPs.

(1) Cellular uptake of silver ions and activation of silver reduction machinery; (2) electron shuttle system involving various cofactors and enzymes; (3 and 4) intra- or extracellular localization of AgNPs; (5) electrostatic interaction between silver ions and cell wall components; and (6) reduction through extracellular enzymes and other organic molecules released in solution.

Shakibai et al. (2003) reported that *A. baumannii* exhibits plasmid-mediated silver resistance rendering bacteria capable of accumulating silver and retaining it by tight binding to a cysteine-rich metalloprotein. Three major gene homologs, namely *silE*, *silP*, and *silS*, of silver resistance machinery have been suggested to play a significant role in the synthesis of AgNPs (Parikh et al., 2008). Silver-binding gene homolog (*silE*) encodes a periplasmic silver-binding protein (silE) which is responsible for silver uptake by presenting histidine sites for silver ion binding. Upon exposure to silver ions, silE-based silver-binding machinery of bacteria gets activated leading to cellular uptake of silver ions. The ions are presented to bacterial silver reduction machinery where biomolecules, generated by silver reduction machinery, bind to the ions, and reduce them to metallic silver nuclei or seed NPs. These particles undergo growth and assembly to form AgNPs of different shapes (spherical or plate-like) followed by their release from the cells via cellular efflux system (Ramanathan et al., 2011). Similarly, Parikh et al. (2008) reported the biosynthesis of AgNPs using the silver-resistant bacterium; *Morganella* sp., where, the three silver-resistant homolog genes (*silE*, *silP*, and *silS*) were recognized in *Morganella* sp. Thus, their presence suggested that this organism forms AgNPs as a mechanism for protection against the toxicity of silver ions. The same was reported for the intracellular biosynthesis of AgNPs by the highly silver-tolerant marine bacterium; *Idiomarina* sp. PR58-8 (Seshadri et al., 2012).

Alternatively, bacteria create an extracellular microenvironment during growth in which silver-specific proteins from silver resistance machinery are released outside the cells. These proteins might reduce Ag^+ ions subsequently forming stable extracellular AgNPs (Parikh et al., 2008). Law et al. (2008) reported also the involvement of the periplasmic c-type cytochrome (MacA) in electron transfer from the inner membrane to the periplasm and the outer membrane c-type cytochrome (OmcF) in surface reduction of Ag^+ ions into AgNPs with the reduction of extracellular electron acceptors. Gaidhani et al. (2013) reported the nitrate reductase independent synthesis of AgNPs in *Acinetobacter*.

3.3 MICROBIAL SYNTHESIS OF GOLD NPS

AuNPs are known to be resistant to chemical oxidation, and stable in a wide range of environmental conditions (Kitching et al., 2015), which widen its applications as it is characterized by reduced environmental toxicity. It has different microbial, environmental, and medical applications (Vincy et al., 2017). Tetrachloroaurate is the main precursor for biosynthesis of gold NPs (AuNPs).

3.3.1 ALGAL PREPARATION OF AuNPs

Lengke et al. (2006) reported the involvement of the cyanobacterial *Plectonema boryanum* metabolites during the synthesis of AuNPs from its salt. Jianping et al. (2007) reported that *C. vulgaris* has strong binding ability toward tetrachloroaurate ions to form algal-bound gold reducing into Au^0. Whereas, approximately, 90% of algal-bound gold attained metallic state, and the crystals of gold were accumulated in the inner and outer parts of cell surfaces with tetrahedral, decahedral, and icosahedral structures. Chakraborty et al. (2009) used living blue-green algae *Lyngbya majuscule*, *Spirulina subsalsa* and green alga *Rhizoclonium hieroglyphicum* to study the biosorption and bioreduction of Au NPs. They did not produce NPs into the solution but desorbed them from the biomass surface. However, the intracellular synthesis of Au-nanorods by the cyanobacterium *Nostoc ellipsosporum* has been reported by Parial et al. (2012a). Moreover, *Phormidium valderianum*, *P. tenue* and *Microcoleus chthonoplastes* were reported in another study for the biosynthesis of AuNPs (Parial et al., 2012b). Where, the changes in pH effected the shape of the produced AuNPs. Senapati et al. (2012) and Geetha et al. (2014) reported the intracellular production of AuNPs using the green microalgae *Tetraselmis kochinensis* and marine cyanobacteria *Gloeocapsa* sp., respectively. The green marine picoeukaryote algae *Picochlorum* sp. has been reported for the intracellular synthesis of AuNPs with an average particles size of 11 nm (El-Sheekh and El-Kassas, 2014b). Whereas, the polysaccharides and protein molecules in the algal cells acted as reducing and stabilizing agents. Moreover, Mubarakali et al. (2013) reported the biosynthesis of AuNPs and silica-gold bionanocomposites by the diatoms (*Navicula atomus* and *Diadesmis gallica*). Moreover, *Plectonema boryanum* (Lengke et al., 2006), *Kappaphycus alvarezii* (Rajasulochana et al., 2010), *C. vulgaris* (Luangpipat et al., 2011), *Chlorella pyrenoidusa* (Oza et al., 2012), and C*alothrix* alga (Kumar et al., 2016) were reported for the biosynthesis of AuNPs.

3.3.2 FUNGAL PREPARATION OF AuNPs

Mukherjee et al. (2001b) reported the intracellular synthesis of AuNPs by *Verticillium* sp. via the biological reduction of $AuCl_4$. Whereby, the AuNPs were observed localized on the surface of the mycelia. The extra- and intracellular biosynthesis of gold NPs by fungus *Trichothecium* sp. has been reported by Absar et al. (2005). It was observed that when the gold ions reacted statically with the *Trichothecium* sp., rapid extracellular formation of spherical rod-like and triangular AuNPs occurred. However, upon shaking, intracellular growth of the AuNPs occurred. In another study, the FTIR analysis proved the involvement of the protein biomolecules in the cell free extract of *Rhizopus oryzae* in bioreduction of Au^{3+} ions into AuNPs which acted also as stabilizing and capping agents (Das et al., 2010). The mycosynthesized AuNPs remained stable for five months. Das et al. (2010) proved the control of the AuNPs shape and average size by varying the precursor salt concentration, the reaction pH and time. Triangular-shaped AuNPs with SPR of 540 nm were prepared using 1000 mg/L $AuCl_4$ and pH 8. While hexagonal AuNPs with SPR of 560 nm were prepared using 1500 mg/L $AuCl_4$ and pH 3. But pentagonal-shaped AuNPs with SPR of 535 nm were prepared applying 2000 mg/L $AuCl_4$ and pH 6. Moreover, a star-shaped AuNPs and SPR of 560 nm were prepared using 2000 mg/L $AuCl_4$ and pH 10. However, using 800 mg/L $AuCl_4$ and pH 10 with reaction time of 10 h produced spherical shaped 50–70 nm AuNPs. Applying the same pH value but with increasing the precursor concentration to 1500 mg/L and reaction time to 15 h the shape changed to spheroidal ones. Extra concentration of precursor to 2500 mg/L produced macro-sized Au^0 of sea urchins like shapes. Not only this but also at 4000 mg/L $AuCl_4$, pH 6 and reaction time of 24 h, 2D-gold-nano-wire network was formed with SPR of 547 nm. Lengthening the time to 30 h led to the formation of 2D Au nano-rods (Das et al., 2010). The change in shape was attributed to the nucleation rate which mainly depended on the precursor/protein ratio. Box-Behnken experimental design and RSM have been applied for the optimization of extracellular mycosynthesis of AuNPs using cell free extract of *Penicillium crustosum* (Barabadi et al., 2014). The elucidated quadratic model proved that the interaction between pH, $AuCl_4$ concentration of $AuCl_4$, and the reaction temperature has a statistical significant effect on the size of the mycosynthesized AuNPs. The optimum conditions for obtaining the lowest average size of AuNPs (53 nm) were elucidated to be pH8, 1.75 mM $AuCl_4$ and 37 °C. The SPR of 527 nm was the main indicator for the production of AuNPs and the absorption peak at 265 nm proved the involvement of fungal protein in the bioreduction reaction (Barabadi et al., 2014). The zeta potential of −26.0 mV confirmed the good stability of the mycosynthesized AuNPs (Barabadi et al., 2014). The extracellular mycosynthesis of polydispersed spherical and triangular-shaped 8–50 nm AuNPs was also reported by *Helminthosporium tetramera*. It was stable for about two whole months (Shelar and Chavan, 2014). The marine endophytic fungus *Cladosporium cladosporioides* isolated from seaweed has been reported for the extracellular mycosynthesis of AuNPs (Manjunath et al., 2017). Whereas, the NADPH-dependent reductase and phenolic compounds proved to be involved in the bioreduction of Au^{3+} ions into AuNPs, with average size of 40–60 nm. The mycosynthesized AuNPs expressed efficient biocidal activity against different pathogenic microbial strains. In another study, the optimum physicochemical parameters for the mycosynthesis of AuNPs using *Aspergillum* sp. WL-Au, were reported to be pH 7.0, 100 mg/mL, and 3 mM biomass and $HAuCl_4$ concentrations, respectively (Qu et al., 2017). Where, the presence of phosphate ions in the phosphate sodium buffer enhanced the intracellular synthesis of the AuNPs and its release. The mycosynthesized AuNPs showed efficient catalytic activity for the reduction of toxic nitro-aromatics 2-nitrophenol, 3-nitrophenol, 4-nitrophenol, *o*-nitroaniline, and *m*-nitroaniline to their corresponding less toxic amino-aromatics in the presence of $NaBH_4$. Moreover, it decolorized 12 different types of azo dyes in presence also of $NaBH_4$ (Qu et al., 2017). Endophytic fungi were earlier reported for extracellular synthesis of spherical shaped AuNPs with average size of 15–30 nm (Nachiyar et al., 2015).

3.3.3 BACTERIAL PREPARATION OF AuNPs

The extracellular bioreduction of gold ions into gold NPs by the thermophilic actinomycete, *Thermomonos* at pH9 and 50 °C has been reported (Sastry et al., 2003). The intracellular recovery of AuNPs by microbial reduction of $AuCl_4$ ions using the anaerobic bacterium *Shewanella algae* has been reported by Konishi et al. (2006). He et al. (2007) reported the extracellular biosynthesis of 10–20 nm AuNPs by *Rhodopseudomonas capsulate* via an NADH-dependent reductase. Whereas, the bioreduction is initiated by the electron transfer from NADH by NADH-dependent reductase as an electron carrier to form NAD^+ and the produced electrons are utilized by Au^{3+} ions for reduction into Au^0 NPs (Figure 3.6).

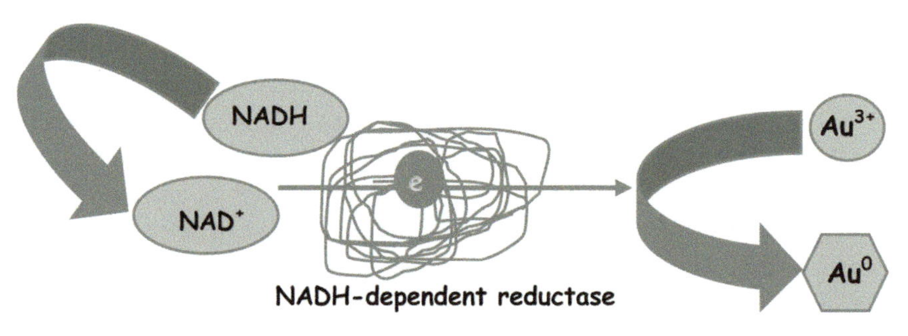

FIGURE 3.6 Bioreduction of Au^{3+} to Au^0 NPs.

He et al. (2007) observed that pH affects the size and shape of the biosynthesized NPs. Whereas, at acidic pH 4, nano-plates of AuNPs were formed, while by increasing the pH, spherical AuNPs in the range of 10–20 nm were observed at pH 7. Kalishwaralal et al. (2009) reported the biosynthesis of gold nano-cubes by *Bacillus lichemiformis*. Wen et al. (2009)

reported the biosynthesis of AuNPs using *Bacillus megatherium* biomass, via the reduction of $AuCl_4^-$ ions, whereas, the reaction time and dodecanethiol (as the capping ligand) were the important parameters in controlling size and morphology of the AuNPs. Kalishwaralal et al. (2010) reported the synthesis and stabilization of spherical-shaped 10–50 nm AuNPs using *Brevibacterium casei*. Whereas, the proteins biomolecules were proved to be responsible for the bioreduction process and acted as capping agents. Moreover, those proteins increased the stability of the biosynthesized AuNPs. The cell free extract of the thermophilic *Geobacillus stearothermophilus* has been reported for the biosynthesis of monodispersed spherical shaped 12 nm AuNPs (Fayaz et al., 2011). Sodium dodecyl sulphate-poly acrylamide gel electrophoresis and FTIR confirmed the involvement of the proteins biomolecules for example NADH- and/or NADPH-dependent reductase enzymes in the bioreduction of Au^{3+} ions into AuNPs. Further, the free amine groups or cysteine residues in the secreted proteins bonded to the NPs increasing their stability and prevented their agglomeration. While, Johnston et al. (2013) reported that the production of a small nonribosomal peptide, delftbactin, as a resistance mechanism of *Delfta acidovorans* to toxic gold ions, via the biosynthesis of inert AuNPs bound to that delftbactin. Consequently, the gold become no longer posed any toxicity problem for the cells. In another older study by Honary et al. (2012a), *Enterobacteriaceae* reported for the extracellular bioreduction of Au^{3+} ions into AuNPs under dark conditions. Where the average size of the biosynthesized NPs differed with the bacterial strain. However, in another study by Singh and Kundu (2014) the main factors affecting the size and shape of the biosynthesized AuNPs using two bacterial strains of *Pseudomonas aeruginosa* and *Rhodopseudomonas capsulata* were the bacterial strains and the pH. Tanzil et al. (2016) proved that the extracellular polymeric substances (EPS), for example, reducing sugars, extracted from electrode-respiring *Geobacter sulfurreducens* biofilms involved in the bioreduction of Au^{3+} ions into AuNPs with average particles size of 20 nm. Further, the FTIR analysis proved the capping of the produced AuNPs by proteins amino-groups (Tanzil et al., 2016).

3.4 MICROBIAL SYNTHESIS OF GOLD–SILVER NPS ALLOY

3.4.1 ALGAL PREPARATION OF Ag–Au NPs

S. platensis is reported for extracellular synthesis of 7–16 nm Ag, 6–10 nm Au, and 17–25 nm (bimetallic 50:50 ratio) of Au core-Ag shell NPs within

120 h at 37 °C and pH 5.6 (Govindaraju et al., 2008). Chakraborty et al. (2009) reported also the extracellular synthesis of gold, silver, and Au/Ag bimetallic NPs by *S. platensis*. Dahoumane et al. (2014) reported the living cells of *Chlamydomonas reinhardtii* for production of stable bimetallic Ag–Au NPs.

3.4.2 FUNGAL PREPARATION OF Ag–Au NPs

F. oxysporum produced an extracellular coenzyme (NADH), which favored the reduction of $HAuCl_4$ and $AgNO_3$, resulting in the formation of highly stable Au–Ag alloy NPs (Senapati et al., 2005). Mushroom such as; *Volvariella volvacea* is reported to produce extracellular Au, Ag and Au–Ag NPs (Daizy, 2009).

3.4.3 BACTERIAL PREPARATION OF Ag–Au NPs

Lactobacillus strains have been described to mediate the biosynthesis of Au, Ag, and Au–Ag NPs (Binoj and Pradeep, 2002). Nair and Pradeep (2002) reported that the exposure of lactic acid bacteria, which are present in the whey of buttermilk, to mixtures of gold and silver ions, produced alloy of Au–Ag NPs. The culture supernatant of *Escherichia coli* was reported for the biosynthesis of 6–12 nm Au–Ag alloy NPs within 20 min at room temperature (Ghorbani, 2017).

3.5 MICROBIAL SYNTHESIS OF COPPER NPS (CuNPs)

Copper NPs have many applications, such as electronic devices (lithium batteries), magnetic phase transitions, gas sensors, industrial cooling and heating systems, mass transfer enhancement, energy storage devices, in the production of cosmetics and pharmaceuticals, biocides, and so on. That is due to its unique electrical, magnetic, thermal, antimicrobial, optical, and catalytic activities (El Zowalaty et al., 2013; Pantidos and Horsfall, 2014). Copper sulfate is the main precursor for biosynthesis of CuNPs.

The nonpathogenic *Pseudomonas stutzeri* reported for the biosynthesis of spherical 8–15 nm CuNPs (Varshney et al., 2010). The extracellular production of CuNPs has been reported by three species of *Penicillium* species: *P. aurantiogriseum, P. citrinum*, and *P. waksmanii* (Honary et al.,

2012b). Where the acidic pH5 produced the smallest particle size and as the concentration of $CuSO_4$ increased the size of the produced CuNPs increased (Honary et al., 2012b). The shape and size of the biosynthesized CuNPs depends also on the utilized bacterial strains. For example, cubic 50–150 nm, 84–130 nm, 10–30 nm, and 15–20 nm CuNPs have been synthesized using *Pseudomonas stutzeri* (Varshney et al., 2011), *Pseudomonas* sp. *(Shobha et al., 2014), Serratia sp.* (Shobha et al., 2014), and *Morganella morganii* (Shobha et al., 2014), respectively. The cell free extract of *Pseudomonas fluorescens* has been reported for the biosynthesis of spherical and hexagonal shaped 49 nm CuNPs (Shantkriti and Rani, 2014). Ghorbani (2016) reported the bioreduction of copper nitrate into CuNPs within an hour using the cell filtrate of *E. coli*. The intracellular synthesis of 20–40 nm CuNPs using *Shewanella oneidensis* has been reported (Kimber et al., 2018). The Gram-negative facultative anaerobe *Shewanella loihica* PV-4 was also reported for intra- and extracellular synthesis of 10–16 nm CuNPs (Lv et al., 2018). Whereas, the microbial synthesis decreased with the increase of salt precursor, but increased with the biomass increment. The low temperature stopped the cell metabolism and consequently the biosynthesis of CuNPs. Moreover, high temperature denaturated microbial enzymes, thus, stop the production of CuNPs. PV-4 used both; vitamins, organic acids and NADH/NADPH as electron donors through soluble proteins and reductase in presence of cytochrome c for electron transfer, to reduce the Cu^{2+} ions into CuNPs. The biosynthesized CuNPs expressed efficient biocidal activity against *E. coli*, even after four cycles of usage as a water desinficatant (Lv et al., 2018). CuNPs destroyed the cell membrane and its cytoplasmic components via the chemical oxidative stress occurred by the production of ROS ($^{\bullet}OH$, H_2O_2 and $^{\bullet}O_2^-$) after binding with thiols. Moreover, CuNPs led to the leakage of the important osmotic solute K^+, which also acts as activator of intracellular enzymes and regulator of intracellular pH (Lv et al., 2018). *Morganella psychrotolerans* reported for the biosynthesis of highly stable CuNPs. However, after three months it started to be transformed to Cu_2O NPs (Pantidos et al., 2018). It is worth to mentioned that such biologically synthesized CuNPs expressed biocidal effect on Gram-positive *Bacillus subtilis* while the commercially available chemically synthesized one did not (Pantidos et al., 2018). Different yeast and fungal strains have been also reported for the biosynthesis of CuNPs as listed in Table 3.4.

The bioactive compounds polysaccharide, proteins, tannins, and steroids found in the extract of the green microalgae *Botryococcus braunii* were reported to be involved in the biosynthesis of spherical and cubical shaped

10–70 nm CuNPs with SPR of 258 nm (Arya et al., 2018). Where these biomolecules acted as bioreductants, stabilizing, and capping agents. The biosynthesized CuNPs expressed efficient biocidal activity against the Gram-negative bacteria *Pseudomonas aeruginosa* MTCC441 and *Escherichia coli* MTCC442, the Gram-positive bacteria *Klebsiella pneumoniae* MTCC109 and *Staphylococcus aureus* MTCC96, and a fungal strain *Fusarium oxysporum* MTCC2087 (Arya et al., 2018).

TABLE 3.4 Some Yeast and Fungal Strains Reported for Biosynthesis of CuNPs

Microbe	Shape and Size	Precursor	References
Hypocrea lixii	Spherical 24.5 nm	$CuCl_2$	José de Andrade et al. (2017)
Stereum hirsutum	Spherical 5–20 nm	$CuCl_2$	Cuevas et al. (2015)
Rhodotorula mucilaginosa	Spherical 10.5 nm	$CuCl_2$	Salvadori et al. (2014)
Penicillium aurantiogriseum	Spherical 89–250 nm	$CuSO_4$	Honary et al. (2012b)
Penicillium citrinum	Spherical 89–295 nm	$CuSO_4$	Honary et al. (2012b)
Penicillium waksmanii	Spherical 79–179 nm	$CuSO_4$	Honary et al. (2012b)

REFERENCES

Abdel-Hadi, A.M., Awad, M.F., Abo-Dahab, N.F., and ElKady, M.F. 2014. Extracellular synthesis of silver nanoparticles by *Aspergillus terreus*: Biosynthesis, characterization and biological activity. Biosci. Biotechnol. Res. Asia. 11(3): 1179–1186.

Abdel Rahim, K., Mahmoud, S.Y., Ali, A.M., Almaary, K.S., Mustafa, A.M.A., and Husseiny, S.M. 2017. Extracellular biosynthesis of silver nanoparticles using *Rhizopus stolonifera*. Saudi J. Biol. Sci. 24: 208–216.

Abdel-Hafez, S.I., Nafady, N.A., Abdel-Rahim, I.R., Shaltout, A.M., and Mohamed, M.A. 2016. Biogenesis and optimization of silver nanoparticles by the endophytic fungus *Cladosporium sphaerospermum*. Int. J. Nano Chem. 2(1): 11–19.

Abirami, M., and Kannabiran, K. 2016. *Streptomyces ghanaensis* VITHM1 mediated green synthesis of silver nanoparticles: Mechanism and biological applications. Front. Chem. Sci. Eng. 10(4): 542–551.

Absar, A., Satyajyoti, S., Khan, M.I., Rajiv, K., and Sastry, M. 2005. Extra-/intracellular biosynthesis of gold nanoparticles by an alkalotolerant fungus, *Trichothecium* sp. J. Biomed. Nanotechnnol. 1: 47–53.

Adio, S.O., Rana, A., Chanabsha, B., BoAli, A.A.K., Essa, M., and Alsaadi, A. 2019. Silver nanoparticle-loaded activated carbon as an adsorbent for the removal of mercury from Arabian gas-condensate. Arab. J. Sci. Eng. 44: 6285–6293.

Ahmad, R.S., Sara, M., Hamid, R.S., Hossein, J, and Nohi, A.A. 2007. Rapid synthesis of silver nanoparticles using culture supernatants of *Enterobacteria*: A novel biological approach. Process Biochem. 42: 919–923.

Ahmad, A., Mukherjee, P., Senapati, S., Mandal, D., Khan, M.I., Kumar, R., and Sastry, M. 2003. Extracellular biosynthesis of silver nanoparticles using the fungus *Fusarium oxysporum*. Colloids Surf. B: Biointerfaces. 28: 313–318.

Amanulla, M. F., Balaji, K., Girilal, M., Yadav, R., Kalaichelvan, P.T., and Venketesan, R. 2010. Biogenic synthesis of silver nanoparticles and their synergistic effect with antibiotics: A study against gram-positive and gram-negative bacteria. Nanomed. Nanotechnol. Biol. Med. 6: 103–109.

Anil Kumar, S., Abyaneh, M.K., Gosavi, S.W., Kulkarni, S.K., Pasricha, R., Ahmad, A., and Khan, M.I. 2007. Nitrate reductase-mediated synthesis of silver nanoparticles from $AgNO_3$. Biotechnol. Lett. 29: 439–445.

Apte, M., Girme, G., Bankar, A., Kumar, A.R., and Zinjarde, S. 2013. 3,4-dihydroxy-L-phenylalanine-derived melanin from *Yarrowia lipolytica* mediates the synthesis of silver and gold nanostructures. J. Nanobiotechnol. 11: 2. https://doi.org/10.1186/1477-3155-11-2.

Arya, A., Gupta, K., Chundawat, T.S., and Vaya, D. 2018. Biogenic synthesis of copper and silver nanoparticles using green alga *Botryococcus braunii* and its antimicrobial activity. Bioinorg. Chem. Appl. 2018. https://doi.org/10.1155/2018/7879403.

Balaji, D.S., Basavaraja, S., Deshpande, R., Bedre Mahesh, D., Prabhakar, B.K., and Venkataraman, A. 2009. Extracellular biosynthesis of functionalized silver nanoparticles by strains of *Cladosporium cladosporioides* fungus. Colloids Surf. B Biointerface. 68: 88–92.

Barabadi, H., Honary, S., Ebrahimi, P., Mohammadi, M.A., Alizadeh, A., and Naghibi, F. 2014. Microbial mediated preparation, characterization and optimization of gold nanoparticles. Braz. J. Microbiol. 45(4): 1493–1501.

Barwal, I., Ranjan, P., Kateriya, S., and Yadav, S.C. 2011. Cellular oxido-reductive proteins of *Chlamydomonas reinhardtii* control the biosynthesis of silver nanoparticles. J. Nanobiotechnol. 9: 56. https://doi.org/10.1186/1477-3155-9-56

Basavaraja, S., Balaji, S.D., Lagashetty, A., Rajasab, A.H., and Venkataraman, A. 2008. Extracellular biosynthesis of silver nanoparticles using the fungus *Fusarium semitectum*. Mater. Res. Bull. 43: 1164–1170.

Becaro, A.A., Jonsson, C.M., Puti, F.C., Siqueira, M.C., Mattoso, L.H.C., Correa, D.S., and Ferreira, M.D. 2015. Toxicity of PVA-stabilized silver nanoparticles to algae and microcrustaceans. Environ. Nanotechnol. Monit. Manag. 3:22–29.

Bekasova, O.D., Brekhovskikh, A.A., Revina, A.A., Dubinchuk, V.T. 2008. Preparation and optical properties of silver nanoparticles in R-phycoerythrin, a protein matrix. Inorg. Mater. 44: 835–841.

Bhainsa, K.C., and D'Souza, S.F 2006. Extracellular biosynthesis of silver nanoparticles using the fungus *Aspergillus fumigatus*. Coll. Surf. B. 47(2): 160–164.

Bhui, D.K., and Misra, A. 2012. Synthesis of worm like silver nanoparticles in methyl cellulose polymeric matrix and its catalytic activity. Carbohydr. Polym. 89: 830–835.

Binoj, N., and Pradeep, T. 2002. Coalescence of nanoclusters and formation of submicron crystallites assisted by *Lactobacillus* strains. Cryst. Grow. Design. 2: 293–298.

Birla, S., Gaikwad, S., Gade, A., and Rai, M., 2013. Rapid synthesis of silver nanoparticles from Fusarium oxysporum by optimizing physicocultural conditions. Sci. World J. 2013. https://doi.org/10.1155/2013/796018.

Brayner, R., Baeberousse, H., Hemadi, M., Djedjat, C., Yepremian, C., Livage, J., Fievet, F., and Coute, A. 2007. Cyanobacteria as bioreactors for the synthesis of Au, Ag, Pd, and Pt nanoparticles via an enzyme-mediated route. J. Nanosci. Nanotechol. 7: 2696–2708.

Bruins, M.R., Kapil, S., and Oehm, F.W. 2000. Microbial. resistance to metals in the environment. Ecotoxicol. Environ. Saf. 45: 198–207.

Chakraborty, N., Banerjee, A., Lahiri, S., Panda, A., Ghosh, A.N. and Pal, R. 2009. Biorecovery of gold using cyanobacteria and an eukaryotic alga with special reference to nanogold formation—a novel phenomenon. J Appl Phycol. 21: 145–152.

Chan, Y.S., and Don, M.M. 2013a. Optimization of process variables for the synthesis of silver nanoparticles by *Pycnoporus sanguineus* using statistical experimental design. J. Korean Soc. Appl. Biol. Chem. 56: 11–20.

Chan, Y.S., and Don, M.M. 2013b. Biosynthesis and structural characterization of Ag nanoparticles from white rot fungi. Mater. Sci. Eng. C. 33:282–288.

Chauhan, R., Kumar, A., and Abraham, J. 2013. A biological approach to the synthesis of silver nanoparticles with *Streptomyces* sp JAR1 and its antimicrobial activity. Sci. Pharm. 81(2): 607–621.

Chen, S.S., and Berns, D.S. 1979. Effect of plastocyanin and phycocyanin on the photosensitivity of chlorophyll-containing bilayer membranes. J. Membr. Biol. 21: 113–127.

Chen, J.C., Lin, Z.H., and Ma, X.X. 2003. Evidence of the production of silver nanoparticles via pretreatment of *Phoma* sp. 3. 2883 with silver nitrate. Lett. Appl. Microbiol. 37: 105–108.

Cuevas, R., Durán, N., Diez, M.C., Tortella, G.R.O., and Rubilar, O. 2015. Extracellular biosynthesis of copper and copper oxide nanoparticles by *Stereum hirsutum*, a native white-rot fungus from Chilean Forests. J. Nanomater. https://doi.org/10.1155/2015/789089.

Dahoumane, S.A., Wijesekera, K., Filipe, C.D.M., and Brennan, J.D. 2014. Stoichiometrically controlled production of bimetallic gold-silver alloy colloids using micro-alga cultures. J. Colloid Interface Sci. 416: 67–72.

Daizy, P. 2009. Biosynthesis of Au, Ag and Au–Ag nanoparticles using edible mushroom extract. Spectrochim. Acta A 73: 374–381.

Dameron, C.T., Reeser, R.N., Mehra, R.K., Kortan, A.R., Carroll, P.J., Steigerwald, M.L., Brus, L.E., and Winge, D.R. 1989. Biosynthesis of cadmium sulfide quantum semiconductor nanocrystallites. Nature. 338(6216): 596–597.

Dar, M.A., Ingle, A., and Rai, M. 2013. Enhanced antimicrobial activity of silver nanoparticles synthesized by *Cryphonectria* sp. evaluated singly and in combination with antibiotics. Nanomedicine. 9(1): 105–110.

Das, S.K., Das, A.R., and Guha, A.K. 2010. Microbial synthesis of multishaped gold nanostructures. Small. 6(9): 1012–1021.

Debabov, V.G., Voeikova, T.A., Shebanova, A.S., Shaitan, K.V., Emel'yanova, L.K., Novikova, L.M., and Kirpichnikov, M.P. 2013. Bacterial synthesis of silver sulfide nanoparticles. Nanotechnol. Russ. 8: 269–276.

Devi, L.S., and Joshi, S.R. 2015. Ultrastructures of silver nanoparticles biosynthesized using endophytic fungi. J. Microsc. Ultrastruct. 3: 29–37.

Devina Merin, D., Prakash, S., and Valentine Bhimb, B. 2010. Antibacterial screening of silver nanoparticles synthesized by marine micro algae. Asian Pac. J. Trop. Dis. 3(10): 797–799.

Durán, N., Marcato, P.D., Alves, O.L., Gabriel, I.H., Souza, D.E., and Esposito, E.J. 2005. Mechanistic aspects of biosynthesis of silver nanoparticles by several *Fusarium oxysporum* strains. J. Nanobiotechnol. 3: 1–7.

El Zowalaty, M., Ibrahim, N.A., Salama, M., Shameli, K., Usman, M., and Zainuddin, N. 2013. Synthesis, characterization, and antimicrobial properties of copper nanoparticles. Int. J. Nanomed. 8: 4467–4479.

Elamawi, R.M., Al-Harbi, R.E., and Hendi, A.A. 2018. Biosynthesis and characterization of silver nanoparticles using *Trichoderma longibrachiatum* and their effect on phytopathogenic fungi. Egypt. J. Biol. Pest. CO. 28(28). https://doi.org/10.1186/s41938-018-0028-1.

Elgorban, A.M., Al-Rahmah, A.N., Sayed, S.R., Hirad, A., Mostafa, A.A., and Bahkali, A.H. 2016. Antimicrobial activity and green synthesis of silver nanoparticles using *Trichoderma viride*. Biotechnol. Biotechnol. Equip. 30(2): 299–304.

El-Naggar, N.E.-A., and Abdelwahed, N.A. 2014. Application of statistical experimental design for optimization of silver nanoparticles biosynthesis by a nanofactory *Streptomyces viridochromogenes*. J. Microbiol. 52: 53–63.

El-Sheekh, M., and El-Kassas, H.Y. 2014a. Application of biosynthesized silver nanoparticles against a cancer promoter *Cyanobacterium, Microcystis aeruginosa*. Asian Pac J. Cancer Prev. 15: 6773–6779.

El-Sheekh, M., and El-Kassas, H.Y. 2014b. Biosynthesis, characterization and synergistic effect of phytogenic gold nanoparticles by marine picoeukaryote *Picochlorum* sp. in combination with antimicrobials. Rend. Fis. Acc. Lincei. 25: 513–521.

Fayaz, A.M., Balaji, K., Girilal, M., Kalaichelvan, P.T., and Venkatesan R. 2009. Mycobased synthesis of silver nanoparticles and their incorporation into sodium alginate films for vegetable and fruit preservation. J Agric. Food Chem. 57: 6246–6252.

Fayaz, A.M., Girilal, M., Rahman, M., Venkatesan, R., and Kalaichelvan, P.T. 2011. Biosynthesis of silver and gold nanoparticles using thermophilic bacterium *Geobacillus stearothermophilus*. Process Biochem. 46: 1958–1962.

Fesharaki, P.J., Nazari, P., Shakibaie, M., Rezaie, S., Banoee, M., Abdollahi, M., and Shahverdi, A.R. 2010. Biosynthesis of selenium nanoparticles using *Klebsiella pneumoniae* and their recovery by a simple sterilization process. Braz. J. Microbiol. 41(2): 461–466.

Fu, M.X., Li, Q.B., Sun, D.H., Lu, Y.H., He, N., Deng, X., Wang, H.X., and Huang, J.L. 2006. Rapid preparation process of silver nanoparticles by bioreduction and their characterizations. Chin. J. Chem. Eng. 14: 114–117.

Gade, A.K., Bonde, P., Ingle, A.P., Marcato, P.D., Durán, N., and Rai, M.K. 2008. Exploitation of *Aspergillus niger* for synthesis of silver nanoparticles. J. Biobased Mater. Bioener. 2: 243–247.

Gaidhani, S., Singh, R., Singh, D., Patel, U., Shevade, K., Yeshvekar, R., Chopade, B.A. 2013. Biofilm disruption activity of silver nanoparticles synthesized by *Acinetobacter calcoaceticus* PUCM 1005. Mater. Lett. 108: 324–327.

Gajbhiye, M., Kesharwani, J., Ingle, A., Gade, A., and Rai, M. 2009. Fungus-mediated synthesis of silver nanoparticles and their activity against pathogenic fungi in combination with fluconazole. Nanomed. Nanotechnol. Biol. Med. 5: 382–386.

Geetha, S., Sathakkathulzariya, J., Aarthi, R., and Blessie, H. 2014. Green synthesis of gold nanoparticle using marine cyanobacteria *Gloeocapsa* sp and the antitumour potential. J. Chem. Pharm. Sci. 4: 172–174.

Gelagutashvili, E. 2013. Binding of heavy metals with C-phycocyanin: A comparison between equilibrium dialysis, fluorescence and absorption titration, Am. J. Biomed. Life Sci. 1: 12–16.

Gericke, M., and Pinches, A. 2006. Biological synthesis of metal nanoparticles. Hydrometallurgy. 83: 132–140.

Ghorbani, H.R. 2016. A simple biological method for copper nanoparticles synthesis. Minerva Biotechnologica. 28(2): 86–88.

Ghorbani, H.R. 2017. Biological synthesis of Au-Ag alloy nanoparticles using *Escherichia coli*. Minerva Biotecnol. 29(1): 30–32.

Ghosh, S., Patil, S., Ahire, M., Kitture, R., Kale, S., Pardesi, K., Cameotra, S.S., Bellare, J., Dhavale, D.D., Jabgunde, A., Chopade, B.A. 2012. Synthesis of silver nanoparticles using *Dioscorea bulbifera* tuber extract andevaluation of its synergistic potential in combination with antimicrobial agents. Int. J. Nanomedicine. 7:483–496.

Glazer, A.N. 1994. Phycobiliproteins a family of valuable, widely used fluorophores. J. Appl. Phycol. 6: 105–112.

Govindaraju, K., Sabjan, K.B., Ganeshkumar, V., and Singaravelu, G. 2008. Silver, gold and bimetallic nanoparticles production using single-cell protein (*Spirulina platensis*) geitler. J. Mater. Sci. 43: 5115–5122.

Graf, P., Mantion, A., Foelske, A., Shkilnyy, A., Masic, A., Thünemann, A.F, and Taubert, A. 2009. Peptide-coated silver nanoparticles: Synthesis, surface chemistry, and pH-triggered, reversible assembly into particle assemblies. Chem. Eur. J. 15: 5831–5844.

Gudikandula, K., and Maringanti, S.C. 2016. Synthesis of silver nanoparticles by chemical and biological methods and their antimicrobial properties. J. Exp. Nanosci. 11(9): 714–721.

Gupta, R.K., Kumar, V., Gundampatid, R.K., Malviya, M., Hasan, S.H., and Jagannadham, M.V. 2017. Biosynthesis of silver nanoparticles from the novel strain of *Streptomyces* Sp. BHUMBU-80 with highly efficient electroanalytical detection of hydrogen peroxide and antibacterial activity. J. Environ. Chem. Eng. 5: 5624–5635.

Gurunathan, S., Kalimuthu, K., Ramanathan, V., Venkataraman, D., Pandian, S.R.K., Muniyandi, J., Hariharan, N., and Eom, S.H. 2009. Biosynthesis, purification and characterization of silver nanoparticles using *Escherichia coli*. Colloids Surface B Biointerface. 74: 328–335.

Hainfeld, J.F., Lin, L., Slatkin, D.N. Dilmanian, F.A., Vadas, T.M., and Smilowitz, H.M. 2014. Gold nanoparticle hyperthermia reduces radiotherapy dose. Nanomedicin. 10(8): 1609–1617.

Hallmann, J., Quadt-Hallmann, A., Mahaffee, W.F., and Kloepper, J.W. 1997. Bacterial endophytes in agricultural crops. Can. J. Microbiol. 43: 895–914.

Haoran, Z., Qingbiao, L., Yinghua, L., Daohua, S., Lin, X., Deng, X., He, N., and Zheng, S. 2005. Biosorption and bioreduction of diamine silver complex by *Corynebacterium*. J. Chem. Technol. Biotechnol. 80: 285–290.

He, S., Guo, Z., Zhang, Y., Zhang, S., Wang, J., and Gu, N. 2007. Biosynthesis of gold nanoparticles using the bacteria *Rhodopseudomonas capsulata*. Mater. Lett. 61: 3984–3987.

Honary, S., Barabadi, H., Fathabad E.G., and Naghibi, F. 2012b. Green synthesis of copper oxide nanoparticles using *Penicillium aurantiogriseum, Penicillium citrinum* and *Penicillium wakasmanii*. Dig. J. Nanomater. Biostruct. 7: 999–1005.

Honary, S., Gharaei-Fathabad, E., Paji, Z.K., and Eslamifar, M. 2012a. A novel biological synthesis of gold nanoparticle by *Enterobacteriaceae* family. Trop. J. Pharm. Res. 11(6): 887–891.

Hosseini-Abari, A., Emtiazi, G., and Ghasemi, S.M. 2013. Development of an eco-friendly approach for biogenesis of silver nanoparticles using spores of *Bacillus athrophaeus*. World J. Microbiol. Biotechnol. 29: 2359–236.

Husseiny, S.M., Salah, T.A., and Anter, H.A. 2015. Biosynthesis of size controlled silver nanoparticles by *Fusarium oxysporum*, their antibacterial and antitumor activities. Beni-Seuf Univ. J. Appl. Sci. 4: 2 2 5–2 31.

Ingle, A., Gade, A., Pierrat, S., Sonnichsen, C., and Rai, M. 2008. Mycosynthesis of silver nanoparticles using the fungus *Fusarium acuminatum* and its activity against some human pathogenic bacteria. Curr. Nanosci. 4: 141–144.

Ingle, A., Rai, M., Gade, A., and Bawaskar, M. 2009. *Fusarium solani*: A novel biological agent for the extracellular synthesis of silver nanoparticles. J. Nanopart. Res. 11: 2079–2085.

Jena, J., Pradhan, N., Dash, B.P., Sukla, L.B., and Panda, P.K. 2012. Biosynthesis and characterization of silver nanoparticles using microalgae *Chlorococcum humicola* and its antibacterial activity. Int. J. Nanomater. Biostruct. 3(1): 1–8.

Jena, J., Pradhan, N., Nayak, R.R., Dash, B.P., Sukla, L.B., Panda, P.K., and Mishra, B.K. 2014. Microalga *Scenedesmus* sp.: A potential low-cost green machine for silver nanoparticle synthesis, J. Microbiol. Biotechnol. 24: 522–533.

Jha, A.K., and Prasad, K. 2010. Biosynthesis of metal and oxide nanoparticles using *Lactobacilli* from yoghurt and probiotic spore tablets. Biotechnol. J. 5: 285–291.

Jianping, X., Jim, Y.L., Daniel, I.C.W., and Yen, P.T. 2007. Identification of active biomolecules in the high-yield synthesis of single-crystalline gold nanoplates in algal solutions. Small. 3(4): 668–672.

Joerger, R., Klaus, T., and Granqvist, C.G. 2000. Biologically produced silver-carbon composite materials for optically functional thin film coatings. Adv. Mater. 12: 407–409.

Johnston, C.W., Wyatt, M.A., Li, X., Ibrahim, A., Shuster, J., Southam, G., and Magarvey, N.A. 2013. Gold biomineralization by a metallophore from a gold-associated microbe. Nat. Chem. Biol. 9: 241–243.

Jolly, L., Vincent, S.J.F., Duboc, P., and Nesser, J.R. 2002 Exploiting exopolysaccharides from lactic acid bacteria. Antonie Van Leeuwenhoek. 82: 367–374.

José de Andrade, C., Maria de Andrade, L., Mendes, M.A., and Oller do Nascimento, C.A. 2017. An overview on the production of microbial copper nanoparticles by bacteria, fungi and algae. Global J. Res. Eng. C. 17(1): 27–33.

Kalimuthu, K., Sureshbabu, R., Venkataraman, D., Bilal, M., and Gurunathan, S. 2008. Biosynthesis of silver nanocrystals by *Bacillus licheniformis*. Colloids Surf. B. 65: 150–153.

Kalishwaralal, K., Deepak, V., Pandian, S.BR.K., Kottaisamy, M., Kanth, S.B.M., Kartikeyan, B., and Gurunathan, S. 2010. Biosynthesis of silver and gold nanoparticles using *Brevibacterium casei*. Colloid. Surf. B. 77: 257–262.

Kalishwaralal, K., Deepak, V., Ramkumarpandian, S., and Gurunathan, S. 2009. Biosynthesis of gold nanocubes from *Bacillus lichemiformis*, Bioresour. Technol. 100: 5356–5358.

Kalishwaralal, K., Deepak, V., Ramkumarpandian, S., Nellaiah, H., and Sangiliyandi, G. 2008. Extracellular biosynthesis of silver nanoparticles by the culture supernatant of *Bacillus licheniformis*. Mater. Lett. 62: 4411–4413.

Kannan, N., Mukunthan, K.S., and Balaji, S., 2011. A comparative study of morphology, reactivity and stability of synthesized silver nanoparticles using *Bacillus subtilis* and *Catharanthus roseus*. Colloids Surf. B. Biointerfaces 86, 378–383.

Kannan, N., and Subbalaxmi, S. 2011. Green synthesis of silver nanoparticles using *Bacillus subtillus* IA751 and its antimicrobial activity. Res. J. Nanosci. Nanotechnol. https://doi.org/10.3923/rjnn.2011.

Karbasian, M., Atyabi, S.M., Siadat, S.D., Momem, S.B., and Norouzian, D. 2008. Optimizing nano-silver formation by *Fusarium oxysporum* (PTCC-5115) employing response surface methodology. Am. J Agric. Biol. Sci. 3: 433–437.

Karthik, L., Kumar, G., Kirthi, A.V., Rahuman, A., and Rao, K.B. 2014. *Streptomyces* sp. LK3 mediated synthesis of silver nanoparticles and its biomedical application. Bioprocess Biosyst. Eng. 37: 261–267.

Karthik, C., and Radha, K.V. 2012. Biosynthesis and characterization of silver nanoparticles using *Enterobacter aerogenes*: a kinetic approach. Dig. J. Nanomater. Biostruct. 7: 1007–1014.

Kathiresan, K., Manivannan, S., Nabeel, M.A., and Dhivya, B. 2009. Studies on silver nanoparticles synthesized by a marine fungus, *Penicillium fellutanum* isolated from coastal mangrove sediment. Colloids Surface B Biointerface. 71: 133–137.

Kimber, R.L., Lewis, E.A., Parmeggiani, F., Smith, K., Bagshaw, H., Starborg, T., Joshi, N., Figueroa, A.I., Laan, G., Cibin, G. Gianolio, D., Haigh, S.J., Pattrick, R.A.D., Turner, N.J., and Lloyd, J.R. 2018. Biosynthesis and characterization of copper nanoparticles using *Shewanella oneidensis*: Application for click chemistry. Small. 14: 1703145.

Kitching, M., Ramani, M., and Marsil, E., 2015. Fungal biosynthesis of gold nanoparticles: Mechanism and scale. Microb. Biotechnol. 8(6): 904–917.

Klaus, T., Joerger, R., Olsson, E., and Granqvist, C.G. 1999. Silver based crystalline nanoparticles, microbially fabricated. Proc. Natl. Acad. Sci. USA. 96: 13611–13614.

Konishi, Y., Tsukiyama, T., Ohno, K., Saitoh, N., Nomura, T., Nagamine, S. 2006. Intracellular recovery of gold by microbial reduction of $AuCl_4$-ions using the anaerobic bacterium Shewanella algae. Hydrometallurgy. 81: 24–29.

Korbekandi, H., Ashari, Z., Iravani, S., and Abbasi, S. 2013. Optimization of biological synthesis of silver nanoparticles using *Fusarium oxysporum*. Iran J. Pharm. Res. 12(3): 289–298.

Korbekandi, H., Iravani, S., and Abbasi, S. 2012. Optimization of biological synthesis of silver nanoparticles using *Lactobacillus casei* subsp. casei. J. Chem. Technol. Biotechnol. 87: 932–937.

Kotakadi, V.S., Gaddam, S.A., Venkata, S.K., Sarma, P.V.G.K., and Sai Gopal, D.V.R. 2016. Biofabrication and spectral characterization of silver nanoparticles and their cytotoxic studies on human CD34 +ve stem cells. 3 Biotech. 6(2): 216. http://doi.org/10.1007/s13205-016-0532-5.

Kowshik, M., Ashtaputre, S., Kharrazi, S., Vogel, W., Urban, J., Kulkarni, S.K., and Paknikar, K.M. 2002. Extracellular synthesis of silver nanoparticles by a silver-tolerant yeast strain MKY3. Nanotechnology. 14: 95–100.

Kowshik, M., Ashtaputre, S., Kharrazi, S., Vogel, W., Urban, J., Kulkarni, S.K., and Paknikarand, K.M. 2003. Extracellular synthesis of silver nanoparticles by a silver-tolerant yeast strain MKY3. Nanotechnology. 14: 95–100.

Kulkarni, R.R., Shaiwale, N.S., Deobagkar, D.N., and Deobagkar, D.D. 2015. Synthesis and extracellular accumulation of silver nanoparticles by employing radiation-resistant *Deinococcus radiodurans*, their characterization, and determination of bioactivity. Int. J. Nanomedicine. 10: 963–974.

Kumar, B., Smita, K., Sánchez, E., Guerra, S., and Cumbal, L. 2016. Ecofriendly ultrasound-assisted rapid synthesis of gold nanoparticles using *Calothrix* algae. Adv. Nat. Sci. Nanosci. Nanotechnology 7: 025013. https://doi.org/10.1088/2043-6262/7/2/025013.

Kumar, C.G., and Mamidyala, S.K. 2011. Extracellular synthesis of silver nanoparticles using culture supernatant of *Pseudomonas aeruginosa*. Colloids Surf. B Biointerface. 84: 462–466.

Kumar, P.S., Balachandran, C., Duraipandiyan, V., Ramasamy, D., Ignacimuthu, S., and Al-Dhabi, N.A. 2015. Extracellular biosynthesis of silver nanoparticle using *Streptomyces* sp. 09 PBT 005 and its antibacterial and cytotoxic properties. Appl. Nanosci. 5: 169–180.

Lakowicz, J.R. 2006. Principles of fluorescence spectroscopy, 3rd edn. New York: Springer Science+Business Media, LLC.

Law, N., Ansari, S., Livens, F.R., Renshaw, J.C., and Lloyd, J.R. 2008. The formation of nano-scale elemental silver particles via enzymatic reduction by *Geobacter sulfurreducens*. Appl. Environ. Microbiol. 4: 7090–7093.

Lengke, M., Ravel, B., Feet, M.E., Wanger, G., Gordon, R.A., and Southam, G. 2006. Mechanisms of gold bioaccumulation by filamentous cynobacteria from gold(III)-chloride complex. Environ. Sci. Technol. 40: 6304–6309.

Lengke, M.F., and Southam, G. 2006. Bioaccumulation of gold by sulfate reducing bacteria cultured in the presence of gold (I)-thiosulfate complex. Geochem. Cosmochim. Acta. 70: 3646–3661.

Lengke, M.F., Fleet, M.E., and Southam, G. 2007. Biosynthesis of silver Nanoparticles by filamentous *Cyanobacteria* from a silver (I) nitrate complex. Langmuir. 23(5): 2694–2699.

LewisOscar, F., Mubarakali, D., Nithya, C., Priyanka, R., Gopinath, V., Alharbi, N.S., and Thajuddin, N. 2015. One pot synthesis and anti-biofilm potential of copper nanoparticles (CuNPs) against clinical strains of *Pseudomonas aeruginosa*. Biofouling. 3(4): 379–391.

Liang, M., Wei, S., Jian-Xin, L., Xiao-Xi, Z., Zhi, H., Wen, L., Zheng-Chun, L., and Jian-Xin, T. 2017. Optimization for extracellular biosynthesis of silver nanoparticles by *Penicillium aculeatum* Su1 and their antimicrobial activity and cytotoxic effect compared with silver ions. Mater. Sci. Eng. C 77: 963–971.

Lima, N., Santos, C., Fernandes, S., Carvalho, J., Dias, N., and Pereira, L. 2014. Synthesis, characterization and antifungal activity of chemically and fungal-produced silver nanoparticles against *Trichophyton rubrum*. J. Appl. Microbiol. 117: 1601–1613.

Lin, Z.Y., Fu, J.K., Wu, J.M., Liu, Y.Y., and Cheng, Hu. 2001. Preliminary study on the mechanism of non-enzymatic bioreduction of precious metal ions. Acta Phys-Chim. Sin. 17: 477–480.

Lin, Z.Y., Zhou, C.H., Wu, J.M., Zhou, J.Z., and Wang, L. 2005. A further insight into the mechanism of Ag^+biosorption by *Lactobacillus* sp strain A09. Spectrochim Acta A Mol. Biomol. Spectrosc. 61: 1195–1200.

Liu, C., Yang, D., Wang, Y., Shi, J., and Jiang, Z. 2012. Fabrication of antimicrobial bacterial cellulose–Ag/AgCl nanocomposite using bacteria as versatile biofactory. J. Nanopart. Res. 14: 1084–1095.

Luangpipat, T., Beattie, I.R., Chisti, Y., and Haverkamp, R.G. 2011. Gold nanoparticles produced in a microalga. J. Nanopart. Res. 13(12): 6439–6445.

Lv, Q. Zhang, B., Xing, X., Zhao, Y., Cai, R., Wang, W., and Gu, Q. 2018. Biosynthesis of copper nanoparticles using *Shewanella loihica* PV-4 with antibacterial activity: Novel approach and mechanisms investigation. J. Hazard. Mater. 347: 141–149.

MacColl, R. 1998. Cyanobacterial phycobilisomes, J. Struct. Biol. 15: 311–334.

Mahdieh, M., Zolanvari, A., Azimee, A.S., and Mahdieh M. 2012. Green biosynthesis of silver nanoparticles by Spirulina platensis. Scientia. Iranica. 19: 926–929.

Maliszewska, I., Szewczyk, K., and Waszak, K. 2009. Biological synthesis of silver nanoparticles. J. Phys. Conf. Ser. 146: 012025. https://doi.org/10.1088/1742-6596/146/1/012025.

Mandal, D., Bolander, M.E., Mukhopadhyay, D., Sarkar, G., and Mukherjee, P. 2006. The use of microorganisms for the formation of metal nanoparticles and their application. Appl. Microbiol. Biotechnol. 69: 485–492.

Manjunath H.M., Joshia, C.G., Danagoudar, A., Poyya, J., Kudva, A.K., and Dhananjaya, B.L. 2017. Biogenic synthesis of gold nanoparticles by marine endophytic fungus *Cladosporium cladosporioides* isolated from seaweed and evaluation of their antioxidant and antimicrobial properties. Process Biochem. 63: 137–144.

Manjunath, H.M., and Joshi, C.G. 2019. Characterization, antioxidant and antimicrobial activity of silver nanoparticles synthesized using marine endophytic fungus—*Cladosporium cladosporioides*. Process Biochem. 82: 199–204.

McFarland, A.D., and Van Duyne, R.P. 2003. Single silver nanoparticles as real-time optical sensors with zeptomole sensitivity. Nano Lett. 3: 1057–1062.

Mock, J., Barbic, M., Smith, R., Schultz, D., and Schultz, S., 2002. Shape effects in plasmon resonance of individual colloidal silver nanoparticles. J. Chem. Phys. 116: 6755–6759.

Mohamedkhair, A.K., Drmosh, Q.A., and Yamani, Z.H. 2019. Silver nanoparticle-decorated tin oxide thin films: Synthesis, characterization, and hydrogen gas sensing. Front. Mater. 6: 188. https://doi.org/10.3389/fmats.2019.00188.

Mohanta, Y.K., and Behera, S.K. 2014. Biosynthesis, characterization and antimicrobial activity of silver nanoparticles by *Streptomyces* sp. SS2. Bioprocess Biosyst. Eng. 37: 2263–2269.

Moharrer, S., Mohammadi, B., Gharamohammadi, R.A., and Yargoli, M. 2012. Biological synthesis of silver nanoparticles by *Aspergillus flavus*, isolated from soil of Ahar copper mine. Indian. J. Sci. Technol. 5(S3): 2443–2444.

Mokhtari, M., Deneshpojouh, S., Seyedbagheri, S., Atashdehghan, R., Abdi, K., Sarkar, S., Minaian, S., Shahverdi, R.H., and Shahverdi, R.A. 2009. Biological synthesis of very small silver nanoparticles by culture suspernatant of *Klebsiella pneumonia*. The effects of visible-light irradiation and the liquid mixing process. Mater. Res. Bull. 44: 1415–1421.

Motamedi Juibari, M., Abbasalizadeh, S., Salehi Jouzani, Gh., and Noruzi, M. 2011. Intensified biosynthesis of silver nanoparticles using a native extremophilic *Ureibacillus thermosphaericus* strain. Mater. Lett. 65: 1014–1017.

Mouxing, F., Qingbiao, L., Daohua, S., Yinghua, L., Ning, H., Xu, D., Huixuan, W., and Jiale, H. 2006. Rapid preparation process of silver nanoparticles by bioreduction and their characterizations. Chin. J. Chem. Eng. 14: 114–117.

Mubarak Ali, D., Arunkumar, J., Harish Nag, K., Ishack, S.S.K.A., Baldev, E., Pandiaraj, D., and Thajuddin N. 2013. Gold nanoparticles from pro and eukaryotic photosynthetic microorganisms—comparative studies on synthesis and its application on biolabelling. Colloids Surf. B. 103: 166–173.

Mubarak Ali, D., Sasikala, M., Gunasekaran, M., and Thajuddin, N. 2011. Biosynthesis and characterization of silver nanoparticles using marine cyanobacterium, *Oscillatoria willei* ntdm01. Dig. J. Nanomater. Biostruct. 6(2): 385–390.

Mukherjee, P., Ahmad, A., Mandal, D., Senapati, S., Sainkar, S.R., Khan, M.I., Parishcha, R., Ajaykumar, P.V., Alam, M., Kumar, R., and Sastry, M. 2001a. Fungus-mediated synthesis of silver nanoparticles and their immobilization in the mycelial matrix: A novel biological approach to nanoparticle synthesis. Nano Lett. 1(10): 515–519.

Mukherjee, P., Ahmad, A., Mandal, D., Senapati, S., Sainkar, S.R., Khan, M.I., Ramani, R., Parischa, R., Ajayakumar, P.V., Alam, M., Sastry, M., and Kumar, R. 2001b. Bioreduction of $AuCl(_4)(^-)$ ions by the fungus, *Verticillium* sp. and surface trapping of the gold nanoparticles formed. Angew. Chem. Int. Ed. Engl. 40: 3585–3588.

Mukherjee, P., Roy, M., Dey, G.K., Mukherjee, P.K., Ghatak, J., Tyagi, A.K., and Kale. S.P. 2008. Green synthesis of highly stabilized nanocrystalline silver particles by a non-pathogenic and agriculturally important fungus *T. asperellum*. Nanotechnology. 19(7): 075103. https://doi.org/10.1088/0957-4484/19/7/075103.

Nachiyar, V., Sunkar, S., Prakash P., and Bavanilatha. 2015. Biological synthesis of gold nanoparticles using endophytic fungi. Der Pharma Chemica. 7(2): 31–38.

Nagy, A.J., and G. Mestl. 1999. High temperature partial oxidation reactions over silver catalysts. Appl. Catal. A Gen. 188: 337–353.

Naik, R.R., Stringer, S.J., Agarwal, G., Jones, S.E., and Stone, M.O. 2002. Biomimetic synthesis and patterning of silver nanoparticles. Nat. Mater. 1: 169–172.

Nair, B., and Pradeep, T. 2002. Coalescense of nanoclusters and formation of submicron crystallites assisted by *Lactobacillus* strains. Crys. Growth Des. 2: 2 93–298.

Nair, P.T.P. 2009. Kinetics of silver nanoparticle biosynthesis: a comparative study in whole cell systems with growing cell systems. New Biotechnol. 25: S205.

Nam, K.T., Lee, Y.J., Krauland, E.M., Kottmann, S.T., and Belcher, A.M. 2008. Peptide mediated reduction of silver ions on engineered biological scaffolds. ACS Nano 2: 1480–1486.

Nithya, R., and Ragunathan, R. 2009. Synthesis of silver nanoparticle using *Pleurotus sajor caju* and its antimicrobial study. Digest J. Nanomater. Biostruct. 4: 623–629.

Omran, B.A., Nassar, H.N., Fatthallah, N.A., Hamdy, A., El-Shatoury, E.H., and El-Gendy, N.Sh. 2018a. Characterization and antimicrobial activity of silver nanoparticles mycosynthesized by *Aspergillus brasiliensis*. J. Appl. Microbiol. 125: 370–382. https://doi.org/10.1111/jam.13776

Omran, B.A., Nassar, H.N., Fatthallah, N.A., Hamdy, A., El-Shatoury, E.H., and El-Gendy, N. Sh. 2018b. Waste upcycling of *Citrus sinensis* peels as a green route for the synthesis of silver nanoparticles. Energ. Sources Part A. 40(2): 227–236. https://doi.org/10.1080/15567036.2017.1410597

Omran, B.A., Nassar, H.N., Younis, S.A., Fatthallah, N.A., Hamdy, A., El-Shatoury, E.H., and El-Gendy, N.S. 2019. Physiochemical properties of *Trichoderma longibrachiatum* DSMZ 16517-synthesized silver nanoparticles for the mitigation of halotolerant sulphate-reducing bacteria. J. Appl. Microbiol. 126: 138–154. https://doi.org/10.1111/jam.14102

Ortega, F.G., Fernández-Baldo, M.A., Fernández, J.G., Serrano, M.J., Sanz, M.I., Diaz-Mochón, J.J., Lorente, J.A., and Raba, J. 2015. Study of antitumor activity in breast cell lines using silver nanoparticles produced by yeast. Int J Nanomedicine. 10: 2021–2031.

Otari, S.V., Patil, R.M., Nadaf, N.H., Ghosh, S.J., and Pawar, S.H. 2012. Green biosynthesis of silver nanoparticles from an actinobacteria *Rhodococcus* sp. Mater. Lett. 72: 92–94.

Othman, A.M., Elsayed, M.A., Elshafei, A.M., and Hassan, M.M. 2017. Application of response surface methodology to optimize the extracellular fungal mediated nanosilver green synthesis. J. Genet. Eng. Biotechnol. 15: 497–504.

Ottoni, C.A., Simões, M.F., Fernandes, S., Gomes dos Santos, J., Sabino da Silva, E., Brambilla de Souza, R.F., and Maiorano, A.E. 2017. Screening of filamentous fungi for antimicrobial silver nanoparticles synthesis. AMB. Expr. 7: 31. https://doi.org/10.1186/s13568-017-0332-2.

Oza, G., Pandey, S., Mewada, A., Kalita, G., and Sharon, M. 2012. Facile biosynthesis of gold nanoparticles exploiting optimum pH and temperature of freshwater alga *Chlorella pyrenoidusa*. Adv. Appl. Sci. Res. 3(3): 1405–1412.

Pal, S., Tak, Y.K., and Song, J.M. 2007. Does the antibacterial activity of silver nanoparticles depend on the shape of the nanoparticle? A study of the gram-negative bacterium *Escherichia coli*. Appl. Environ. Microbiol. 73(6): 1712–1720.

Pantidos, N., and Horsfall, L.E. 2014. Biological synthesis of metallic nanoparticles by bacteria, fungi and plants. J. Nanomed. Nanotechnol. 5: 233. https://doi.org/10.4172/2157-439. 1000233.

Pantidos, N., Edmundson, M.C., and Horsfall, L. 2018. Room temperature bioproduction, isolation and anti-microbial properties of stable elemental copper nanoparticles. New Biotechnol. 40: 275–281.

Parial, D., Patra, H.K., Dasgupta, A.K., and Pal, R. 2012b. Screening of different algae for green synthesis of gold nanoparticles. Eur. J. Phycol. 47(1): 22–29.

Parial, D., Patra, H.K., Roychoudhury, P., Dasgupta, A.K., and Pal, R. 2012a. Gold nanorod production by cyanobacteria—a green chemistry approach. J. Appl. Phycol. 24: 55–60.

Parikh, R.Y., Singh, S., Prasad, B.L.V., Patole, M.S., Sastry, M., and Shouche, Y.S. 2008. Extracellular synthesis of crystalline silver nanoparticles and molecular evidence of silver resistance from *Morganella* sp.: Towards understanding biochemical synthesis mechanism. ChemBioChem 9: 1415–1422.

Patel, V. Berthold, D., Puranik, P., Gantar, M. 2015. Screening of *cyanobacteria* and microalgae for their ability to synthesize silver nanoparticles with antibacterial activity. Biotechnol. Rep. 5: 112–119.

Pereira, L., Dias, N., Carvalho, J., Fernandes, S., Santos, C., and Lima, N. 2014. Synthesis, characterization and antifungal activity of chemically and fungal-produced silver nanoparticles against *Trichophyton rubrum* J. Appl. Microbiol. 117: 1601–1613.

Prakash, A., Sharma, S., Ahmad, N., Ghosh, A., and Sinha, P. 2011. Synthesis of AgNPs by *Bacillus cereus* bacteria and their antimicrobial potential. J. Biomater. Nanobiotechnol. 2: 155–161.

Prakasham, R.S., Buddana, S., Yannam, S., and Guntuku, G. 2012. Characterization of silver nanoparticles synthesized by using marine isolate *Streptomyces albidoflavus*. J. Microbiol. Biotechnol. 22: 614–621.

Prasad, K., Jha, A.K., Prasad, K., and Kulkarni, A.R. 2010. Can microbes mediate nano-transformation? Indian J. Phys. 84(10): 1355–1360.

Priyabrata, M., Ahmad, A., Deendayal, M., Satyajyoti, S., Sainkar, S.R., Khan, M.I., Parishcha, R., Ajaykumar, P.V., Alam, M., Kumar, R., and Sastry, M. 2001. Fungus mediated synthesis of silver nanoparticles and their immobilization in the mycelial matrix: a novel biological approach to nanoparticle synthesis. Nano Lett. 1: 515–519.

Pugazhenthiran, N., Anandan, S., Kathiravan, G., Prakash, U.N.K. Crawford, S., and Ashok-kumar, M. 2009. Microbial synthesis of silver nanoparticles by *Bacillus* sp. J. Nanopart. Res. 11: 1811–815.

Pulit-Prociak, J., and Banach, M. 2016. Silver nanoparticles—a material of the future…? Open Chem. 14: 76–91.

Qu, Y., Pei, X., Shen, W., Zhang, X., Wang, J., Zhang, Z., Li, S., You, S., Ma, F., and Zhou, J. 2017. Biosynthesis of gold nanoparticles by *Aspergillum* sp. WL-Au for degradation of aromatic pollutants. Physica E. 88: 133–141.

Rahimi, G., Alizadeh, F., and Khodavandi, A. 2016. Mycosynthesis of silver nanoparticls from Candida albicans and its antibacterial activity against *Escherichia coli* and *Staphylococcus aureus*. Trop. J. Pharm. Res. 15(2): 371–375.

Rai, M., Yadav, A., and Gade, A. 2009. Silver nanoparticles as a new generation of antimicrobials. Biotechnol Adv 27: 76–83.

Rajasulochana, P., Dhamotharan, R., Murugakoothan, P., Murugesan, S., and Krishnamoorthy, P. 2010. Biosynthesis and characterization of gold nanoparticles using the alga *Kappaphycus alvarezii*. Int. J. Nanosci. 9: 511–516.

Rajeshkumar, S., Malarkodi, C., Paulkumar, K., Vanaja, M., Gnanajobitha, G., and Annadurai, G. 2013. Intracellular and extracellular biosynthesis of silver nanoparticles by using marine bacteria *Vibrio alginolyticus*. Nanosci. Nanotechnol. 3: 21–25.

Ramanathan, R., O'Mullane, A.P., Parikh, R.Y., Smooker, P.M., Bhargava, S.K., and Bansal, V. 2011. Bacterial kinetics-controlled shape-directed biosynthesis of silver nanoplates using *Morganella psychrotolerans*. Langmuir. 27: 714–719.

Rashmi, S., and Preeti, V. 2009. Biomimetic synthesis and characterization of protein capped silver nanoparticles. Biores. Technol. 100: 501–504.

Ravindra. B.K., and Rajasab, A.H. 2014. A comparative study on biosynthesis of silver nanoparticles using four different fungal species. Int. J. Pharm. Pharm. Sci. 6(1): 372–376.

Renugadevi, K., Aswini, R.V. 2012. Microwave irradiation assisted synthesis of silver nanoparticles using *Azadirachta indica* leaf extract as a reducing agent and in vitro evaluation of its antibacterial and anticancer activity. Int. J. Nanomater. Biostruct. 2(2): 5–10.

Rita Evelyne, J., and Subbiayh, R. 2014. Biosynthesis of silver nanoparticles from *Streptomyces olivaceous* and its antimicrobial activity. Int. J. Pharm. Res. Health Sci. 2: 166–172.

Roy, S., Mukherjee, T., Chakraborty, S., and Kumar das, T. 2013. Biosynthesis, characterisation and antifungal activity of silver nanoparticles by the fungus *Aspergillus foetidus* MTCC8876. Digest J. Nanomater. Biostruct. 8: 197–205.

Sabry, S.A., Ghozlan, H.A., and Abou-Zeid, D.M. 1997. Metal tolerance and antibiotic resistance patterns of a bacterial population isolated from sea water. J. Appl. Microbiol. 82: 245.

Sadhasivam, S., Shanmugam, P., and Yun, K. 2010. Biosynthesis of silver nanoparticles by *Streptomyces hygroscopicus* and antimicrobial activity against medically important pathogenic microorganisms. Colloids Surf. B: Biointerfaces. 81: 358–362.

Saifuddin, N., Wong, C.W., and Nur Yasumira, A.A. 2009. Rapid biosynthesis of silver nanoparticles using culture supernatant of bacteria with microwave irradiation. E-J. Chem. 6: 61–70.

Salunkhe, G.R., Ghosh, S., Santoshkumar, R.J., Khade, S., Vashisth, P., Kale, T., Chopade, S., Pruthi, V., Kundu, G., Bellare, J.R., and Chopade, B.A. 2014. Rapid efficient synthesis and characterization of silver, gold and bimetallic nanoparticles from the medicinal plant *Plumbago zeylanica* and their application in biofilm control. Int. J. Nanomedicine. 9: 2635–2653.

Salvadori, M.R., Ando, R.A., Nascimento, C.A.O., and Corrêa, B. 2014. Intracellular biosynthesis and removal of copper nanoparticles by dead biomass of yeast isolated from the wastewater of a mine in the Brazilian Amazonia. PLOS One. 9(1): e87968.

Samadi, N., Golkaran, D., Eslamifar, A., Jamalifar, H., Fazeli, M.R., and Mohseni, F.A. 2009. Intra/extracellular biosynthesis of silver nanoparticles by an autochthonous strain of *Proteus mirabilis* isolated from photographic waste. J. Biomed. Nanotechnol. 5: 247–253.

Samundeeswari, A., Dhas, S.P., Nirmala, J., John, S.P., Mukherjee, A., and Chandrasekaran, N. 2012. Biosynthesis of silver nanoparticles using actinobacterium *Streptomyces albogriseolus* and its antibacterial activity. Biotechnol. Appl. Biochem. 59: 503–507.

Sanghi, R., and Verma, P. 2009. Biomimetic synthesis and characterisation of protein capped silver nanoparticles. Bioresour. Technol. 100: 501–504.

Sarkar, S., Jana, D., Samanta, K., and Mostafa, G., 2007. Facile synthesis of silver nano particles with highly efficient antimicrobial property. Polyhedron. 26, 4419–4426.

Sastry, M., Ahmad, A., Khan, M.I., and Kumar, R. 2003. Biosynthesis of metal nanoparticles using fungi and actinomycete. Curr. Sci. 85: 162–170.

Sathishkumar, M., Sneha, K., Won, S., Cho, C., Kim, S., and Yun, Y., 2009. Cinnamon zeylanicum bark extract and powder mediated green synthesis of nano-crystalline silver particles and its bactericidal activity. Colloids Surf. B. Biointerfaces. 73: 332–338.

Sedlakova-Kadukova, J., Velgosova, O., Vosatka, M., Lukavsky, J., Dodd, Willner, J., and Fornalczyk, A. 2017. Control over the biological synthesis of ag nanoparticles by selection of the specific algal species. Arch. Metall. Mater. 62(3): 1439–1442.

Selvakannan, P.R., Swami, A., Srisathiyanarayanan, D., Shirude, P.S., Pasricha, R., Mandale, A.B., and Sastry, M. 2004. Synthesis of aqueous Au core–Ag shell nanoparticles using tyrosine as a pH-dependent reducing agent and assembling phase-transferred silver nanoparticles at the air–water interface. Langmuir. 20: 7825–7836.

Senapati, S., Ahmed, A., Khan, M.I., Sastry, M., and Kumar, R. 2005. Excellent biosynthesis of bimetallic Au–Ag alloy nanoparticles. Small. 1: 517–520.

Senapati, S., Syed, A., Moeez, S., Kumar, A., Ahmad, A. 2012. Intracellular synthesis of gold nanoparticles using alga Tetraselmis kochinensis. Mater. Lett. 79: 116–118.

Seshadri, S., Prakash, A., and Kowshik, M. 2012. Biosynthesis of silver nanoparticles by marine bacterium, *Idiomarina* sp. PR58-8. Bull. Mater. Sci. 35: 1201–1205.

Shahverdi, A.R., Fakhimi, A., Shahverdi, H.R., and Sara, M. 2007b. Synthesis and effect of silver nanoparticles on the antibacterial activity of different antibiotics against *Staphylococcus aureus* and *Escherichia coli*. Nanomed. Nanotechnol. Biol. Med. 3: 168–171.

Shahverdi, A.R., Minaeian, S., Shahverdi, H.R., Jamalifar, H., and Nohi, A.A. 2007a. Rapid synthesis of silver nanoparticles using culture supernatants of *Enterobacteria*: A novel biological approach. Process Biochem. 42: 919–923.

Shakibai, M.R., Dhakephalkar, P.K., Kapdnis, B.P., and Chopade, B.A. 2003. Silver resistance in Acinetobacter baumannii BL54 occurs through binding to a Ag-binding protein. Iranian. J. Biotechnol. 1: 41–46.

Shaligram, N.S., Bule, M., Bhambure, R., Singhal, R.S., Singh, S.K., Szakacs, G., and Pandey, A. 2009. Biosynthesis of silver nanoparticles using aqueous extract from the compact in producing fungal strain. Process Biochem. 44: 939–943.

Shanthi, S., Jayaseelan, B.D., Velusamy, P., Vijayakumar, S., Chih, C.T., and Vaseeharan, B. 2016. Biosynthesis of silver nanoparticles using a probiotic *Bacillus licheniformis* Dahb1 and their antibiofilm activity and toxicity effects in *Ceriodaphnia cornuta*. Microb Pathog. 93: 70–77.

Shantkriti, S., and Rani, P. 2014. Biological synthesis of Copper nanoparticles using *Pseudomonas fluorescens*. Int. J. Curr. Microbiol. Appl. Sci. 3(9): 374–383.

Sharma, G., Nakuleshwar, D.J., Kumar, M., and Mohammad, I.A. 2015. Biological synthesis of silver nanoparticles by cell-free extract of *spirulina platensis*. J. Nanotechnol.4: 1–6.

Shelar, G.B., and Chavan, A.M. 2014. Extracellular biological synthesis, characterization and stability of gold nanoparticles using the fungus *Helminthosporium tetramera*. Int. J. Pure App. Biosci. 2(3): 281–285.

Sherif, H., Taher, S., and Hend, A., 2015. Biosynthesis of size controlled silver nanoparticles by Fusarium oxysporum, their antibacterial and antitumor activities. Beni-Suef Univ. J. Basic Appl. Sci. 4, 225–231.

Shobha, G., Moses, V., and Ananda, S. 2014. Biological synthesis of copper nanoparticles and its impact—A review. Int. J. Pharm. Sci. Invent. 3: 28–38.

Si, S., and Mandal, T.K. 2007. Trytophan-based peptides to synthesize gold and silver nanoparticles: A mechanistic and kinetic study. Chem. Eur. J. 13: 3160–3168.

Singh, P.K., and Kundu, S. 2014. Biosynthesis of gold nanoparticles using bacteria. Proc. Natl. Acad. Sci., India, Sect. B Biol. Sci. 84(2): 331–336.

Singh, D., Rathod, V., Fatima, L., Kausar, A., Vidyashree, N.A., and Priyanka, B. 2014. Biologically reduced silver nanoparticles from *Streptomyces* sp. VDP-5 and its antibacterial efficacy. Int. J. Pharm. Sci. Res. 4: 31–36.

Singh, R., Shedbalkar, U.U., Wadhwani, S.W., and Chopade, B.A. 2015. Bacteriagenic silver nanoparticles: Synthesis, mechanism, and applications. Appl. Microbiol. Biotechnol. 99: 4579–4593.

Singh, R., Wagh, P., Wadhwani, S., Gaidhani, S., Kumbhar, A., Bellare, J., and Chopade, B.A. 2013. Synthesis, optimization, and characterization of silver nanoparticles from *Acinetobacter calcoaceticus* and their enhanced antibacterial activity when combined with antibiotics. Int. J. Nanomedicine. 8: 4277–4290.

Singh, T., Jyoti, K., Patnaik, A., Singh, A., Chauhan, R., and Chandel, S.S. 2017. Biosynthesis, characterization and antibacterial activity of silver nanoparticles using an endophytic fungal supernatant of *Raphanus sativus*. J. Gen. Eng. Biotechnol. 15: 31–39.

Sintubin, L., Verstraete, W., and Boon, N. 2012. Biologically produced nanosilver: Current state and future perspectives. Biotechnol. Bioeng. 109: 2422–2436.

Sintubin, L., Windt, W.D., Dick, J., Mast, J., Ha, D., Verstraete, W., and Boon, N. 2009. Lactic acid bacteria as reducing and capping agent for the fast and efficient production of silver nanoparticles. Appl. Microbiol. Biotechnol. 84: 741–749.

Sneha, K., and Yun, Y-S. 2013. Recovery of microbially synthesized gold nanoparticles using sodium citrate and detergents. Chem. Eng. J. 214: 253–261.

Subashini, J., Khanna, V.G., and Kannabiran, K. 2014. Anti-ESBL activity of silver nanoparticles biosynthesized using soil *Streptomyces* species. Bioprocess Biosyst. Eng. 37: 999–1006.

Sudha, S.S., Rajamanickam, K., and Rengaramanujam, J. 2013. Microalgae mediated synthesis of silver nanoparticles and their antimicrobial activity against pathogenic bacteria. Ind. J. Exp. Biol. 52: 393–399.

Sukanya, M., Saju, K., Praseetha, P., and Sakthivel, G. 2013. Therapeutic potential of biologically reduced silver nanoparticles from actinomycete cultures. J. Nanosci. 2013. http://dx.doi.org/10.1155/2013/94071.

Syafiuddin, A., Salmiati, Salim, M.R., Kueh, A.B.H., Hadibaratad, T., and Nur, H. 2017. A review of silver nanoparticles: research trends, global consumption, synthesis, properties, and future challenges. J. Chin. Chem. Soc. 64: 732–756.

Syed, A., Saraswati, S., Kundu, G.C., and Ahmad, A. 2013. Biological synthesis of silver nanoparticles using the fungus *Humicola* sp. and evaluation of their cytoxicity using normal and cancer cell lines Spectrochim. Acta A: Mol. Biomol. Spectrosc. 114: 144–147.

Tanzil, A.H., Sultana, S.T., Saunders, S.R., Dohnalkova, A.C., Shi, L., Davenport, E., Ha, P., and Beyenal, H. 2016. Production of gold nanoparticles by electrode-respiring *Geobacter sulfurreducens* biofilms. Enzyme Microb. Technol. 95: 69–75.

Thakkar, K.N., Mhatre, S.S., and Parikh, R.Y. 2011. Biological synthesis of metallic nanoparticles. Nanomedicine 6(2): 257–262.

Tyupa, D.V., Kalenov, S.V., Baurina, M.M., Yakubovich, L.M., Morozov, A.N., Zakalyukin, R.M., Sorokin, V.V., and Skladnev, D.A. 2016. Efficient continuous biosynthesis of silver nanoparticles by activated sludge micromycetes with enhanced tolerance to metal ion toxicity. Enzyme Microb. Technol. 95: 137–145.

van den Hondel, C.A., Punt, P.J., and van Gorcom, R.F. 1992. Production of extracellular proteins by the filamentous fungus *Aspergillus*. Antonie. Van Leeuwenhoek. 61(2): 153–160.

van Hullebusch, E., Zandvoort, M., and Lens P. 2003. Metal immobilization by biofilms: Mechanisms and analytical tools. Rev. Environ. Sci. Biotechnol. 2: 9–33.

Varshney, R., Seema, B., Gaur, M.S., and Pasricha, R. 2010. Characterization of copper nanoparticles synthesized by a novel microbiological method. J. Met. 62: 102–104.

Varshney, R., Seema, B., Gaur, M.S., and Pasricha, R. 2011. Copper nanoparticles synthesis from electroplating industry effluent. Nano Biomed. Eng. 3: 115–119.

Vaseeharan, B., Ramasamy, P., and Chen, J.C. 2010. Antibacterial activity of silver nanoparticles (AgNPs) synthesized by tea leaf extracts against pathogenic *Vibrio harveyi* and its protective efficacy on juvenile *Feneropenaeus indicus*. Lett. Appl. Microbiol. 50(4): 352–356.
Vidyasagar, G., Shankaravva, B., Begum, R., and Imrose, R.R. 2012. Antimicrobial activity of silver nanoparticles synthesized by *Streptomyces* species JF714876. Int. J. Pharm. Sci. Nanotechnol. 5: 1638–1642.
Vigneshwaran, N., Ashtaputre, N.M., Varadarajan, P., Nachane, R.P., Paralikar, K.M., and Balasubramanya, R.H. 2007. Biological synthesis of silver nanoparticles using the fungus *Aspergillus flavus*. Mater. Lett. 61(6): 1413–1418.
Vigneshwaran, N., Kathe, A.A., Varadarajan, P.V., Nachane, R.P., and Balasubramanya, R.H. 2006. Biomimetics of silver nanoparticles by white rot fungus *Phaenerochaete chrysosporium*. Colloids Surface B Biointerface. 53(1): 55–59.
Vijayaraghavan, K., and Kamala Nalini, S.P. 2010. Biotemplates in the green synthesis of silver nanoparticles. Biotechnol. J. 5: 1098–1110.
Vincy, W., Mahathalana, T.J., Sukumaran, S., and Jeeva, S. 2017. Algae as a source for synthesis of nanoparticles—A review. Int. J. Latest Trends Eng. Technol. Special Issue—International Conference on Nanotechnology: The Fruition of Science, pp. 5–9.
Wang, H., Chen, H., Wang, Y., Huang, J., Kong, T., Lin, W., Zhou, Y., Lin, L., Sin, D, and Li, Q. 2012. Stable silver nanoparticles with narrow size distribution non-enzymatically synthesized by *Aeromonas* sp. SH10 cells in the presence of hydroxyl ions. Curr. Nanosci. 8: 838–846.
Wen, L., Lin, Z., Gu, P., Zhou, J., Yao, B., Chen, G., and Fu, J. 2009. J. Nanopart. Res., 2008. Extracellular biosynthesis of monodispersed gold nanoparticles by a SAM capping route. 11: 279–288.
Xie, J., Lee, J.Y., Wang, D.I.C., and Ting, Y.P. 2007. Silver nanoplates: from biological to biomimetic synthesis. ACS Nano. 1: 429–439.
Xue, B., He, D., Gao, S., Wang, D., Yokoyama, K., and Wang, L. 2016. Biosynthesis of silver nanoparticles by the fungus *Arthroderma fulvum* and its antifungal activity against genera of *Candida*, *Aspergillus* and *Fusarium*. Int J Nanomedicine. 11: 1899–1906.
Yin, Y., Liu, J., and Jiang, G. 2012. sunlight-induced reduction of ionic Ag and Au to metallic nanoparticles by dissolved organic matter. ACS Nano. 6: 7910–7919.
Zaheer, Z. 2018. Biogenic synthesis, optical, catalytic, and in vitro antimicrobial potential of Ag-nanoparticles prepared using Palm date fruit extract. J. Photochem. Photobiol., B Biol. 178: 584–592.
Zhang, H., Li, Q., Lu., Y., Sun, D., Lin, X., Deng, X., He, N., and Zheng, S. 2005. Biosorption and bioreduction of diamine silver complex by *Corynebacterium*. J. Chem. Technol. Biotechnol. 80: 285–290.
Zhang, H.R., Li, Q.B., Wang, H.X., Sun, D.H., Lu, Y.H., and He, N. 2007. Accumulation of silver(I) ion and diamine silver complex by *Aeromonas* SH10 biomass. Appl. Biochem. Biotechnol. 143:54–62.

CHAPTER 4

Biosynthesis of Metal Oxide Nanoparticles

4.1 INTRODUCTION

The European Union has issued legislation on waste electrical/electronic equipment and restriction of hazardous substances, which come in parallel with the worldwide enlargement demands, industrial requirements, and forthcoming technologies that depend mainly on nanoparticles. Thus, there is an imperative need to develop an eco-friendly approach for nanomaterials synthesis that does not depend on toxic chemicals. Plants and microorganisms (i.e., algae, bacteria, yeast, and fungi) with their components are acting as biofactories for metal oxides (Yadav et al., 2019).

Metal oxides nanoparticles have many applications in the petroleum industry such as photodegradation of petroleum hydrocarbons pollutants and biocidal effect on macrofoulants and biocorrosion causing aerobic and anaerobic microorganisms, such as sulfate reducing bacteria (SRB). SRB are anaerobic bacteria reducing sulfate to corrosive sulfide causing severe corrosion of metals (Wade et al., 2011). Metal oxides can be applied in the oil reservoir, enhanced oil recovery, drilling, well stimulations, well cementing, deasphalting of heavy crude oil, and also as catalysts for some reactions in the refining processes, antiscaling, and corrosion inhibitors (Yu et al., 2010; Nassar et al., 2011; Rezvani et al., 2012; Nasr-El-Din et al., 2013; William et al., 2014; Murugesan et al., 2016; Khalil et al. 2017; Agista et al., 2018; Esmaeilnezhad et al., 2019). Moreover, some of the metal oxides nanoparticles, such as magnetite and hematite have some environmental applications. For example, oil spill collectors accelerating the bioremediation rate of petroleum hydrocarbons polluted environment (Saed et al., 2014; Liang et al., 2015; Abdullah et al., 2018). Magnetite can accelerate the rate of some bioupgrading process in the petroleum industry, such as the biodesulfurization and biodenitrogenation processes (Shan et al., 2005; Zakaria et al., 2015; El-Gendy and Speight, 2016; El-Gendy and Nassar, 2018).

Zinc oxide is an inorganic compound. It appears as a white powder and is nearly soluble in water. Its most common crystal structures are wurtzite (hexagonal) and zinc blende. Zinc oxide, with its unique physical and chemical properties, such as high chemical stability, high electrochemical coupling efficiency, broad range of radiation absorption, and high photo-stability, is a multifunctional material (Segets et al. 2009). In materials science, zinc oxide is classified as a semiconductor. Its high thermal stability makes it attractive for potential use in electronics, laser technology, and optoelectronics (Wang et al., 2008). The piezo- and pyro-electric properties of ZnO mean that it can be used as a sensor, converter, energy generator, and photo-catalyst in hydrogen production. Because of its hardness and rigidity, it is an important material in the ceramics industry. While its low toxicity, biocompatibility, and biodegradability make it a material of interest for biomedicine and in proecological systems (Ludi and Niederberger, 2013). The powder ZnO is widely used as an additive in numerous materials and products including ceramics, glass, cement, rubber (e.g., car tires), lubricants, paints, ointments, adhesives, plastics, sealants, pigments, foods (source of Zn nutrient), batteries, ferrites, and free retardants (Sabir et al., 2014). El-Gendy et al. (2016) used ZnO NPs for photodegradation of chlorophenols. The ZnO NPs have been reported for good stimulation and increasing the good productivity via the improvement of rheological characteristics of fracturing fluid (Nasr-El-Din et al., 2013). ZnO NPs have been reported not only as biocides and mitigate microbial influenced corrosion in the petroleum industry. But it is also used as catalysts in the petrochemical industry, and it is also used as selective sensors for ammonia gas and H_2S removal from drilling fluids (Mirhendi et al., 2013).

The variety of structures of nanometric zinc oxide means that ZnO can be classified among new materials with potential applications in many fields of nanotechnology. ZnO NPs provide one of the greatest assortments of varied particle structures among all known materials. ZnO can occur in one-dimensional (1D), two-dimensional (2D), and three-dimensional (3D) structures. One-dimensional structures make up the largest group, including nanoribbons (Pan et al., 2001), nanohelixes, nanosprings and nanorings (Kong et al., 2004), nanobelts (Huang et al., 2006), nanorods (Frade et al., 2012), nanocombs (Xu et al., 2012), and nanowires (Nikoobakht et al., 2013). Zinc oxide can be obtained in 2D structures, such as nanoplate/nanosheet and nanopellets (Chiu et al., 2010). Examples of 3D structures of zinc oxide include flower, dandelion, snowflakes, coniferous urchin-like, etc. (Polshettiwa et al., 2009). The antimicrobial activity of the ZnO NPs involves the release of reactive oxygen species (ROS) from the surface of ZnO which causes fatal damage to microorganisms. According to Varaprasad et al. (2016), ROS are known to cause severe oxidative stress by damaging DNA, cell membranes, and cellular proteins. The rupture of the cell wall is because of the surface activity of ZnO NPs, which causes the decomposition of the cell wall and subsequently the cell membrane damage leading to the leakage of cell contents, and ultimate cell death. Krupa et al. (2019) reported the efficient antimacrofouling activity of the green synthesized ZnO NPs via coconut water. ZnO NPs are usually synthesized using zinc acetate dihydrate and zinc nitrate.

Titanium dioxide (TiO_2) is the most widely used white pigment, for example, in paints. It has high brightness and a very high refractive index. TiO_2 is also an effective opacifier, making substances more opaque. It has a wide range of industrial applications that include the food industry, photocatalytic media, gas sensors, paints, and in cosmetic industry as a white pigment, water treatment, air purification, solar energy, UV absorber, semiconductor, and agricultural industries. Moreover, more extensive applications include paints, papers, plastics, inks, toothpastes, and adding the TiO_2 to skimmed milk makes it appear brighter, more opaque, and more palatable. Almost every sunscreen contains titanium dioxide. It is a physical blocker for UVA (ultraviolet light with a wavelength of 315–400 nm) and UVB (ultraviolet light with a wavelength of 280–315 nm) radiation. It is chemically stable and not decolorized under UV light. Titanium dioxide can be added to the surface of cements, tiles, and paints to give the material sterilizing, deodorizing, and antifouling properties. This is because the photocatalytic properties of TiO_2 mean that, in the presence of water, hydroxyl free radicals are formed which can convert organic molecules to CO_2 and water and also destroy microorganisms. An attractive feature of TiO_2 photocatalytic disinfection is its potential to be activated by visible light, for example, sunlight. TiO_2NPs are the most commonly used NPs to deactivate pathogens in water. TiO_2 can kill both Gram-negative and Gram-positive bacteria. The antibacterial activity of TiO_2 is related to ROS production, especially hydroxyl free radicals and peroxide formed under UV irradiation via oxidative and reductive pathways, respectively (Li et al., 2008). Inactivation of microorganisms depends on several factors, for example, concentration of TiO_2, type of microorganism, intensity and wavelength of light, degree of hydroxylation, pH, temperature, oxygen availability, ROS, and retention time (Markowska-Szczupak et al., 2011). TiO_2 NPs have been used in the preparation of catalysts applied in the oxidative desulfurization process in petroleum refining (Rezvani et al., 2012). TiO_2, Fe_2O_3, CuO, and ZnO NPs are used as additives in well cementing in the petroleum industry (Hurnaus and Plank, 2015). Titanium chloride is the most commonly used metal salt used for the synthesis of TiO_2 NPs. TiO_2 and Fe_3O_4 NPs have been reported for deasphatling of heavy crude oil (Nassar et al., 2011). Hematite NPs are also reported to be used in the oxidative desulfurization process in petroleum refining (Khalil et al., 2015).

The fairly stable, cost-effectiveness, and readily available copper oxide nanoparticles CuO and Cu_2O NPs compared to other expensive noble metals, like Au, Pt, and Ag, is a p-type semiconductor with the optical bandgap energy (E_g, eV) of ~1.2–1.8 and 2.2–2.6 eV, respectively (Murali et al., 2015) and have potential antimicrobial activity. It can be applied as gas sensors, antifouling agents, catalysts for the water–gas shift reaction. It can be applied also in the oxidation of the CO automobile exhaust gases and in the preparation of organic and inorganic nanostructure composites (Gnanavel et al., 2017; Pansambal et al., 2017; Buazar et al., 2019). Moreover, Cu_2O NPs have been applied in antifouling paints and coatings for petroleum pipelines (Yadav et al., 2019). The anabolic and catabolic activities of a model SRB, *Desulfovibrio vulgaris* Hidenborough have been reported to be inhibited by the high concentration of CuO NPs (>50 mg/L) (Chen et al., 2019b). CuO and ZnO NPs have been reported to be used together with xanthan gum to improve rheological, thermal, and electrical properties of water-based drilling fluid (William et al., 2014). CuO NPs are reported to be applied for the adsorbents used for adsorptive desulfurization in petroleum refining (Khodadadi et al., 2012). Copper nitrate, copper (II) sulfate pentahydrate, copper acetate, or copper chloride dihydrate are the widely used precursors for the green synthesis of copper oxides.

Several plants' parts and microorganisms have been reported for the biosynthesis of metal oxide nanoparticles, such as zinc oxide, iron oxides (IOs), copper oxides, and titanium dioxide. This chapter will summarize the most published reports concerning this with some examples for the application of such biosynthesized metal oxides in the petroleum sector. The most important characterization instruments will be illustrated. The dependence of the obtained type, size, and shape of metal oxides NPs on the applied biological source and the physicochemical conditions of its biosynthesis process will be extensively discussed. The biocomponents and mechanisms involved in the biosynthesis of metal oxide NPs will be also explained.

IO nanomaterials exist in many forms in nature. Magnetite (Fe_3O_4), maghemite (γ-Fe_2O_4), and hematite (α-Fe_2O_3) are the most common forms used as nanoadsorbents (Chan and Ellis, 2004). Generally, due to the small size of nanosorbent materials, their separation and recovery from contaminated water are great challenges for water treatment. However, magnetic Fe_3O_4 and γ-Fe_2O_4 can be easily separated and recovered from the system with the assistance of an external magnetic field. Therefore, they have been successfully used as sorbent materials in the removal of various heavy metals from water systems. Moreover, Fe_3O_4-NPs have several interesting characteristics as being with superparamagnetic behavior, biocompatible, and biodegradable. Furthermore, the study of Fe_3O_4 nanofluid on thermal conductivity and viscosity with the presence of external magnetic source and electric field has also gained popularity in heat transfer applications.

Magnetic iron NPs were exploited for numerous applications, such as catalysts for organic syntheses, water splitting, magnetic storage media, biosensors, separation process, and environmental remediation, besides, many biomedical usages like cellular therapy, tissue repair, magnetic resonance imaging (MRI), hyperthermia, catalysis, and drug delivery. Fundamental magnetic properties of nanoparticles critically define their potential applications, such as hard magnets for data storage and soft magnetic materials for magnetic switches. The properties of magnetic ferrite nanoparticles can be adjusted by size surface, shape, assembly, coupling, and doping. For instance, IO nanoparticles over 28 nm are ferrimagnetic and widely used for magnetic separation and as ferrofluids for liquid seals. While IO nanoparticles below 28 nm are superparamagnetic at room temperature, and they are heavily explored for biomedical applications, such as drug delivery, cancer therapy via magnetic hyperthermia as contrast agents for MRI, and the emerging technique of magnetic particle imaging (Bao et al., 2016). IO nanoparticles smaller than 4 nm become primarily paramagnetic and can be used as positive MRI contrast agents. Recently, it has been shown that nonspherical IO nanoparticles could improve their usefulness for biomedical applications. Naturally present superparamagnetic Fe_3O_4 nanoscale aggregates in paraffin wax deposits are reported to control the physicochemical properties of crude oils (Lesin et al., 2010). Functionalized magnetic nanoparticles are reported to be used for reservoir mapping, dewatering of heavy crude oil emulsions, emulsion separation which would solve many problems in upstream and downstream processes, for selective extraction for crude oil impurities which might cause corrosion and/or deposit formation, and for the preparation of magnetic nanofluids for drilling, completion, and enhanced oil recovery (Kapusta et al., 2011; Cocuzza et al., 2012; Peng et al., 2012; Ali et al., 2015; Simonsen et al., 2018; 2019). Ferrous sulfate $FeSO_4$ and mixture of ferrous $FeCl_2$ and ferric $FeCl_3$ are the most commonly used metal salts used to produce IOs nanoparticles.

4.2 PHYTOSYNTHESIS OF NANOMETAL OXIDES

4.2.1 PHYTOSYNTHESIS OF ZINC OXIDE NANOPARTICLES (ZnO NPS)

Many plant parts have been applied for the green synthesis of ZnO NPs. Clark and Macquarrie (2002) reported the phytosynthesis of ZnO NPs using *Acalypha indica* fresh leaf aqueous extract and zinc acetate dihydrate ($ZnC_4H_6O_4$) (99% purity) as precursors. Within 2 h and at alkaline pH 12, the precipitates of ZnO NPs were taken out and washed repetitively with distilled water followed by ethanol to remove the impurities of the obtained product, and then dried overnight at 60 °C in a vacuum oven overnight. Hot water extract of *Aloe barbadensis* miller leaf has been reported for the biopreparation of wurtzite structure, polydispersed spherical, and hexagonal-shaped 25–40 nm ZnO NPs using zinc nitrate as precursor (Sangeetha et al., 2011). It is worth mentioning that the average size of the prepared ZnO NPs increased with the extract concentration (Sangeetha et al., 2011). The plant *A. indica* has also the potential to be utilized for phytosynthesis of ZnO NPs. Fifty milliliter of *A. indica* leaves' hot water extract was allowed to react with 5 g of zinc nitrate at temperature 60 °C (Gnanasangeetha and Tambavani, 2013). This whole mixture was then boiled until it was reduced to a deep yellow color paste. This paste was then collected in a ceramic container and heated in an oven at 400 °C for 2 h.

Bougainvillea glabra is a lush evergreen subtropical vine (Ali et al., 2013). It belongs to the family Nyctaginaceae. It is native to Latin America (Brazil), commonly grown in gardens, porches, and boundary walls (Shah et al., 2006). Among different species of ornamental plants, *B. glabra* is a well-known medicinal plant that grows on various lands and climates (Gillis, 1976). The *B. glabra* leaf water extract was used for the preparation of ZnO NPs (Samzadeh-Kermani et al., 2016). The XRD study revealed the prepared ZnO NPs that were crystalline in nature. Hexagonal-shaped ZnO NPs were revealed from SEM images. The biologically prepared ZnO NPs showed strong antibacterial activity against Gram-positive and Gram-negative bacteria.

Bhumi and Savithramma (2014) used the aqueous leaf extract of *Catharanthus roseus* and zinc acetate for the preparation of ZnO NPs, whereas the pH was adjusted to 12 using NaOH. Jafarirad et al. (2014) reported the green synthesis of ZnO NPs and CuO NPs using the aqueous extract of *Rosa sahandina*. Koli (2015) synthesized ZnO NPs using aqueous extract of *Cheilocostus speciosus* (Costus speciosus) leaves. It expressed an efficient antibacterial effect and good antidiabetic property. Later, the green synthesis of ZnO NPs was successfully achieved by using *Ruta graveolens* stem hot water extract (Lingaraju et al., 2016). The UV/Vis absorption spectrum showed an absorption band at 355 nm due to ZnO NPs. Sharp intense diffraction peaks appearing at 2 h of 31.75°, 34.40°, 36.26°, 47.53°, 56.61°, 62.90°, 67.95°, and 69.06° correspond with those from (100), (101), (102), (110), (103), (200), (112), and (201) orientations, respectively. The crystallite size was found to be 28 nm using Debye–Scherer's formula. SEM images showed that the particles were spherical in shape with a size of 28 nm. Significant antibacterial activity was observed against four bacterial strains namely Gram-negative *Klebsiella aerogenes* (National collection of industrial microorganisms (NCIM) 2098), *Escherichia coli* (NCIM-5051) and *Pseudomonas desmolyticum* (NCIM-2028) and a Gram-positive *Staphylococcus aureus* (NCIM-5022). *Limonia acidissima* L. leaf extract was reported to mediate the synthesis of ZnO NPs and AgNPs (Patil and Taranath 2016). The presence of alcohol, phenol, carboxylic acids, and alkaloids in *L. acidissima* leaf extract was responsible for reducing, capping, and stabilization of AgNPs and ZnO NPs. The results revealed that 400 μg/mL AgNPs showed the maximum zone of inhibition 15.16, 15.5, and 13.33 mm against *S. aureus*, *S. typhi*, and *P. aeruginosa*, respectively. While ZnO NPs showed less activity in comparison to AgNPs owing to the agglomeration of NPs. Moreover, highly stable and spherical ZnO NPs were produced by using zinc acetate and *Ixora coccinea* leaf hot water extract (Yedurkar et al. 2016). An absorption peak was obtained at 340 nm using UV/Vis spectrophotometer. Dynamic light scattering (DLS) analysis showed an average particle size of 145.1 nm whereas a high zeta potential value confirmed the stability of the formed ZnO NPs. The SEM analysis revealed spherical morphology of NPs and energy dispersive X-ray (EDX) analysis confirmed the formation of highly pure ZnO NPs. Hence, the leaf extract of *I. coccinea* act as a reducing as well as surface stabilizing agent for the synthesis of ZnO NPs. The potential of *Parthenium hysterophorus* leaf water extract for the synthesis of ZnO NPs was evaluated (Datta et al. 2017). The green synthesized NPs were characterized using UV/Vis spectroscopy revealing a maximum absorbance peak at 400 nm. Spherical and cylindrical-shaped particles were revealed from SEM and TEM analysis with average particle size ranging from 16 to 45 nm. Jeyabharathi et al. (2017) reported the phytosynthesis of spherical-shaped ZnO NPs using the aqueous extract of *Amaranthus caudatus*, which acted as a reducing and capping agent. Those green ZnO expressed a more efficient antibacterial effect on the Gram-positive *Staphylococcus epidermidis* than the Gram-negative *Enterobacter aerogene*. The green ZnO NPs were found to influence the normal development of zebrafish embryos in a dose-dependent manner, it was proved at high concentrations (>10 mg/mL). Santhoshkumar et al. (2017) reported the green synthesis of ZnO NPs using *Passiflora caerulea* leaf hot water extract. Whereas the synthesis of ZnO NPs was evident from the color change from pale white to yellow. A characteristic peak to ZnO NPs appeared at 380 nm using UV/Visible spectrophotometry. X-ray diffraction (XRD) peaks obtained at 31.8°, 34.44°, 36.29°, 47.57°, 56.61°, 67.96°, and 69.07° corresponded to the lattice plane of (100), (002), (101), (102), (110), (112), and (201) suggesting the face-centered cubic (fcc) crystal structure of the prepared NPs, respectively. Using Scherer's equation, the average size of the synthesized NPs was found to be 37 nm. While the SEM analysis of ZnO NPs demonstrated that the size was approximately in the range of 30–50 nm. Chaudhuri and Malodia (2017) also reported the biosynthesis of ZnO NPs using *Calotropis gigantean* leaf hot water extract. The combination of 200 mM zinc acetate salt and 15 mL of leaf extract was ideal for the synthesis of nearly 20 nm ZnO NPs. The UV/Vis absorption band showed a peak near 350 nm, which is characteristic of ZnO NPs. The DLS data showed a single peak at 11 nm (100%) and the polydispersity index (PdI) of 0.245. The XRD analysis showed that they were highly crystalline with an average size of 10 nm. The SEM images showed that the particles were spherical in nature. The presence of zinc and oxygen was confirmed by EDX and the atomic % of zinc and oxygen were 33.31 and 68.69, respectively. Two-dimensional and 3D images of ZnO NPs were obtained by atomic force microscopy (AFM) studies, which indicated that those were monodisperse having size ranges between 1.5 and 8.5 nm. The presence of phytochemicals (eicosatrienoic acid methyl ester, hexatriacontaine, trimethyl undecatriene, and trifluoroacetic acid), volatile essential oil (phytol), and flavonoids (varinging, quercitrin, hesperitin, and kaempferol) aided in the biosynthesis process of ZnO NPs. The *C. gigantean* leaf water extract also contains acalyphamide, 2-methylanthraquinone, tri-*o*-methyl ellagic acid, sitosterol, glucoside, stigmasterol, quinine, tannins, resins, and essential oils (Chaudhuri and Malodia, 2017).

Different fruit peels' extracts were used as reducing agents to reduce zinc nitrate as a source of the zinc ions into its nanoparticle state (Nava et al. 2017). *Citrus paradisi* (grapefruit) and *Citrus aurantifolia* (lemon) peel hot water extract was used to biologically synthesize ZnO NPs. The formation mechanism was explained on the basis that the biomolecules including flavonoids, limonoids, and carotenoids present in the peel extracts act as ligation agents. The aromatic hydroxyl groups from the biomolecules formed complexing agents with the precursor salts and ligate with zinc ions. This, in turn, started the nucleation process, which then led to the reduction and shaping of the NPs. That system undergoes direct decomposition when calcinated at 400 °C which led to the release of the ZnO NPs (Nava et al., 2017).

Zare et al. (2017) reported the capability of aqueous extract of *Cuminum cyminum* (cumin) for the green synthesis of spherical/oval highly stable ZnO NPs with an average size of 7 nm and characteristic absorption peak in the range of 370 nm. The optimum operating conditions were pH 8, 70 °C, 1 mM zinc nitrate, and 10 mL extract. It showed the higher antibacterial effect on Gram-negative bacteria than the Gram-positive ones, because of the resistive thick layers of peptidoglycan in the cell walls. It expressed also an efficient antifungal effect. While the green synthesis of ZnO NPs using *Chelidonium majus* hot water extract and the assessment of their cytotoxic and antimicrobial properties were performed by Dobrucka et al. (2018). The study showed that the biosynthesized ZnO NPs with an average size of 10 nm, were of excellent antimicrobial activities, and demonstrated high efficiency in the treatment of epithelial cells of human nonsmall cell lung cancer A549. 2-hydroxy-1, 4-napthoquionone, Linalool α-terpineol, EtherphenylvinylA 1,3-indandione, Eugenol, Oxirane-tetradecyl, and Hexadecanoic acid the main constituents of *Lawsonia inermis* plant extract are reported to be responsible for the bioreduction, capping, and stabilization of ZnO NPs (Jayarambabu et al., 2018). The green synthesis of ZnO NPs by *Pongamia pinnata* leaf hot water extract was studied by Malaikozhundana and Vinodhini (2018). *P. pinnata* is a forest tree belonging to the family Leguminosae and is commonly used for biodiesel production (Meera et al. 2003). Different plant parts such as leaves, root, bark, flowers, and seeds of *P. pinnata* contain a number of furano-flavonoid compounds including karanjin, pongamol, pongapin, glabrin, karanja chromene, karanjone, and pongaglabrone are the principal furano-flavonoids present in the plant. The green synthesized ZnO NPs were crystalline in nature with a mean size of 21.3 nm and the zeta potential measurement demonstrated that the ZnO NPs were negatively charged (−12.45 mV) and were moderately stable. In another study, upon mixing the hot water extract of green leaves of *Tabernaemontana divaricate* with zinc nitrate hexahydate, a yellow-colored paste was formed. That turned to white ZnO NPs after heating for 2 h, at 450 °C (Raja et al., 2018). The absorption peak at 376 nm with an energy gap of 3.26 eV encouraged ZnO NPs photocatalytic degradation efficiency on methylene blue dye under sunlight irradiation, due to the production of electrons and holes, upon sunlight irradiations. Where, the produced superoxide radicals by the reaction of an electron with oxygen and hydroxyl radicals generated by the reaction of holes with hydroxide ions/water, degraded the dye molecules. The diffraction peaks of 2θ = 31.73, 34.33, 36.00, 47.31, 56.37, 62.48, 66.08, 67.61, and 68.70 which correspond to (100), (002), (101), (102), (110), (103), (112), (201), and (004) confirmed the production of hexagonal wurtzite crystalline structured ZnO NPs, respectively. The Fourier-transform infrared (FTIR) analysis proved the involvement of phenolic acid, flavonoids, and vitamins extract constituents in phytosynthesis, capping, chelating, and stabilization of 20–50 nm ZnO NPs (Scheme 4.1).

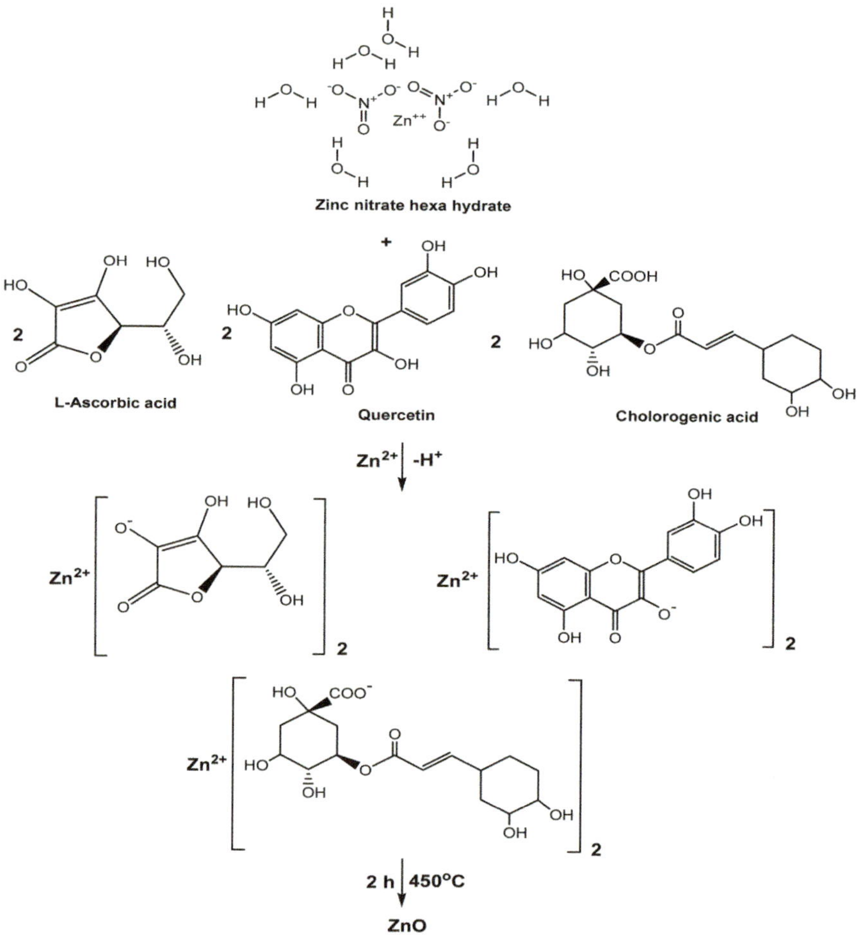

SCHEME 4.1 Phytosynthesis mechanism of ZnO NPs using hot water extract of *Tabernae-montana divaricate.*

Singh et al. (2018) reported the green and sustainable synthesis of ZnO quantum dots (QDs) using zinc acetate (precursor) and *Eclipta alba* leaf hot water extract as a reducing agent. The optimum conditions were found to be 40 °C, pH 7, 5 mL zinc acetate (5 mM), 7 mL leaf extract, and reaction time of 75 min. The transmission electron microscopy (TEM) depicted homogeneous distribution of spherical ZnO QDs with average particle size of 6 nm. The selected area electron diffraction (SAED) analysis revealed the crystalline nature of ZnO QDs with a hexagonal wurtzite phase and lattice constants a = b = 0.32 nm, and c = 0.52 nm. Those ZnO QDs expressed significant antibacterial activity against *E. coli*. Raja et al. (2018) reported also the efficient biocidal effect of ZnO NPs against *Salmonella paratyphi*, *E. coli*, and *Staphylococcus aureu*. Abbasi et al. (2019) reported the phytosynthesis of crystalline hexagonal-shaped 35 nm ZnO NPs using hot water extract of *Linum usitatissimum* roots. While Bayrami et al. (2019) reported the ultrasonic-assisted phytosynthesis of spherical-shaped 21 nm ZnO NPs using aqueous leaf extract of whortleberry (*Vaccinium arctostaphylos* L.). The prepared ZnO NPs expressed an efficient biocidal effect on *S. aureus* and *E. coli* and photodegradation on Rhodamine B (RhB) dye. In another study, the hot water extract of Cucurbita seed has been reported for the phytosynthesis of crystalline wurtzite structure, rod, rectangular, and hexagonal-shaped ZnO NPs with the average size of 35 nm (Velsankar et al., 2019). The prepared ZnO NPs expressed an efficient biocidal effect against *E. coli*, *Bacillus pumilus*, and *Salmonella typhi* bacteria, and *Aspergillus flavus* and *Aspergillus niger* fungi (Velsankar et al., 2019). That biocidal effect was attributed to the electrostatic interaction between the positively charged ZnO NPs and the negatively charged microbial cells. These ZnO NPs penetrated the cell wall with their lipids, proteins, and sugars layer, which leads to cell membrane elongation and rupture. The formation of the easily cell wall penetrating Zn^{2+} (Figure 4.1) with the formation of ROS H_2O_2 (hydrogen peroxide), OH^- radical, O_2^{-2} ions lead to an oxidative stress mechanism, impair and attack DNA and cell lines. That consequently, hinders the bacterial growth and development. Moreover, it leads to cell wall disruption, exertion of cytoplasmic contents, and liberation of the interior cell contents leading to cell damage, and finally, cell death occurs (Velsankar et al., 2020).

4.2.2 PHYTOSYNTHESIS OF TITANIUM DIOXIDE NANOPARTICLES (TiO_2 NPS)

Leaf extract of *C. roseus* was reported to phytosynthesize irregular-shaped titanium dioxide NPs, with average size ranged from 25 to 110 nm

(Velayutham et al., 2012). *Psidium guajava* leaf hot water extract mediated the biosynthesis of TiO_2 NPs (Santhoshkumar et al., 2014). *P. guajava* leaves are commonly used as a popular medicine for diarrhea, wounds, ulcers, rheumatic pain, and they are also can be chewed to relieve toothache. Quercetin is a major flavonoid present in *P. guajava* leaves. XRD pattern of the synthesized TiO_2 NPs showed the presence of both anatase and rutile forms which can be denoted at 2θ peaks at 27.57°, 36.21°, 41.37°, 54.45°, 56.76°, and 69.12° for (110), (101), (111), (211), (220), and (112) reflections, respectively. Field emission scanning electron microscope (FESEM) revealed the presence of spherical-shaped particles with an average size of 32.58 nm. Abdul Jalil et al. (2016) reported two methods for the green synthesis of TiO_2 NPs with good optical properties, using *Curcuma longa* plant aqueous extract. One produced 91.37 nm anatase colloidal solutions and 76.36 nm nanopowder of titanium dioxide, with average crystallite size of 22.881 and 43.088 nm, respectively, and the other method produced just 92.6 nm pure anatase nanopowder, with average crystallite size of 45.808 nm. The nanopowder had three crystalline forms: anatase, rutile, and brookite. All the prepared TiO_2 expressed more effective antifungal activity relative to that of industrial synthetic TiO_2 NPs.

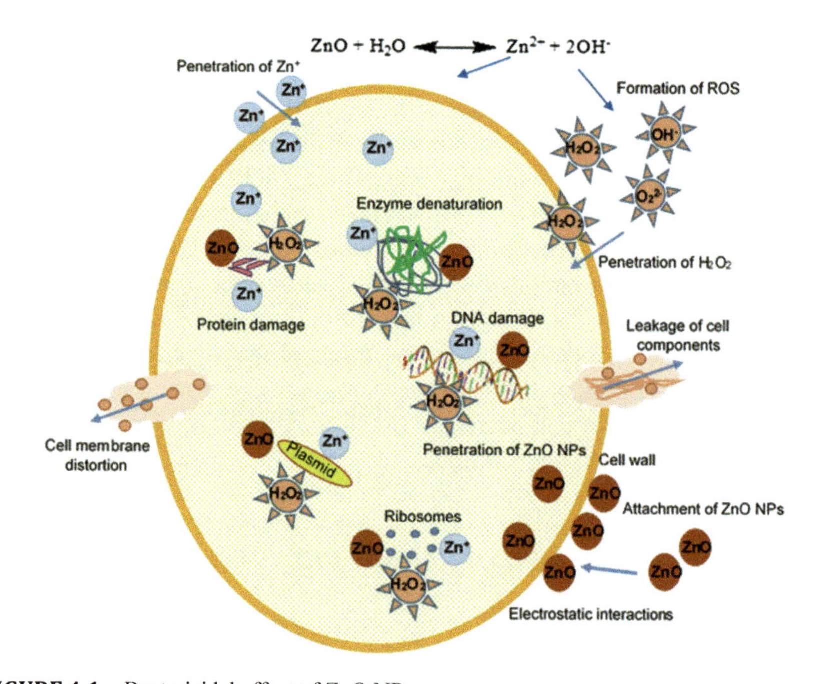

FIGURE 4.1 Bactericidal effect of ZnO NPs.

The green synthesis of TiO_2 NPs was attained also by using *Trigonella foenum-graecum* leaf hot water extract (Subhapriya and Gomathipriya, 2018). The XRD displayed the existence of TiO_2 NPs which was confirmed by the incidence of peaks at 25.28° corresponds to 101 reflections of anatase form. The high-resolution scanning electron microscope (SEM) perceptions revealed spherical-shaped TiO_2 with average size of 20–90 nm. Goutam et al. (2018) reported the green synthesis of TiO_2 NPs using *Jatropha curcas* leaf hot water extract. Aqueous leaf extract of *Artemisia haussknechtii* has been also reported for phytosynthesis of TiO_2 NPs (Alavi and Karimi, 2018). *Jatropha* belongs to the family, *Euphorbiaceae*, and widely cultivated as a petro plant for biodiesel production in many parts of the world. DLS revealed the average size particles of ≈75 nm. The EDX spectrum revealed the purity of the prepared TiO_2 NPs as both titanium (Ti) and oxygen (O) were present in the sample with weight percentage, 34% and 23%, and atomic percentage, 13% and 26%, respectively. The XRD confirmed the crystallinity of the prepared TiO_2 NPs. The FESEM images evidenced the successful green synthesis of spherical-shaped TiO_2 NPs with a diameter ranging from 10 to 20 nm. The phytochemicals such as polyphenols, flavonoids, alkaloids, antioxidants, terpenoids, steroids, free amino acids, and tannins present in the leaves of *J. curcas* perhaps proved to play a key role, as capping and stabilizing agents for the prepared particles. The surface area of the green synthesized TiO_2 NPs was 27.038 m^2/g. Results confirmed that the green synthesized TiO_2 NPs had mesoporous nature with a large surface area and thus, contained a number of active sites on the surface, which enhanced the adsorption phenomenon, consequently it could play a significant role in the removal of pollutants from wastewater. The green synthesized TiO_2 NPs showed 82.26% and 76.48% removal of chemical oxygen demand (COD) and chromium (Cr^{+6}), simultaneously, from real tannery waste water (TWW) after its solar photocatalytic treatment and thus, convincingly demonstrated the remarkable potential for wastewater treatment using the green synthesized TiO_2 NPs. The photocatalytic degradation of TWW obeyed the first-order kinetics.

4.2.3 PHYTOSYNTHESIS OF COPPER OXIDES (CuO AND Cu_2O) NANOPARTICLES

Many reports have been published for the phytosynthesis of copper oxide NPs, *Magnolia kobus* (Lee et al., 2011), *A. barbadensis* Miller (Gunalan et al., 2012), *Gloriosa superba* L. plant extract (Naika et al., 2015), *Ormocarpum cochinchinense* leaves (Gnanavel et al., 2017), *Saraca indica* leaves (Prasad et al., 2017), *Acanthospermum hispidum* L. extract (Pansambal et al., 2017), *Ixora coccinea* extract (Yedurkar et al., 2017), and *Hibiscus rosa-sinensis* flower extracts (Rajendran et al., 2018). Eugenol, eugenic acid, caryophyllene, urosolic acid, luteolin, rosmarinic acid, aesculin, limatrol, linalool, apigenin, isothymusin, carotene, and ascorbic acid present in *Ocimum tenuiflorum* leaf extract (Sumitha et al., 2016), and cysteine proteases present in the latex of *Calotropis procera* L. (Harne et al., 2012) were responsible for the synthesis of spherical-shaped CuO NPs. Moreover, rod-shaped 140 nm CuO NPs were prepared by *Carica papaya* leaves extract (Sankar et al., 2014). While, the aldehyde group present in reducing sugars of *Manihot esculenta* leaf extract played an important role in the phytosynthesis of Cu_2O NPs (Ramesh et al., 2011). The water-soluble carbohydrates present in *Tridax procumbens* leaf extract were responsible for the preparation of a mixture of hexagonal and cubic Cu_2O NPs, within only 10 min, using Fehling's solution as precursor (Gopalakrishnan et al., 2012). In another study, the hot water extract of green tea (*Camellia sinensis*) leaves was used to prepare truncated cubes and spheres-shaped Cu_2O NPs with SPR 466 nm and average particles size of 34.36 nm (Riya and George, 2015). Whereas, the carbohydrates with aldehydes group acted as the main reducing, capping, and stabilizing agents.

$$RCHO + 2Cu^{2+} \longrightarrow 2H_2O + RCOOH + Cu_2O + 4H^+$$

Moreover, the produced Cu_2O NPs expressed efficient bactericidal effect on pathogenic *E. coli*, *S. aureus*, *K. pneumoniae*, *P. mirabilis*, and *B. cereus* (Gopalakrishnan et al., 2012).

Borah et al. (2016) reported the phytosynthesis of octahedral 4–10 nm Cu_2O NPs using aqueous extract of *Syzygium jambos* (L.) leaves. Ramesh and Gopalakrishnan (2016) reported the phytogenesis of nanoneedles Cu_2O NPs within only 15 min using aqueous extract of *Arachis hypogaea* leaves. Borah et al. (2017) reported the phytosynthesis of spherical-shaped 7.33 nm Cu_2O NPs, within only 10 min, using extract of banana pulp.

Even plant wastes such as banana peels (Ghosh et al., 2017) have been applied for the preparation of Cu_2O NPs. In another study by Nwanya et al. (2019), red-colored cubic 10–26 nm Cu_2O NPs were prepared by the aqueous extract of *Zea mays L.* dry husk. Upon annealing such Cu_2O at 600 °C for 2 h, they were completely thermally oxidized into spherical-shaped 30–90 nm CuO NPs. Whereas the CuO NPs produced at an annealing temperature of 300 °C is recommended as a biocide for *E. coli* and *S. aureus*. While the Cu_2O NPs is recommended as biocide for *Pseudomonas aeruginosa* and *Bacillus licheniformis*. Moreover, a mixture of Cu_2O and CuO NPs expressed more efficient photocatalytic degradation properties on different organic dyes than pure CuO NPs. This was attributed to the presence of more electron-hole pairs and a decrease in the electron-hole recombination rate. Since, CuO is a p-type semiconductor with a narrow bandgap of 1.3–1.7 eV (Gnanavel et al., 2017). Thus, its conduction band and valence band are lower than those of Cu_2O (2.15 eV). Therefore, the copper oxides NPs mixture forms a heterostructure type II staggered band structure and favored charge transfer which consequently positively enhanced the photocatalytic efficiency. Lignin in alkaline medium has been also reported for the phytosynthesis of Cu_2O NPs (Li et al., 2016). Whereas copper hydroxide is formed in an alkaline medium and the Cu^{2+} ions have been reduced by the strong reducing lignin aliphatic hydroxyl groups into Cu(I) of Cu_2O NPs. While the hydroxyl groups of lignin are oxidized to carbonyls. At the same time, lignin acts as a strong stabilizing agent for the produced Cu_2O NPs via its sufficient adsorption properties for metal ions by its many oxygenous groups (Scheme 4.2).

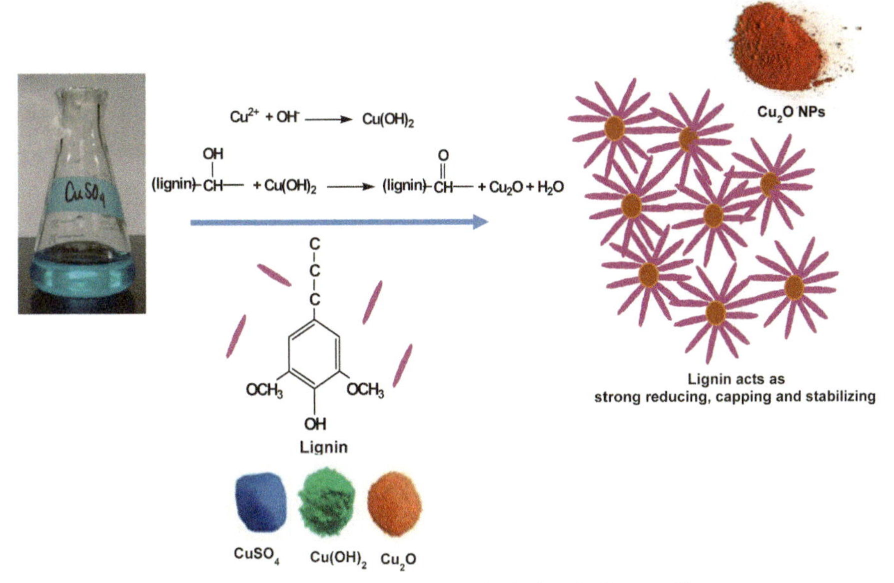

SCHEME 4.2 Phytosynthesis of Cu_2O NPs via lignin in alkaline medium.

Seaweeds have been also reported for the biosynthesis of copper oxide nanoparticles. For example, brown seaweed *Bifurcaria bifurcate* extract was reported to produce spherical-shaped Cu_2O and CuO NPs with average particles size of 5–45 nm (Abboud et al., 2014). While the spherical

Cu-cored Cu_2O NPs with average particles size of 53 nm were reported to be prepared using an aqueous extract of the red seaweed *Kappaphycus alvarezii* (Khanehzaei et al., 2014).

Odoom-Wubah et al. (2015) reported the phytosynthesis of flower like CuO microstructures with average size of average size 1.372.0 mm and consisted of several 2D nanorods, using aqueous leaf extract of *Cinnamomum camphora,* at room temperature. Hot water extract of *Drypetes sepiaria* leaf has been reported for the phytosynthesis of cupric oxide CuO NPs (Narasaiah et al., 2017). Whereas the deep blue color of copper nitrate solution changed to brownish-black with surface plasmon resonance of 298 nm. The XRD analysis confirmed the characteristic diffraction peaks at 2θ of 32.37, 35.19, 38.49, 48.14, 58.01, 61.31, 66.01, 67.66, 72.37, and 75.19, assigning to (110), (11–1), (111), (20–2), (202), (113), (022), (220), (31–2), and (004) planes, respectively, of monoclinic structure of CuO with the average crystalline size of 25 nm. While the TEM analysis proved the biomolecular capped spherical-shaped 18.77 nm CuO NPs. The prepared CuO NPs expressed potential photocatalytic activity on Congo red dye, producing Lueco Congo red, within 14 min and in presence of a low concentration of $NaBH_4$. *P. guajava* hot water leaf extract has been reported for the phytosynthesis of 19.19 nm elongated spheroids CuO NPs (Sreeju et al., 2017). The phytosynthesized CuO NPs proved to be a cost-effective catalyst for wastewater treatment. It expressed efficient biocidal activity against *E. coli* and *S. aureus*. That CuO NPs performed good catalytic degradation properties toward different dyes; methylene blue, methyl orange, methyl red, and eosin yellow, in presence of $NaBH_4$. Moreover, it efficiently catalyzed the reduction of 2-nitrophenol, 3-nitrophenol, and 4-nitrophenol into their corresponding less toxic amino-compounds, in the presence of excess $NaBH_4$ (Sreeju et al. 2017). While Yedurkar et al. (2017) reported the production of spherical-shaped CuO NPs using hot water extract of *I. Coccinea* leaf. Whereas the DLS analysis proved monodispersed and highly stable CuO NPs with average particle size of 167.1 nm, PDI of 0.345, and with Zeta potential of −21 mV. In another study by Yugandhar et al. (2017), the hot water extract of *Syzygium alternifolium* stem bark phytosynthesized polydispersed nonagglomerated spherical-shaped 5–13 nm CuO NPs, with efficient microbicidal effect on *E. coli* and *T. harzianum*. While Sathiyavimal et al. (2018) reported the green synthesis of brownish-black nanorods 50 nm CuO NPs using *Sida acuta* leaves extract, which expressed efficient photocatalytic degradation properties on toxic organic pollutants and efficient antimicrobial activity against Gram-positive and Gram-negative bacteria. The EDX diffractive analysis proved the high purity of the produced CuO NPs. The saponin-rich *Sapindus mukorossi* fruit extract has been also used for the phytosynthesis of CuO nanowires (Sundar et al., 2018). The TEM analysis proved a uniform 1D nanostructure CuO nanowires that were 800 nm in length and 50–100 nm in width that is made up of several small nanowires with a width of about 10 nm (Sundar et al., 2018). The hot water extract of the brown algae *Cystoseira trinodis* has been used for the photosynthesis of CuO NPs, using $CuSO_4$, as precursor salt (Gu et al., 2018). Within only 90 min and under ultrasonic conditions (400 w, 20kH), spherical-shaped 9 nm CuO NPs have been produced. Which expressed efficient photodegradation of methylene blue under UV and solar irradiation and efficient biocidal activity against the Gram-negative bacteria *E. coli* (ATCC 27853), *Enterococcus faecalis* (ATCC 9212), *Salmonella typhimurium* (ATCC 14028), and Gram-positive bacteria *S. aureus* (ATCC 25923), *Bacillus subtilis* (ATCC 12228), and *Streptococcus faecalis* (ATCC 29212) (Gu et al., 2018).

The components of the walnut (*Juglans regia*) leaf extract, malic acid, 3-o-caffeoylquinic acids, quercetin-o-pentoside, sucrose, disaccharide, α-tocopherol, tocopherol isomer, and other phenolic compounds, have been also reported for the green synthesis of CuO NPs (Asemani and Anarjan, 2019). Whereas the optimization of genesis process has been performed applying response surface methodology, based on Box–Behnken experimental design. That produced a monodispersed, highly stable, spherical, and crystalline, 80 nm CuO NPs with SPR 226 nm, using 1 g copper salt, and 14 mL walnut leaf extract applying 490 °C of furnace temperature. Beetroot (*Beta vulgaris* L) hot water extract was also reported for the phytogenesis of CuO NPs from copper (II) sulfate pentahydrate (Chandrasekaran et al., 2019). Whereas the AFM showed bimodal distributions of CuO NPs with average size distribution from 11.4 to 63.9 nm. The prepared CuO NPs expressed higher antibacterial potentials on Gram-negative *E. coli*, *Pseudomonas*, and *Salmonella* than the tested Gram-positive *Staphylococcus*. Which was attributed to the thick rigid peptidoglycan layer in the Gram-positive bacterial cell walls that perform as a barrier for the penetration of CuO NPs. However, the overall high biocidal efficiency was attributed to the small size of the prepared NPs, 33.47 nm, as measured by TEM, and their high surface to volume ratio. Thus, bind to the cell membrane with high affinity, penetrate into the cell wall, and bind with sulfur and carboxyl group of amino acids. CuO NPs would also produce ROS, finally destroying DNA, inactivating essential enzymes and/or proteins, leading to cell mortality (Chandrasekaran et al., 2019). Khatami et al. (2019) reported the efficient antifungal effect of 80 nm sized CuO

NPs phytosynthesized using tea extract. Chen et al. (2019a) reported the phytosynthesis of CuO NPs using the aqueous extract of papaya leaf. The phytosynthesized spherical-shaped 2.29 nm CuO NPS proved to be an efficient biocide against the soilborne *Ralstonia solanacearum*. Chen et al. (2019a) reported that CuO NPs prevent biofilm formation, reduce swarming motility and disturb ATP production. Whereas the absorption of multiple CuO NPs by the bacterial cells causes nanomechanical damage to the cytomembrane and downregulation of genes involved in pathogenesis and motility. In another study by Buazar et al. (2019), the X-ray photoelectron spectroscopy proved the purity of the prepared CuO NPs using hot aqueous wheat (*Triticum aestivum*) seed extract. Where the main characteristic peaks of CuO are 939.87 eV and 954.57 eV, which correspond to the Cu $2p_{3/2}$ and Cu $2p_{1/2}$, respectively (Prasad et al., 2017). The CuO NPs appeared as a single sharp peak with SPR at 300 nm, only after 25 min of reaction. The FTIR spectroscopy proved the formation of CuO and the involvement of extract-derived electron-rich biomolecules in reducing Cu cations from divalent oxidation state into Cu^0 metallic, then directly converted to CuO NPs because of the superior chemical reactivity of bare nanoscale copper metal surface (Scheme 4.3). Such biomolecules are also involved in capping/stabilizing the produced CuO NPs (Scheme 4.3). The XRD pattern proved also the monoclinic structure of the produced CuO NPs with a crystallite size of 20.76 nm. The SEM and TEM proved the stability of the monodispersed spherical-shaped 22 ± 1.5 nm CuO NPs for 3 months. The phytosynthesized CuO NPs expressed efficient reusability in photocatalytic degradation of 4-nitrophenol, in the presence of $NaBH_4$ (Buazar et al., 2019).

4.2.4 PHYTOSYNTHESIS OF iron oxides NANOPARTICLES

Senthil and Ramesh (2012) synthesized Fe_3O_4 NPs by the reduction of ferric chloride ($FeCl_3$) solution using *T. procumbens* leaf hot water extract containing carbohydrates with an aldehyde group as a reducing agent. The possible reduction mechanism leading to the formation of Fe_3O_4 NPs from the iron precursor, $FeCl_3$ was proposed as follows. First, $FeCl_3$ was hydrolyzed to form ferric hydroxide and released H^+ ions in the proper pH value and temperature. After that, ferric hydroxide was partially reduced by the plant extract containing carbohydrates (glucose) to form Fe_3O_4 NPs; the aldehyde group was oxidized to the corresponding acid. In the XRD spectrum, four distinct peaks were clearly distinguished. The peaks were perfectly indexed to crystalline Fe_3O_4 not only in their peak positions but also

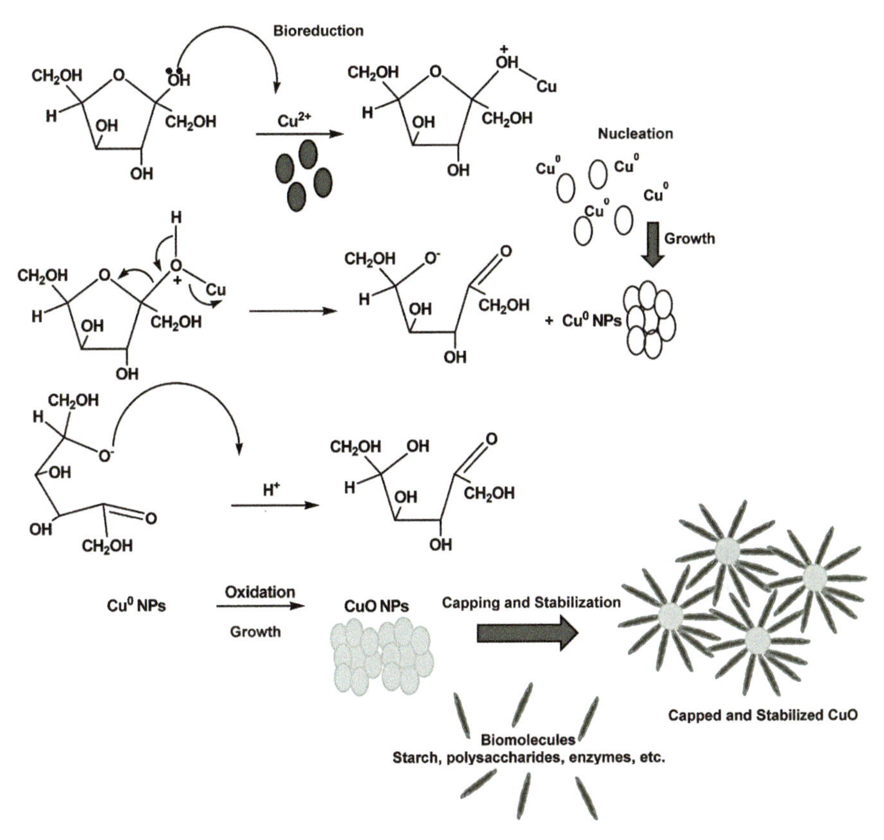

SCHEME 4.3 Phytosynthesis of CuO NPs by plant extracts.

in their relative intensities. The peaks with 2θ values of 29.1°, 35.4°, 56.7°, and 61.2° corresponded to the crystal planes of (200), (311), (511), and (440) of crystalline Fe_3O_4, respectively. Using Scherer's equation, the crystallite size was found to be in the range of 80–100 nm. The SEM micrographs revealed irregular sphere-shaped particles with rough surfaces. All the prepared NPs were well separated and no agglomeration was noticed. The prepared IO NPs exhibited antibacterial activity against *P. aeruginosa.* The biosynthesis of magnetic iron oxide (Fe_3O_4) nanoparticles by the reduction of $FeCl_3$ solution, using seaweed (*Sargassum muticum*) aqueous extract, has been reported (Mahdavi et al., 2013). The sulfated polysaccharides in the extract acted as reducing agent and efficient stabilizer, whereas the sulfate, hydroxyl, and aldehyde groups might have reduced the Fe^{3+} and stabilized the nanoparticles. The decrease in pH during the phytosynthesis process denoted the contribution of the OH groups in the reduction process.

Initially, $FeCl_3$ hydrolyzed to form ferric hydroxide with the release of H^+ ions then the formed ferric hydroxide was partially reduced by the extract constituents to Fe_3O_4-NPs, while the aldehyde groups were oxidized to the corresponding acids. The TEM revealed a size of approximately 18 ± 4 nm. The XRD showed the crystalline nature of the cubic-shaped NPs. Moreover, the magnetization measurements recorded with vibration sample magnetometer (VSM) revealed a specific saturation magnetization value of 22.1emu/g with a negligible coercivity Hc of hysteresis loop (82.3 Oe) and consequently no remanence Mr (2.75emu/g). All of those results indicated the superparamagnetic nature of the green synthesized Fe_3O_4-NPs. That was attributed to the small size of the NPs (<25 nm) and they have been considered to have a single magnetic domain. However, the magnetic property of these green synthesized nanoparticles was lower than those prepared by the coprecipitation method, since all the metal-reducing and stabilizer agents acted as a capping agent and provided a robust coating on the metal nanoparticles in a single step. But those green agents made the functional bioactivity of Fe_3O_4-NPs (antimicrobial) comparably higher than particles that were synthesized by chemical method.

The Fe_3O_4 NPs were also reported to be successfully synthesized using pomegranate (*Punica granatum*) leaf water extract (Rao et al. 2013). These NPs were utilized for modification of two strains (NCIM 3589 and NCIM 3590) of heat-killed yeast cells *Yarrowia lipolytica*, which were further employed as biosorbents to remove hexavalent chromium. Moreover, Lunge et al. (2014) synthesized 2–25 nm magnetite nanoparticles (MNPs) using tea wastewater extract. SEM showed cuboid/pyramid shaped crystal structures of Fe_3O_4 NPs. TEM showed the particle size in the range of 5–25 nm. XRD pattern was identical to magnetite. The prepared MNPs were tested for removal of arsenic As (III) and As (V) from aqueous solution. They exhibited high adsorption capacity for arsenic. The estimated cost of As (III) removal from the water was estimated to be negligible. The adsorption data obeyed Langmuir adsorption isotherm with a high adsorption capacity of 188.69 mg/g for arsenic (III), and 153.8 mg/g for arsenic (V). The mean sorption energy (E) calculated indicated physicochemical sorption process. A pseudo second-order kinetic model fitted best for As (III) adsorption on the prepared Fe_3O_4 NPs and the derived activation energy was 64.27 kJ/mol. Thermodynamic calculations revealed the endothermic nature of adsorption.

An environmentally friendly effective technique was demonstrated by Thakur and Karak (2014) to prepare IO/reduced graphene oxide nanohybrid at room temperature by using banana peel ash and *Colocasia esculenta* leaf water extract. The diffraction peaks of the pure IO nanoparticles observed at 2θ 30.15°, 36.2°, 43.32°, 53.89°, 57.13°, and 62.29° were assigned to (220), (311), (400), (422), (511), and (440) crystal planes, respectively. Makarov et al. (2014) reported the synthesis of 30 nm Fe_3O_4NPs using *Hordeum vulgare* and *Rumex acetos* leaf water extract. The role of pH was considerable in the stability of the prepared Fe_3O_4NPs. It was found that the stability of *H. vulgare* synthesized iron nanoparticles was increased at pH 3.0. Similarly, amorphous iron NPs with a diameter of 10–40 nm were produced by extract of *R. acetosa.* Where it was highly stable at low pH (pH = 3.7) as compared to *H. vulgare* (pH 5.8). Factorial design of experiments was applied to optimize the microwave-assisted phytosynthesis of Fe_3O_4-NPs by the aqueous extract of Turmeric (*C. longa* L.) leaves (Herlekar and Barve, 2014). The optimum conditions were found to be the 8% plant material, 1 molar $FeCl_3$ solution, and 9 h of contact time, which expressed the highest removal of Orthophosphate (PO_4) and COD from domestic sewage. The iron nanoparticles showed characteristic peak of magnetite at 377 nm. The SEM, EDX, and XRD analyses revealed well dispersed crystalline spherical-shaped magnetite with sizes ranged between 17.9 and 28.7 nm. Recently, Ehrampoush et al. (2015) reported that the size of IO NPs synthesized from *Citrus unshiu* tangerine peel water extract was dependent on its concentration. It was observed that the size of the NPs decreased with increasing the concentration of tangerine peel extract from 2% to 6%. Average size particle was recorded using DLS and was found to be 200 nm. However, the SEM image showed that the IO NPs synthesized by tangerine were relatively uniform, spherical-shaped particles with an average size of 50 nm. While Muthukumar and Matheswaran (2015) tailored IO NPs using *Amaranthus spinosus* leaf water extract. These NPs were spherical with rhombohedral phase structure, smaller in size with a large specific surface area, and less aggregation than those produced with sodium borohydride.

El-Kassas et al. (2016) reported the utilization of two seaweeds *Padina pavonica* (Linnaeus) Thivy and *Sargassum acinarium* (Linnaeus) extracts as reductants of $FeCl_3$ to phytogenic Fe_3O_4-NPs, with surface plasmon band around 402 nm and 415 nm and particle sizes of 10–19.5 nm and 21.6–27.4 nm, respectively. FTIR analyses proved that the sulfated polysaccharides in the algae extracts acted as dual functions, reducing the $FeCl_3$ and stabilizing the phytogenic Fe_3O_4-NPs. Then, the phytogenic Fe_3O_4-NPs were immobilized in calcium alginates beads and used for lead removal, which expressed adsorption capacity of 91% and 78% after 75 min, respectively. Yew et al. (2016) reported the application of seaweed (*K. alvarezii*) aqueous extract as a green reducing and stabilizing agents during the green synthesis of magnetite Fe_3O_4. The prepared magnetite was separated by the aid of an external permanent magnet. The prepared Fe_3O_4 NPs were found to be spherical shaped with an average size of 14.7 nm. The XRD analysis showed the diffraction peaks of Fe_3O_4 NPs at $2\theta = 30.56°$, 35.86°, 43.46°, 54.01°, 57.38°, 63.00°, and 74.46°, which were assigned to the crystal planes of (200), (311), (400), (422), (511), (440), and (533), respectively.

The Fe_3O_4-NPs were successfully prepared using the Andean blackberry leaf hot water extract (Kumar et al., 2016). The formation of Fe_3O_4-NPs was preliminarily indicated by the change in the coloration from yellow to black. It was revealed from TEM that the Fe_3O_4-NPs were spherical in shape and aggregated with size ranging from 40 to 70 nm. The SAED pattern recorded for the spherical Fe_3O_4 NPs clearly showed the ring like electron diffraction patterns, typical of the polycrystalline Fe_3O_4 NPs structure. The average particle size distribution of Fe_3O_4-NPs was 54.5 ± 24.6 nm. The watermelon rind aqueous extract was used as a solvent, capping, and reducing agent in the synthesis of highly crystalline spherical-shaped Fe_3O_4-NPs, with a fcc structure and average size of 20 nm. The VSM proved saturation magnetization (Ms) value of about 14.2 emu/g, the coercive force (Hc) of 285.58 G and magnetic remanence (Mr) of 2.62 emu/g. The green synthesized magnetic NPs proved to have catalytic activity for the synthesis of ethyl 6-methyl-4-(4-nitrophenyl)-2-oxo-1,2,3,4-tetrahydropyrimidine-5-carboxylate, in the presence of ethanol as a solvent. Shojaee and Mahdavi Shahri (2016) reported the green synthesis of cubic iron magnetic nanoparticles Fe_3O_4-NPs with the average size of 35 nm, using the aqueous extract of Shanghai White tea (*Camelia sinensis*) as a reducing and capping agent. The reduction potential of caffeine and other polyphenols in the extract was found to be sufficient to reduce metals, whereas the decrease in pH during the phytosynthesis process implied the contribution of the OH groups in the reduction process. The FTIR analysis suggested that the hydroxyl and carbonyl groups of the main constituents of the white tea: epicatechin, epigallocatechin, gallic acid, and epigallocatechin gallate played a key role during the formation of Fe_3O_4-NPs, via the following suggested mechanism: complexation with Fe salts, then simultaneous reduction of Fe and finally capping with the oxidized polyphenols/caffeine.

According to Ali et al. (2017) (α-Fe_2O_3), NPs were successfully synthesized from the *Citrus reticulum* (mandarin) peel hot water extract using ferric sulfate at pH 8. Then, the obtained IO was calcined at 800 °C for 2 h to obtain pure hematite nanoparticles. The prepared α-Fe_2O_3 NPs were porous, quasispherical, rough shaped particles with average particles size of 20–63 nm, weak ferromagnetic properties, and bandgap (Eg) of 2.38 eV (Ali et al., 2017). The *C. reticulum* (CR) peels' extract is mainly composed of 5-hydroxymethylfurfural; 2,3-dihydro-3,5-dihyroxy-6- methyl-4H-pyran-4-one (DDMP); 2-methoxy-4-vinylphenol (i.e., 4-vinylguaiacol); n-hexadecanoic acid; 5-methyl-2-furancarboxaldehyde; d-limonene; 2,4-dihydroxy-2,5-dimethyl-3(2H) furan-3-one; and 3,5-dihdroxy-2-methy-4H-pyran-4-one (Omran et al., 2013). Thus, the presence of phenolic compounds and the flavonoid DDMP was suggested to act as reducing and stabilizing agents. While the 4-vinylguaiacol was suggested to can act as encapsulating and complexing agent. A color change from pale yellow to blackish-brown indicates the formation of IO nanoparticles occurred within few seconds. The formation of the IO NPs was depicted using UV/Vis spectrophotometer (Figure 4.2).

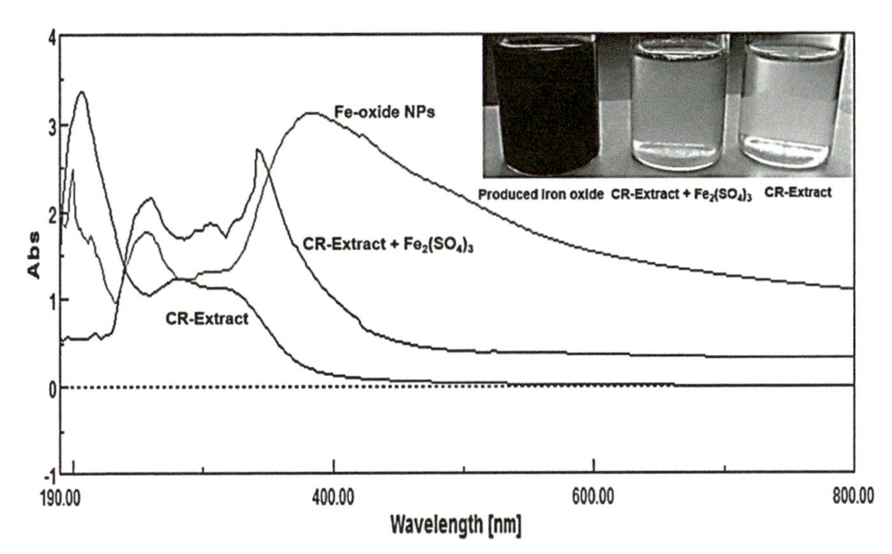

FIGURE 4.2 UV/Vis spectrum and steps of green synthesis of iron-oxide NPs.

Where the disappearance of the sharp peak at $\lambda_{342\ nm}$ in the UV-spectrum of the $Fe_2(SO_4)_3$/CR-extract mixture with the appearance of the broad peak in the range of $\lambda_{318\ nm}$–$\lambda_{608\ nm}$ in the UV spectrum of the IO with maximum absorbance at $\lambda_{384\ nm}$ was attributed to the oxidation of zero-valent iron to IO NPs. The thermal transition during the calcination process of the prepared IO was investigated with differential scanning calorimetric-thermal gravimetric analysis (TGA/DSC). Which depicted, three distinct weight loss steps: 30–115 °C, 150–250 °C, and 250–350 °C of ≈7%, 10.4%, and 6%, respectively, with an approximate total weight loss of 24% (Figure 4.3). The first mass loss might was attributed to the evaporation of water. The second mass loss was attributed to the volatilization of some organic compounds left from *C. reticulum* extract. The third minor mass loss might was attributed to the minor decomposition of the prepared IO.

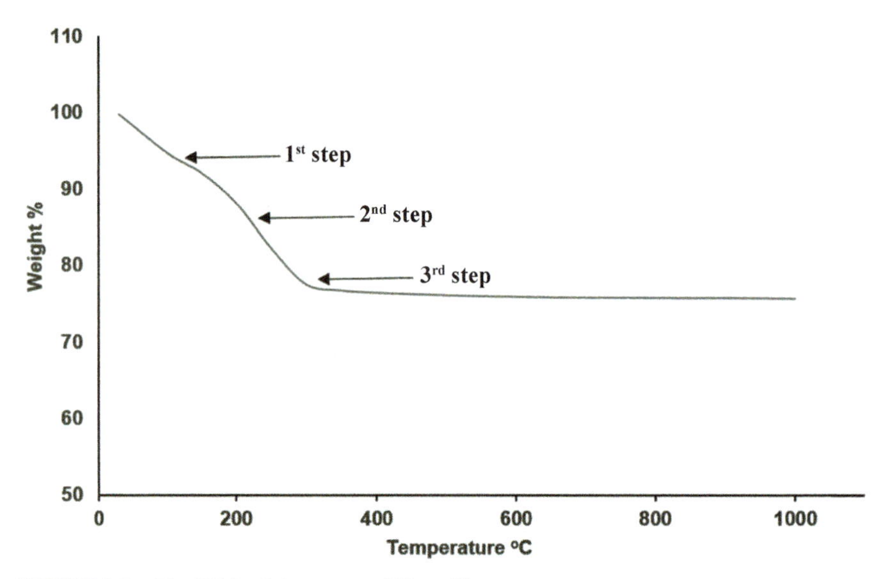

FIGURE 4.3 The TGA of the prepared Fe-oxide.

The XRD pattern (Figure 4.4) showed a pure crystalline α-Fe_2O_3 (JCPDS card number: 04-s008-7623), with the average crystalline size of 57.59 nm, as calculated from Scherer's equation. The FTIR-analysis showed the dominant peaks characteristic for the Fe-O stretching of α-Fe_2O_3 at 539.53 and 471.16 cm^{-1}, the broad peak near 3440.88 cm^{-1} assigned for the OH stretching, and the peak at 1635 cm^{-1} assigned for H–O–H. That proved the role of the extract constituents in the green synthesis of IO NPs. Ali et al. (2017) proposed two possible mechanisms for the green synthesis of hematite. Whereas the reduction would occur upon two steps: first, via the formation of a complex by breaking the –OH bond of the phenols in the CR-extract and forming a partial bond with the Fe^{3+} ions. Then, secondly, the breakage of the partial bond would occur, and then the electrons transferred to reduce the metal ions to nanoparticles and thus the phenols itself got oxidized to ortho-quinone. Then, the produced highly unstable Fe° transformed to IOs nanoparticles in a short span of time before it got finally capped with the components of the CR-extract.

$$nFe^{3+} + 3Ar{-}(OH)_n \longrightarrow nFe^0 + 2nAr{=}O + 2nH^+$$

where Ar is the phenyl group, and n is the number of hydroxyl groups oxidized by Fe^{3+}.

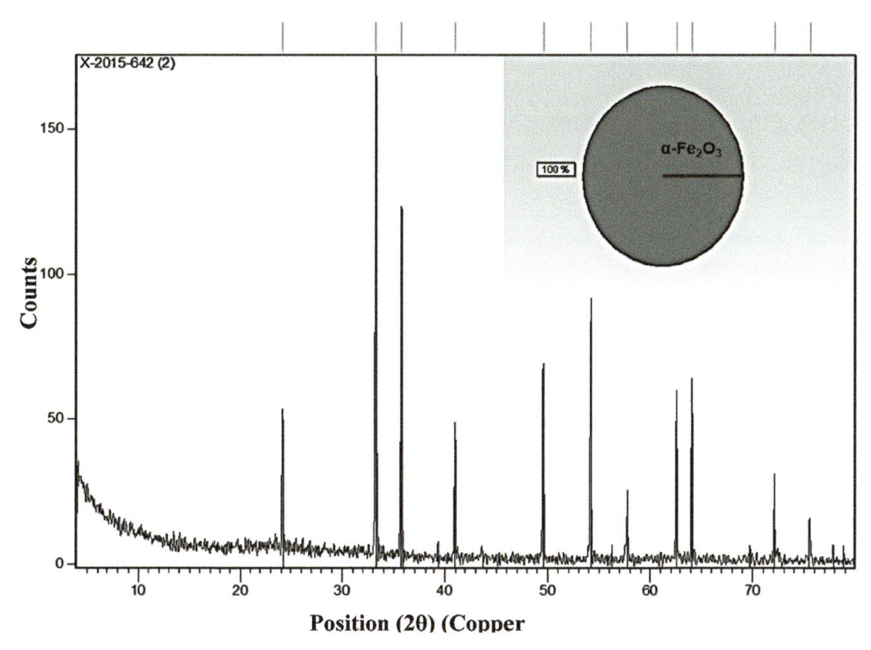

FIGURE 4.4 The XRD of the green synthesized iron oxide (α-Fe_2O_3).

While the suggested other mechanism was the ferric sulfate would have the first dissolved in water producing $[Fe(OH)_2(OH)]_2^+$ or $[Fe(OH)_4(OH)_2]^+$ via the deprotonation of the coordinated water molecules. The hydrolyzed iron species would have formed a complex with the deprotonated phenolic compounds in the CR-extract within the performed hydrothermal process. Aggregations of the relatively hydrophobic phenols would have been occurred, and at elevated temperatures, phase transition of $[Fe(OH)_2]_2^+$ to Fe_2O_3 occurred, whereas at higher temperatures, the phenolic compounds would have been decomposed, leaving pores in the α-Fe_2O_3, as was confirmed by the SEM and TEM analysis. The SEM micrograph of the green synthesized α-Fe_2O_3 (Figure 4.5a) revealed aggregates of irregular sphere-shaped particles with rough surfaces. While, the HRTEM micrograph (Figure 4.5b) proved aggregates of porous, irregular, rough, and quasispherical-shaped particles of α-Fe_2O_3. The particles size ranged between approximately 20 to

63 nm, which nearly coincided with that calculated from the XRD analysis. The prepared α-Fe_2O_3 NPs showed good photocatalytic decontamination of polluted water from anionic and cationic dyes and dicholorophenols under visible light irradiation (Figure 4.6). The primary photocatalytic oxidation mechanism is believed to proceed by the formation of the powerful oxidizing hydroxyl radicals and/or the adsorbed O_2 would have captured the photogenerated electrons, producing $O_2^{2-\bullet}$ that can directly degrade the pollutant. Moreover, the $O_2^{2-\bullet}$ can also react with the photogenerated holes and then form active $OH^\bullet$ radicals and peroxides which also photodegrade, the pollutants (Ali et al., 2017).

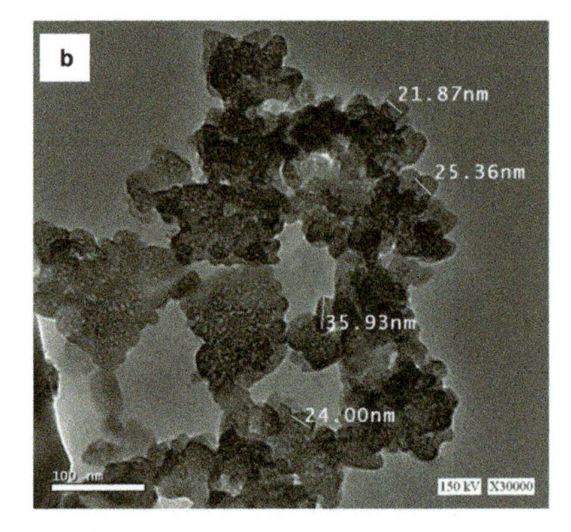

FIGURE 4.5 SEM (a) and TEM (b) micrographs of the pure calcined green synthesized α-Fe_2O_3.

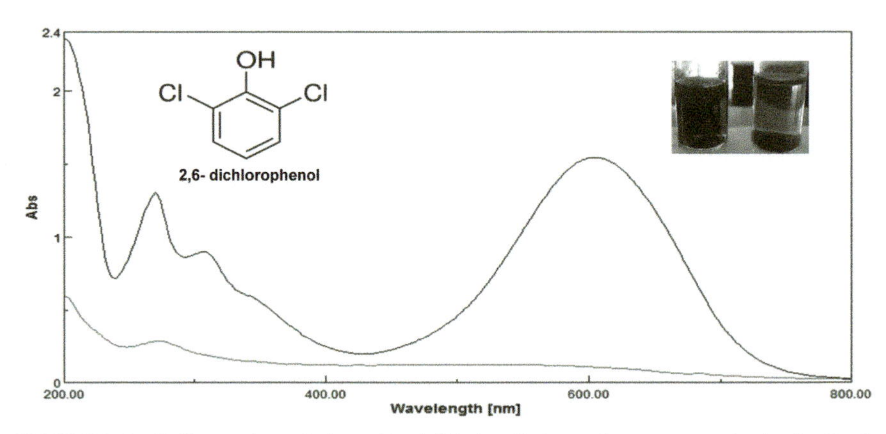

FIGURE 4.6 Pollutant degradation with visible irradiation using green synthesized α-Fe_2O_3 NPs.

Other studies using agroindustrial wastes proved that the green synthesized Fe_3O_4-NPs which would be capped with green substrates have an encouraging prospective in differ applications. Whereas the phytogenic Fe_3O_4-NPs prepared by extract of tea residue, coffee waste hydrochar, and corn Zea mays were reported for arsenic removal (Lunge et al., 2014), azo dye Acid Red 17 removal (Khataee et al., 2017), and antimicrobial and drug delivery applications (Patra et al., 2017), respectively.

Lagenaria siceraria leaves hot aqueous extract was found to be capable of phytosynthesis of cubic shaped and stabilized magnetized Fe_3O_4 NPs nanoparticles with a size range of 30–100 nm, with an enhanced antioxidant property and antibacterial effect against the Gram-negative *E. coli* and the Gram-positive *S. aureus* (Kanagasubbulakshmi and Kadirvelu, 2017). Aqueous ferric chloride hexahydrate and ferrous chloride tetrahydrate with 2/1 molar ratio and sodium hydroxide and leaf hot water extract of *Platanus Orientalis* L. were used to prepare highly stable magnetized quasispherical Fe_3O_4 with average size of 7.69 ± 1.55 nm and zeta potential of −24.80 mV. Whereas the extract acted simultaneously as an efficient stabilizer and capping agent (Nurbas et al., 2017). Rajiv et al. (2017) reported the preparation of IO NPs using cost-effective and environment-friendly method via *Lantana camera* leaf hot water extract. UV/Vis absorption spectrum showed a wide absorption peak at 370 nm. XRD confirmed the crystalline nature of the phytosynthesized IO nanoparticles. The nanoparticle size was found to be between 10 and 20 nm. EDX revealed the purity of the prepared IO NPs that were composed of 65.88% of iron and 34.12% of oxygen, respectively. Inhibition zone of $\approx 20.10 \pm 1$ mm was observed against *Pseudomonas* sp. at 100 $\mu g\ ml^{-1}$ of IO NPs. Synthesis of hematite nanoparticles (α-Fe_2O_3) NPs was accomplished by exploiting *Anacardium occidentale* leaf hot water extract as a reducing and capping agent (Rufus et al., 2017). The XRD pattern confirmed the crystalline nature of the synthesized hematite nanoparticles. The main characteristic peaks of α-Fe_2O_3 at 2θ values of 24.4°, 33.2°, 35.8°, 41.1°, 49.6°, 54.2°, 57.8°, 62.6°, 64.1°, 72.1°, 75.6°, 77.9°, 80.9°, 83.2°, 85.2°, and 88.7° corresponding to the reflections from the planes (012), (104), (110), (113), (024), (116), (018), (214), (300), (1010), (220), (036), (128), (0210), (134), and (226), respectively. Moreover, the irregular-shaped particles were revealed from the TEM image. The prepared α-Fe_2O_3 NPs showed weak ferromagnetism and enhanced the thermal conductivity of conventional base fluids: water and ethylene glycol at room temperature. It showed an efficient catalytic degradation to methyl red and eosin yellow dyes in the presence of excess sodium borohydride ($NaBH_4$). It expressed also sufficient biocidal activity against the pathogenic Gram-negative *E. coli* and Gram-positive *S. aureus* (Rufus et al., 2017). Tharunya et al. (2017) reported a single step and completely green biosynthetic method for the reduction of ferrous sulfate solution with superparamagnetic Fe_3O_4-NPs using fig (*Ficus carica*) fruit ethanolic extract. The *F. carica* is known to contain polyphenols and flavonoids that can act both as a reducing agent and a capping agent. While aqueous leaf extract of *Leucas aspera* was used to prepare magnetic Fe_3O_4 NPs with average size <20 nm (Veeramanikandan et al., 2017). However, it was aggregated to form irregular rhombic-shaped aggregates with panoramic view and rough surfaces of a wide size range 117 μm–1.29 mm, as depicted by SEM analysis. Such Fe_3O_4 NPs expressed efficient antibacterial and antioxidant activities (Veeramanikandan et al., 2017). The rich phytochemical constituents (polysaccharides, flavonoids, alkaloids, glycosides, phenols, saponins, tannins, phytosterols, anthraquinones, coumarins, and reducing sugars) of the extract proved to play a vital role in the phytosynthesis process. Cheera et al. (2018) reported the green synthesis of spherical-shaped Fe_3O_4 NPs with an average diameter of 16–20 nm, using *Acacia nilotica* leaf aqueous extract as a reducing and capping agent. The magnetic parameters, namely, saturation magnetization (Ms), remant magnetization (Mr), and coercivity (Hc) values were 11.1 emu/g, 1.92 emu/g, and 342 Oe, respectively, and proved the ferromagnetic behavior of the prepared nanoparticles. The phytogenic NPs exhibited supreme catalytic activity through reduction of Congo red dye within 30 min, in the presence of $NaBH_4$, at room temperature. Facile synthesis of IO NPs using the hot water leaf extract of *Eichhornia crassipes* was reported by Jagathesan and Rajiv (2018). *E. crassipes* is one of the aquatic weeds. It belongs to Pontederiaceae family which is resistant to all eradication methods. It contains secondary metabolites such as phenols, sterols, flavonoids, terpenoids, anthoquinones, and phenalenone compounds (Vanathi et al., 2014). Broad absorption peak at 379 nm was observed using UV/Vis spectrophotometer. The EDX analysis confirmed the formation of pure IO NPs. Purity level of the particles was analyzed. *E. crassipes* mediated IO NPs had 77.08% of iron and 22.97% of oxygen, respectively. SEM revealed that the synthesized NPs were rod-shaped and arranged without aggregation.

4.3 MICROBIAL SYNTHESIS OF METAL OXIDES NANOPARTICLES

Microbial synthesis of metal oxides NPs can be performed either intracellularly or extracellularly (Hulkoti and Taranath, 2014; Zikalala et al., 2018; Yusof et al., 2019). The intracellular mechanisms involve five steps (Figure 4.7) (Zhang et al., 2011; Iravani, 2014): (1) the trapping, via the transportation of metal ions into the cell wall by electrostatic attraction between positively charged metal ions and negative charges on microbial cell walls of proteins, enzymes, and polysaccharides. The small metal ions can diffuse throughout the cell wall into the cytoplasmic membrane. (2) The reduction of metal ions to a metal atom by the enzymes found in the cell wall via, for example, the electron transfers from NADH by NADH-dependent reductase that acts as an electron carrier. (3) Then, the nuclei start to grow to form NPs and accumulate in the cytoplasm and the periplasmic space of the cell wall. (4) The stabilization of the prepared NPs occurred via the proteins or peptides and amino acids, such as cysteine, tyrosine, and tryptophan exist inside the cells. (5) Finally, to get the purified NPs, the cells are subjected to ultra-sonication.

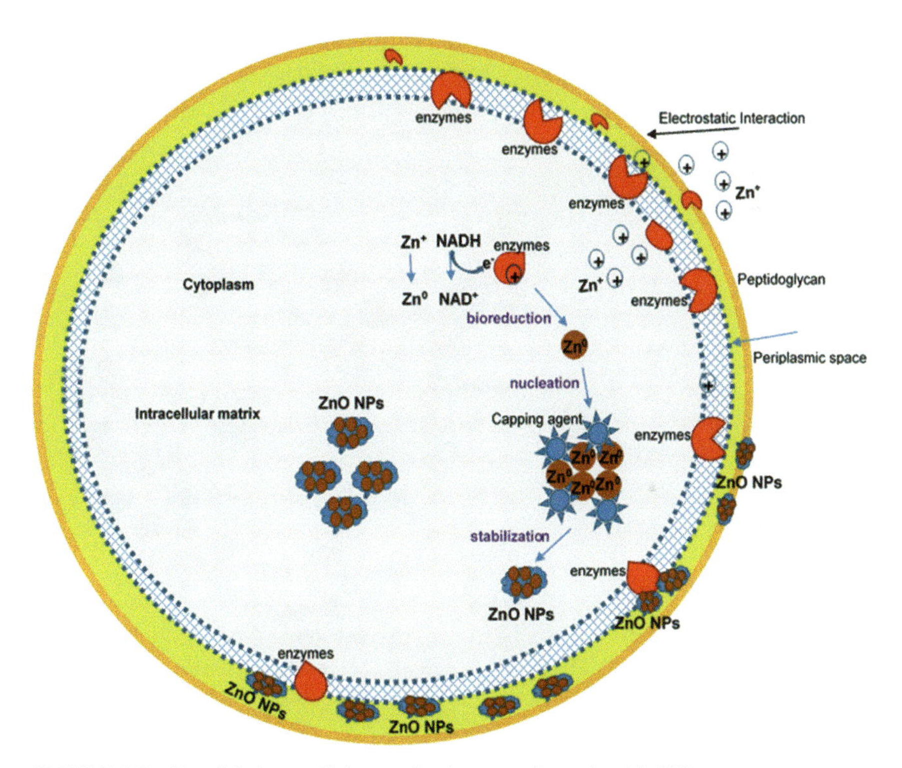

FIGURE 4.7 Possible intracellular synthesis steps of metal oxide NPs.

While the extracellular biosynthesis of metal oxides NPs can be performed by the interaction of the secreted proteins and/or some organic compounds released by the cells or located on the microbial cell wall, with the metal ions. Extracellular synthesis of metal oxide NPs can be also done via enzymes located in the cell membrane and/or released by the cells. For example, nitrate reductase enzyme which is involved in the nitrogen cycle and catalyzes the conversion of nitrate to nitrite is also involved in the bioreduction of metal ions. Which can be initiated by the electron transfer from NADH by NADH-dependent reductase that acts as an electron carrier (Jain et al., 2013; Kundu et al., 2014; Balraj et al., 2017). Consequently, the metal ions captured electron and reduced to zero-charged metal, which subsequently forms metal oxide NPs (Figure 4.8). These bioreductants can also act as capping and stabilizing agents, preventing the agglomeration of the biosynthesized metal oxide NPs (Omran et al., 2020). It is very important to keep in mind that the physicochemical parameters of the biosynthesis process, such as pH, initial precursor concentration, temperature, reaction time, cell age and type, stirring rate, and illumination, and affect the size and shape of the biosynthesized NPs (Omran et al., 2020).

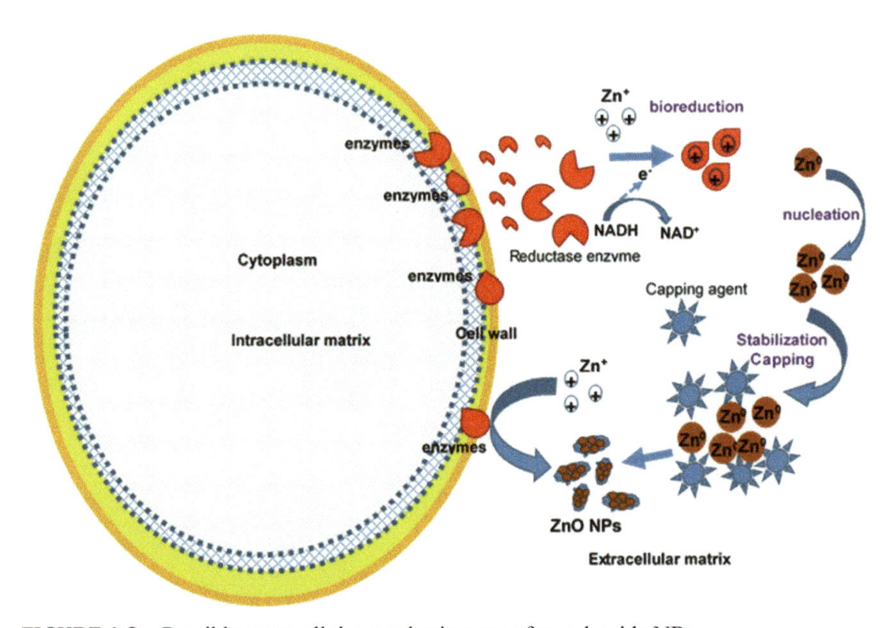

FIGURE 4.8 Possible extracellular synthesis steps of metal oxide NPs.

Bharde et al. (2006) proved that the extracellular hydrolysis of the anionic iron complexes $K_3[Fe(CN)_6]$ and $K_4[Fe(CN)_6]$ by cationic proteins secreted by the fungi *Fusarium oxysporum* and *Verticillium* sp., at room temperature, led to the formation of crystalline magnetite NPs with the size range of 20–50 nm and 10–40 nm, respectively. The FTIR analysis revealed the presence of proteins in the quasispherical and cube-shaped magnetite particles, respectively. Those Fe_3O_4 exhibited a signature of a ferrimagnetic transition with a negligible amount of spontaneous magnetization at low temperature. That study suggested the bioremediation of iron cyanide complexes via the synthesis of MNP. Moreover, *F. oxysporum* and *Verticillium* sp. produced quasispherical magnetite NPs from a mixture of 2:1 molar $FeCl_3$ and $FeCl_2$. Abdeen et al. (2016) reported a biophysical method for the preparation of spherical-shaped superparamagnetic iron NPs and ferromagnetic Fe_3O_4 NPs, with saturation magnetization amounts of 112 and 68 emu/g, small remanence, and coercivity and average particle size of 8 ± 2 and 50 ± 1, respectively. Whereas *A. niger* was used for the decomposition of $FeSO_4$ and $FeCl_3$ to FeS and Fe_2O_3, respectively. Then, the produced particles were exposed to supercritical conditions of ethanol for 1 h at 300 °C and pressure of 850 psi. Finally, the system was cooled down to room temperature and the magnetic particles were collected by permanent magnets. The obtained values for magnetization and coercivity promoted the usage of Fe NPs for different applications in the petroleum industry and many other medical applications such as hyperthermia for cancer treatment and contrast agent for MRI and drug delivery.

The pH, as well as the partial pressure of gaseous hydrogen (r-H_2) or redox potential of the culture solution, has been reported to play an important role in the biosynthesis of spherical-shaped cuprous oxide Cu_2O and anatase TiO_2 NPs using *Lactobacillus* and baker's yeast, that is, *Saccharomyces cerevisiae* (Jha et al., 2009; Prasad et al., 2010). The average size of the biosynthesized Cu_2O and TiO_2 NPs recorded 10–20 nm and 8–35 nm under low r-H_2, respectively. Where the oxidoreductases are pH-sensitive enzymes, and at low pH values, oxidases are activated and high oxidation potentials occur (Figure 4.9).

Jha and Prasad (2010) reported the *Lactobacillus*-mediated biosynthesis of 10–70 nm anatase TiO_2 NPs. *Fusarium* spp. has been reported for the mycosynthesis of triangular-shaped >100 nm ZnO NPs (Velmurugan et al., 2010). In another study, a probiotic bacteria *Lactobacillus* spp. has been used for biogenesis of hexagonal ZnO crystal structure, spherically shaped with a diameter of about 30.3–38 nm (Salman et al., 2018).

The crystalline 70 nm ZnO NPs have been also biosynthesized by using *Lactobacillus salivarius* L3 (Hu et al., 2013). *Lactobacillus plantarum* VITES07 has been also reported for the microbial synthesis of spherical-shaped 7–19 nm ZnO NPs (Selvarajan and Mohanasrinivasan, 2013). *Lactobacillus johnsonii* was also reported for the extracellular biosynthesis of 4–9 nm irregular shaped TiO_2 and spherical-shaped ZnO (Al-Zahrani et al., 2018). *B. subtilis* reported to synthesize spherical to oval-shaped 66–77

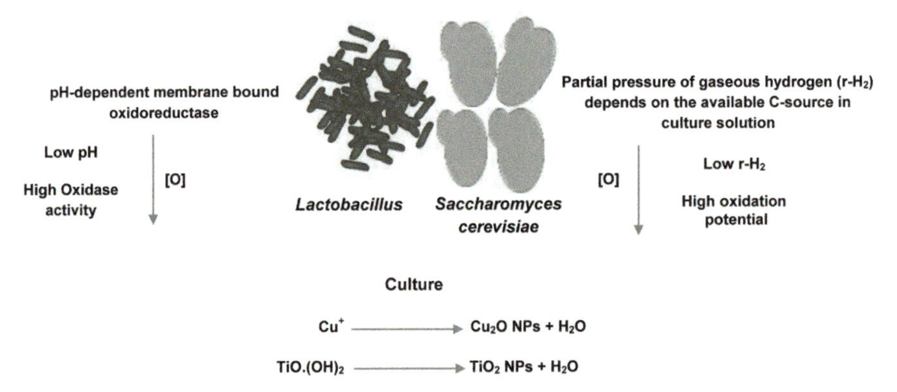

FIGURE 4.9 Possible mechanism for the biosynthesis of Cu_2O and TiO_2 NPs.

nm anatase TiO_2 NPs using $TiO(OH)_2$ as a precursor (Kirthi et al., 2011). The mycosynthesis of spherical-shaped 100–140 nm ZnO NPs by *Aspergillus aeneus* has been reported (Jian et al., 2013). The cell-free filtrate of *Aspergillus flavus* TFR7 mycosynthesized 12–15 nm TiO_2 NPs (Raliya et al., 2015). Sarkar et al. (2014) reported the mycosynthesis of multishaped (spherical, triangular, and hexagonal) 45–150 nm ZnO NPs using *Alternaria alternata* (Fr.) Keissl (1912). In another study, smooth and elongated shaped ZnO NPs have been prepared using *Pichia fermentas* JA2 (Chauha et al., 2015). Shamsuzzaman et al. (2017) reported the application of *Candida albicans* in the biosynthesis of quasispherical 25 nm ZnO NPs. In an earlier study, *Aeromonas hydrophila* has been used for bacterio-synthesis of ZnO and TiO_2 NPs (Jayaseelan et al., 2012; 2013). The absorption peak of 374 nm with bandgap of 3.32 ev proved the biosynthesis of ZnO NPs. Moreover, the XRD analysis confirmed the crystalline hexagonal ZnO phase (wurtzite structure) with diffraction peaks at 31.75°, 34.37°, 47.60°, 56.52°, 66.02°, and 75.16° with reflections of (100), (002), (101), (102), (110), (112), and (202), respectively (Jayaseelan et al., 2012). However, the XRD pattern of the synthesized TiO_2 NPs revealed intense peaks at 27.47, 31.77, 36.11, 41.25, 54.39, 56.64, and 69.53 corresponding to 110, 100, 101, 111, 211, 220, and 301 Bragg's reflection, respectively, proving rutile TiO_2 NPs (Jayaseelan et al., 2013). The AFM and the field emission scanning electron microscope (FSEM) proved smooth and spherical-shaped ZnO and TiO_2 NPs with average particles size of 57.72 nm and 40.5 nm, respectively. Whereas the alcohols, phenols, primary amines, lactones, and aliphatic amines in *A. hydrophila* are involved in the biosynthesis of nanoparticles (Jayaseelan et al., 2012; 2013). The biosynthesized ZnO and TiO_2 NPs reported to have an efficient bactericidal and fungicidal effect of

ZnO NPs against *A. hydrophila, E. coli, S. aureus, P. aeruginosa, E. faecalis, Streptococcus pyogenes, A. flavus, A. niger,* and *C. albicans* (Jayaseelan et al., 2012; 2013). The bactericidal effect of the biosynthesized ZnO might be attributed to the generation of H_2O_2 on the surface of ZnO NPs (Yamamoto, 2001). Upon the entrance of ZnO NPs through the cells' mitochondria, it induces oxidative stress causing cell death due to apoptosis (Xia et al., 2007). Moreover, ZnO NPs cause distorts and damage to the bacterial cell membrane, leading to the leakage of intracellular contents, and consequently cell death (Liu et al., 2009). It is important to mention the phototoxicity of TiO_2 NPs on bacterial cells under sunlight and/or UV irradiations. Where the oxidative attack of such irradiated NPs to microbial cells occurred and ROS (H_2O_2,·OH and ·O_2) produced on the surface of the irradiated TiO_2 NPs promoting the peroxidation of the polyunsaturated phospholipids component of the cell membrane causing direct cell damage of the cell wall (Jayaseelan et al., 2013). The proteins secreted by *S. cerevisiae* reported to be the main reason for the biosynthesis of quasispherical 10 nm ZnO NPs and in forming also a "corona" around the NPs (Sharan et al., 2015). Nontoxic hexagonal structure spherical-shaped ZnO NPs was mycosynthesized by soil fungus *Aspergillus terreus* VIT 2013 (Sangappa et al., 2013). The extracellular biosynthesis of ZnO nanorods with a diameter size ranged between 60 and 95 nm has been reported using *Fusarium Solani* (Venkates et al., 2013). The mycelial cell filtrate of the endophytic fungus *Cochliobolus geniculatus* has been reported for the mycosynthesis of polydispersed, nonagglomerated, crystalline, quasispherical, and wurtzite structure 2–6 nm ZnO NPs with bandgap of 3.28 eV and SPR of 378 nm (Kadam et al., 2019). Several other publications reported for the microbial synthesis of ZnO NPs (Table 4.1).

The Gram-negative *Serratia* sp. has been reported for the intracellular biosynthesis of polydispersed 10–30 nm CuO NPs (Hasan et al., 2007). The proposed mechanism is the biosynthesis of the CuNPs inside the cells then followed by cell death and lysis. Once the CuNPs leaked out, they immediately oxidized to CuO NPs. The maximum production occurred during the stationary phase of the bacterial cell growth where the oxidative stress due to the depletion of their energy source and accumulation of wastes are peaked. Moreover, the exposure to Cu^{2+} increased the stress as the cells were trying to overcome the toxicity of metal ions and the occurrence of the osmotic stress also led to cell death and lysis (Hasan et al., 2007).

Zarasvand and Rai et al. (2016) reported the biosynthesis of polydispersed 400 nm CuO NPs using *Shewanella indica.* The biosynthesized CuO NPs expressed efficient biocidal activity against the SRB *Desulfovibrio marinisediminis* GSR3, recording a minimum inhibitory concentration of 100 μg/mL.

TABLE 4.1 Some Examples for Microbially Synthesized ZnO NPs

Microbes	Shape and Size	Application	References
Thermophilic *Bacillus haynesii* CDL3	Hexagonal zincite Spherical-shaped 50 nm	Antibacterial activity	Rehman et al. (2019)
Aspergillus niger	Spherical ~61 nm	Antimicrobial activity	Kalpana et al. (2018)
Bacillus megaterium NCIM2326	Rod and cubic 47–95 nm	Antimicrobial activity	Saravanan et al. (2018)
Halomonas elongate IBRC-M 10214	Multi-shaped ~18.11 nm	Antimicrobial activity	Taran et al. (2018)
Lactobacillus paracasei LB3	~Spherical 1179 nm	Antimicrobial activity	Król et al. (2018)
Pichia kudriavzevii	Hexagonal wurtzite 10–61 nm	Antimicrobial activity	Moghaddam et al. (2017)
Sphingobacterium thalpophilum	Triangular 40 nm	Antimicrobial activity	Rajabairavi et al. (2017)
Staphylococcus aureus	Needle-shaped Acicular 10–50 nm	Antimicrobial activity	Rauf et al. (2017)
Streptomyces sp.	Spherical 20–50 nm	Antimicrobial activity	Balraj et al. (2017)
Aspergillus fumigatus JCF	Spherical 60–80 nm	Antimicrobial activity	Rajan et al. (2016)
Aspergillus fumigatus TFR8	Oblate spherical 2.9 nm	Exopolysaccharide producer enhancer by *Bacillus subtilis* strain JCT1	Raliya et al. (2014)
Bacillus licheniformis MTCC9555	Flower 250 nm	Photodegradation of dye	Tripathi et al. (2014)
Rhodococcus pyridinivorans NT2b	Spherical 100–120 nm	Antibacterial activity	Kundu et al. (2014)
Aspergillus terreus	Spherical ~54.8 nm	Antifungal activity	Baskar et al. (2013)
Lactobacillus sporogens	Hexagonal ~145.7 nm	Antimicrobial activity	Mishra et al. (2013)
Lactobacillus sporogens	Hexagonal 5–15 nm	Photodegradation of hydrocarbon pollutants	Prasad and Jha (2009)

The biosynthesized CuO NPs using the cell-free filtrate of *Bacillus* sp. FU4 have been reported as efficient biocide for *E. coli* ATCC 25922 and *S. aureus* ATCC 43300 (Taran et al., 2017). Orthogonal array of Taguchi experimental design has been applied for optimizing the biosynthesis process. Spherical shaped 2–41 nm CuO NPs have been obtained at optimum conditions of

0.01 M $CuSO_4$, 96 h biomass age and 96 h reaction time. The Analysis of variance (ANOVA) proved that the concentration of $CuSO_4$ was the most important factor relative to biomass age and reaction time (Taran et al., 2017). In another study by Taran et al. (2018), Taguchi experimental design has been also applied for optimizing the biosynthesis process of anatase TiO_2 NPs using the supernatant of *Halomonas elongate* IBRC-M 10214 growth culture. Spherical shaped 104.63 nm TiO_2 NPs have been obtained at optimum conditions of 0.1 M $TiO.(OH_2)$, 37 °C and 96 h reaction time. The analysis of variance (ANOVA) proved that the concentration of the precursor $TiO.(OH_2)$ was the most important factor relative to the reaction temperature and time (Taran et al., 2018). The zeta potential value of −31.1 mV and -30.9 mV proved the high stability of the biosynthesized CuO NPs and ZnO NPs by actinomycetes and mycelial cell filtrate of *C. geniculatus*, respectively (Nabila and Kannabiran, 2018; Kadam et al., 2019). The cell-free filtrate of *Trichoderma asperellum* has been also reported for the mycosynthesis of spherical-shaped 110 nm CuO NPs (Saravanakumar et al., 2019). Several other publications reported the microbial synthesis of CuO and TiO_2 NPs (Table 4.2).

TABLE 4.2 Some Examples for Microbially Synthesized CuO and TiO_2 NPs

Microbes	Shape and Size	Application	References
Actinomycete isolate VITBN4	CuO Spherical 61.7 nm	Antimicrobial activity	Nabila and Kannabiran (2018)
Morganela morganii	CuO Spherical < 10 nm	Antibacterial activity	Ghasemi et al. (2017)
Bacillus amyloliquefaciens	Anatase TiO_2 Spherical 22.11–97.28 nm	Photocatalytic degradation of sulfonated dyes	Khan and Fulekar (2016)
Bacillus mycoides	Anatase TiO_2 Spherical 40–60 nm	Quantum Dot sensitized solar cells	Órdenes-Aenishanslins et al. (2014)
Aeromonas hydrophila	Rutile TiO_2 Spherical 28–54 nm	Antibacterial activity	Jayaseelan et al. (2013)

REFERENCES

Abbasi, B.H., Zahir, A., Ahmad, W., Nadeem, M., Giglioli-Guivarc'h, N., and Hano, C. 2019. Biogenic zinc oxide nanoparticles-enhanced biosynthesis of lignans and neolignans in cell suspension cultures of *Linum usitatissimum* L. Artif. Cell Nanomed. Biotechnol. 47(1): 1367–1373.

Abboud, Y., Saffaj, T., Chagraoui, A., Ei-Bouari, A., Brouzi, K., Tanane, O., Ihssane, B. 2014. Biosynthesis, characterization and antimicrobial activity of copper oxide nanoparticles (CONPs) produced using brown alga extract (*Bifurcaria bifurcata*). Appl. Nanosci. 4: 571–576.

Abdeen, M., Sabry, S., Ghozlan, H., El-Gendy, A.A., and Carpenter, E.E. 2016. Microbial-physical synthesis of Fe and Fe_3O_4 magnetic nanoparticles using *Aspergillus niger* YESM1 and supercritical condition of ethanol. J. Nanomater. 2016, 9174891. http://dx.doi.org/10.1155/2016/9174891\

Abdul Jalill, R.D.H., Nuaman, R.S., and Abd, A.N. 2016. Biological synthesis of titanium dioxide nanoparticles by *Curcuma longa* plant extract and study its biological properties. World Sci. News. 49(2): 204–222.

Abdullah, M.M.S., Atta, A.M., Allohedan, H.A., Alkhathlan, H.Z., Khan, M., and Ezzat, A.O. 2018. Green synthesis of hydrophobic magnetite nanoparticles coated with plant extract and their application as petroleum oil spill collectors. Nanomaterials. 8: 855. doi:10.3390/nano8100855.

Agista, M.N., Guo, K., and Yu, Z. 2018. A state-of-the-art review of nanoparticles application in petroleum with a focus on enhanced oil recovery. Appl. Sci. 8: 871–900.

Alavi, M., and Karimi, N. 2018. Characterization, antibacterial, total antioxidant, scavenging, reducing power and ion chelating activities of green synthesized silver, copper and titanium dioxide nanoparticles using *Artemisia haussknechtii* leaf extract. Artif. Cell Nanomed. Biotechnol. 46(8): 2066–2081.

Ali, H.R., Nassar, H.N., and El-Gendy, N.Sh. 2017. Green synthesis of α-Fe_2O_3 using *Citrus reticulum* peels extract and water decontamination from different organic pollutants. Energy Source A. 39(13): 1425–1434. https://doi.org/10.1080/15567036.2017.1336818

Ali, N., Baloch, M. A., and Chachar, Q. 2013. Influence of naphthalene acetic acid (NAA) on sprouting and rooting potential of stem cuttings of bougainvillea. Sci. Int. 25: 299–304.

Ali, N., Zhang, B., Zhang, H., Zaman, W., Li, X., Li, W., and Zhang, Q. 2015. Interfacially active and magnetically responsive composite nanoparticles with raspberry like structure; synthesis and its applications for heavy crude oil/water separation. Colloids Surf. A. 472: 38–49.

Al-Zahrani, H., El-Waseif, A., and El-Ghwas, D. 2018. Biosynthesis and evaluation of TiO_2 and ZnO nanoparticles from in vitro stimulation of *Lactobacillus johnsonii*. J. Innov. Pharm. Biol. Sci. 5(1):16–20.

Asemani, M., and Anarjan, N. 2019. Green synthesis of copper oxide nanoparticles using Juglans regia leaf extract and assessment of their physico-chemical and biological properties. Green Process Synth. 8: 557–567.

Balraj, B., Senthilkumar, N., Siva, C., Krithikadevi, R., Julie, A., Potheher, I.V., and Arulmozhi, M. 2017. Synthesis and characterization of zinc oxide nanoparticles using marine *Streptomyces* sp. with its investigations on anticancer and antibacterial activity. Res. Chem. Intermed. 43: 2367–2376.

Bao, Y., Wen, T., Samia, A.C.S., Khandhar, A., and Krishnan, K.M. 2016. Magnetic nanoparticles: material engineering and emerging applications in lithography and biomedicine. J. Mater. Sci. 51: 513–553.

Baskar, G., Chandhuru, J., Fahad, K.S., and Praveen, A.S. 2013. Mycological synthesis, characterization and antifungal activity of zinc oxide nanoparticles. Asian J. Pharm. Technol. 3: 142–146.

Bayrami, A., Alioghli, S., Pouran, S.R., Habibi-Yangjeh, A., Khataee, A., and Ramesh, S., 2019. A facile ultrasonic-aided biosynthesis of ZnO nanoparticles using *Vaccinium arctostaphylos* L. leaf extract and its antidiabetic, antibacterial, and oxidative activity evaluation. Ultrason. Sonochem. 55: 57–66.

Bharde, A., Rautaray, D., Bansal, V., Ahmad, A., Sarkar, I., Yusuf, S.M., Sanyal, M., and Sastry, S. 2006. Extracellular biosynthesis of magnetite using fungi. Small. 2(1): 135–141.

Bhumi, G., and Savithramma, N. 2014. Biological synthesis of zinc oxide nanoparticles from *Catharanthus roseus* (l.) G. Don. leaf extract and validation for antibacterial activity. Int. J. Drug Dev. Res. 6(1): 208–214.

Borah, R., Saikia, E., Jyoti Bora, S., and Chetia, B. 2016. On-water synthesis of phenols using biogenic Cu_2O nanoparticles without using H_2O_2. RSC Adv. 6: 100443–100447.

Borah, R., Sankar, E.S., Bora, J., and Chetia, B. 2017. Banana pulp extract mediated synthesis of Cu_2O nanoparticles: an efficient heterogeneous catalyst for the ipso-hydroxylation of arylboronic acids. Tetrahedron Lett. 58: 1211–1215.

Buazar, F., Sweidi, S., Badri, M., and Kroushawi, F. 2019. Biofabrication of highly pure copper oxide nanoparticles using wheat seed extract and their catalytic activity: a mechanistic approach. Green Process Synth. 8: 691–702.

Chan, H.B.S., and Ellis, B.L. 2004. Carbon-encapsulated radioactive 99mTc nanoparticles. Adv. Mater. 16: 144–149.

Chandrasekaran, R., Yadav, S.A., and Sivaperumal, S. 2019. Phytosynthesis and characterization of copper oxide nanoparticles using the aqueous extract of *Beta vulgaris* L and evaluation of their antibacterial and anticancer activities. J. Clust. Sci.

Chaudhuri, S.K., and Malodia, L. 2017. Biosynthesis of zinc oxide nanoparticles using leaf extract of *Calotropis gigantea*: characterization and its evaluation on tree seedling growth in nursery stage. Appl. Nanosci. 7: 501–512.

Chauhan, R., Reddy, A., and Abraham, J. 2015. Biosynthesis of silver and zinc oxide nanoparticles using *Pichia fermentans* JA2 and their antimicrobial property. Appl. Nanosci. 5: 63–71.

Cheera, P.K., Sreenivasulu, V., Govinda, S., Himageerish, K.K., Deepa, T., Vasantha, J.N.V.V., and Venkateswarlu, P. 2018. Biosynthesis of the Fe_3O_4 nanoparticles using *Acacia nilotica* leaf extract and their effect on degradation of Congo red dye in aqueous solution. Trends Textile Eng. Fashion Technol. 1(3). TTEFT.000513.2018.

Chen, J., Mao, S., Xu, Z., and Ding, W. 2019a. Various antibacterial mechanisms of biosynthesized copper oxide nanoparticles against soilborne *Ralstonia solanacearum*. RSC Adv. 9: 3788–3799.

Chen, Z., Gaoa, S.-h., Jina, M., Suna, S., Lu, J., Yang, P., Bonda, P.L., Yuan, Z., and Guo, J. 2019b. Physiological and transcriptomic analyses reveal CuO nanoparticle inhibition of anabolic and catabolic activities of sulfate-reducing bacterium. Environ. Int. 125: 65–74.

Chiu, W.S., Khiew, P.S., Cloke, M., Isa, D., Tan, T.K., Radiman, S., Abd-Shukor, R., Abd-Hamid, M.A., Huang, N.M., Lim, H.N., Chia, C.H. 2010. Photocatalytic study of two-dimensional ZnO nanopellets in the decomposition of methylene blue. Chem. Eng. J. 158, 345–352.

Clark, J., and Macquarrie, D. 2002. Handbook of Green Chemistry and Technology, Blackwell Publishing, Oxfordshire.

Cocuzza, M., Pirri, C., Rocca, V., and Verga, F. 2012. Current and future nanotech applications in the oil industry. Am. J. Appl. Sci. 9(6): 784–793.

Datta, A., Patra, C., Bharadwaj, H., Kaur, H., Dimri, N., and Khajuria, R. 2017. Green synthesis of zinc oxide nanoparticles using *Parthenium hysterophorus* leaf extract and evaluation of their antibacterial properties. J. Biotechnol. Biomater. 7: 3–8.

Dobrucka, R., Dlugaszewska, J., and Kaczmarek, M. 2018. Cytotoxic and antimicrobial effects of biosynthesized ZnO nanoparticles using of *Chelidonium majus* extract. Biomed. Microdevices. 20: 5. https://doi.org/10.1007/s10544-017-0233-9

Ehrampoush, M.H., Miria, M., Salmani, M.H., and Mahvi, A.H. 2015. Cadmium removal from aqueous solution by green synthesis iron oxide nanoparticles with tangerine peel extract. J. Environ. Health Sci. Eng. 13: 84–91.

El-Gendy, N.Sh., and Nassar, H.N. 2018. Biodesulfurization in petroleum refining. Wiley & Sons, Hoboken, NJ, and Scrivener Publishing LLC, Beverly, MA, USA. https://doi.org/10.1002/9781119224075

El-Gendy, N.Sh., and Speight, J.G. 2016. Handbook of Refinery Desulfurization. CRC Press, Taylor & Francis Group, Boca Raton, FL, USA. https://doi.org/10.1201/b19102

El-Gendy, N.Sh., El-Salamony, R.A., and Younis, S.A. 2016. Green synthesis of fluorapatite from waste animal bones and the photo-catalytic degradation activity of a new ZnO/green biocatalyst nano-composite for removal of chlorophenols. J. Water Process Eng. 12: 8–19. https://doi.org/10.1016/j.jwpe.2016.05.007

El-Kassas, H.Y., Aly-Eldeen, M.A., and Gharib, S.M. 2016. Green synthesis of iron oxide (Fe_3O_4) nanoparticles using two selected brown seaweeds: characterization and application for lead bioremediation. Acta Oceanol. Sin. 35(8): 89–98.

Esmaeilnezhad, E., Karimian, M., and Choi, H.J. 2019. Synthesis and thermal analysis of hydrophobic iron oxide nanoparticles for improving in-situ combustion efficiency of heavy oils. J. Ind. Eng. Chem. 71: 402–409.

Frade, T., Jorge, M.E., and Gomes, A. 2012. One-dimensional ZnO nanostructured films: effect of oxide nanoparticles. Mater. Lett. 82: 13–15.

Ghasemi, N., Sheini, F.J., and Zekavati, R. 2017. CuO and Ag/CuO nanoparticles: biosynthesis and antibacterial properties. Mater. Lett. 196: 78–82.

Ghosh, P.R., Fawcett, D., Sharma, S.B., and Poinern, G.E. J. 2017. Production of high-value nanoparticles via biogenic processes using aquacultural and horticultural food waste. Materials. 10: 852. doi:10.3390/ma10080852.

Gillis, W.T. 1976. Floral development of *Bougainvillea spectabilis* Wild. *Boerhaavia diffusa* L. and *Mirabilis jalapa* L. (Nyctaginaceae). Bot. J. Linean Soc. 20: 34–41.

Gnanasangeetha, D., and Tambavani, D.S. 2013. Biogenic production of zinc oxide nanoparticles using *Acalypha indica*. J. Chem. Biol. Phys. 4(1): 238–246.

Gnanavel, V., Palanichamy, V., and Roopan, S.M. 2017. Biosynthesis and characterization of copper oxide nanoparticles and its anticancer activity on human colon cancer cell lines (HCT-116). J. Photochem. Photobiol. B. 171: 133–138.

Gopalakrishnan, K., Ramesh, C., Ragunathan, V., and Thamilselvan, M. 2012. Antibacterial activity of Cu_2O nanoparticles on *E. coli* synthesized from *Tridax procumbens* leaf extract and surface coating with polyaniline. Dig. J. Nanomater. Biostruct. 7(2): 833–839.

Goutam, S.P., Saxena, G., Singh, V., Yadav, A.K., Bharagava, R.N., and Thapa, K.B. 2018. Green synthesis of TiO_2 nanoparticles using leaf extract of Jatropha curcas L. for photocatalytic degradation of tannery wastewater. Chem. Eng. J. 336: 386–396.

Gu, H., Chen, X., Chen, F., Zhou, X., Parsaee, Z. 2018. Ultrasound-assisted biosynthesis of CuO-NPs using brown alga *Cystoseira trinodis*: characterization, photocatalytic AOP, DPPH scavenging and antibacterial investigations. Ultrason. Sonochem. 41: 109–119.

Gunalan, S., Sivaraj, R., and Venckatesh, R. 2012. *Aloe barbadensis* Miller mediated green synthesis of mono-disperse copper oxide nanoparticles: optical properties. Spectrochim. Acta, A. 97: 1140–1144.

Harne, S., Sharma, A., Dhaygude, M., Joglekar, S., Kodam, K., and Hudlikar, M. 2012. Novel route for rapid biosynthesis of copper nanoparticles using aqueous extract of *Calotropis procera* L. latex and their cytotoxicity on tumor cells. Colloids Surf. B. 95: 284–288.

Hasan, S.S., Singh, S., Parikh, R.Y., and Dharne, M.S. 2007. Bacterial synthesis of copper/copper oxide nanoparticles. J. Nanosci. Nanotechnol. 8: 1–6.

Herlekar, M., and Barve, S. 2014. Optimization of microwave assisted green synthesis protocol for iron oxide nanoparticles and its application for simultaneous removal of multiple pollutants from domestic sewage. Int. J. Adv. Res. 3(4): 331–345.

Hu, W.-F., Liu, X.-Y., Zhang, M., Shen, M., Pang, X., Zhou, H.-Y., and Zhu, J.-F. 2013. Biosynthesis of ZnO nanoparticles by *Lactobacillus salivarius* L3. Modern Food Sci. Technol. 29(9): 2192–2198.

Huang, Y., He, J., Zhang, Y., Dai, Y., Gu, Y., Wang, S., and Zhou, C. 2006. Morphology, structures and properties of ZnO nanobelts fabricated by Zn-powder evaporation without catalyst at lower temperature. J. Mater. Sci. 41: 3057–3062.

Hulkoti, N.I., and Taranath, T.C. 2014. Biosynthesis of nanoparticles using microbes—a review. Colloids Surf. B Biointerfaces. 121: 474–483.

Hurnaus, T., and Plank, J. 2015. Crosslinking of guar and HPG based fracturing fluids using ZrO_2 nanoparticles. SPE International Symposium on Oilfield Chemistry, The Woodlands, TX. https://doi.org/10.2118/173778-MS .

Iravani, S. 2014. Bacteria in nanoparticle synthesis: current status and future prospects. Int. Scholarly Res. Not. 359316. http://dx.doi.org/10.1155/2014/359316.

Jafarirad, S., Mehrabi, M., and Rassul pur, E. 2014. Biological synthesis of zinc oxide and copper oxide nanoparticles. Proceedings of International Conference on Chemistry, Biomedical and Environment Engineering, Antalya, 62–64. http://dx.doi.org/10.17758/IAAST.A1014056

Jagathesan, G., and Rajiv, P. 2018. Biosynthesis and characterization of iron oxide nanoparticles using *Eichhornia crassipes* leaf extract and assessing their antibacterial activity. Biocatal. Agric. Biotechnol. 13: 90–94.

Jain, N., Bhargava, A., Tarafdar, J.C., Singh, S.K., and Panwar, J. 2013. A biomimetic approach towards synthesis of zinc oxide nanoparticles. Appl. Microbiol. Biotechnol. 97: 859–869.

Jayarambabu, N., Rao, K.V., and Rajendar, V. 2018. Biogenic synthesis, characterization, acute oral toxicity studies of synthesized Ag and ZnO nanoparticles using aqueous extract of *Lawsonia inermis*. Mater. Lett. 211: 43–47.

Jayaseelan, C., Abdul Rahuman, A., Kirthi, A.V., Marimuthu, S., Santhoshkumar, T., Bagavan, A., Gaurav, K., Karthik, L., and Bhaskara Rao, K.V. 2012.Novel microbial route to synthesize ZnO nanoparticles using *Aeromonas hydrophila* and their activity against pathogenic bacteria and fungi. Spectrochim. Acta A. 90: 78–84.

Jayaseelan, C., Abdul Rahuman, A., Roopan, S.M., Kirthi, A.V., Venkatesan, J., Kim, S.-K., Iyappan, M., and Siva, C. 2013. Biological approach to synthesize TiO_2 nanoparticles using *Aeromonas hydrophila* and its antibacterial activity. Spectrochim. Acta A. 107: 82–89.

Jeyabharathi, S., Kalishwaralal, K., Sundar, K., and Muthukumaran, A. 2017. Synthesis of zinc oxide nanoparticles (ZnONPs) by aqueous extract of *Amaranthus caudatus* and evaluation of their toxicity and antimicrobial activity. Mater. Lett. 209: 295–298.

Jha, A.K., and Prasad, K. 2010. Biosynthesis of metal and oxide nanoparticles using *Lactobacilli* from yoghurt and probiotic spore tablets. Biotechnol. J. 5: 285–291.

Jha, A.K., Prasad, K., and Kulkarni, A.R. 2009. Synthesis of TiO_2 nanoparticles using micro-organisms. Colloids Surf. B. 71: 226–229.

Kadam, V.V., Ettiyappan, J.P., and Balakrishnan, R.M. 2019. Mechanistic insight into the endophytic fungus mediated synthesis of protein capped ZnO nanoparticles. Mater. Sci. Eng. B. 243: 214–221.

Kalpana, V.N., Kataru, B.A.S., Sravani, N., Vigneshwari, T., Panneerselvam, A., and Devi Rajeswari, V. 2018. Biosynthesis of zinc oxide nanoparticles using culture filtrates of *Aspergillus niger*: antimicrobial textiles and dye degradation studies. OpenNano. 3: 48–55.

Kanagasubbulakshmi, S., and Kadirvelu, K. 2017. Green synthesis of iron oxide nanoparticles using *Lagenaria siceraria* and evaluation of its antimicrobial activity. Def. Life Sci. J. 2(4): 422–427.

Kapusta, S., Balzano, L., and te Riele, P.M. 2011. Nanotechnology applications in oil and gas exploration and production. International Petroleum Technology Conference. Society of Petroleum Engineers, Bangkok. .

Khalil, M., Jan, B.M., Tong, C.W., and Berawi, M.A. 2017. Advanced nanomaterials in oil and gas industry: design, application and challenges. Appl. Energy. 191: 287–310.

Khalil, M., Lee, R.L., and Liu, N. 2015. Hematite nanoparticles in aquathermolysis: a desulfurization study of thiophene. Fuel. 145: 214–220.

Khan, R., and Fulekar, M.H. 2016. Biosynthesis of titanium dioxide nanoparticles using *Bacillus amyloliquefaciens* culture and enhancement of its photocatalytic activity for the degradation of a sulfonated textile dye Reactive Red 31. J. Colloid. Interface Sci. 475: 184–191.

Khanehzaei, H., Ahmad, M.B., Shameli, K., and Ajdari, Z. 2014. Synthesis and characterization of Cu@Cu2O core shell nanoparticles prepared in seaweed *Kappaphycus alvarezii* media. Int. J. Electrochem. Sci. 9: 8189–8198.

Khataee, A., Kayan, B., Kalderis, D., Karimi, A., Akay, S., and Konsolakis, M. 2017. Ultrasound-assisted removal of Acid Red 17 using nanosized Fe_3O_4-loaded coffee waste hydrochar. Ultrasound Sonochem. 35: 72–80.

Khatami, M., Varma, R.S., Heydari, M., Peydayesh, M., Sedighi, A., Askari, H.A., Rohani, M., Baniasadi, M., Arkia, S., Seyedi, F., and Khatami, S. 2019. Copper oxide nanoparticles greener synthesis using tea and its antifungal efficiency on *Fusarium solani*. Geomicrobiol. J. 36(9): 777–781.

Khodadadi, A., Torabi angajia, M., Talebizadeh rafsanjania, A., and Yonesib, A. 2012. Adsorptive desulfurization of diesel fuel with nano copper oxide (CuO). Proceedings of the 4th International Conference on Nanostructures (ICNS4), 12–14.

Kirthi, A.V., Abdul Rahuman, A., Rajakumar, G., Marimuthu, S., Santhoshkumar, T., Jayaseelan, C., Elango, G., Abduz Zahir, A., Kamaraj, C., and Bagavan, A. 2011. Biosynthesis of titanium dioxide nanoparticles using bacterium *Bacillus subtilis*. Mater. Lett. 65: 2745–2747.

Koli, A. 2015. Biological synthesis of stable zinc oxide nanoparticles and its role as anti-diabetic and anti-microbial agents. Int. J. Acad. Res. 2(4): 139–143.

Kong, X., Ding, Y., Yang, R., and Wang, Z.L. 2004. Single-crystal nanorings formed by epitaxial self-coiling of polar-nanobelts. Science. 303: 1348–1351.

Król, A., Railean-Plugaru, V., Pomastowski, P., Złoch, M., and Buszewski, B. Mechanism study of intracellular zinc oxide nanocomposites formation. Colloids Surf. A Physicochem. Eng. Aspects. 553: 349–358.

Krupa, N.D., Grace, A.N., and Raghavan, V. 2019. Process optimisation for green synthesis of ZnO nanoparticles and evaluation of its antimacrofouling activity. IET Nanobiotechnol. 13(5): 510–514.

Kumar, B., Smita, K., Cumbal, L., Debut, A., Galeas, S., and Guerrero, V.H. 2016. Phytosynthesis and photocatalytic activity of magnetite (Fe_3O_4) nanoparticles using the Andean blackberry leaf. Mater. Chem. Phys. 179: 310–315.

Kundu, D., Hazra, C., Chatterjee, A., Chaudhari, A., Mishra, S. 2014. Extracellular biosynthesis of zinc oxide nanoparticles using *Rhodococcus pyridinivorans* NT2: multifunctional textile finishing, biosafety evaluation and in vitro drug delivery in colon carcinoma. J. Photochem. Photobiol. B. Biol. 140: 194–204.

Lee, H. J., Lee, G., Jang, N.R., Yun, J.H., Song, J.Y., and Kim B.S. 2011. Biological synthesis of copper nanoparticles using plant extract. Nanotechnology. 1(1): 371–372.

Lesin, V.I., Koksharov, Yu A., and Khomutov, G.B. 2010. Magnetic nanoparticles in petroleum. Pet. Chem. 50 (2): 102–105.

Li, M., Mahendra, S., Lyon, D.Y., Brunet, L., Liga, M.V., Li, D., and Alvarez, P.J.J. 2008. Antimicrobial nanomaterials for water disinfection and microbial control: potential applications and implications. Water Res. 42: 4591–4602.

Li, P., Lv, W., and Ai, S. 2016. Green and gentle synthesis of Cu_2O nanoparticles using lignin as reducing and capping reagent with antibacterial properties. J. Exp. Nanosci. 11(1): 18–27.

Liang, J., Du, N., Song, S., and Hou, W. 2015. Magnetic demulsification of diluted crude oilin-water nanoemulsions using oleic acid-coated magnetite nanoparticles. Colloids Surf. A. 466: 197–202.

Lingaraju, K., Naika, H.R., Manjunath, K., Basavaraj, R.B., Nagabhushana, H., Nagaraju, G., and Suresh, D. 2016. Biogenic synthesis of zinc oxide nanoparticles using *Ruta graveolens* (L.) and their antibacterial and antioxidant activities. Appl. Nanosci. 6: 703–710.

Liu, Y., He, L., Mustapha, A., Li, H., Hu, Z.Q., and Lin, M. 2009. Antibacterial activities of zinc oxide nanoparticles against *Escherichia coli* O157:H7. J. Appl. Microbiol. 107: 1193–1201.

Ludi, B., and Niederberger, M. 2013. Zinc oxide nanoparticles: chemical mechanism and classical and non-classical crystallization. Dalton Trans. 42: 12554–12568.

Lunge, S., Singh, S., and Sinha, A. 2014. Magnetic iron oxide (Fe_3O_4) nanoparticles from tea waste for arsenic removal. J. Magn. Magn. Mater. 356: 21–31.

Mahdavi, M., Namvar, F., Ahmad, M.B., and Mohamad, R. 2013. Green biosynthesis and characterization of magnetic iron oxide (Fe_3O_4) nanoparticles using seaweed (*Sargassum muticum*) aqueous extract. Molecules. 18: 5954–5964.

Makarov, V.V., Makarova, S.S., Love, A.J., Sinitsyna, O.V., Dudnik, A.O., Yaminsky, I.V., Taliansky, M.E., and Kalinin, N.O. 2014. Biosynthesis of stable iron oxide nanoparticles in aqueous extracts of *Hordeum vulgare* and *Rumex acetosa* plants. Langmuir. 30(20): 5982–5988.

Malaikozhundana, B., and Vinodhini, J. 2018. Nanopesticidal effects of *Pongamia pinnata* leaf extract coated zinc oxide nanoparticle against the Pulse beetle, *Callosobruchus maculatus*. Mater. Today Commun. 14: 106–115.

Markowska-Szczupak A., Ulfig, K., and Morawski, A.W. 2011. The application of titanium dioxide for deactivation of bio particulates: an overview. Catal. Today. 169: 257–69.

Meera, B., Kumar, S., and Kalidhar, S.B. 2003. A review of the chemistry and biological activity of *Pongamia pinnata*. J. Med. Aromat. Plant Sci. 25(5): 441–446.

Mirhendi, M., Emtiazi, G., and Roghanian, R. 2013. Antibacterial activities of nano magnetite ZnO produced in aerobic and anaerobic condition by *Pseudomonas stutzeri*. Jundishapur J. Microbiol. 6(10): e10254. doi:10.5812/jjm.10254.

Mishra, M., Paliwal, J.S., Singh, S.K., Selvarajan, E., Subathradevi, C., and Mohanasrinivasan, V. 2013. Studies on the inhibitory activity of biologically synthesized and characterized zinc oxide nanoparticles using *Lactobacillus sporogens* against *Staphylococcus aureus*. J. Pure Appl. Microbiol. 7: 1263–1268.

Moghaddam, A.B., Moniri, M., Azizi, S., Rahim, R.A., Ariff, A.B., Saad, W.Z., Namvar, F., and Navaderi, M. 2017. Biosynthesis of ZnO nanoparticles by a new Pichia kudriavzevii yeast strain and evaluation of their antimicrobial and antioxidant activities. Molecules. 22: 1–18.

Murali, D.S., Kumar, S., Choudhary, R.J., Wadikar, A.D. Jain, M.K., and Subrahmanyam, A. 2015. Synthesis of Cu_2O from CuO thin films: Optical and electrical properties. AIP Adv. 5: 047143. https://doi.org/10.1063/1.4919323.

Murugesan, S., Monteiro, O.R., and Khabashesku, V.N. 2016. Extending the lifetime of oil and gas equipment with corrosion and erosion resistant Ni-B-nanodiamond metal-matrix-nanocomposite coatings. Offshore Technology Conference, Houston, TX. https://doi.org/10.4043/26934-MS.

Muthukumar, H., and Matheswaran, M. 2015. *Amaranthus spinosus* leaf extract mediated FeO nanoparticles: physicochemical traits, photocatalytic and antioxidant activity. ACS Sustain. Chem. Eng. 3: 3149–3156.

Nabila, M.I., and Kannabiran, K. 2018. Biosynthesis, characterization and antibacterial activity of copper oxide nanoparticles (CuO NPs) from actinomycetes. Biocatal. Agric. Biotechnol. 15: 56–62.

Naika, H.R., Lingaraju, K., Manjunath, K., Kumar, D., Nagaraju, G., Suresh, D., and Nagabhushana, H. 2015. Green synthesis of CuO nanoparticles using *Gloriosa superba* L. extract and their antibacterial activity. J Taibah Univ. Sci. 9(1): 7–12.

Narasaiah, P., Mandal, P.K., and Sarada, N.C. 2017. Biosynthesis of copper oxide nanoparticles from Drypetes sepiaria Leaf extract and their catalytic activity to dye degradation. IOP Conference Series: Materials Science and Engineering. 263 (2017) 022012. doi:10.1088/1757–899X/263/2/022012.

Nasr-El-Din, H.A., Gurluk, M.R., and Crews, J.B. 2013. Enhancing the performance of viscoelastic surfactant fluids using nanoparticles. EAGE Annual Conference and Exhibition Incorporating SPE Europec, London. https://doi.org/10.2118/164900-MS.

Nassar, N.N., Hassan, A., and Pereira-Almao, P. 2011. Metal oxide nanoparticles for asphaltene adsorption and oxidation. Energy Fuels. 25: 1017–1023.

Nava, O.J., Soto-Robles, C.A., Gomez-Gutierrez, C.M., Vilchis-Nestor, A.R., Castro-Beltran, A., Olivas, A., and Luque, P.A. 2017. Fruit peel extract mediated green synthesis of zinc oxide nanoparticles. J. Mol. Struct., 1147: 1–6.

Nikoobakht, B., Wang, X., Herzing, A., and Shi, J. 2013. Scable synthesis and device integration of self-registered one-dimensional zinc oxide nanostructures and related materials. Chem. Soc. Rev. 42: 342–365.

Nurbas, M., Ghorbanpoor, H., and Avci, H. 2017. An eco-friendly approach to synthesis and characterization of magnetite (Fe_3O_4) nanoparticles using *Platanus orientalis* L. leaf extract. Dig. J. Nanomater. Bios. 12(4): 993–1000.

Nwanya, A.C., Razanamahandry, L.C., Bashir, A.K.H., Ikpo, C.O., Nwanya, S.C., Botha, S., Ntwampe, S.K.O., Ezema, F.I., Iwuoha, E.I., and Maaz, M. 2019. Industrial textile effluent treatment and antibacterial effectiveness of *Zea mays L.* Dry husk mediated bio-synthesized copper oxide nanoparticles. J. Hazrard. Mater. 375: 281–289.

Odoom-Wubah, T., Chen, X., Huang, J., and Li, Q. 2015. Template-free biosynthesis of flower-like CuO microstructures using *Cinnamomum camphora* leaf extract at room temperature. Mater. Lett. 161: 387–390.

Omran, B.A., Fatthallah, N.A., El-Gendy, N.Sh., El-Shatoury, E.H., and Abouzeid, M.A., 2013. Green biocides against sulphate reducing bacteria and macrofouling organisms. J. Pure Appl. Microbiol. 7(3): 2219–2232.

Omran, B.A., Nassar, H.N., Younis, S.A., El-Salamony, R.A., Fatthallah, N.A., Hamdy, A., El-Shatoury, E.H., and El-Gendy, N.Sh. 2020. Novel mycosynthesis of cobalt oxide nanoparticles using *Aspergillus brasiliensis* ATCC 16404—optimization, characterization and antimicrobial activity. J. Appl. Microbiol. 128: 438–457. https://doi.org/10.1111/jam.14498

Órdenes-Aenishanslins, N.A., Saona, L.A., Durán-Toro, V.M., Monrás, J.P., Bravo, D.M., and Pérez-Donoso, J.M. 2014. Use of titanium dioxide nanoparticles biosynthesized by *Bacillus mycoides* in quantum dot sensitized solar cells. Microb. Cell Fact. 13: 90. http://www.microbialcellfactories.com/content/13/1/90

Pan, Z.W., Dai, Z.R., and Wang, Z.L. 2001. Nanobelts of semiconducting oxides. *Science,* 291: 1947–1949.

Pansambal, S., Deshmukh, K., Savale, A., Ghotekar, S., Pardeshi, O., Jain, G., Aher, Y., and Pore, D. 2017. Phytosynthesis and biological activities of fluorescent CuO nanoparticles using *Acanthospermum hispidum* L. extract. J. Nanostruct. 7(3): 165–174.

Patil, B.N., and Taranath, T.C. 2016. *Limonia acidissima* L. leaf mediated synthesis of silver and zinc oxide nanoparticles and their antibacterial activities. Microb. Pathog. 115: 227–232.

Patra, J.K., Ali, M.S., Oh, I.-G., and Baek, K.-H. 2017. Proteasome inhibitory, antioxidant, and synergistic antibacterial and anticandidal activity of green biosynthesized magnetic Fe_3O_4 nanoparticles using the aqueous extract of corn (*Zea mays* L.) ear leaves. Artif. Cells Nanomed. Biotechnol. 45: 349–356.

Peng, J., Liu, Q., Xu, Z., and Masliyah, J. 2012. Novel magnetic demulsifier for water removal from diluted bitumen emulsion. Energy Fuels. 26: 2705–2710.

Polshettiwar, V., Baruwati, B., and Varma, R.S. 2009. Self-assembly of metal oxides into three-dimensional nanostructures: synthesis and application in catalysis. ACS Nano. 9: 728–736.

Prasad, K., and Jha, A.K. 2009. ZnO nanoparticles: synthesis and adsorption study. Nat Sci. 1: 129–35.

Prasad, K., Jha, A.K., Prasad, K., and Kulkarni, A.R. 2010. Can microbes mediate nano-transformation? Indian J. Phys. 84(10): 1355–1360.

Prasad, K.S., Patra, A., Shruthi, G., and S. Chandan. 2017. Aqueous extract of *Saraca indica* leaves in the synthesis of copper oxide nanoparticles: finding a way towards going green. J. Nanotechnol. 7502610. https://doi.org/10.1155/2017/7502610.

Raja, A., Ashokkumar, S., Marthandam, R.P., Jayachandiran, J., Khatiwada, C.P., Kaviyarasu, K., Raman, R.G., and Swaminathan, M. 2018. Eco-friendly preparation of zinc oxide nanoparticles using Tabernaemontana divaricata and its photocatalytic and antimicrobial activity. J. Photochem. Photobiol. B. 181: 53–58.

Rajabairavi, N., Raju, C.S., Karthikeyan, C., Varutharaju, K., Nethaji, S., Hameed, A.S.H., Haja, A.S., Appakan, S. 2017. Biosynthesis of novel zinc oxide nanoparticles (ZnO NPs) using endophytic bacteria *Sphingobacterium thalpophilum*. Springer Proc. Phys. 189: 245–524.

Rajan, A., Cherian, E., and Baskar, G. 2016. Biosynthesis of zinc oxide nanoparticles using *Aspergillus fumigatus* JCF and its antibacterial activity. Int. J. Mod. Sci. Technol. 1: 52–57.

Rajendran, A., Siva, E., Dhanraj, C., and Senthilkumar, S. 2018. A green and facile approach for the synthesis copper oxide nanoparticles using *Hibiscus rosa-sinensis* flower extracts and its antibacterial activities. J. Bioprocess. Biotech. 8(3): 324. doi: 10.4172/2155–9821.1000324.

Rajiv, P., Bavadharani, B., Kumar, M.N., and Vanathi, P. 2017. Synthesis and characterization of biogenic iron oxide nanoparticles using green chemistry approach and evaluating their biological activities. Biocatal. Agric. Biotechnol. 12: 45–49.

Raliya, R., Biswas, P., and Tarafdar, J.C. 2015. TiO_2 nanoparticle biosynthesis and its physiological effect on mung bean (*Vigna radiata* L.). Biotechnol. Rep. 5: 22–26.

Raliya, R., Tarafdar, J.C., Mahawar, H., Kumar, R., Gupta, P., Mathur, T., Kaul, R.K., Kumar, P., Kalia, A., Gautam, R., Singh, S.K., Gehlot, H.S. 2014. ZnO nanoparticles induced exopolysaccharide production by *B. subtilis* strain JCT1 for arid soil applications. Int. J. Biol. Macromol. 65: 362–368.

Ramesh, C., and Gopalakrishnan, K. 2016. Biogenic synthesis of Cu_2O nanoparticles using aqueous solutions of *Arachis hypogaea* leaf extracts. MCASJR. 3: 79–87.

Ramesh, C., Hari Prasad, M., and Ragunathan, V. 2011. Antibacterial behaviour of Cu_2O nanoparticles against Escherichia coli; reactivity of Fehling's solution on *Manihot esculenta* leaf extract. Curr. Nanosci. 7(5): 770–775.

Rao, A., Bankar, A., Kumar, A.R., Gosavi, S., and Zinjarde, S. 2013. Removal of hexavalent chromium ions by *Yarrowia lipolytica* cells modified with phyto-inspired Fe^0/Fe_3O_4 nanoparticles. J. Contam. Hydrol. 146: 63–73.

Rauf, M.A., Owais, M., Rajpoot, R., Ahmad, F., Khan, N., and Zubair, S. 2017. Biomimetically synthesized ZnO nanoparticles attain potent antibacterial activity against less susceptible: *S. aureus* skin infection in experimental animals. RSC Adv. 7: 36361–36373.

Rehman, S., Jermy, B.R., Akhtar, S., Borgio, J.F., Abdul Azeez, S., Ravinayagam, V., Al Jindan, R., Alsalem, Z.H., Buhameid, A., and Gani, A. 2019. Isolation and characterization of a novel thermophile; *Bacillus haynesii*, applied for the green synthesis of ZnO nanoparticles. Artif. Cells Nanomed. Biotechnol. 47(1): 2072–2082.

Rezvani, M.A., Shojaie, A.F., and Loghmani, 2012. Synthesis and characterization of novel nanocomposite, anatase sandwich type polyoxometalate, as a reusable and green nano catalyst in oxidation desulfurization of simulated gas oil. Catal. Commun. 25: 36–40.

Riya, L., and George, M. 2015. Green synthesis of cuprous oxide nanoparticles. Int. J. Adv. Res. Sci. Eng. 4: 315–322.

Rufus, A., Sreeju, N., Vilas, V., and Philip, D. 2017. Biosynthesis of hematite (α-Fe_2O_3) nano-structures: size effects on applications in thermal conductivity, catalysis, and antibacterial activity. J. Mol. Liq. 242: 537–549.

Sabir, S., Arshad, M., and Chaudhari, S.K. 2014. Zinc oxide nanoparticles for revolutionizing agriculture: synthesis and applications. Sci. World J. 2014: 925494. doi:10.1155/2014/925494.

Saed, D., Nassar, H.N., El-Gendy, N.Sh., Zaki, T., Moustafa, Y.M., and Badr, I.H.A. 2014. The enhancement of pyrene biodegradation by assembling MFe_3O_4 nano-sorbents on the surface of microbial cells. Energy Source A. 36(17): 1931–1937. https://doi.org/10.1080/15567036.2014.889782

Salman, J.A.S., Kadhim, A.A., and Haider, A.J. 2018. Biosynthesis, characterization and antibacterial effect of Zno nanoparticles synthesized by *Lactobacillus* spp. J Glob. Pharma Technol. 10(3): 348–355.

Samzadeh-Kermani, A., Izadpanah, F., and Mirzaee, M. 2016. The improvements in the size distribution of zinc oxide nanoparticles by the addition of a plant extract to the synthesis. Cogent Chem. 2: 1150389–115096.

Sangappa, M., Vandana, S.P., Bharath, A.U., and Thiagarajan, P. 2013. Mycobiosynthesis of novel non toxic zinc oxide nanoparticles by a new soil fungus *Aspergillus terreus* VIT 2013. J. Chem. Pharm. Res. 5(12): 1155–1161

Sangeetha, G., Rajeshwari, S., and Venckatesh, R. 2011. Green synthesis of zinc oxide nanoparticles by *Aloe barbadensis* miller leaf extract: structure and optical properties. Mater. Res. Bull. 46: 2560–2566.

Sankar, R., Manikandan, P., Malarvizhi, V., Fathima, T., Shivashangari, K.S., and Ravikumar, V. 2014. Green synthesis of colloidal copper oxide nanoparticles using *Carica papaya* and its application in photocatalytic dye degradation. Spectrochim. Acta Mol. Biomol. Spectrosc. 121: 746–750.

Santhoshkumar, J., Kumar, S. V., and Rajeshkumar, S. 2017. Synthesis of zinc oxide nanoparticles using plant leaf extract against urinary tract infection pathogen. Resour. Effic. Technol. 3: 459–465.

Santhoshkumar, T., Rahuman, A.A., Jayaseelan, C., Rajakumar, G., Marimuthu, S., Kirthi, A.V., Velayutham, K., Thomas, J., Venkatesan, J., and Kim, S. 2014. Green synthesis of titanium dioxide nanoparticles using *Psidium guajava* extract and its antibacterial and antioxidant properties. Asian Pac. J. Trop. Dis. 7(12): 968–976.

Saravanakumar, K., Shanmugam, S., Varukattu, N.B., Ali, D.M., Kathiresan, K., and Wang, M.-H. 2019. Biosynthesis and characterization of copper oxide nanoparticles from indigenous fungi and its effect of photothermolysis on human lung carcinoma. J. Photochem. Photobiol. B. 190: 103–109.

Saravanan, M., Gopinath, V., Chaurasia, M.K., Syed, A., Ameen, F., and Purushothaman, N. 2018. Green synthesis of anisotropic zinc oxide nanoparticles with antibacterial and cytofriendly properties. Microb. Pathog. 115: 57–63.

Sarkar, J., Ghosh, M., Mukherjee, A., Chattopadhyay, D., and Acharya, K. 2014. Biosynthesis and safety evaluation of ZnO nanoparticles. Bioprocess Biosyst. Eng. 37: 165–171.

Sathiyavimal, S., Vasantharaj, S., Bharathi, D., Saravanan, M., Manikandan, E., Kumar, S.S., and Pugazhendhi, A. 2018. Biogenesis of copper oxide nanoparticles (CuONPs) using *Sida acuta* and their incorporation over cotton fabrics to prevent the pathogenicity of Gram negative and Gram positive bacteria. J. Photochem. Photobiol. B. Biol. 188: 126–134.

Segets, D., Gradl, J., Taylor, R.K., Vassilev, V., and Peukert, W. 2009. Analysis of optical absorbance spectra for the determination of ZnO nanoparticle size distribution, solubility, and surface energy. ACS Nano. 3: 1703–1710.

Selvarajan, E., and Mohanasrinivasan, V. 2013. Biosynthesis and characterization of ZnO nanoparticles using *Lactobacillus plantarum* VITES07. Mater. Lett. 112: 180–182.

Senthil, M., and Ramesh, C. 2012. Biogenic synthesis of Fe_3O_4 nanoparticles using *Tridax procumbens* leaf extract and its antibacterial activity on *Pseudomonas aeruginosa*. Dig. J. Nanomater. Biostruct. 7: 1655–1661.

Shah, S.T., Zamir, R., Muhammad, T., and Ali, H. 2006. Mass propagation of *bougainvillea spectabilis* through shoot tip culture. Pak. J. Bot. 38: 953–959.

Shamsuzzaman, M.A., Khanam, H., and Aljawfi, R.N. 2017. Biological synthesis of ZnO nanoparticles using *C. albicans* and studying their catalytic performance in the synthesis of steroidal pyrazolines. Arab. J. Chem. 10: S1530–S1536.

Shan, G., Xing, J., Zhang, H., and Liu, H. 2005. Biodesulfurization of dibenzothiophene by microbial cells coated with magnetite nanoparticles. Appl. Environ. Microbiol. 71(8): 4497–4502.

Sharan, C., Khandelwal, P., and Poddar, P. 2015. Biomilling of rod-shaped ZnO nanoparticles: a potential role of *Saccharomyces cerevisiae* extracellular proteins. RSC Adv. 5: 1883–1889.

Shojaee, S., and Mahdavi Shahri, M. 2016. Green synthesis and characterization of iron oxide magnetic nanoparticles using Shanghai White tea (*Camelia sinensis*) aqueous extract. J. Chem. Pharm. Res. 8: 138–143.

Shojaee, S., and Shahri, M.M. 2016. Green synthesis and characterization of iron oxide magnetic nanoparticles using Shanghai White tea (*Camelia sinensis*) aqueous extract. J. Chem. Pharm. Res. 8(5): 138–143.

Simonsen, G., Strand, M., and Øye, G. 2018. Potential applications of magnetic nanoparticles within separation in the petroleum industry. J. Pet. Sci. Eng. 165: 488–495.

Simonsen, G., Strand, M., Norrman, J., and Øye, G. 2019. Amino-functionalized iron oxide nanoparticles designed for adsorption of naphthenic acids. Collids Surf. A. 568: 147–156.

Singh, A.K., Pal, P., Gupta, V., Yadav, T.P., Gupta, V., and Singh, S.P. 2018. Green synthesis, characterization and antimicrobial activity of zinc oxide quantum dots using *Eclipta alba*. Mater. Chem. Phys. 203: 40–48.

Sreeju, N., Rufus, A., and Philip, D. 2017. Studies on catalytic degradation of organic pollutants and anti-bacterial property using biosynthesized CuO nanostructures. J. Mol. Liq. 242: 690–700.

Subhapriya, S., and Gomathipriya, P. 2018. Green synthesis of titanium dioxide (TiO_2) nanoparticles by *Trigonella foenum-graecum* extract and its antimicrobial properties. Microb. Pathog. 116: 215–220.

Sumitha, S., Vidhya, R.P., Lakshmi, M.S., and Prasad, K.S. 2016. Leaf extract mediated green synthesis of copper oxide nanoparticles using *Ocimum tenuiflorum* and its characterization. Int. J. Chem. Sci. 14(1): 435–440.

Sundar, S., Venkatachalam, G., and Kwon, S.J. 2018. Biosynthesis of copper oxide (CuO) nanowires and their use for the electrochemical sensing of dopamine. Nanomaterials. 8: 823. doi:10.3390/nano8100823.

Taran, M., Rad, M., and Alavi, M. 2017. Antibacterial activity of copper oxide (CuO) nanoparticles biosynthesized by *Bacillus* sp. FU4: optimization of experiment design. Pharm. Sci. 23: 198–206.

Taran, M., Rad, M., and Alavi, M. 2018. Biosynthesis of TiO_2 and ZnO nanoparticles by *Halomonas elongata* IBRC-M 10214 in different conditions of medium. BioImpacts. 8: 81–89.

Thakur, S., and Karak, N. 2014. One-step approach to prepare magnetic iron oxide/reduced graphene oxide nanohybrid for efficient organic and inorganic pollutants removal. Mater. Chem. Phys. 144: 425–432.

Tharunya, P., Subha, V., Kirubanandan, S., Sandhaya, S., and Renganathan, S. 2017. Green synthesis of superparamagnetic iron oxide nanoparticle from *Ficus carica* fruit extract, characterization studies and its application on dye degradation studies. Asian J. Pharm. Clin. Res. 10(3): 125–128.

Tripathi, R.M., Bhadwal, A.S., Gupta, R.K., Singh, P., Shrivastav, A., and Shrivastav, B.R. 2014. ZnO nanoflowers: novel biogenic synthesis and enhanced photocatalytic activity. J. Photochem. Photobiol. B. Biol. 141: 288–295. jphotobiol.2014.10.001.

Vanathi, P., Rajiv, P., Narendhran, S., Rajeshwari, S., Rahman, P.K.S.M., and Venckatesh, R. 2014. Biosynthesis and characterization of phytomediated zinc oxide nanoparticles: a green chemistry approach. Mater. Lett. 134: 13–15.

Varaprasad, K., Raghavendra, G.M., Jayaramudu, T., and Seo, J. 2016. Nano zinc oxide-sodium alginate antibacterial cellulose fibres. Carbohydr. Polym. 135: 349–55.

Veeramanikandan, V, Madhu, G.C., Pavithra, V., Jaianand, K., and Balaji, P. 2017. Green synthesis, characterization of iron oxide nanoparticles using *Leucas aspera* leaf extract and evaluation of antibacterial and antioxidant studies. Int. J. Agr. Innov. Res. 6(2): 2319–1473.

Velayutham, K., Rahuman, A.A., Rajakumar, G., Santhoshkumar, T., Marimuthu, S., Jayaseelan, C, Bagavan, A., Kirthi, A.V., Kamaraj, C., Zahir, A.A., and Elango, G. 2012. Evaluation of *Catharanthus roseus* leaf extract-mediated biosynthesis of titanium dioxide nanoparticles against *Hippobosca maculata* and *Bovicola ovis*. Parasitol. Res. 111(6): 2329–2337.

Velmurugan, P., Shim, J., You, Y., Choi, S., Kamala-Kannan, S., Lee, K.J., Kim, H.J., and Oh, B.-T. 2010. Removal of zinc by live, dead, and dried biomass of *Fusarium* spp. Isolated from the abandoned-metal mine in South Korea and its perspective of producing nanocrystals. J. Hazard. Mater. 182: 317–324.
Velsankar, K., Sudhahar, S., Maheshwaran, G., and Krishna Kumar, M. 2019. Effect of biosynthesis of ZnO nanoparticles via *Cucurbita* seed extract on *Culex tritaeniorhynchus* mosquito larvae with its biological applications. J. Photochem. Photobiol. 200: 111650.
Velsankar, K., Sudhahar, S., Parvathy, G., and Kaliammal, R. 2020. Effect of cytotoxicity and antibacterial activity of biosynthesis of ZnO hexagonal shaped nanoparticles by *Echinochloa frumentacea* grains extract as a reducing agent. Mater. Chem. Phys. 239: 121976.
Venkatesh, K.S., Palani, N.S., Krishnamoorthi, S.R., Thirumal, V., and Ilangovan, R. 2013. Fungus mediated biosynthesis and characterization of zinc oxide nanorods. AIP Conference Proceedings 1536(93). https://doi.org/10.1063/1.4810116.
Wade, S.A., Mart, P.L., and Trueman, A.R. 2011. Microbiologically influenced corrosion in maritime vessels. Corros. Mater. 36: 68–79.
Wang, Z.L. 2008. Splendid one-dimensional nanostructures of zinc oxide: a new nanomaterial family for nanotechnology. ACS Nano. 2: 1987–1992.
William, J.K.M., Ponmani, S., Samuel, R., Nagarajan, R., and Sangwai, J.S. 2014. Effect of CuO and ZnO nanofluids in xanthan gum on thermal, electrical and high pressure rheology of water-based drilling fluids. J. Pet. Sci. Eng. 117: 15–27.
Xia, T., Kovochich, M., and Nel, A.E. 2007. Impairment of mitochondrial function by particulate matter (PM) and their toxic components: implications for PM-induced cardiovascular and lung disease. Front. Biosci. 12: 1238–1246.
Xu, T., Ji, P., He, M., and Li, J. 2012. Growth and structure of pure ZnO micro/nanocombs. J. Nanomater. 2012: 797935. doi:10.1155/2012/797935.
Yadav, M., Jain, A., and Malhotra, P. 2019. A review on the sustainable routes for the synthesis and applications of cuprous oxide nanoparticles and their nanocomposites. Green Chem. 21: 937–955.
Yamamoto, O. 2001. Influence of particle size on the antibacterial activity of zinc oxide. Int. J. lnorg. Mater. 3: 643–646.
Yedurkar, S., Maurya, C., and Mahanwar, P. 2016. Biosynthesis of zinc oxide nanoparticles using *Ixora Coccinea* leaf extract—a green approach. Open J. Synth. Theory Appl. 5: 1–14.
Yedurkar, S.M., Maurya, C.B., and Mahanwa, P.A. 2017. A biological approach for the synthesis of copper oxide nanoparticles by *Ixora Coccinea* leaf extract. J. Mater. Environ. Sci. 8(4): 1173–1178.
Yew, Y.P., Shameli, K., Miyake, M., Kuwano, N., Bahiyah Bt Ahmed Khairudin, N., Eva Bt Mohamad, S., and Lee, K.X. 2016. Green synthesis of magnetite (Fe_3O_4) nanoparticles using seaweed (*Kappaphycus alvarezii*) extract. Nanoscale Res. Lett. 11: 276. doi: 10.1186/s11671–016-1498–2.
Yu, J., Berlin, J., Lu, W., Zhang, L., Kan, A., Zhang, P., Walsh, E.E., Work, S., Chen, W., Tour, J., Wong, M., and Tomson, M.B. 2010. Transport study of nanoparticles for oilfield application. SPE International Conference on Oilfield Scale, Aberdeen. https://doi.org/10.2118/131158-MS.
Yugandhar, P., Vasavi, T., Devi, P.U.M., and Savithramma, N. 2017. Bioinspired green synthesis of copper oxide nanoparticles from *Syzygium alternifolium* (Wt.) Walp: characterization and evaluation of its synergistic antimicrobial and anticancer activity. Appl. Nanosci. 7: 417–427.

Yusof, H.M., Mohamad, R., Zaidan, U.H., and Abdul Rahman, N. 2019. Microbial synthesis of zinc oxide nanoparticles and their potential application as an antimicrobial agent and a feed supplement in animal industry: a review. J. Anim. Sci. Biotechnol. 10: 57. https://doi.org/10.1186/s40104-019-0368-z.

Zakaria, B.S., Nassar, H.N., Saed, D., El-Gendy, N.Sh. 2015. Enhancement of carbazole denitrogenation rate using magnetically decorated *Bacillus clausii* BS1. Pet5. Sci. Technol. 33(7): 802–811. https://doi.org/10.1080/10916466.2015.1014966

Zarasvand, K.A., and Rai, V.R. 2016. Inhibition of a sulfate reducing bacterium, *Desulfovibrio marinisediminis* GSR3, by biosynthesized copper oxide nanoparticles. 3 Biotech. 6: 84. doi 10.1007/s13205–016-0403–0.

Zare, E., Pourseyedi, S., Khatami, M., and Darezereshki, E. 2017. Simple biosynthesis of zinc oxide nanoparticles using nature's source, and it's in vitro bio-activity. J. Mol. Struct. 1146: 96–103.

Zhang, X., Yan, S., Tyagi, R.D., and Surampalli, R.Y. 2011. Synthesis of nanoparticles by microorganisms and their application in enhancing microbiological reaction rates. Chemosphere. 82: 489–494.

Zikalala, N., Matshetshe, K., Parani, S., and Oluwafemi, O.S. 2018. Biosynthesis protocols for colloidal metal oxide nanoparticles. Nano-Struct. Nano-Objects. 16: 288–299.

CHAPTER 5

Nanobiotechnology and Mitigation of Microbiologically Influenced Corrosion in Petroleum Industry

5.1 INTRODUCTION

Chemical corrosion and bio corrosion of pipelines and water systems are serious problems in petroleum industry. It has been reported that microbiologically influenced corrosion (MIC) contributes to approximately 20% of the annual metal damage in oil and gas industry (Sharad et al., 2016; Abdul Rasheed et al., 2019a, b). Zakaria et al. (2012) reported that MIC is responsible for; approximately 75% of corrosion in productive oil wells, more than 50% of failures of buried pipelines and it is also the main cause of corrosion occurrence in drilling and pumping equipment and storage tanks. In an earlier study, MIC was estimated to represent 40% of the internal pipelines corrosion in oil and gas industry (Zhu et al., 2003; Rajasekar et al., 2010). Moreover, MIC-localized corrosion reported to cause from 70–95% of the pipeline internal leaks (Javaherdashti, 2008). It has been estimated that 50% of the worldwide total corrosion-related problem cost is caused by MIC (Lin and Ballim, 2012). MIC is mainly caused by the presence and adherence of the microbes on the pipelines, which consequently leads to metal deterioration and damage as a result of microbial activities (Rajasekar et al., 2007; Aruliah and Ting, 2014; Agarry et al., 2015; Gu et al., 2019). For enhancing corrosion and metal deterioration, microbial population creating such biofilms need energy and carbon sources, electron donors and acceptors, and water (Omran et al., 2019). It has been reported that MIC biofilms are easily formed on rough surfaces than smooth ones and on 304 stainless steel than on 316 (Song et al., 2018). Moreover, the pitting corrosion occurred into the 2507 duplex stainless steel in petroleum and chemical industry, and circulating cooling water systems is mainly due to sulfate-reducing bacteria (SRB) attack (Liang et al., 2014). It has been reported that 77% of the occurring corrosion in US production wells is attributed to the SRB attack

(Iverson, 2001). SRB also mainly attacks the oilfield water injection systems (Song et al., 2018). Further, from the environmental and economic point of view, MIC has many negative impacts. It would lead to reservoir souring, oil contamination, and sudden pipeline failure, which would cause sudden oil and/or gas leakage, explosion, and fire occurrence (Vanaei et al., 2017). *Desulfovibrio* sp. as the main example of SRB in the oil field has been reported to produce approximately 10 g/L toxic and corrosive H_2S, during its active division phase (Turkiewicz et al., 2013). Biofilm extracellular polysaccharide substance (EPS) has sticky properties that aid the microbial adherence to the metal surfaces, producing a slime layer, which protects the microbial biofilm constituents form the antimicrobial action of the injected biocides (Aruliah and Ting, 2014). The EPS can contribute up to 90% of the biofilm (Bhola et al., 2010). EPS is composed of proteins, peptides, lipids, carbohydrates, humic acids, exogenous DNA, and polysaccharides (Kang et al., 2014). The biofilm is first formed by the adherence and attachment of the planktonic aerobic microorganisms, which produce such EPS forming a locally oxygen free shelter below the sessile microorganisms that enhances the growth of the sessile SRB (Wadood et al., 2015; El-Shamy et al., 2016; Gu et al., 2019; Jia et al., 2019). However, it should be known that there are planktonic and sessile SRB. There are also microorganisms other than the SRB contributing to MIC, which are methanogens, iron-reducing bacteria (IRB) (e.g., *Pseudomonas* sp., *Geobacillus* sp., and *Shewanella* sp. reduce ferric Fe^{3+} to ferrous Fe^{2+}), acid-producing bacteria (e.g., *Thiobacillus* sp., *Acetobacter* sp., and *Clostridium* sp.), metal-oxidizing bacteria (e.g., *Gallionella*, *Sphaerotilus*, and *Leptothrix*), and the slime-producing bacteria (Al-Abbas et al., 2012, 2013; Sharad et al., 2016; Vigneron et al., 2016; Arruda de Queiroz et al., 2018). Leptospirilli such as *Leptospirillum ferrooxidans* would represent as much as one-half of the ferrous-iron-oxidizing population (Sand et al., 1992). But SRB are known to be the main contributor of MIC (Wang et al., 2010; Esquivel et al., 2011; Xu et al., 2013; Loto, 2017). The rate of MIC of steel by SRB, for example, *Desulfotomaculum nigrificans*, is reported to be approximately six times than the control sample in formation water without SRB, since SRB accelerate both the anodic and cathodic reactions (Liu et al., 2019). Besides this, the pits' depth in the presence of SRB can reach approximately 7.7 times than those of control, which is usually attributed to the presence of SRB heterogeneous biofilm that created the serious galvanic effect and consequently accelerated the localized corrosion (Liu et al., 2019). The MIC induced by a consortium of SRB and IRB is reported to be 10 times higher than that induced by SRB alone, as ferrous ions produced via the ferric reduction would inhibit the development of the protecting iron sulfide film (Valencia-Cantero and Peña-Cabriales, 2014). In another study, the MIC of steel subjected to bacterial consortium of IRB, SRB, and *Bacillus* sp. is reported to be approximately five times higher than that subjected to a consortium of IRB and SRB alone (Valencia-Cantero and Peña-Cabriales, 2003). Moreover, *Pseudomonas* sp. has been reported to enhance the rate of MIC as it stops the repassivation; thus, the pits continue to grow (Lan et al., 2017), which might be attributed to the production of organic acids and/or enhancement of cathodic reduction due to the catalase excreted by that genus (Lan et al., 2017). The MIC would consequently occur through cathodic depolarization, concentration and/or galvanic cell formation, and direct electron transfer (Enning and Garrelfs, 2014; El-Shamy et al., 2016). Biofilm causes clogging and plugging in the pipelines (Little and Lee, 2014; Jia et al., 2019). The metabolic products of biofilm microorganisms, such as sulfides and organic acids, together with the EPS, change the interface chemistry, such as pH value and dissolved oxygen concentration, which first leads to localized pitting and/or crevice corrosion and then consequently enhances the corrosion rate. Thus, such biofilms should be continuously removed to enhance the flow in the pipelines and prevent the formation of biocorrosion and metal wall perforations. Otherwise, sudden failure and breakage in pipelines would occur, which would consequently cause, as mentioned before, the leakage of petroleum leading to environmental pollution and open firing. Besides this, biofouling as a development of the biofilm formation usually occurs. Once the biofilm microorganisms, which are also known as microfoulants, produce the thick gelatinous slime layer, macrofoulants such as mussels, alga, barnacles, etc., can gather onto biofilm forming a huge and massive well-adhered and compact biodeposit (Telegdi et al., 2016). Such biofoulants together with the other insoluble biocorrosion products and precipitates plug the pipelines, decrease the heat exchange, and cause many other problems in oil and gas industry, which consequently lower the economic function effectiveness. Accordingly, it should be taken into consideration that upon choosing biocide for real field application, it should have a wide biocidal effect on different expected microbial strains creating the biofilm.

The mitigation of MIC is of an urgent need for many industrial sectors, not only for the oil and gas industry, as it attacks the cooling systems and most of the industrial infrastructure. Thus, it is mandatory to find a solution to decrease the capital cost and prevent the sudden industrial failure, production downtime, and/or sudden shutdown caused by MIC. Expensive and nonecofriendly chemical corrosion inhibitors and biocides are usually applied in petroleum industry to overcome such a massive problem (Smith et al., 2010; Muñoz-Bonilla and Fernández-García, 2012; Wei et al., 2013; Yeole et al., 2015; Abd-Elaal et al., 2017; Nkemnaso et al., 2018; Sharma et al., 2018; Jia et al., 2019). Upon the biofilm formation, the resistibility of sessile microbial population against the injected biocides increases. Besides this, the repeated application of biocide would create a microbial biocide acclimatization. So, this would lead to biocide dosage escalation, which would consequently have a lot of negative impacts on nontarget microorganisms and ecosystem (Omran et al., 2013). Thus, the application of cost-effective and eco-friendly green corrosion inhibitors and biocides is a worldwide mandate now. Moreover, the application of metal/metal oxide nanoparticles (NPs) alone or as biocide enhancer in mitigation of MIC is expected to solve a lot of aforementioned problems in oil and gas industry. This chapter summarizes the worldwide efforts in the application of nanobiotechnology to overcome the problem of biocorrosion and formation of corrosive biofilms.

5.2 BIOSYNTHESIS OF SOME NANOPARTICLES UNDER LIMITED AERATION AS A TOOL FOR BIOREMEDIATION

Thiobacillus ferrooxidans under anaerobic condition reduces ferric iron, while under aerobic condition re-oxidizes the ferrous iron (Sand, 1989). Then, the dissolved sulfides, H_2S, HS^-, and S^-, produced by the SRB would react with the dissolved Fe(II) precipitating FeS NPs (Fortin et al., 1996). Labrenz and Banfield (2004) reported the formation of small-sized sphalerite (ZnS) deposits (2–5 nm) in a naturally occurring biofilm of SRB *Desulfobacteriaceae* at low temperature of 8 °C and slightly alkaline media pH 7.2–8.6. Thus, SRB can be applied to bioremediate heavy metal polluted water. SRB enhance the reaction of the produced sulfides with toxic metal ions precipitating them as stable metal sulfides NPs, which can be easily removed from the polluted water (Panda et al., 2016; Ayangbenro et al., 2018).

Lengke and Southam (2006) reported the bioaccumulation of gold by SRB collected from a gold mine in South Africa cultured in the presence of gold(I)-thio sulfate complex. That intracellularly produced small-sized AuNPs < 10 nm. The Gram-negative *Pseudomonas stutzeri* reported to produce in biofilm 12-nm ZnS at pH12 and 27 °C and 44-nm ZnS at pH 7 and 37 °C within two weeks, under anaerobic conditions, on a piece of zinc metal (Mirhend et al., 2013).

Desulfovibrio alaskensis G20 and *D. desulfuricans* 8307 are anaerobic SRB, which have been reported for the formation of palladium and platinum NPs within 2 h at room temperature (see Figure 5.1) (Capeness et al., 2015). Thus, it can be recommended for bioremediation of water contaminated with Pd^{2+} and Pt^{2+} ions. *Desulfovibrio alaskensis* G20 and *D. desulfuricans* 8307 have also been reported for the formation of 10-nm nickel sulfide NiS NPs within only 30 min under anaerobic condition and at room temperature (Capeness et al., 2015).

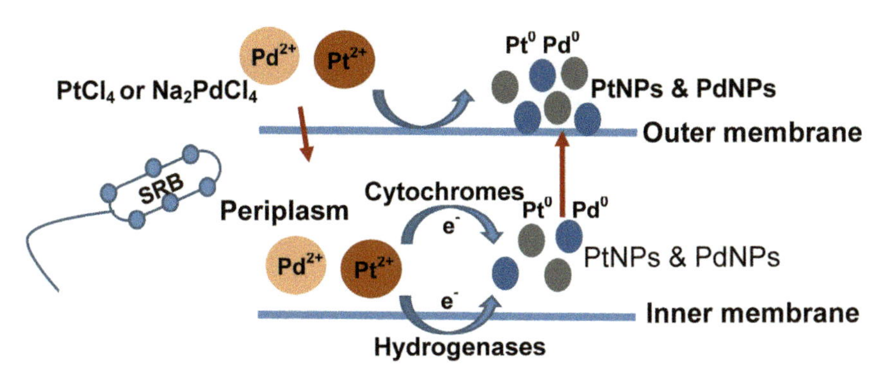

FIGURE 5.1 A hypothetical mechanism for biosynthesis of metal NPs using SRB.

In situ bioprecipitation of heavy metals by the integration of Fe^0 and SRB for heavy metal contaminated aquifer sediment and groundwater remediation on a microcosm level has been reported, where glycerol was used as the C-source (Kumar et al., 2014a). The effect of different sizes of Fe^0 proved that sulfate reduction occurred only in microcosms amended with micro and granular Fe^0, where glycerol acted as the electron donor for the sulfate reduction, while nanosized Fe^0 exhibited a bactericidal effect on the SRB. However, Cd, Ni, As, and Zn were removed by sorption, and the highest efficiency was observed in nano Fe^0 microcosm. This study proved the importance of controlling the pH, oxidation redox potential, and the size of F^0 for achieving the highest heavy metal remediation and sulfate reduction (Kumar et al., 2014a).

The uptake of NPs by SRB-EPS biofilm is now a tool for bioremediation in wastewater treatment plants, taking into consideration that such an EPS biofilm is released under stressed conditions (Mathur et al., 2017). Although the presence of TiO_2 NPs has been reported to negatively affect the SRB activity and also act as an antibacterial agent for different Gram-positive and negative bacteria, especially under UV irradiation, due to its photocatalytic characteristics (Cai et al., 2013). However, there was a recorded occurrence of synergetic effect between SRB consortia (Mathur et al., 2017), where it showed higher tolerance to ≤1 μg/mLTiO_2 NPs, without any decrease in sulfate reduction efficiency, even under UV irradiation. Also, a higher release of EPS and increment in specific growth rate and biofilm thickness and density than the individual cultures of *Desulfovibrio desulfuricans* and *Desulfotomaculum nigrificans* were also observed (Mathur et al., 2017).

Take the advantage that the enzymatic system of SRB can perform magnetite reduction and produce iron sulfide and also the dissolved Fe(III) complexes and Fe (oxyhydr)oxides can perform as electron acceptors for SRB growth via a reduction mechanism. Thus, this would help in the biosynthesis of different forms of iron oxides NPs by SRB, for example, the biotransformation of maghemite into magnetite (Zhou et al., 2019), which can occur via the direct bioreduction of water Fe(III) by SRB and/or Fe (oxyhydr)oxides, by extracellular electron transfer, as reported by Zhou et al. (2019).

5.3 NANOMATERIALS FOR MITIGATION OF CORROSIVE BIOFILM

The planktonic corrosive microorganisms upon their adherence and attachment to metal surfaces start to produce the biofilm (see Figure 5.2). The sessile cells composing the biofilm are known to be more resistible to biocides than planktonic cells, due to the polysaccharide covering, which decreases the permeability of the injected biocide (Santos da Silva et al., 2019). It would require the injection of 10 times or even more of biocide concentration and/or amount to kill the sessile cells than the planktonic ones (Omran et al., 2013; Xu and Gu, 2015), which would be very problematic to environment and ecosystem. In an attempt to overcome such problem, besides the biofilm formation and biocorrosion, the incorporation of biocidal agents in the form of nanometals and/or nanometal oxides into polymeric nanocomposite, and its applications as paintings and/or coatings with antimicrobial activities are reported to be very promising (Kubacka et al., 2014).

Fahmy et al. (2018) reported the phytosynthesis of FeNPs using different plant extracts (e.g., tea, henna, red spinach, curry leaf, cloves, curcuma, holy basil, Indian gooseberry, tridax daisy, spinach, pomegranate fruit and peels, *Eucalyptus*, *Dodonaea viscosa, Gardenia jasminoides* and *Colocasia esculenta*, and *Sargassum muticum*) and the biosynthesis of FeNPs using *Acinetobacter* spp. and *Aspergillus oryzae*, where those green-synthesized FeNPs expressed efficient biocidal activity on a wide range of microorganisms creating the MIC biofilms.

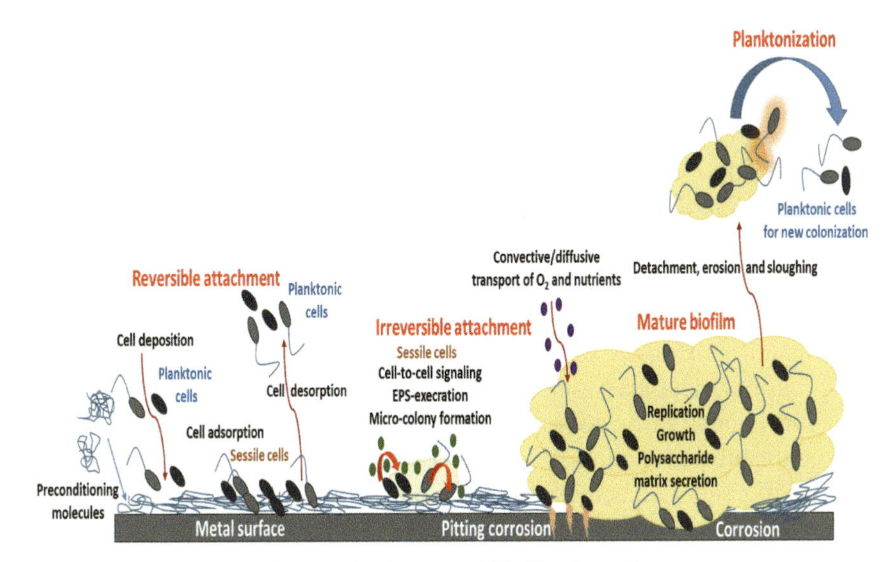

FIGURE 5.2 Schematic diagram for the steps of biofilm formation.

ZnO NPs have been reported as inhibitors for the formation of *Pseudomonas aeruginosa* biofilm (Lee et al., 2014), whereas Zn^{2+} and ZnO NPs at concentration <3 mM inhibited virulence factors via the inhibition of the production of pseudomonas quinolone signal, pyocyanin, and siderophore pyochelin, but they did not affect the production of rhamnolipids. However, it increased the production of siderophore pyoverdine and the cells' hydrophilicity, without any elucidation of bactericidal effect on planktonic *P. aeruginosa* (Lee et al., 2014). Nevertheless, ZnO NPs reported for inhibition of the biofilm formation and the cell growth of *Escherichia coli* and *Staphylococcus aureus* (Applerot et al., 2012). Moreover, the biosynthesized 16.35-nm ZnO by *Pseudomonas stutzeri*, under aerobic conditions at pH7, were reported to express an efficient biocidal activity against *Escherichia coli*, *Staphylococcus aureus*, and *P. stutzeri*, which are usually found in microbial biofilm in petroleum industry (Mirhend et al., 2013).

Moreover, the TiO_2 and ZnO NPs have been reported as efficient biocides for the heterotrophic and pathogenic bacterial population. As such, metal oxide NPs damage the protein, DNA, and cell wall (Zhang, 2010; Saadat, 2011; Asrari and Cheraghpour, 2018). Moreover, the photocatalytic biocidal activity of TiO_2 has also been reported (Kubacka et al., 2014). Antimicrobial titania (TiO_2) incorporated nanocomposites have also been successfully reported (Kubacka et al., 2007, 2012, 2014; Saadatmand and Yazdanshenas, 2012). CuNPs incorporated into a polyamide layer have been reported in another study as an efficient antimicrobial thin-film nanocomposite membrane (Cherif et al., 2018), while ZnO NPs reported as antibiofilms and inhibitors for the synthesis of exo-polysaccharides synthesis by *Pseudomonas aeruginosa* (Ishwarya et al., 2018).

Sadek et al. (2019) applied gamma radiation to reduce the particles size of the biosynthesized ZnO and CuO NPs using the extract of *Sargassum muticum*. The biosynthesized NPs recorded average size of 15 and 20 nm, respectively. The dispersed biosynthesized nanometal oxides in polyethylene oxide polymer expressed a significant biocidal effect on biofilm forming microorganisms *Proteus mirabilis*, *Pseudomonas aeruginosa*, and *Staphylococcus aureus*. The prepared nanocomposite also decreased the adherence of such microbial population to carbon mild steel coupon C1010. Thus, it lowered the biocorrosion. Besides this, low concentration of such a nanocomposite (150 ppm) also acted as a mix-type corrosion inhibitor (i.e., cathodic and anodic) with anticorrosion activity of 92.3% on mild steel coupon C1010 immersed in acidic solution (1-M HCl). That recorded efficient potential was attributed to the presence of active functional groups such as the NH_2, CO, and CHO, from the aglal extract on the biosynthesized NPs, which acted as stabilizing and capping agents during the phytosynthesis process. Those groups enhanced the adsorption of the prepared nanocomposite on the mild steel coupon C1010. Moreover, the small size of the CuO NPs enhanced their penetration into the microbial cells, which consequently destroyed its cell membrane, enzymatic system, and DNA, leading to cell rupture and death.

Jayabalan et al. (2019) reported the extracellular biosynthesis of ZnO NPs by *Pseudomonas putida* (MCC 2989) mainly via the excreted reductant nitrate reductase and polysaccharides. The biosynthesized spherical and hexagonal shaped ZnO NPs found to be with a crystalline wurtzite structure, a bandgap of 4 eV, and an average size of 25–45 nm. The X-ray photoelectron spectroscopy (XPS) proved the presence of Zn2p 1/2, Zn2p 3/2, O1S, and C1S, which were clearly appeared at the bonding energies of 1044.49 eV, 1021.52 eV, 530 eV, and 284.83 eV, respectively, confirming the biosynthesis of ZnO NPs. The biosynthesized ZnO NPs inhibited the biofilm formation by *Bacillus cereus* (MCC, 2039), and *Enterococcus faecalis* (MCC, 2041). That was attributed to the induction of reactive oxygen species (ROS) by ZnO NPs, which might have destroyed the EPS (Ishwarya et al., 2018). The biosynthesized ZnO NPs also expressed an efficient biocidal activity against the Gram-negative *Pseudomonas otitidis* (MCC 2509), *Pseudomonas oleovorans* (MCC 2566), *Acinetobacter baumannii* (MCC 2366), and the Gram-negative *Bacillus cereus* (MCC, 2039) and *Enterococcus faecalis* (MCC, 2041). The efficient biocidal activity was related to the constituents of the bacterial cell wall, whereas the Gram-positive bacteria, which have 20–80-nm-thick cell wall with a multilayer of peptidoglycan polymer containing teichoic acid and lipoteichoic acid, would chelate Zn^{2+} ions and transport them into the cell, producing the ROS, destroying the enzymatic system, proteins, and DNA, and causing cell lysis and death (Kumar et al., 2011; Hood and Skaar, 2012). In the case of the Gram-negative bacteria with 7–8-nm-thin cell wall of peptidoglycan polymer containing porins, the small-sized ZnO NPs would easily penetrate the cell membranes and consequently destruct the cell membrane with the leakage of the cell components (Ganesh et al., 2019). The toxicity of ZnO is due to its dissolution to Zn^{2+} ions, which can inhibit enzymes such as the alkaline phosphatise, polymerases, and carboxy peptidase. It can interact with the sulphhydryl group of enzymes. It can also strongly bind to Cystein, Histidine, and Aspartate side chains of proteins. Moreover, upon the entrance of ZnO NPs into the cell, disturbance in ionic strength, osmotic pressure, and cell functions, such as growth, metabolic activity, and replication occurs and sometimes completely inhibited. Usually, the smaller the size of NPs, the higher their biocidal activity and the occurrence of cell destruction (Saravanakumar et al., 2015; Barapatre et al., 2016; Narendhran and Sivaraj, 2016).

5.4 NANOMATERIALS AS BIOCIDES FOR CORROSIVE SULFATE-REDUCING BACTERIA

The main contributor for the MIC is the SRB (Sharad et al., 2016; Khouzani et al., 2019; Santos da Silva et al., 2019). Most of the SRB are known to be obligate anaerobes; however, some are known to be facultative anaerobes, for example, *Vibrio* and *Photobacterium* species (Xiaohong et al., 2019). However, the most popular strains in oil and gas industry are the obligate anaerobes, Gram-negative *Desulfovibrio,* and Gram-positive *Desulfotomaculum*. SRB can manage to stay alive in an aerobic environment until the environment becomes suitably anaerobic for them to grow. SRB can be found everywhere in the oil and gas production facilities from deep inside a well, to all the way, to the treatment facilities, till the transport network and storage infrastructure. SRB can survive over a wide range of pH (4–9.5). They can grow in a wide range of salinities. SRB can tolerate pressure up to 500 atm. Most of the SRB exist in temperature ranges of 25–60 °C. SRB obtain their energy from organic nutrients and/or from hydrogen, iron, and other inorganic compounds, while the cellular biomass would be plagiarized from inorganic carbon (Macedo de Souza et al., 2017). Thus, it could be both heterotrophic and autotrophic (see Figure 5.3).

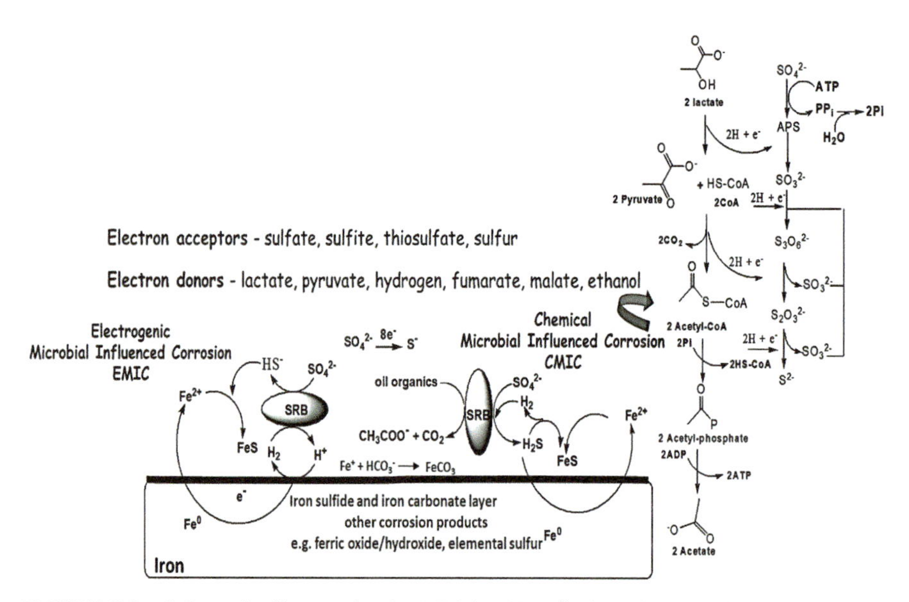

FIGURE 5.3 Schematic diagram for the MIC by SRB. *Adenosine phosphosulfate—APS, Adenosine triphosphate—ATP, Adenosine diphosphate—ADP, pyrophosphate—PPi, and phosphate—Pi.

SRB in an anaerobic respiration use sulfate instead of oxygen and oxidize organic substances or molecular H_2 to organic acids and CO_2 by the reduction of sulfate to hydrogen sulfide (Little et al., 1992). The interaction between SRB metabolic products and ferrous metal produces aggressive corrosive environment such as H_2S and some other metal sulfates and sulfides (Khowdiary et al., 2017; Abdul Rasheed et al., 2019a, b). The production of FeS in the MIC biofilm by SRB and the consequent acceleration of the initial corrosion rate are due to anodic depolarization and cathodic depolarization (see Scheme 5.1) by hydrogenase (Lee et al., 2003; Enning and Garrelfs, 2014).

$$4Fe \rightarrow 4Fe^{2+} + 8e^-$$
$$SO_4^{2-} + 8H^+ + 8e^- \rightarrow HS^- + 3H_2O + OH^-$$
$$HS^- + H^+ \leftrightarrow H_2S$$
$$Fe^+ + S^{2-} \rightarrow FeS$$
$$3Fe^{2+} + 6OH^- \rightarrow 3Fe(OH)_2$$

SCHEME 5.1 Anodic and cathodic depolarization by SRB.

FeS and EPS slime layers produce a dark-colored sludge, which hampers the oil flow (Valencia-Cantero and Peña-Cabriales, 2014). Several metal/metal oxide NPs have been added to coating polymers to act as both corrosion inhibitors and biocides for SRB and mitigate the corrosive biofilm formation, for example, AgNPs (Want et al., 2010), TiO_2 (Ma et al., 2009; Wang et al., 2010), CuO (Fathy et al., 2013), and ZnO (Fathy et al., 2013); these coatings mainly prevent the microbial adhesion toward the metal surface. However, much research is still required for the commercialization of those dual-effect coatings. Even metal/metal oxide NPs can be added to biocides to enhance their efficiency. For example, the biocidal efficiency of nanopowder cationic thiol surfactants against SRB was enhanced by AgNPs (Azzam et al., 2012). Only 600 mg/L of CuNPs/glutaraldehyde has been reported to completely inhibit the SRB growth within only 3 h (Aiad et al., 2014a).

The anatase TiO_2 NPs with an average size of 20–40 nm were used to prepare a Ni–P–Cr–TiO_2 composite coating (Ma et al., 2009). It showed a unique anticorrosive potential in 3.5% NaCl media and also prevented the formation of biofilm and MIC by SRB. That was attributed to the photocatalytic activity of TiO_2, which led to the production of •OH free radicals, decomposing the cell wall and membrane and consequently leading to cell death and lysis.

AgNPs have been green synthesized within 5 min, using sunlight and cationic surfactant as reducing and capping agents, respectively (Aiad et al., 2014b). Such green-synthesized capped AgNPs showed efficient antimicrobial activity against the Gram-positive *Sarcina lutea*, *Bacillus pumilus*, and *Micrococcus luteus*, the Gram-negative *Pseudomonas aeruginosa*, the yeast *Candida albicans*, the fungus *Penicillium chrysogenum*, and SRB *Desulfomonas pigra*. Further study proved that the highly stable hexagonal and spherical shaped green-synthesized AgNPs capped by cationic surfactant with zeta potential of +30 mV expressed high biocidal efficiency against different fungal strains, Gram-positive and Gram-negative bacteria, and SRB (Shaban et al., 2015). That was attributed to the fact that the positive charges of the capped AgNPs increased their adsorption to the negatively charged cell membranes. Later, in another study by Shaban et al. (2019), it was proved that more crystalline and smaller dispersed spherical shaped 28–58-nm AgNPs with a more control on size, shape, and increased stability (zeta potential +66 mV) can be occurred upon the increase of the Schiff base cationic surfactant hydrophobic tail. Moreover, the increase in the biocidal efficiency of such surfactants against fungi, SRB, and Gram-positive and negative bacteria occurred by the incorporation of AgNPs. However, with the increase in the tail length, the activity of the Ag-capped cationic surfactant decreased, which is known as the cutoff effect. That was parallel to the decrease of critical micelle concentration values, which occurred with the increase in the surfactant hydrophobic tail length; thus, their adsorption on the cell wall decreased. Consequently, the increment of the nanocolloid surface activity led to an increase in the accumulation of the AgNPs on the cell wall due to the physical interaction between the positively charged AgNPs and the negatively charged cell walls. The smaller the size of the surfactant as a biocide and AgNPs are, the more the penetration throughout the cell wall and membrane would occur, which consequently leads to cell damage (Shaban et al., 2019).

Khowdiary et al. (2017) reported the efficient biocidal activity of AgNPs loaded on cationic surfactants of nanohybrids size 34–74 nm against SRB *Desulfomonas pigra*. That was attributed to the double biocidal efficiency of the AgNPs and the cationic surfactant via adsorption onto the cell membrane, due to the amphipathic properties and high surface activity of the surfactant, which increased by the increment of the length of the polyethylene glycol chain, then penetrated into the cells, disturbing its enzymatic system and metabolism, and then led to cell mortality. Moreover, Ag^+ ions have been reported to prevent the bacterial adhesion toward the metal surface, which is the required first step of biofilm formation (Abdul Rasheed et al., 2019a). This would open a new field of research concerning the production of biosurfactants with good biocidal efficiency and its hybridization with green-synthesized metal/metal oxide NPs (see Tables 5.1–5.4), for enhancing their capabilities in SRB mitigation and biofilm formation. Thus, cost effectiveness and ecofriendliness overcome the MIC problem.

TABLE 5.1 Some Examples for the Green Synthesis of TiO_2 NPs

Extract	Type, Shape, and Size	References
Aloe barbadensis leaf	Anatase spherical 20 nm	Rajkumaria et al. (2019)
East African highland banana (Musa AAA) sap	Anatase spherical 6.1 nm	Wagutu et al. (2019)
Jatropha curcas L. sap	Anatase spherical 5.5 nm	Wagutu et al. (2019)
Diospyros ebenum leaf	Anatase spherical 24–33 nm	Sun et al. (2019)
Prunusdomestica domestica L. (Plum) peels	Anatase cylindrical 200 nm	Ajmal et al. (2019)
Prunus persia L. (Peach) peels	Anatase cylindrical 47.1 p–63.21 nm	Ajmal et al. (2019)
Actinidia deliciosa (Kiwi) peels	Anatase cylindrical 54.17–85.13 nm	Ajmal et al. (2019)
Tamarindus indica leaf	Anatase spherical 20–40 nm	Hiremath et al. (2019)

TABLE 5.1 *(Continued)*

Extract	Type, Shape, and Size	References
Trigonella foenum graceum leaf	Anatase spherical 20–90 nm	Subhapriya and Gomathipriya (2018)
Jatropha curcas L. leaf	Anatase spherical 13 nm	Goutam et al. (2018)
Lippia citriodora leaf	Anatase/Rutile spherical 20–40 nm	Senthilkumar et al. (2018)
Bengal gram beans (*Cicer arietinum* L.)	Anatase/Rutile spherical 14 nm	Kashale et al. (2016)
Euphorbia heteradena Jaub root	Rutile spherical 20 nm	Nasrollahzadeh and Sajadi (2015)
Psidium guajava leaf	Anatase spherical 32.58 nm	Santhoshkumar et al. (2014)
Jatropha curcas L. *Latex*	Anatase spherical 20–50 nm	Hudlikar et al. (2012)
Catharanthus roseus leaf	Anatase/Rutile irregular 65 nm	Velayutham et al. (2012)
Annona squamosal fruit peels	Rutile spherical 23 nm	Roopan et al. (2012)

TABLE 5.2 Some Examples for the Green Synthesis of Copper Oxide NPs

Extract	Type, Shape, and Size	References
Callistemon viminalis flower	CuO cuboids and hierarchical platelet 14.9–42.4 nm	Sone et al. (2020)
Mint leaf	CuO cubic 22–25 nm	Aziz et al. (2020)
Psidium guajava leaf	CuO spherical 2–6 nm	Singh et al. (2019)
Nerium oleander leaf	CuO shpherical 21 nm	Sebeia et al. (2019)
Moringha oriefera	CuO spherical 2–100 nm	Revathi et al. (2019)
Aloe barbadensis Miller Aloe vera leaf	CuO spherical 15–30 nm	Rajkumaria et al. (2019)
Zea mays L. Dry husk	Cu_2O cubic 10–26 nm	Nwanya et al. (2019)
Cistus incanus leaf	CuO spherical 15–25 nm	Jing et al. (2019)
Aloe barbadensis leaf	CuO spherical and cubic 5–30 nm	Batool et al. (2019)
Arbutus unedo leaf	CuO spherical 50 nm	Yu et al. (2018)
Bauhinia tomentosa leaf	CuO shpherical 22–40 nm	Sharmila et al. (2018)
Malus domestica leaf	CuO spherical 18–20 nm	Jadhav et al. (2018)
Rosa canina fruit	CuO spherical 15–25 nm	Hemmati et al. (2018)
Psidium guajava leaf	CuO elongated spheroids 17–20 nm	Sreeju et al. (2017)
Fortunella japonica fruit	CuO spherical 5–10 nm	Singh et al. (2017)
Drypetes sepiaria leaf	CuO spherical 18.77 nm	Narasaiah et al. (2017)
Arka (*Calotropis gigantea*) leaf	Cu_2O various shapes 5–15 nm	Behera and Giri (2016)
Acalypha indica leaf	CuO spherical 15–30 nm	Sivaraj et al. (2014)
Aloe barbadensis Miller Aloe vera leaf	CuO spherical 15–30 nm	Gunalan et al. (2012)

TABLE 5.3 Some Examples for the Green Synthesis of ZnO NPs

Extract	Type, Shape and Size	References
Echinochloa frumentacea grains	Hexagonal 35–85 nm	Velsankar et al. (2020)
Artemisia annua	Spherical 20 nm	Wang et al. (2020)
Allium sativum (garlic)	Spheriodal <100 nm	Arciniegas-Grijalba et al. (2019)
Silybum marianum L. seed	Spherical 19.5 nm	Arvanag et al. (2019)
Berberis aristata leaf	Needle like 20–40 nm	Chandra et al. (2019)
Rhamnus virgata	Hexagonal and triangular 20–30 nm	Iqbal et al. (2019)
Olea europaea olive leaf	Cubic 41.0 nm	Ogunyemi et al. (2019)
Matricaria chamomilla chamomile flower	Cubic 51.2 nm	Ogunyemi et al. (2019)
Lycopersicon esculentum M. red tomato fruit	Cubic 51.6 nm	Ogunyemi et al. (2019)
Punica granatum fruit peels	Spherical 32.98 nm	Sukri et al. (2019)
Punica granatum fruit peels	Hexagonal 81.84 nm	Sukri et al. (2019)
Coriandrum sativum leaf	Hexagonal wurtzite 58.2–216 nm	Yashni et al. (2019)
Eryngium foetidum L. (Culantro)	Spherical 8 nm	Begum et al. (2018)
Ficus racemose leaf	Hexagonal wurtzite 15 nm	Birusanti et al. (2018)
Citrus aurantifolia peels	Pyramids 50 nm	Çolak and Karaköse (2017)
Trifolium pretense flowers	Sperical 60–70 nm	Dobrucka and Długaszewska (2016)
Ixora coccinea roots	Spherical 80–130 nm	Yedurkar et al. (2016)
Vitex negundo flowers	Hexagonal 10–13 nm	Ambika and Sundrarajan (2015)
Nephelium lappaceum peels	Needle like 50 nm	Yuvakkumar et al. (2014)
Polygala tenuifolia roots	Spherical 33–73.5 nm	Nagajyothi et al. (2013)
Punica granutum peels	Spherical and square like 50–100 nm	Sharma et al. (2010)

TABLE 5.4 Some Examples for the Green Synthesis of Metal NPs

Extract	Type, Shape, and Size	References
Abutilon indicum leaves	Spherical AgNPs 7–17 nm	Ashokkumar et al. (2015)
Alternanthera dentate leaves	Spherical AgNPs 5–100 nm	Kumar et al. (2014b)
Citrus sinensis peels	Spherical AgNPs 10–35	Kaviya et al. (2011)
Memecylon edule leaves	Triangular, circular, hexagonal AgNPs 20–50 nm	Elavazhagan and Arunachalam (2011)
Nelumbo nucifer leaves	Spherical, triangular AgNPs 25–80 nm	Santhoshkuma et al. (2011)

TABLE 5.4 *(Continued)*

Extract	Type, Shape, and Size	References
Trigonella-foenum graecum seeds	Spherical AuNPs 15–25 nm	Aromal and Philip (2012)
Cassia fistula stem	Spherical AuNPs 55–98 nm	Daisy and Saipriya (2012)
Mirabilis jalapa flowers	Spherical AuNPs 100 nm	Vankar and Bajpai (2010)
Mentha piperita leaves	Spherical Ag, Au NPs 90–150	MubarakAli et al. (2011)
Acalypha indica leaves	Spherical Ag, Au NPs 20–30 nm	Krishnaraj et al. (2010)
Red peanut skin	Spherical FeNPs 10.6 nm	Pan et al. (2020)
Mango peel	Amorphous Zero valent FeNPs	Desalegn et al. (2019)
Azadirachta indica (neem) leaves	Spherical FeNPs 50–100 nm	Pattanayak and Nayak (2013)
Camellia sinensis (green tea)	Spherical FeNPs 5–10 nm	Hoag et al. (2009)
Hawthorn fruit	Spherical 60-nm AgNPs, 200-nm CuNPs	Długosz et al. (2020)
Sumaq (Rhuscoriaria L.) fruits	Semi spherical CuNPs 22–27 nm	Ismail (2020)
Duranta erecta fruit	Spherical CuNPs 70 nm	Ismail et al. (2019)
Camellia sinensis (green tea)	Spherical CuNPs 67–99 nm	Mandava et al. (2017)
Syzygium cumini leaf	Spherical CuNPs 4.88 nm	Fernando and Gurulakshmi (2016)

Omran et al. (2019) reported the efficient biocidal effect of the mycosynthesized nonagglomerated multishaped, spherical, hexagonal, triangular, and cuboid 5–11-nm AgNPs, on a halotolerant planktonic mixed culture of SRB, which was isolated from an Egyptian oil field, where a complete inhibition of SRB occurred at 2000-mg/L AgNPs. As can be seen from the high-resolution transmission electron microscope (HERTM) analysis of the SRB cultures without the biocidal treatment with AgNPs, well-shaped colonies of SRB appeared, which were surrounded by some form of stable extracellular material for example the EPS (see Figure 5.4a). At different higher magnification power (see Figure 5.4b), the SRB appeared as a long and thin rod- and/or vibrio-shaped flagellated cells, for motility with a regular outlined cell wall and regularly distributed cytoplasm. The presence of the produced FeS NPs within the SRB cell was also obvious. Intact cells also appeared with double-membrane-layered cell wall structure and/or amorphous dense coat and had electron-translucent cytoplasm free of conspicuous reserve granules (see Figure 5.4c). But after the treatment with 2000-mg/L AgNPs, HRTEM analysis showed a clear evidence of an alteration in cell morphology, damaged colonies of SRB, overlaid with AgNPs (see Figure 5.4d). There was an obvious disruption of SRB cell membranes, a lysis in cell wall, and cuts in its flagella (see Figure 5.4e). Moreover, shredding in bacterial cells and a cytoplasmic extraction with a release in cell components also occurred (see Figure 5.4f). In addition, a loss in the stable extracellular martial (see Figure 5.4d) and disruption in the double-membrane-layered cell wall structure (see Figure 5.4e) were also observed. Thus, it is recommendable to use such green-synthesized AgNPs in water disinfection from SRB and/or as additives to coatings applied on petroleum pipelines. Such a biocidal effect was attributed to the small-sized mycosynthesized AgNPs and increased surface-to-volume ratio, which promoted their interaction with the bacterial cell membrane. According to Coutinho et al. (1993), the planktonic or sessile SRB cells are always surrounded by stable extracellular martial denoted as glycocalyx (i.e., a cell coat), which is a polysaccharide layer that extends from the outermost S-layer of the Gram-negative bacterial surface that usually protect it from the biocides. But the mycosynthesized AgNPs as seen from the HERTM micrographs successfully attacked such a protecting layer.

Although of the aforementioned biocidal effect of AgNPs on SRB, however, Chen et al. (2018) reported the stimulated cell proliferation and growth enhancement effect of low concentration of AgNPs ($\leq$100 mg/L) on sulfate reducing bacterium *Desulfovibrio vulgaris* in the sewer system. However, it did not affect the sulfate reduction. Thus, it is important to investigate the possible molecular mechanism that is responsible for the stimulating effect of AgNPs on *D. vulgaris* and the enzymatic system involved in AgNPs resistance and detoxification.

Zarasvand and Rai et al. (2016) reported the biosynthesis of polydispersed 400-nm CuO NPs using *Shewanella indica*. The biosynthesized CuO NPs expressed efficient biocidal activity against the SRB *Desulfovibrio marinisediminis* GSR3, recording a minimum inhibitory concentration of 100 μg/mL. The anabolic and catabolic activities of a model SRB, *Desulfovibrio vulgaris* Hidenborough, have been reported to be inhibited by high concentration of CuO NPs (>50 mg/L) (Chen et al., 2019a), whereas the cell proliferation and sulfate reduction rate were decreased. That was attributed to the overproduction of ROS and the inhibition of energy conversion, cell mobility, and iron starvation (see Figure 5.5).

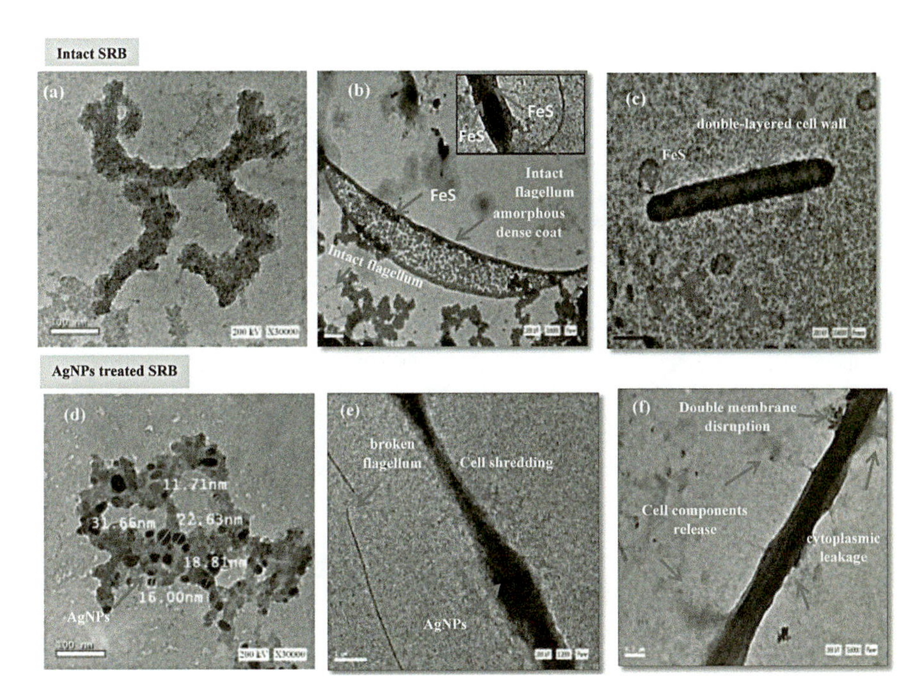

FIGURE 5.4 Biocidal effect of mycosynthesized AgNPs on SRB isolated from an oil field.

ZnO NPs have been reported for their inhibitory action on SRB, anaerobic biotransformation of organic compounds, and enzymes involved in degradation of dyes such as phenol oxidase, lignin peroxidase, veratryl alcohol oxidase, and azo reductase in wastewater treatment (Rasool and Lee, 2018), where sulfate and dyes were electron acceptors, and glucose acted as an electron donor. The presence of even low concentrations of ZnO NPs (10 mg/L) inhibited the oxidation of glucose, that is, the cosubstrate and also inhibited the chemical oxygen demand removal. But it did not affect the sulfate reduction. However, high concentration of ZnO NPs, that is, 100–200 mg/L, decreased the sulfate reduction by 57.8% and 44.1%, respectively, compared with the control (without ZnO NPs), which recorded approximately 78%. That was attributed to the inhibitory effect of the adsorbed ZnO NPs on SRB, as depicted by the scanning electron microscope. As ZnO NPs penetrated the cells, it dissolute into Zn^{+}, and then damaged and deformed the SRB cells at 100 mg/L ZnO NPs with the occurrence of a complete cell lysis at 200 mg/L ZnO NPs (Rasool and Lee, 2018). In an earlier study by Das et al. (2017), the 100-mg/L Fe_3O_4 NPs reduced the biocorrosion of iron by *Halanaerobium* sp., and the sulfide production rate was greatly lowered.

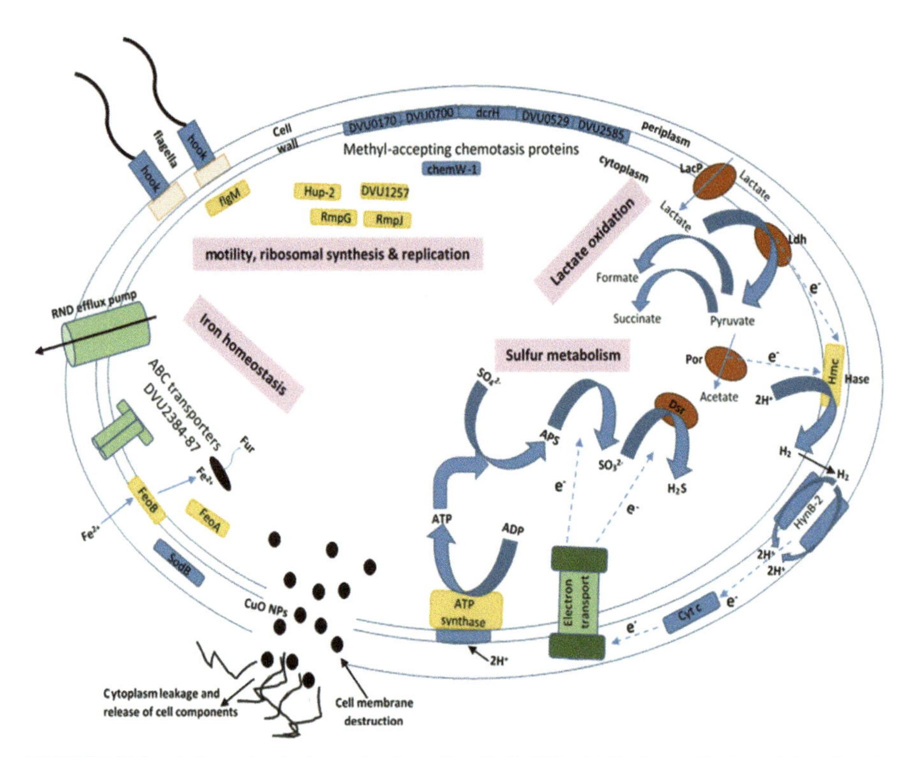

FIGURE 5.5 A hypothetical mechanism for CuO NPs inhibition effect on *D. vulgaris*. The yellow and blue blocks; represent the encoding genes that were up- or down-regulated, respectively. Abbreviations: LacP—lactate permease; Ldh—lactate dehydrogenase; Por—pyruvate-ferrodoxin oxidoreductase; Hase—hydrogenase; Cyt c—cytochrome c; APS—adenosine 5′-phosphosulfate; Dsr—dissimilatory sulfite-reductase; dcrH—methyl-accepting chemotaxis protein; cheW-1—chemotaxis protein; flgM—negative regulator of flagellin synthesis; Hup-2—DNA-binding protein; Rmp—50S ribosomal protein; Feo—ferrous iron transport protein; SodB—superoxide dismutase; hmc—operon protein; and HynB-2—periplasmic [NiFe] hydrogenase.

To overcome and/or lower any expected toxicity on nontarget living organisms, nanometal oxides with high biocidal activity would be incorporated into green nontoxic antimicrobial polymer, such as chitosan (Rasool et al., 2018; Abdul Rasheed et al., 2019b). This would keep the functionality and increase the stability and activity of such metal oxide NPs. The Ag and ZnO NPs, which can be green synthesized (Omran et al., 2018; Arciniegas-Grijalba et al., 2019), phosphate, and hydroxyapatite, which can also be prepared from natural sources, such as animal bones (El-Gendy and Nassar, 2020) can be incorporated into chitosan to prepare green nanocomposite with enhanced antibacterial activities (Yan et al., 2015). Even AgNPs incorporate hydroxyapatite nanocomposite reported to act as an efficient antibacterial coating (Zhang et al., 2017). Zhai et al. (2019) reported the biocidal and antibacterial effect of Zn–Ni–chitosan coatings on SRB and *Escherichia coli* with a recorded decrease in the MIC of the coated carbon steel.

The spherical shaped interlinked chitosan and 10% ZnO NP nanocomposite (CZNC-10) with average size of 100–150 nm and zeta potential of +29 mV was reported as an efficient biocide for SRB at a very low concentration of 250 μg/mL (Abdul Rasheed et al., 2019b), where its effect on biocorrosion of carbon steel coupons in simulated inject seawater, revealed a great damage on biofilm within four days of incubation, with a corrosion inhibition efficiency (%IE) of approximately 73.4%. After seven days, an obvious decrease in coupons' surface roughness (approximately 51%), healthy SRB cells, and extracellular polymeric substance (EPS) on the coupons surface, with a recorded %IE of approximately 71%. Semiquantitative analysis for the prediction of SRB and the biocidal efficiency of CZNC-10 was performed by measuring the sulfur content in the biofilm by energy-dispersive X-ray spectroscopy (EDX). That revealed 70% reduction in the S content, after 28 days of incubation, which consequently proved the mortality of SRB cells. Moreover, the XPS proved a decrease in the peak intensities of FeS, S 2p, and S 2s, indicating the inhibition of SRB activity. Besides that, it also confirmed the decrease in the biotic reduction of sulfate into FeS in the presence of CZNC-10. The roughness is attributed to the formation of pits due to the formation of H_2S, which was followed by iron sulfide precipitation and production of elemental sulfur. However, the profilometry analysis proved that in the absence of CZNC-10, the pit size reached approximately 31.4 μm width with 1.6 μm depth and decreased to approximately 20 μm and 0.7 μm, respectively, in presence of CZNC-10. The electrochemical impedance spectroscopy analysis proved an increase of approximately 2.8 times in charge transfer resistance (R_{ct}) on carbon steel coupon, after 28 days of incubation. The inhibition of the SRB activity and %IE of approximately 61.3%, keeping nearly the aspect ratios of the large pits (approximately 2.1×10^3–2.2×10^3), after 28 days, are advantageous as it means that the biocide did not change the corrosion form, as this would eliminate any unexpected performance in the system attacked by 250 μg/mL, that is, the minimum inhibitory concentration. The chitosan NPs can attack the bacterial cell membrane via its $-NH_2$ group, increasing its permeability, which consequently led to the protein leakage and cell damage (Qi et al., 2004), while the ZnO NPs generated its expected ROS, leading to oxidative stress, which consequently damaged the bacterial enzymatic system, proteins, and DNA (Rasool et al., 2018).

Ituen et al. (2020) reported the phytosynthesized stable, monodispersed, nonagglomerated, and round-shaped 45-nm AgNPs, with an average zeta potential of –43.2 mV and polydispersity index of 0.32, using the aqueous extract of tangerine (*Citrus reticulata*) peels. Such phytosynthesized AgNPs was tested for their efficiency in X80 steel corrosion mitigation both chemically in 1-M HCl and microbially by a Gram-negative SRB *Desulfovibrio* sp, isolated from produced water samples. The phytosynthesized AgNPs expressed efficient biocidal activity against SRB with a minimum inhibitory concentration of 2.88 mg/L, better than that of the peel extract alone, which was 6.56 mg/L. That was attributed to the penetration of the AgNPs into the SRB cells, which destroyed their enzymes and proteins and consequently altered its activity, and then finally led to its mortality. Upon the injection of the phytosynthesized AgNPs in SRB culture with X80 steel coupons, at the minimum inhibitory concentration (2.88 mg/L), the corrosion rate decreased from 0.47 mpy to 0.1 mpy at 303 K with inhibition efficiency of approximately 76.2%. That was attributed to the mortality of the corrosive SRB, where its count decreased by 3 log within 14 days and also due to the adsorption of the phytosynthesized AgNPs on the metal coupon's surface. Moreover, the addition of 50 ppm D-tyrosine enhanced the inhibition of microbial corrosion to reach 83.3% at 303 K. The main advantage of that study was the recorded efficient corrosion inhibition of the phytosynthesized AgNPs under the acidic condition, where it was proved to be a physisorption mechanism, as it increased with the increase in NP concentrations and decreased with the increase in temperature, recording 93.9% and 83.3% at 303 and 333 K, respectively. That followed the Langmuir isotherm with a free energy change of adsorption ΔG_{ads} of –19.59. From the thermodynamic analysis at different temperatures and NP concentrations, the adsorption was proved to be exothermic and resulted in orderliness of the bulk solution because of interactive adsorption of the NPs onto the steel surface. From the electrochemical measurements, it was proved that the adsorbed film of the phytosynthesized AgNPs insulated the coupons' surface resulting in an increase in the local dielectric, where, kinetically, the adsorption of the NPs involved the displacement of water molecules by the NPs, which consequently resulted in a decrease in the active corrosion sites and reduced the corrosion rate. The phytosynthesized AgNPs were found to be a mixed-type corrosion inhibitor with anodic predominance. It increased the charge transfer and film resistances but reduced the corrosion current. Surface analysis proved that the phytosynthesized AgNPs protected the steel coupons, where the pitting and the localized corrosion damage were greatly reduced. The EDX analysis presumptively proved that the C, N, and O sites of the peels extract were involved in the adsorption and protection of the coupons surface, while the Fourier transform infrared spectroscopy analysis confirmed the involvement of –OH, –NH, and C=O functional groups of the extract in the surface protection (Ituen et al., 2020). Thus, as the phytosynthesized AgNPs proved to be a more thermally stable acid and microbial corrosion inhibitor than the peels extract itself, it would be recommended to be applied in oil and gas industry.

REFERENCES

Abd-Elaal, A.A., Elbasiony, N.M., Shaban, S., and Zaki, E.G. 2017. Studying the corrosion inhibition of some prepared nonionic surfactants based on 3-(4-hydroxyphenyl) propanoic acid and estimating the influence of silver anoparticles on the surface parameters. *J. Mol. Liquids*. 249: 304–317.

Abdul Rasheed, P., Jabbar, K.A., Mackey, H.R., and Mahmoud, K.A. 2019a. Recent advancements of nanomaterials as coatings and biocides for the inhibition of sulfate reducing bacteria induced corrosion. *Curr. Opin. Chem.* Eng. 25: 35–42.

Abdul Rasheed, P., Jabbara, K.A., Rasool, K., Pandey, R.P., Sliem, M.H., Helal, M., Samara, A., Abdullah, A.M., and Mahmoud, K.A. 2019b. Controlling the biocorrosion of sulfate-reducing bacteria (SRB) on carbon steel using ZnO/chitosan nanocomposite as an eco-friendly biocide. *Corros. Sci.* 148: 397–406.

Agarry, S.E., Salam, K.K., Arinkoola, A.O., and Soremekun, I.O. 2015. Microbiologically influenced corrosion of mild steel in crude oil environment. *Eur. J. Eng. Technol.* 3(6): 40–52.

Aiad, I., El-Sukkary, M.M., Soliman, E.A., El-Awady, M.Y., and Shaban, S.M. 2014b. In situ and green synthesis of silver nanoparticles and their biological activity. *J. Ind. Eng. Chem.* 20: 3430–3439.

Aiad, I., Hassanien, A.A., El-Sayed, W.N., Tawfik, S.M., and Nazeif, A. 2014a. Synergistic effect of copper sulfate with some traditional biocides for killing the sulfate reducing bacteria in oil fields. *Middle East J. Appl. Sci.* 4: 373–384.

Ajmal, N., Saraswat, K., Bakht, M.F., Riadi, Y., Ahsan, M.J., and Noushad, M. 2019. Cost-effective and eco-friendly synthesis of titanium dioxide (TiO_2) nanoparticles using fruit's peel agro-waste extracts: Characterization, in vitro antibacterial, antioxidant activities. *Green Chem. Lett. Rev.* 12(3): 244–254.

Al-Abbas, F.M., Bhola, R., Spear, J.R., Olson, D.L., and Mishra, B. 2013. Electrochemical characterization of microbiologically influenced corrosion on linepipe steel exposed to facultative anaerobic *Desulfovibrio* sp. *Int. J. Electrochem. Sci.* 8: 859–871.

Al-Abbas, F.M., Spear, J.R., Kakpovbia, A., Balhareth, N.M., Olson, D.L., and Mishra, B. 2012. Bacterial attachment to metal substrate and its effects on microbiologically-influenced corrosion in transporting hydrocarbon pipelines. *J. Pipeline Eng.* 11(1): 63–72.

Ambika, S., and Sundrarajan, M. 2015. Green biosynthesis of ZnO nanoparticles using *Vitex negundo* L. extract: Spectroscopic investigation of interaction between ZnO nanoparticles and human serum albumin. *J. Photochem. Photobiol.* 149: 143–148.

Applerot, G., Lellouche, J., Perkas, N., Nitzan, Y., Gedanken, A., and Banin, E. 2012. ZnO nanoparticle coated surfaces inhibit bacterial biofilm formation and increase antibiotic susceptibility. *RSC Adv.* 2: 2314–2321.

Arciniegas-Grijalba, P.A., Patiño-Portela, M.C., Mosquera-Sánchez, L.P., Guerra Sierra, B.E., Muñoz-Florez, J.E., Erazo-Castillo, L.A., and Rodríguez-Páez, J.E. 2019. ZnO-based nanofungicides: Synthesis, characterization and their effect on the coffee fungi *Mycena citricolor* and *Colletotrichum* sp. *Mater. Sci. Eng. C.* 98: 808–825.

Aromal, S.A., and Philip, D. 2012. Green synthesis of gold nanoparticles using *Trigonella foenum-graecum a*nd its size dependent catalytic activity. *Spectrochim. Acta A.* 97: 1–5.

Arruda de Queiroz, G., Andrade, J.S., Malta, T.B.S., Vinhas, G., and Gomes de Andrade Lima, M.A. 2018. Bioflm formation and corrosion on carbon steel API 5lx60 in clayey soil. *Mater. Res.* 21(3): e20170338. doi: http://dx.doi.org/10.1590/1980-5373-MR-2017-0338.

Aruliah, R., and Ting, Y.-P. 2014. Characterization of corrosive bacterial consortia isolated from water in a cooling tower. *ISRN Corrosion.* 2014: 803219. doi: 10.1155/2014/803219.

Arvanag, F.M., Bayrami, A., Habibi-Yangjeh, A., and Pouran, S.R. 2019. A comprehensive study on antidiabetic and antibacterial activities of ZnO nanoparticles biosynthesized using *Silybum marianum* L Seed extract. *Mater. Sci. Eng. C.* 97: 397–405.

Ashokkumar, S., Ravi, S., Kathiravan, V., and Velmurugan, S. 2015. Synthesis of silver nanoparticles using *A. indicum* leaf extract and their antibacterial activity. *Spectrochim. Acta Part A: Mol. Biomol. Spectrosc.* 134: 34–39.

Asrari, E., and Cheraghpour, N. 2018. Dioxide titanium nanoparticles efficiency on removing water heterotrophic bacteria population. *Environ. Qual.* 28: 43–49.

Ayangbenro, A.S., Olanrewaju, O.S., and Babalola, O.O. 2018. Sulfate-reducing bacteria as an effective tool for sustainable acid mine bioremediation. *Front. Microbiol.* 9: 1986. 10.3389/fmicb.2018.01986.

Aziz, W.J., Abid, M.A., and Hussein, E.H. 2020. Biosynthesis of CuO nanoparticles and synergistic antibacterial activity using mint leaf extract. *Mater. Technol.* 35: 447–451. https://doi.org/10.1080/10667857.2019.1692163

Azzam, E.M.S., Sami, R.M., and Kandile, N.G. 2012. Activity inhibition of sulfate reducing bacteria using some cationic thiol surfactants and their nanostructures. *Am. J. Biochem.* 2: 29–35.

Barapatre, A., Ramadil, K., and Jha, H., 2016. Synergistic antibacterial and antibiofilm activity of silver nanoparticles biosynthesized by lignin-degrading fungus. *Biores. Bioprog.* 3: 1–13.

Batool, M., Qureshi, M.Z., Hashmi, F., Mehboob, N., and Daoush, W.M. 2019. Adsorption of Congo red (acid red 28) azodye on biosynthesized copper oxide nanoparticles. *Asian J. Chem.* 31(3): 707–713.

Begum, S., Ahmaruzzaman, M., Adhikar, P.P. 2018. Ecofriendly bio-synthetic route to synthesize ZnO nanoparticles using *Eryngium foetidum* L. and their activity against pathogenic bacteria. *Mater. Lett.* 228: 37–41.

Behera, M., and Giri, G. 2016. Inquiring the photocatalytic activity of cuprous oxide nanoparticles synthesized by a green route on methylene blue dye. *Int. J. Ind. Chem.* 7: 157–166.

Bhola, R., Bhola, S.M., Mishra, B., and Olson, D.L. 2010. Microbiologically influenced corrosion and its mitigation: A review. *Mater. Sci. Res. India.* 7(2): 407–412

Birusanti, A.B., Mallavarapu, U., Nayakanti, D., and Espenti, C.S. 2018. Plant-mediated ZnO nanoparticles using *Ficus racemosa* leaf extract and their characterization, antibacterial activity. *Asian J. Pharm. Clin. Res.* 11(9): 463–467.

Cai, Y., Strømme, M., and Welch, K. 2013. Photocatalytic antibacterial effects are maintained on resin-based TiO_2 Nanocomposites after cessation of UV irradiation. *Plos One.* 8(10): e75929. https://doi.org/10.1371/journal.pone.0075929.

Capeness, M.J., Edmundson, M.C., and Horsfall, L.E. 2013. Nickel and platinum group metal nanoparticle production by *Desulfovibrio alaskensis* G20. *New Biotechnol.* 32(6): 727–731.

Chandra, H., Patel, D., Kumari, P., Jangwan, J.S., and Yadav, S. 2019. Phyto-mediated synthesis of zinc oxide nanoparticles of *Berberis aristata*: Characterization, antioxidant activity and antibacterial activity with special reference to urinary tract pathogens. *Mater. Sci. Eng. C.* 102: 212–220.

Chen, Z., Gaoa, S.-h., Jina, M., Suna, S., Lu, J., Yang, P., Bonda, P.L., Yuan, Z., and Guo, J. 2019. Physiological and transcriptomic analyses reveal CuO nanoparticle inhibition of anabolic and catabolic activities of sulfate-reducing bacterium. *Environ. Int.* 125: 65–74.

Chen, Z., Lu, J., Gao, S.-H., Jin, M., Bond, P.L., Yang, P., Yuan, Z., and Guo, J. 2018. Silver nanoparticles stimulate the proliferation of sulfate reducing bacterium *Desulfovibrio vulgaris*. *Water Res.* 129: 163–171.

Cherif, A.Y., Arous, O., Mameri, N., Zhu, J., Said, A.A., Vankelecom, I., Simoens, K., Bernaerts, K., and Van der Bruggen, B. 2018. Fabrication and characterization of novel antimicrobial thin film nano-composite membranes based on copper nanoparticles. *J. Chem. Technol. Biotechnol.* 93: 2737–2747.

Çolak, H., and Karaköse, E. 2017. Green synthesis and characterization of nanostructured ZnO thin films using Citrus aurantifolia (Lemon) peel extract by spin-coating method. *J. Alloys Compounds.* 690: 658–662.

Coutinho, C.M.L.M., Magalhaes, F.C. and Araujo-Jorge, T.C. 1993. Scanning electron microscope study of biofilm formation on different flow rates over metal surfaces using sulphate-reducing bacteria. *Biofouling.* 7: 19–29.

Daisy, P., and Saipriya, K. 2012. Biochemical analysis of *Cassia fistula* aqueous extract and phytochemically synthesized gold nanoparticles as hypoglycemic treatment for diabetes mellitus. *Int. J. Nanomed.* 7: 1189–1202.

Das, K.R., Kerkar, S., Meena, Y., and Mishra, S. 2017. Effects of iron nanoparticles on iron-corroding bacteria. *3 Biotech.* 7(6): 385. doi: 10.1007/s13205-017-1018-9.

Desalegn, B., Megharaj, M., Chen, Z., and Naidu, R. 2019. Green synthesis of zero valent iron nanoparticle using mango peel extract and surface characterization using XPS and GC-MS. *Heliyon.* 5(5): e01750.

Długosz, O., Chwastowski, J., and Banach, M. 2020. Hawthorn berries extract for the green synthesis of copper and silver nanoparticles. *Chem. Papers.* 74: 239–252.

Dobrucka, R., and Długaszewska, J. 2016. Biosynthesis and antibacterial activity of ZnO nanoparticles using *Trifolium pratense* flower extract. *Saudi J. Biol. Sci.* 23: 517–523.

Elavazhagan, T., Arunachalam, K.D. 2011. *Memecylon edule* leaf extract mediated green synthesis of silver and gold nanoparticles. *Int. J. Nanomed.* 6: 1265–1278.

El-Gendy, N. Sh., and Nassar, H.N. 2020. Sustainable photo- and bio-catalysts for wastewater treatment. Chapter 5, in "Photocatalysts in advanced oxidation processes for wastewater treatment." In: Elvis Fosso-Kankeu, Ed., *Sadanand Pandey and Suprakas Sinha Ray.* Wiley, Hoboken, NJ, USA. https://doi.org/10.1002/9781119631422.ch5

El-Shamy, A.M., Zohdy, K.M., and El-Dahan, H.A. 2016. Control of corrosion and microbial corrosion of steel pipelines in salty environment by polyacrylamide. *Ind. Chem.* 2: 120. doi: 10.4172/2469-9764.1000120.

Enning, D., and Garrelfs, J. 2014. Corrosion of iron by sulfate-reducing bacteria: New views of an old problem. *Appl. Environ. Microbiol.* 80: 1226–1236.

Enning, D., and Garrelfs, J. 2014. Corrosion of iron by sulfate-reducing bacteria: New views of an old problem. *Appl. Environ. Microbiol.* 80(4): 1226–1236.

Esquivel, R.G., Olivares, G.Z., Gayosso, M.J.H., and Trejo, A.G. 2011. Cathodic protection of XL 52 steel under the influence of sulfate reducing bacteria. *Mater. Corros.* 62: 61–67.

Fahmy, H.M., Mohamed, F.M., Marzouq, M.H., Mustafa, A.B. Alsoudi, A.M., Ali, O.A., Mohamed, M.A., and Mahmoud, F.A. 2018. Review of green methods of iron nanoparticles synthesis and applications. *BioNanoScience*. 8: 491–503. https://doi.org/10.1007/s12668-018-0516-5

Fathy, M., Badawi, A., Mazrouaa, A.M., Mansour, N.A., Ghaz,y E.A., and Elsabee, M.Z. 2013. Styrene N-vinylpyrrolidone metal nanocomposites as antibacterial coatings against sulfate reducing bacteria. *Mater. Sci. Eng. C.* 33: 4063–4070.

Fernando, J., and Gurulakshmi, P. 2016. Green synthesis and characterization of copper nanoparticles and their applications. *J. Nanosci. Technol.* 2(5): 234–236.

Fortin, D., Davis, B, and Beveridge, T.G. 1996. Role of *Thiobacillus* and sulfate-reducing bacteria in iron biocycling in oxic and acidic mine tailings. *FEMS Microbiol. Ecol.* 21: 11–24.

Ganesh, M., Lee, S.G., Jayaprakash, J., Mohankumar, M., and Jang, H.T. 2019. *Hydnocarpus alpina* Wt extract mediated green synthesis of ZnO nanoparticle and screening of its anti-microbial, free radical scavenging, and photocatalytic activity. *Biocatal. Agri. Biotech.* 19: 101129. https://doi.org/10.1016/j.bcab.2019.101129.\

Goutam, S.P., Saxena, G., Singh, V., Yadav, A.K., Bharagava, R.N., and Thapa, K.B. 2018. Green synthesis of TiO_2 nanoparticles using leaf extract of *Jatropha curcas* L. for photocatalytic degradation of tannery wastewater. *Chem. Eng. J.* 336: 386–396.

Gu, T., Jia, R., Unsal, T., and Xu, D. 2019. Toward a better understanding of microbiologically influenced corrosion caused by sulfate reducing bacteria. *J. Mater. Sci. Technol.* 35: 631–636.

Gunalan, S., Sivaraj, R., and Venckatesh, R. 2012. *Aloe barbadensis Miller* mediated green synthesis of mono-disperse copper oxide nanoparticles: Optical properties. *Spectrochim Acta A. Mol. Biomol. Spectrosc.* 97: 1140–1144.

Hemmati, S., Mehrazin, L., Hekmati, M., Izadi, M., and Veisi, H. 2018. Biosynthesis of CuO nanoparticles using *Rosa canina* fruit extract as a recyclable and heterogeneous nanocatalyst for C-N Ullmann coupling reactions. *Mater. Chem. Phys.* 214: 527–532.

Hiremath, S., Antony Raja, M.A.L., Chandra Prabhab, M.N., and C. Vidya. 2019. *Tamarindus indica* mediated biosynthesis of nano TiO_2 and its application in photocatalytic degradation of Titan yellow. *J. Environ. Chem. Eng.* 6: 7338–7346.

Hoag, G.E., Collins, J.B., Holcomb, J.L., Hoag, J.R., Nadagouda, M.N., and Varma, R.S. 2009. Degradation of bromothymol blue by 'greener' nano-scale zero-valent iron synthesized using tea polyphenols. *J. Mater. Chem.* 19: 8671–8677.

Hood, M.I., and Skaar, E.P. 2012. Nutritional immunity: Transition metals at the pathogen–host interface. *Nat. Rev. Microbiol.*. 10: 525–537.

Hudlikar, M., Joglekar, S., Dhaygude, M., and Kodam, K. 2012. Green synthesis of TiO_2 nanoparticles by using aqueous extract of *Jatropha curcas* L. latex. *Mater Lett.* 75: 196–199.

Iqbal, J., Abbasi, B.A., Mahmood, T., Kanwal, S., Ahmad, R., and Ashraf, M. 2019. Plant-extract mediated green approach for the synthesis of ZnONPs: Characterization and evaluation of cytotoxic, antimicrobial and antioxidant potentials. *J. Mol. Struct.* 1189: 315–327.

Ishwarya, R., Vaseeharan, B., Kalyani, S., Banumathi, B., Govindarajan, M., Alharbi, N.S., Kadaikunnan, S., Al-Anbr, M.N., Khaled, J.M., and Benelli, G. 2018. Facile green synthesis of zinc oxide nanoparticles using *Ulva lactuca* seaweed extract and evaluation of their photocatalytic, antibioflm and insecticidal activity. *J. Photochem. Photobiol., B.* 178: 249–258.

Ismail, M., Gul, S., Khan, M.I., Khan, M.A., Asiri, A.M., and Khan, S.B. 2019. Green synthesis of zerovalent copper nanoparticles for efficient reduction of toxic azo dyes Congo red and methyl orange. *Green Process Synth.* 8: 135–143.

Ismail, M.I.M. 2020. Green synthesis and characterizations of copper nanoparticles. *Mater. Chem. Phys.* 240: 122283.

Ituen, E., Ekemini, E., Yuanhua, L., and Singh, A. 2020. Green synthesis of *Citrus reticulata* peels extract silver nanoparticles and characterization of structural, biocide and anticorrosion properties. *J. Mol. Struct.* 1207: 127819.

Iverson, W.P. 2001. Research on the mechanisms of anaerobic corrosion. *Int. Biodeter. Biodegr.* 47(2): 63–70.

Jadhav, M.S., Kulkarni, S., Raikar, P., Barretto, D.A., Vootla, S.K., and Raikar, U.S. 2018. Green biosynthesis of CuO & Ag–CuO nanoparticles from *Malus domestica* leaf extract and evaluation of antibacterial, antioxidant and DNA cleavage activities. *New J. Chem.* 42: 204–213.

Javaherdashti, R. 2008. *Microbiologically Influenced Corrosion: An Engineering Insight.* Springer-Verlag, London, UK.

Jayabalan, J., Mani, G., Krishnan, N., Pernabas, J., Devadoss, J.M., and Jang, H.T. 2019. Green biogenic synthesis of zinc oxide nanoparticles using *Pseudomonas putida* culture and its in vitro antibacterial and anti-bioflm activity. *Biocatal. Agri. Biotechnol.* 21: 101327. https://doi.org/10.1016/j.bcab.2019.101327.

Jia, R., Unsal, T., Xu, D., Lekbach, Y., and Gu, T. 2019. Microbiologically influenced corrosion and current mitigation strategies: A state of the art review. *Int. Biodeter. Biodegrad.* 137: 42–58.

Jing, C., Yan, C.-J., Yuan, X.-T., and Zhub, L.-P. 2019. Biosynthesis of copper oxide nanoparticles and their potential synergistic effect on alloxan induced oxidative stress conditions during cardiac injury in Sprague–Dawley rats. *J. Photochem. Photobiol. B. Biol.* 198: 111557.

Kang, F., Alvarez, P.J., and Zhu, D. 2014. Microbial extracellular polymeric substances reduce Ag^+ to silver nanoparticles and antagonize bactericidal activity. *Environ. Sci. Technol.* 48(1): 316–322.

Kashale, A.A., Gattu, K.P., Ghule, K., Ingole, V.H., Dhanayat, S., Sharma, R., Chang, J.-Y., and Ghule, A.V. 2016. Biomediated green synthesis of TiO_2 nanoparticles for lithium ion battery application. *Composites B.* 99: 297–304.

Kaviya, S., Santhanalakshmi, J., Viswanathan, B., Muthumary, J., and Srinivasan, K. 2011. Biosynthesis of silver nanoparticles using *Citrus sinensis* peel extract and its antibacterial activity. *Spectrochem. Acta A. Mol. Biomol. Spectrosc.*79: 594–598.

Khouzani, M.K., Bahrami, A., Hosseini-Abari, A., Khandouzi, M., and Taheri, P. 2019. Microbiologically influenced corrosion of a pipeline in a petrochemical plant. *Metals.* 9: 459. doi: 10.3390/met9040459.

Khowdiary, M.M., El-Henawy, A.A., Shawky, A.M., Sameeh, M.Y., and Negm, N.A. 2017. Synthesis, characterization and biocidal efficiency of quaternary ammonium polymers silver nanohybrids against sulfate reducing bacteria. *J. Mol. Liq.* 230: 163–168.

Krishnaraj, C., Jagan, E.G., Rajasekar, S., Selvakumar, P., Kalaichelvan, P.T., and Mohan, N. 2010. Synthesis of silver nanoparticles using *Acalypha indica* leaf extracts and its antibacterial activity against water borne pathogens. *Colloid Surf. B.* 76: 50–56.

Kubacka, A., Diez, S., Rojo, D., Bargiela, R., Ciordia, S., Zapico, I., Albar, J.P., Barbas, C., Martins dos Santos, V.A.P., Fernández-García, M., and Ferrer, M. 2014. Understanding the antimicrobial mechanism of TiO_2-based nanocomposite films in a pathogenic bacterium. *Sci. Rep.* 4: 4134. doi: 10.1038/srep04134.

Kubacka, A., Ferrer, M., and Fernández-García, M. 2012. Kinetics of photocatalytic disinfection in TiO_2-containing polymer thin films: UV and visible light performance. *Appl. Catal. B.* 121/122: 230–248.

Kubacka, A., Serrano, C., Ferrer, M., Lünsdorf, H., Bielecki, P., Cerrada, M.L., Fernández-García, M., and Fernández-García, M. 2007. High-performance dual-action polymer-TiO_2 nanocomposite films via melting processing. *Nano Lett.* 7: 2529–2534.

Kumar, A., Pandey, A.K., Singh, S.S., Shanker, R., and Dhawan, A., 2011. Cellular uptake and mutagenic potential of metal oxide nanoparticles in bacterial cells. *Chemosphere*. 83: 1124–1132.

Kumar, D.A., Palanichamy, V., and Roopan, S.M. 2014b. Green synthesis of silver nanoparticles using *Alternanthera dentate* leaf extract at room temperature and their antimicrobial activity. *Spectrochim Acta Part A: Mol Biomol Spectrosc.* 127: 168–171.

Kumar, N., Omoregie, E.O., Rose, J., Masion, A., Lloyd, J.R., Diels, L., and Bastiaens, L. 2014a. Inhibition of sulfate reducing bacteria in aquifer sediment by iron nanoparticles. *Water Res.* 51: 64–72.

Labrenz, M., and Banfield, J.F. 2004. Sulfate-reducing bacteria-dominated biofilms that precipitate ZnS in a subsurface circumneutral-pH mine drainage system. *Microb. Ecol.* 47(3): 205–217.

Lan, G., Chen, C., Liu, Y., Lu, Y., Du, J., Taob, S., and Zhang, S. 2017. Corrosion of carbon steel induced by a microbial enhanced oil recovery bacterium *Pseudomonas* sp. SWP-4. *RSC Adv.* 7: 5583–5594.

Lee, J.-H., Kim, Y.-G., Cho, M.H., and Lee, J. 2014. ZnO nanoparticles inhibit *Pseudomonas aeruginosa* biofilm formation and virulence factor production. *Microbiol. Res.* 169: 888–896.

Lee, W., Lewandowski, Z., Morrison, M., Characklis, W.G., Avci, R., and Nielsen, P. 1993. Corrosion of mild steel underneath anaerobic biofilms containing sulfate-reducing bacteria part II. At high dissolved oxygen concentration. *Biofouling*. 7(3): 217–239.

Lengke, M., and Southam, G. 2006. Bioaccumulation of gold by sulfate-reducing bacteria cultured in the presence of gold(I)-thio sulfate complex. *Geochim. Et Cosmochim. Acta.* 70(14): 3646–3661.

Liang, C.H., Wang, H,.and Huang, N.B. 2014. Effects of sulphate-reducing bacteria on corrosion behaviour of 2205 duplex stainless steel. *J. Iron Steel Res. Int.* 21: 444–450.

Lin, J., and Ballim, R. 2012. Biocorrosion control: Current strategies and promising alternatives—A review. *Afr. J. Biotechnol.* 11(91): 15736–15747.

Little, B.J., and Lee, J.S. 2014. Microbiologically influenced corrosion: An update. *Int. Mater. Rev.* 59: 384–393.

Little, B.J., Wagner, P., and Mansfeld, F. 1992. An overview of microbiologically influenced corrosion. *Electrochim. Acta* 37(12): 2185–2194.

Liu, H., Meng, G., Li, W., Gu, T., and Liu, H. 2019. Microbiologically influenced corrosion of carbon steel beneath a deposit in CO_2-saturated formation water containing *Desulfotomaculum nigrificans*. *Front. Microbiol.* 10: 1298. doi: 10.3389/fmicb.2019.01298.

Loto, C.A. 2017. Microbiological corrosion: mechanism, control and impact: A review. *Int. J. Adv. Manuf. Technol.* 92: 4241–4252.

Ma, J., Shi, Y., Di, J., Yao, Z., and Liu, H. 2009. Effect of TiO_2 nanoparticles on anticorrosion property in amorphous Ni–P–Cr composite coating in artificial seawater and microbial environment. *Mater. Corros.* 60(4): 274–279.

Macedo de Souza, P., Regina de Vasconcelos Goulart, F., Marques, J.M., Bizzo, H.R., Blank, A.F., Groposo, C., Paula de Sousa, M., Vólaro, V., Alviano, C.S., Moreno, D.S.A., and Seldin, L. 2017. Growth inhibition of sulfate-reducing bacteria in produced water from the petroleum industry using essential oils. *Molecules*. 22: 648. doi:10.3390/molecules22040648

Mandava, K., Kadimcharla, K., Keesara, N.R., Fatima, S.N., Bommena, P., and Batchu, U.R. 2017. Green synthesis of stable copper nanoparticles and synergistic activity with antibiotics. *Indian J. Pharm. Sci.* 79(5): 695–670.

Mathur, A., Bhuvaneshwari, M., Babu, S., Chandrasekaran, N., and Mukherje, A. 2017. The effect of TiO2 nanoparticles on sulfate-reducing bacteria and their consortium under anaerobic conditions. *J. Environ. Chem. Eng.* 5: 3741–3748.

Mirhendi, M., Emtiazi, G., and Roghanian, R. 2013. Antibacterial activities of nano magnetite ZnO produced in aerobic and anaerobic condition by *Pseudomonas stutzeri*. *Jundishapur J. Microbiol.* 6(10): e10254. doi: 10.5812/jjm.10254.

MubarakAli, D., Thajuddin, N., Jeganathan, K., Gunasekaran, M., 2011. Plant extract mediated synthesis of silver and gold nanoparticles and its antibacterial activity against clinically isolated pathogens. *Colloid Surf. B.* 85: 360–365.

Muñoz-Bonilla, A., and Fernández-García, M. 2012. Polymeric materials with antimicrobial activity. *Prog. Polym. Sci.* 37: 281–339.

Nagajyothi, P.C., Minh An, T.N., Sreekanth, T.V.M., Lee, J. Il, Joo, D.L., Lee, K.D. 2013. Green route biosynthesis: characterization and catalytic activity of ZnO nanoparticles. *Mater. Lett.* 108: 60–163.

Narasaiah, P., Mandal, B.K., and Sarada, N.C. 2017. Biosynthesis of copper oxide nanoparticles from *Drypetes sepiaria* Leaf extract and their catalytic activity to dye degradation. *IOP Conf. Series: Materials Science and Engineering.* 263: 022012 doi:10.1088/1757-899X/263/2/022012.

Narendhran, S., and Sivaraj, R., 2016. Biogenic ZnO nanoparticles synthesized using L. aculeate leaf extract and their antifungal activity against plant fungal pathogens. *Bull. Mater. Sci.* 39: 1–5.

Nasrollahzadeh, M., and Sajadi, S.M. 2015. Synthesis and characterization of titanium dioxide nanoparticles using *Euphorbia heteradena jaub* root extract and evaluation of their stability. *Ceram. Int.* 41: 14435–14439.

Nkemnaso, O.C., and Chibueze, I.K., 2018. Advances in industrial biofilm control with nanotechnology—A review. *Int. J. Chem. Biomol. Sci.* 4(4): 41–59.

Nwanya, A.C., Razanamahandry, L.C., Bashir, A.K.H., Ikpo, C.O., Nwanya, S.C., Botha, S., Ntwampe, S.K.O., Ezema, F.I., Iwuoha, E.I., and Maaz, M. 2019. Industrial textile effluent treatment and antibacterial effectiveness of *Zea mays L.* Dry husk mediated bio-synthesized copper oxide nanoparticles. *J. Hazr. Mater.* 375: 281–289.

Ogunyemi, S.O., Abdallah, Y., Zhang, M., Fouad, H., Hong, X., Ibrahim, E., Masum, M. M.I., Hossain, A., Mo, J., and Li, B. 2019. Green synthesis of zinc oxide nanoparticles using different plant extracts and their antibacterial activity against *Xanthomonas oryzae* pv. Oryzae. Artif. Cell Nanomed. Biotechnol. 47(1): 341–352.

Omran, B.A., Fatthallah, N.A., El-Gendy, N.Sh., El-Shatoury, E.H., and Abouzeid, M.A., 2013. Green Biocides against sulphate reducing bacteria and macrofouling organisms. *J. Pure Appl. Microbiol.* 7(3): 2219–2232.

Omran, B.A., Nassar, H.N., Fatthallah, N.A., Hamdy, A., El-Shatoury, E.H., and El-Gendy, N. Sh. 2018. Waste upcycling of *Citrus sinensis* peels as a green route for the synthesis of silver nanoparticles. *Energ. Sources Part A.* 40(2): 227–236. https://doi.org/10.1080/15567036.2017.1410597

Omran, B.A., Nassar, H.N., Younis, S.A., Fatthallah, N.A., Hamdy, A., El-Shatoury, E.H., and El-Gendy, N.Sh. 2019. Physiochemical properties of Trichoderma longibrachiatum DSMZ 16517-synthesized silver nanoparticles for the mitigation of halotolerant sulphate-reducing bacteria. *J. Appl. Microbiol.* 126: 138–154. https://doi.org/10.1111/jam.14102

Pan, Z., Lin, Y., Sarkar, B., Owens, G., and Chen, Z. 2020. Green synthesis of iron nanoparticles using red peanut skin extract: Synthesis mechanism, characterization and effect of conditions on chromium removal. *J Colloid Interface Sci.* 558: 106–114.

Panda, S., Mishra, S., and Akcil, A. 2016. Bioremediation of acidic mine effluents and the role of sulfidogenic biosystems: A mini-review. *Euro-Mediterr J. Environ. Integr.* 1: 8. doi: 10.1007/s41207–016-0008–3.

Pattanayak, M., and Nayak, P.L. 2013. Green synthesis and characterization of zero valent iron nanoparticles from the leaf extract of *Azadirachta indica* (neem). *World J. Nano Sci. Technol.* 2: 6–9.

Qi, L.X.Z., Jiang, X., Hu, C., and Zou, X. 2004. Preparation and antibacterial activity of chitosan nanoparticles. *Carbohydr. Res.* 339: 2693–2700.

Rajasekar, A., Babu, T.G., Maruthamuthu, S., Pandian, S.K., Mohanan, S., and Palaniswamy, N. 2007. Biodegradation and corrosion behaviour of Serratia marcescens ACE2 isolated from an Indian diesel-transporting pipeline. *World J. Microbiol. Biotechnol.* 23: 1065–1074.

Rajasekar, A., Maruthamuthu, S., and Rahman, P K.S.M. 2010. Microbial communities in petroleum pipeline and its relationship with bio-corrosion, Indo-UK Project, Teeside University.

Rajkumaria, J., Magdalane, C.M., Siddhardha, B., Madhavan, J., Ramalingam G., Al-Dhabie, N.A., Arasue, M.V., Ghilane, A.K.M., Duraipandiayan, V., and Kaviyarasuf, K. 2019. Synthesis of titanium oxide nanoparticles using *Aloe barbadensis* mill and evaluation of its antibiofilm potential against *Pseudomonas aeruginosa* PAO1. *J. Photoch. Photobio. B. Biol.* 201: 111667.

Rasool, K., and Lee, D.S. 2016. Effect of ZnO nanoparticles on biodegradation and biotransformation of co-substrate and sulphonated azo dye in anaerobic biological sulfate reduction processes. *Int. Biodeter. Biodegr.* 109: 150–156.

Rasool, K., Nasrallah, G., Younes, N., Pandey, R., Abdul Rasheed, P., and Mahmoud, K.A. 2018. Green" ZnO-interlinked chitosan nanoparticles for the efficient inhibition of sulfate-reducing bacteria in inject seawater. *ACS Sustain. Chem. Eng.* 6: 3896–3906.

Revathi, B., Rajeshkumar, S., Anitha, R., and Lakshmi, T. 2019. Biosynthesis of copper oxide nanoparticles using herbal formulation and its characterization. *Int. J. Res. Pharm. Sci.* 10(3): 2117–2119.

Roopan, S.M., Bharathi, A., Prabhakarn, A., Abdul Rahuman, A., Velayutham, K., Rajakumar, G., Padmaja, R.D., Lekshmi, M., and Madhumitha, G. 2012. Efficient phyto-synthesis and structural characterization of rutile TiO_2 nanoparticles using *Annona squamosa* peel extract. *Spectrochim Acta A. Mol. Biomol. Spectrosc.* 98: 86–90.

Saadat, M. 2011. Study of TiO_2 nanoparticles toxicity and bactericide on *Sodomounose aeromonse*, *J. Comp. Pathol. Iran.* 8: 497–502.

Saadatmand, M., Yazdanshenas M. 2012. Investigation of anti-microbial properties of chitosan-TiO_2 nanocomposite and its use on sterile gauze pads. *Med. Lab. J.* 6: 59–72.

Sadek, R.F., Farrag, H.A., Abdelsalam, S.M., Keiralla, Z.M.H., Raafat, A.I., and Araby, E. A powerful nanocomposite polymer prepared from metal oxide nanoparticles synthesized via brown algae as anti-corrosion and anti-biofilm. 2019. *Front. Mater.* 6: 140. doi: 10.3389/fmats.2019.00140.

Sand, W. 1989. Ferric iron reduction by *Thiobacillus ferrooxidans* at extremely low pH-values. *Biogeochemistry*. 7: 195–201.

Sand, W., Rohde, K., Sobotke, B., and Zenneck, C. 1992. Evaluation of *Leptospirillum ferrooxidans* for leaching. *Appl. Environ. Microbiol.* 58(1): 85–92.

Santhoshkumar, T., Abdul Rahuman, A., Jayaseelan, C., Rajakumar, G., Marimuthu, S., Kirthi, A.V., Velayutham, K., Thomas, J., Venkatesan, J., and Kim, S.-K. 2014. Green synthesis of titanium dioxide nanoparticles using *Psidium guajava* extract and its antibacterial and antioxidant properties. *Asian Pac. J. Trop. Med.* 7(12): 968–976.

Santhoshkumar, T., Rahuman, A.A., Rajakumar, G., Marimuthu, S., Bagavan, A., and Jayaseelan, C. 2011. Synthesis of silver nanoparticles using *Nelumbo nucifera* leaf extract and its larvicidal activity against malaria and filariasis vectors. *Parasitol Res.* 108: 693–702.

Santos da Silva, P., Ferreira de Senna, L., Gonçalvesa, M.M.M., and Baptista do Lago, D.C. 2019. Microbiologically-influenced corrosion of 1020 carbon steel in artifcial seawater using Garlic oil as natural biocide. *Mater. Res.* 22(4): e20180401. doi: http://dx.doi.org/10.1590/1980–5373-MR-2018-0401.

Saravanakumar, A., Ganesh, M., Jayaprakash, J., and Jang, H.T. 2015. Biosynthesis of silver nanoparticles using *Cassia tora* leaf extract and its antioxidant and antibacterial activities. *J. Ind. Eng. Chem.* 28: 277–281.

Sebeia, N., Jabli, M., and Ghith, A. 2019. Biological synthesis of copper nanoparticles, using *Nerium oleander* leaves extract: Characterization and study of their interaction with organic dyes. *Inorg. Chem. Commun.* 105: 36–46.

Senthilkumar, S., Ashok, M., Kashinath, L., Sanjeeviraja, C., and Rajendran, A. 2018. Phytosynthesis and characterization of TiO_2 nanoparticles using *Diospyros ebenum* leaf extract and their antibacterial and photocatalytic degradation of crystal violet. *Smart Sci.* 6: 1–9.

Shaban, S.M., Aiad, I., El-Sukkary, M.M., Soliman, E.A., and El-Awady. M.Y. 2015. Preparation of capped silver nanoparticles using sunlight and cationic surfactants and their biological activity. *Chin. Chem. Lett.* 26: 1415–1420.

Shaban, S.M., Aiad, I., Yassin, F.A., and Mosalam, A. 2019. The tail effect of some prepared cationic surfactants on silver nanoparticle preparation and their surface, thermodynamic parameters, and antimicrobial activity. *J. Surfact. Deterg.* 22: 1445–1460.

Sharad, A.A., Faja, O.M., Younis, K.M., ALwan, M.G., Drais, A.A., Bloh, A.H., Usup, G., Sahrani, F.K., and Ahmad, A. 2016. Isolation and identification of anaerobic cultivable iron reducing bacteria from crude oil and detection of biofilms formation on carbon steel surfaces. *Am. Eurasian J. Sustain. Agric.* 10(2): 55–64.

Sharma, D., Rajput, J., Kaith, B., Kaur, S.M., and Sharma, S. 2010. Synthesis of ZnO nanoparticles and study of their antibacterial and antifungal properties. *Thin Solid Films.* 19: 1224–1229.

Sharma, M., Liu, H., Chen, S., Cheng, F., Voordouw, G., and Gieg, L. 2018. Effect of selected biocides on microbiologically influenced corrosion caused by *Desulfovibrio ferrophilus* IS5. *Sci. Rep.* 8: 16620. doi: 10.1038/s41598–018-34789–7.

Sharmila, G., Pradeep, R.S., Sandiya, K., Santhiya, S., Muthukumaran, C., Jeyanthi, J., Kumar, N. M., and Thirumarimurugan, M. 2018. Biogenic synthesis of CuO nanoparticles using *Bauhinia tomentosa* leaves extract: Characterization and its antibacterial application. *J. Mol. Struct.* 1165: 288–292.

Singh, J., Kumar, V., Kim, K.-H., and Rawat, M. 2019. Biogenic synthesis of copper oxide nanoparticles using plant extract and its prodigious potential for photocatalytic degradation of dyes. *Environ. Res.* 177: 108569.

Singh, S., Kumar, N., Kumar, M., Jyoti, Agarwal, A., and Mizaikoff, B. 2017. Electrochemical sensing and remediation of 4-nitrophenol using bio-synthesized copper oxide nanoparticles. *Chem. Eng. J.* 313: 283–292.

Sivaraj, R., Rahman, P. K.S.M., Rajiv, P., Narendhran, S., and Venckatesh, R. 2014. Biosynthesis and characterization of *Acalypha indica* mediated copper oxide nanoparticles and evaluation of its antimicrobial and anticancer activity. *Spectrochim Acta A. Mol. Biomol. Spectrosc.* 129: 255–258.

Smith, K., Gould, K. A., Ramage, G., Gemmell, C., Hinds, J., and Lang, S. 2010. Influence of tigecycline on expression of virulence factors in biofilm-associated cells of methicillin-resistant Staphylococcus aureus. *Antimicrob. Agents Chemother.* 54: 380–387.

Sone, B.T., Diallo, A., Fuku, X.G., Gurib-Fakim, A., and Maaza, M. 2020. Biosynthesized CuO nano-platelets: Physical properties & enhanced thermal conductivity nanofluidics. *Arab. J. Chem.* 13(1): 160–170.

Song, W., Chen, X., He, C., Li, X., Liu, C. 2018. Microbial corrosion of 2205 duplex stainless steel in oilfield produced water. *Int. J. Electrochem.* Sci. 13: 675–689.

Sreeju, N., Rufus, A., and Philip, D. 2017. Studies on catalytic degradation of organic pollutants and anti-bacterial property using biosynthesized CuO nanostructures. *J. Mol. Liq.* 242: 690–700.

Subhapriya, S. and Gomathipriya, P. 2018. Green synthesis of titanium dioxide (TiO_2) nanoparticles by *Trigonella foenum-graecum* extract and its antimicrobial properties. *Microb. Pathog.* 116: 215–220.

Sukri, S.N.A.M., Shameli, K., Wong, M.M.-T., Teow, S.-Y., Chew, J., Ismail, N.A. 2019. Cytotoxicity and antibacterial activities of plant-mediated synthesized zinc oxide (ZnO) nanoparticles using *Punica granatum* (pomegranate) fruit peels extract. *J. Mol. Struct.* 1189: 57–65.

Sun, Y., Wang, S., and Zheng, J. 2019. Biosynthesis of TiO_2 nanoparticles and their application for treatment of brain injury—An in-vitro toxicity study towards central nervous system. *J. Photochem. Photobiol. B: Biol.* 194: 1–5.

Telegdi, J., Trif, L., and Románszki, L. 2016. Smart antibiofouling composite coatings for naval applications. In: M.F. Montemor (Ed.), *Smart Composite Coatings and Membranes: Transport, Structural, Environmental and Energy Applications*, Woodhead Publishing, pp. 123–155, Ch. 9.

Turkiewicz, A., Brzeszcz, J., and Kapusta, P. 2013. The application of biocides in the oil and gas industry. *NATFA GAZ.* 2: 103–111.

Valencia-Cantero, E., and Peña-Cabriales, J.J. 2014. Effects of iron-reducing bacteria on carbon steel corrosion induced by thermophilic sulfate-reducing consortia. *J. Microbiol. Biotechnol.* 24(2): 280–286.

Valencia-Cantero, E., Peña-Cabriales, J.J., Martínez-Romero, E. 2003. The corrosion effects of sulfate- and ferric-reducing bacterial consortia on steel. *Geomicrobiol. J.* 20: 157–169.

Vanaei, H.R., Eslami, A., and Egbewande, A. 2017. A review on pipeline corrosion, in-line inspection (ILI), and corrosion growth rate models. *Int. J. Pressure Vessels Piping.* 149: 43–54.

Vankar, P.S., and Bajpai, D. 2010. Preparation of gold nanoparticles from *Mirabilis jalapa* flowers. *Ind. J. Biochem. Biophys.* 47, 157–160.

Velayutham, K., Abdul Rahuman, A., Rajakumar, G., Santhoshkumar, T., Marimuthu, S., Jayaseelan, C., Bagavan, A., Kirthi, A.V., Kamaraj, C., Zahir, A.Z., and Elango, G. 2012. Evaluation of *Catharanthus roseus* leaf extract-mediated biosynthesis of titanium dioxide nanoparticles against *Hippobosca maculata* and *Bovicola ovis*. *Parasitol. Res.* 111: 2329–2337.

Velsankar, K., Sudhahar, S., Parvathy, G., and Kaliammal, R. 2020. Effect of cytotoxicity and Antibacterial activity of biosynthesis of ZnO hexagonal shaped nanoparticles by

Echinochloa frumentacea grains extract as a reducing agent. *Mater. Chem. Phys.* 239: 121976.

Vigneron, A., Alsop, E.B., Chambers, B., Lomans, B.P., Head, I.M., Tsesmetzis, N. 2016. Complementary microorganisms in highly corrosive biofilms from an offshore oil production facility. *Appl. Environ. Moicrobiol.* 82(8): 2545–2554.

Wadood, H.Z., Rajasekar, A., Ting, Y.-P., and Sabari, A.N. 2015. Role of Bacillus subtilis and Pseudomonas aeruginosa on corrosion behaviour of stainless steel. *Arab. J. Sci. Eng.* 40: 1825–1836.

Wagutu, A.W., Yano, K., Sato, K., Park, E., Iso, Y. and Isobe, T. 2019. *Musa* AAA and *Jatropha curcas* L. sap mediated TiO_2 nanoparticles: Synthesis and characterization. *Sci. Afr.* 6: e00203.

Wan, D., Yuan, S., Neoh, K.G., and Kang, E.T. 2010. Surface functionalization of copper via oxidative graft polymerization of 2,20-bithiophene and immobilization of silver nanoparticles for combating biocorrosion. *ACS Appl. Mater. Interf.* 2: 1653–1662.

Wang, D., Cui, L., Chang, X., and Guan, D. 2020. Biosynthesis and characterization of zinc oxide nanoparticles from *Artemisia annua* and investigate their effect on proliferation, osteogenic differentiation and mineralization in human osteoblast-like MG-63 Cells. *J. Photochem. Photobiol. B. Biol.* 202: 111652.

Wang, H., Wang, Z., Hong, H., and Yin, Y. 2010. Preparation of cerium-doped TiO_2 film on 304 stainless steel and its bactericidal effect in the presence of sulfate-reducing bacteria (SRB). *Mater. Chem. Phys.* 124: 791–794.

Wei, H., Ding, D., Wei, S., and Guo, Z. 2013. Anticorrosive conductive polyurethane multiwalled carbon nanotube nanocomposites. *J. Mater. Chem. A.* 1: 10805–10813.

Xiaohong, L., Hui, X., Wenjun, Z., Yongqian, L., Xuexi, T., Jizhou, D., Zhibo, Y., Jing, W., Fang, G., and Guoqing, D. 2019. Analysis of cultivable aerobic bacterial community composition and screening for facultative sulfate-reducing bacteria in marine corrosive steel. *Chin. J. Oceanol. Limn.* 37(2): 600–614.

Xu, D., and Gu, T. 2015. The War against problematic bioflms in the oil and gas industry. *J. Microb. Biochem. Technol.* 7: e124. doi: 10.4172/1948-5948.1000e124.

Xu, D., Huang, W., Ruschau, G., Hornemann, J., Wen, J., and Gu, T. 2013. Laboratory investigation of MIC threat due to hydrotest using untreated seawater and subsequent exposure to pipeline fluids with and without SRB spiking. *Eng. Fail Anal.* 28: 149–159.

Yan, Y., Zhang, X., Li, C., Huang, Y., Ding, Q., and Pang, X. 2015. Preparation and characterization of chitosan-silver/hydroxyapatite composite coatings onTiO_2 nanotube for biomedical applications. *Appl. Surf. Sci.* 332: 62–69.

Yashni, G., Al-Gheethi, A.A., Mohamed, R.M.S.R., and Amir Hashim, M.K. 2019. Green synthesis of ZnO nanoparticles by *Coriandrum sativum* leaf extract: Structural and optical properties. *Desalin. Water Treat.* 167: 245–257.

Yedurkar, S., Maurya, C., and Mahanwar, P. 2016. Biosynthesis of zinc oxide nanoparticles using *Ixora coccinea* leaf extract—A green approach. *Open J. Synth. Theory Appl.* 5: 1–14.

Yeole, K., Mahajan, L., and Mhaske, S. 2015. Poly(o-anisidine)-MWCNT nanocomposite: synthesis, characterization and anticorrosion properties. *Polym. Composit.* 36: 1477–1485.

Yu, Y., Fei, Z., Cui, J., Miao, B., Lu, Y., and Wu, J. 2018. Biosynthesis of copper oxide nanoparticles and their in vitro cytotoxicity towards nasopharynx cancer (kb cells) cell lines. *Int. J. Pharmacol.* 14: 609–614.

Yuvakkumar, R., Suresh, J., and Hong, S.I. 2014. Green synthesis of zinc oxide nanoparticles. *Adv. Mater. Res.* 952: 59–71.

Zakaria, A.E., Gebreil, H.M., and Abdelaal, N.M. 2012. Control of microbiologically induced corrosion in petroleum industry using various preventive strategies. *Arab J. Nucl. Sci. Appl.* 45(2): 460–478.

Zarasvand, K.A., and Rai, V.R. 2016. Inhibition of a sulfate reducing bacterium, Desulfovibrio marinisediminis GSR3, by biosynthesized copper oxide nanoparticles. *3 Biotech*. 6: 84. doi: 10.1007/s13205–016-0403–0.

Zhai, X., Ren, Y., Wang, N., Guan, F., Agievich, M., Duan, J., and Hou, B. 2019. Microbial corrosion resistance and antibacterial property of electrodeposited Zn–Ni–chitosan coatings. *Molecules*. 24: 1974. doi: 10.3390/molecules24101974.

Zhang, L. 2010. Mechanistic investigation into antibacterial behavior of suspensions of ZnO nanoparticles against *E. coli*. *J. Nanopart. Res.* 12(5): 1625–1636.

Zhang, X., Chaimayo, W., Yang, C., Yao, J., Miller, B.L., and Yates, M.Z. 2017. Silver-hydroxyapatite composite coatings with enhanced antimicrobial activities through heat treatment. *Surf. Coat. Technol.* 325: 39–45.

Zhou, Y., Gao, Y., Xie, Q., Wang, J., Yue, Z., Wei, L., Yang, Y., Li, L., and Chen, T. 2019. Reduction and transformation of nanomagnetite and nanomaghemite by a sulfate-reducing bacterium. *Geochim. Cosmochim. Acta.* 256: 66–81.

Zhu, X.Y., Lubeck, J., and Kilbane II, J.J. 2003. Characterization of microbial communities in gas industry pipelines. *Appl. Environ. Microbiol.* 69(9): 5354–5363.

CHAPTER 6

Application of Nanobiotechnology in Petroleum Refining

6.1 INTRODUCTION

There is a worldwide continuous increase in energy consumption. As reported by the U.S. Energy Information Administration (2016), it has been estimated to increase by 48% within 2012–2040. The transportation sector alone consumes approximately 25% of energy and estimated to increase by 1.4%/year within 2012–2040. The worldwide consumption of energy is mainly from nonrenewable sources (petroleum, natural gas, and coal), which approximately represent 83%, and nearly half of which comes from crude oil (Yi et al., 2019). The rest of energy demands are covered by renewable sources, such as biofuels, hydropower, solar energy, wind energy, etc. (Misra et al., 2018). However, there is a worldwide decrease in the high grade oil reserves, which are characterized by low viscosity and low asphaltene, sulfur, and nitrogen contents (Alves et al., 2015). This consequently led to the increase in the generation of polluting gases, which adds to the problem of global climate change and adversely affected the human health. The presence of sulfur in motor fuels has critical harming effect on the catalytic converters in motorized engines, increasing the combustion-related emissions, that is, the particulate matters (PMs), CO, CO_2, SOx, and NOx, which, consequently, increases global warming and air and water pollution (Srivastava, 2012; Abd Al-Khodor and Albayati, 2020). Moreover, high concentration of sulfur in fuels dramatically decreases the efficiency and lifetime of emission gas treatment systems in cars (Mužic and Sertić-Bionda, 2013). Incomplete combustion of fossil fuels causes emission of aromatic sulfur and nitrogen compounds; oxidation of these compounds in the atmosphere would lead to the aerosol of sulfuric and nitric acids (Porto et al., 2018). For example, NOx emission is significantly increased by 66%, corresponding to an increase in sulfur content of gasoline from 40 to 150 mg/L. It has been reported that approximately 73% of the produced SO_2 is from anthropogenic origin and is due to the combustion of petroleum and its derivatives. The NOx and CO_2 are thought by many to be the primary causes of "chemical smog" as well as "greenhouse gas" (GHG) accumulation. It has also been reported that sulfur is the main cause of emissions of PM. All of those aforementioned harmful emissions affect the stratospheric ozone, increasing the hole in the Earth's protective ozone layer (Larentis et al., 2011). It has been reported that approximately 2% sulfur in diesel fuel can be directly converted to PM emissions. The PM and SOx are known to be carcinogenic. The visible, dark black component of smoke is carbon that has incompletely burned. The soot resulted from the use of the lower quality fuel (i.e., less refined one), which contains a large amount of mutagenic and carcinogenic polyaromatic hydrocarbons (PAHs). Diesel exhaust is considered the most carcinogenic exhaust and accounts for approximately 25% of all smoke and soot in the atmosphere. It has been reported that relatively high concentration of SO_2 (>100 ppm) expresses harmful effects on the human respiratory system, where it can cause mortality within short time exposure to 400–500 ppm. Besides, very low concentrations of 1–2 ppm SO_2 would be enough to express sever damage to plant (Schmidt et al., 1973). Moreover, upon the emissions of SO_2 and NO_2 in the atmosphere, they react with hydrogen producing the weak sulfurous acid, strong sulfuric acid, and nitric acid, which is the main cause of acid rain and haziness that reduces the average temperature of the affected area and leads to climate change (Sadare et al., 2017). About 25% of the acidity of rain is due to the presence of nitric acid (HNO_3), and approximately 75% is related to the presence of sulfuric acid (H_2SO_4) (see Scheme 6.1). Acid rain has many negative impacts on the ecosystem and environment; for example, it causes soil pollution, destroys green area, kills forests, and damages crops, leather, cars, and buildings. It also poisons lakes and rivers leading to a devastating effect on their fauna and flora and falling in fish population. Also, the presence of high levels of sulfate in water affects the human health, as it causes diarrhea and dehydration. Acid rains cause degradation of many soil minerals producing metal ions that are then washed away in the runoff, causing the release of toxic ions, such as Al^{3+}, into the water streams; moreover, loss of important minerals, such as Ca^{2+} (see Scheme 6.1), from the soil would kill trees and damage crops and cause solid erosion. Acid rains have also a negative impact on building and monuments.

Thus, to overcome the problem of climate change, the worldwide environmental legislation has put limits for the sulfur and nitrogen levels in the transportation fuels to decrease both the tailpipe and evaporative combined emissions of nonmethane organic gas, NOx, and SOx emissions

(Sadare et al., 2017; Porto et al., 2018). Consequently, refineries must have economically feasible techniques to remove sulfur and nitrogen from crude oil and refinery streams to the extent needed to mitigate these unwanted effects.

$$2NO\ (g) + O_2\ (g) \longrightarrow 2NO_2\ (g)$$

$$3NO_2\ (g) + H_2O \longrightarrow 2HNO_3\ (aq) + NO\ (g)$$

$$2NO_2\ (g) + H_2O \longrightarrow HNO_3\ (aq) + HNO_2\ (aq)$$

$$2SO_2\ (g) + O_2\ (g) \longrightarrow 2SO_3\ (g)$$

$$SO_2\ (g) + H_2O \longrightarrow H_2SO_3\ (aq)$$

$$SO_3\ (g) + H_2O \longrightarrow H_2SO_4\ (aq)$$

$$H_2SO_4\ (aq) + CaCO_3\ (s) \longrightarrow Ca^{2+}\ (aq) + SO_4^{2-}\ (aq) + H_2O + CO_2$$

SCHEME 6.1 Acidity of acid rains and degradation of soil minerals.

Crude oil is generally composed of 80%–87% carbon and 10%–15% hydrogen. Then come sulfur that represents approximately 0.05%–6% and nitrogen and oxygen representing 0.1%–2% and 0.05%–1.5%, respectively. Moreover, it contains traces of heavy metals, for example, V, Ni, Fe, Al, Na, Ca, Cu, and U. However, in some recent publications, sulfur, which is the third most abundant constituent in crude oil, has been reported to range between 0.03% and 15% (wt.%) based on its origin and geographical source (El-Gendy and Nassar, 2018). Alkhalili et al. (2017) reported that the acidic mercaptans and thiols compromise to about 1%–15% of the total sulfur compounds in crude oil, whereas the nonacidic sulfides and thiophenic compounds contribute to about 50%–80%. Then come to lower extent the disulfides representing 7%–15% of the total sulfur content. Sometimes, thiophenes (i.e., thiophene, benzothiophene, dibenzothiophene, and their derivatives) compromise from 50% to as much as 95% of the sulfur compounds in crude oils and its fractions (Mohebali and Ball, 2016). Speight (2014) reported that the nonbasic nitrogen compounds (e.g., carbazole) and the basic ones (e.g., quinoline, Qn) represent approximately 70%–75% and 25%–30% of the total nitrogen content in crude oil, respectively. The sulfur and nitrogen contents increase with the increase in the boiling point of the distillate fractions (Misra et al., 2018). It is worth to note that, if the American Petroleum Institute (API) gravity of the crude oil is less than 10, it would be considered as heavy crude oil, while the sour crude oil is the one that has S-content of more than 0.5% (Abd Al-Khodor and Albayati, 2020). Generally, there is an inversely proportional relationship between the S-content and the API value; the lower the API value, the higher the S-content of the crude oil. Nevertheless, as much as the oil has low viscosity and sulfur content, as much easier as it will be its extraction and refining processes (Lateef et al., 2019). Moreover, light crude oils mean higher yield of hydrocarbons and lower environmental pollution.

In general, the presence of sulfur and nitrogen compounds is undesirable, as they are corrosive, causing a lot of problems in pipelines and pumping and refining equipment (Sadare et al., 2017). Moreover, the high sulfur- and nitrogen-content crude oil would lead to a lot of drawbacks in the refining process, for example, deactivation of some catalysts in downstream processing and upgrading of hydrocarbons (Srivastava, 2012; Bhadra and Jhung, 2019). It has been reported that the sulfur compounds in heavy crude oils are mainly of three-membered-ring polycyclic sulfur compounds or higher, for example, dibenzothiophene and its alkyl-substituted derivatives such as 4-methyldibenzothiophene and 4,6-dimethyldibenzothiophene (Song et al., 2003; El-Gendy and Speight, 2016; Alkhalili et al., 2017; El-Gendy and Nassar, 2018). These compounds are the most refractory compounds in the hydrodesulfurization (HDS) process (Sadare et al., 2017; Lin et al., 2020). Moreover, the presence of S-compounds contributes to coke formation and to catalyst poisoning during the refining of crude oil, thus reducing the process yields (Speight and El-Gendy, 2017). Besides, the presence of N-compounds usually inhibits the HDS process, and it has been reported to increase in the following order: carbazole (CAR) < quinoline < indole (Laredo et al., 2001). However, in another study by Tao et al. (2017), the inhibiting effect of quinolone was much higher than that of indole. It should be known that, although the concentration of nitrogen is relatively low in petroleum compared to that of sulfur, its presence is of a much greater significance in refinery operations than might be expected from such small amounts. Like S-compounds, most of the N-containing compounds in petroleum are currently removed by the conventional chemical and physical refinery processes (El-Gendy and Nassar, 2018).

Briefly, hydrotreatment of heavy crude oil is the most important process in petroleum refining, which also includes HDS and hydrodenitrogenation (HDN). It requires high temperature and pressure, hydrogen, and efficient catalyst (Speight, 2014; Garcia-Montoto et al., 2020). As the molecular weight of the S- and N-compounds increases, more harsh, energy-consuming, and expensive conditions are required for hydrotreatment. At elevated temperature (290–455 °C) and pressure (8–10 MPa) in the presence of cobalt- and nickel-promoted Mo/γ-Al_2O_3, HDS and HDN occur with the production of hydrogen sulfide and ammonia, respectively, with a simultaneous, saturation of aromatics, that is, hydrodearomatization (Misra et al., 2018). For reaching the ultra-low S- and N-content fuel hydrotreatment, although it tends to improve diesel quality by raising its cetane number, however, it tends to decrease the gasoline quality by lowering its octane number (Song and Ma, 2003; Sadare et al., 2017; Abid et al., 2019). Not only this, but increased operational and capital costs as well as more carbon dioxide emissions also occur (Martínez et al., 2017; Lateef et al., 2019).

HDN under high temperature and pressure is the most widely method for the removal of nitrogen polyaromatic heterocyclic compounds from petroleum, but it is like HDS, which is expensive, hazardous, and needs more severe conditions, more hydrogen, and a higher energy consumption (Speight and El-Gendy, 2017). Moreover, the NH_3 produced during the HDN can also inhibit the HDS catalysts. Besides, hydrogenation of aromatic rings is required prior to the attack of the C–N bonds for nitrogen removal. Nitrogen compounds can be responsible for the poisoning of cracking catalysts, and they also contribute to gum formation in such products as domestic fuel oil (Speight, 2014). CAR as an example for the nonbasic nitrogenous PAHs can directly impact the refining processes in two ways: (1) during the cracking process, CAR can be converted into basic derivatives, which can be adsorbed to the active sites of the cracking catalysts; and (2) it directly inhibits the HDS catalysts. Thus, the removal of CAR and other nitrogen compounds would significantly increase the extent of catalytic cracking and, consequently, the gasoline yield. It has been reported that, by 90% reduction in the nitrogen content, a 20% increase in gasoline yield occurs. That has a major economic improvement in low-margin high-volume refining processes (Benedik et al., 1998). Basic N-compounds are more inhibitory for catalysts than the nonbasic ones. But, they can potentially be converted into basic compounds during the refining/catalytic cracking process (Asumana et al., 2011). Thus, they are also inhibitory to catalysts. Moreover, metals such as nickel and vanadium are potent inhibitors for catalysts, and in petroleum, these metals are typically associated with N-compounds (Mogollon et al., 1998). Thus, removal of nitrogen is preferable before the HDS process (Singh et al., 2011a). Nitrogenous compounds lower the oxidative desulfurization efficiency due to its competitive parallel oxidation (Souza et al., 2009). Also, polycyclic aromatic nitrogen heterocyclic (PANH) compounds are extracted during the extractive desulfurization; due to its basicity, it also affects the extraction and catalytic oxidative desulfurization (Huh et al., 2009; Jia et al., 2009). It should be known that polycyclic aromatic sulfur heterocyclic (PASH) and PANH compounds are like PAHs have carcinogenic, mutagenic activities and acute toxicity (Wang and Krawiec, 1994). This would cause a lot of environmental issues upon the occurrence of any oil spill, as these compounds persist for a long period of time and can be accumulated in organisms' tissues (El-Gendy and Nassar, 2018).

From the economic point of view, it is important to maximize the sulfur and nitrogen removal keeping the fuel-energy content. There are a lot of existing investments to apply bioprocess, for example, biodesulfurization (BDS) and biodenitrogenation (BDN) as complementary to the existing refining technologies due to their milder operating conditions, and for being cleaner and selective processes with low emissions and no production of undesirable by-products (Singh et al., 2012; Kitashov et al., 2019).

The most challenging goal for the industrial development of BDS technology is the catalytic removal of recalcitrant organosulfur compounds, for example, dibenzothiophene and its derivatives, using microbial cells or enzymes under aerobic or anaerobic conditions. Anaerobic BDS desulfurizes the fuels producing hydrogen sulfide same as the HDS process (see Figure 6.1). Thus, the produced H_2S gas can be treated with existing refinery desulfurization plants (e.g., Claus process). The other advantage of anaerobic BDS is the minimal production of undesired colored and gum forming products, which would be formed during the bio-oxidation of hydrocarbons. However, due to the low reaction rates, safety and cost concerns, in addition to the lack of identification of the specific enzymes responsible for the anaerobic BDS and isolation of anaerobic microorganisms effective enough for practical petroleum anaerobic desulfurization, this process has not been developed up till now (Speight and El-Gendy, 2017; El-Gendy and Nassar, 2018). Consequently, aerobic BDS has been the focus of most of the research in BDS (Mohebali and Ball, 2016).

Complete mineralization and Kodama pathways are not recommended for aerobic BDS of fuels, as oxidation of the hydrocarbon skeleton reduces the fuel value (see Figure 6.2). However, the most recommendable biodesulfurizing microorganisms to be applied in petroleum bioupgrading are those which desulfurize DBT via the 4S-pathway, without affecting its hydrocarbon skeleton, producing 2-hydroxybiphenyl (2-HBP) and/or 2,2′-bihydroxybiphenyl (2,2′-BHBP) (see Figure 6.2).

DBT → H_2 → SH → H_2 → Biphenyl + H_2S

FIGURE 6.1 Possible anaerobic biodesulfurization pathway.

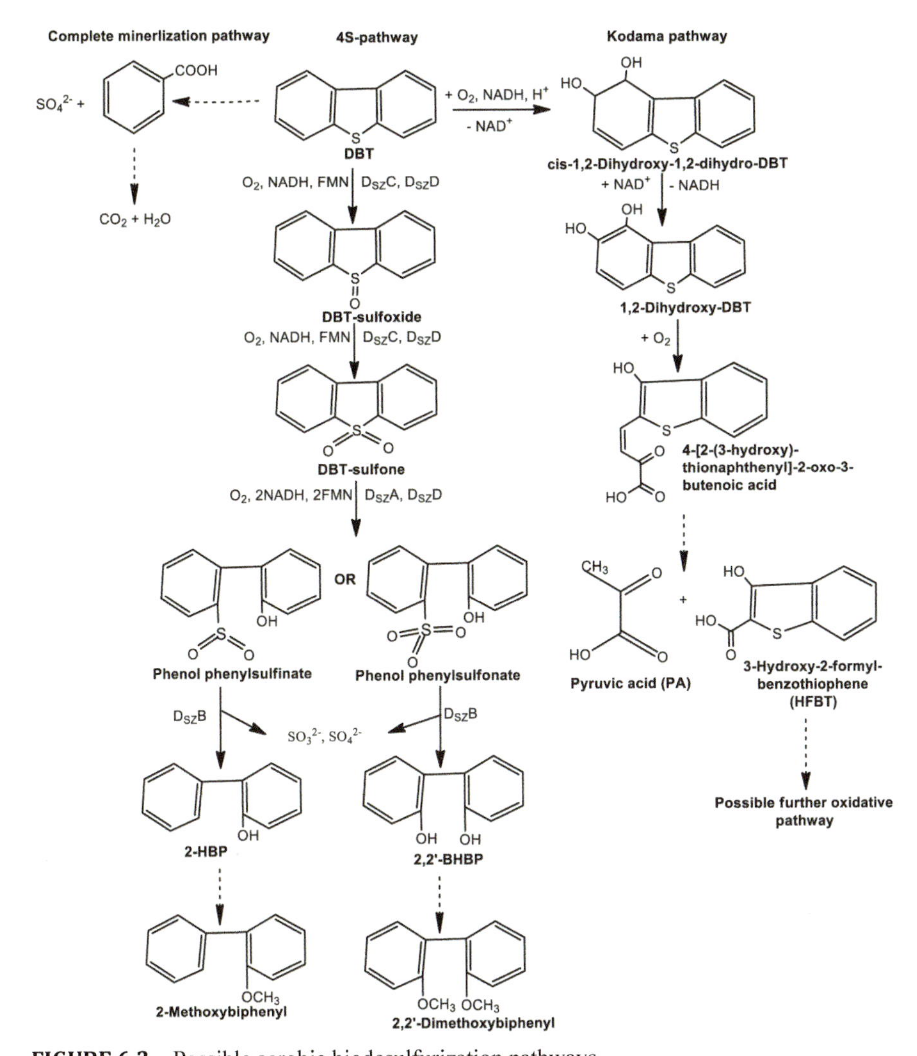

FIGURE 6.2 Possible aerobic biodesulfurization pathways.

However, those phenolic compounds are reported to express feedback inhibition and thus decrease the rate of BDS (El-Gendy and Nassar, 2018).

Research in the field of BDS managed to isolate microorganisms, for example, *Mycobacterium* sp. (Chen et al., 2009), *Achromobacter* sp. (Bordoloi et al., 2014), *Rhodococcus erythropolis* HN2 (El-Gendy et al., 2014), and *Pseudomonas aeruginosa* (Al-Jailawi et al., 2015), which are capable to detoxify such phenolic compounds via methoxylation into their corresponding 2-methoxybiphenyl and 2,2′-dimethoxy-1,1′-biphenyl as final-end products (see Figure 6.2).

The milder and safer process conditions of aerobic BDS lead to significant reductions in GHG emissions and energy requirements relative to HDS processes (Sadare et al., 2017; Lateef et al., 2019), where the capital and operating costs to set up a BDS-process is reported to be 50% and 10%–20% lower than that of a HDS process, with approximately 80% reduction in GHG emissions (El-Gendy and Speight, 2016). Thus, it is very recommendable as alternative and/or complementary to conventional desulfurization processes. The first aerobic BDS pilot plant was built in the mid-1990s by ENCHIRA Biotechnology Corporation (formerly Energy Biosystems Corporation) (Singh et al., 2012). The maximum reduction in the sulfur content achieved with different microorganisms using heavy crude oil as the sulfur source is in the range of 47%–82% (Setti et al., 1993; El-Gendy et al., 2006; Torkamani et al., 2008a, b, 2009; Agarwal and Sharma, 2010; Bhatia and Sharma, 2010, 2012; Li and Jiang, 2013; Adlakha et al., 2016).

Although BDS is characterized by lower operational cost and produces valuable by-products, the main disadvantage of BDS is its lower reaction rates compared to HDS (Yi et al., 2019). It has been predicted that the BDS catalyst must have S-removal capability within the range of 1–3 mmol DBT/g dry cell weight/h in real oil feed. Thus, the efficiency of the present biocatalysts is desired to be enhanced by 500-fold to reach a commercially feasible process (Kilbane, 2006). Alves et al. (2015) performed a cost analysis study, comparing two BDS process designs, upstream and downstream, in a conventional HDS process. The BDS costs and emission estimations were made considering the BDS of DBT as a model for S-compounds, while HDS estimations were made based on crude oil HDS. The BDS downstream HDS configuration is found to be the best alternative to be applied in oil refinery, from the point of lower energy consumption, GHG emissions, and operational costs, to obtain almost S-free fuels. Moreover, there are still some unsolved aspects such as stability, lifetime, inhibition effects, high costs of biocatalyst, reactor design, quantity of water in the media, separation of aqueous organic phases, low enzyme activity, low mass transfer rates, and biocatalyst recovery. Also, large amounts of biomass are needed (typically 2.5 g biomass per g sulfur), and biological systems must be kept alive to function, which can be difficult under the variable input conditions found in refineries (Nehlsen et al., 2005). Nevertheless, the cost of the culture medium represents 30%–40% of the total amount (El-Gendy and Nassar, 2018). Moreover, the desulfurization rate stops before the complete removal of sulfur compounds, as growing cells may be deactivated by the accumulation of 2-hydroxybiphenyl (2-HBP) (Irani et al., 2011). For a cost-effective BDS process, the reaction time and catalyst longevity should be of 1 and 400 h, respectively (Pacheco et al., 1999). Different research studies reported the desired rate of BDS for its commercialization to be 20 μmol DBT/min/g DCW (Nazari et al., 2017), 3-mM S/g DCW/h (Singh, 2015) and, in other reports, 1.2–3 mM/g DCW/h (El-Gendy and Nassar, 2018), whereas the maximum rate achieved till date is 320 μM S/g DCW/h (Kilbane, 2006; Singh, 2015). El-Gendy and Nassar (2018) recommended different aspects to overcome such drawbacks. For example, (1) isolation of new biodesulfurizing microorganism with the capabilities of utilizing wide range of organosulfur compounds; (2) isolation of thermophilic microorganisms to be applied as a downstream process after HDS and safe time and energy consumed in cooling the feed to the BDS process; (3) isolation of halotolerant microorganisms to decrease the requirement of fresh water; (4) isolation of more hydrophobic microorganisms to overcome the problem of oil/water phase ratio and desulfurize more amount of feed per cycle; (5) isolation of new biodesulfurizing microorganisms with the ability to overcome the feedback inhibition caused by 2-HBP, 2,2′-BHBP and sulfate ions; (6) applying genetic engineering to increase the copy number of desulfurization genes and/or producing recombinant microorganisms with hydrocarbons and inhibitors tolerance and higher BDS efficiency; (7) applying biodesulfurizing microorganisms with the capability of producing biosurfactants instead of using chemical surfactants, to overcome the problem of mass transfer, increase the contact between the cells and the organosulfur compounds, and lower the operational cost; (8) immobilization and coating cells with nanoparticles (NPs) to enhance the rate of BDS, increase the stability and lifetime of the biocatalyst, overcome the toxic effect of the produced products, solvents, and hydrocarbons, and, moreover, solve the problem of biocatalyst recovery; (9) applying statistical optimization and mathematical modeling to lower the cost of process optimization; and (10) finally producing valuable product during the BDS process. Thus, in view of industrial application of BDS, cell immobilization is considered to be one of the most promising approaches (Huang et al., 2012).

Although selective BDS of petroleum and its distillates has been practically investigated, there is a little information about BDN of oil feed without affecting its calorific value. It has been estimated that BDN of petroleum would be beneficial for deep denitrogenation, where the classical hydroprocessing methods are costly and nonselective (Vazquez-Duhalt et al., 2002). It will also eliminate the contribution of fuel nitrogen to NOx emissions. However, the economics of nitrogen-removal processes are affected by the amount of the associated hydrocarbon lost from the fuel during the BDN process.

Generally, the currently well-established CAR-BDN pathway resembles that of DBT-Kodama pathway (see Figure 6.3). It is obvious that is economically unfeasible due to the degradation of hydrocarbon skeleton, which consequently leads to a loss in the fuel value. However, most of the CAR degrading microorganisms produce 2′-aminobiphenyl-2,3-diol as the first step in CAR-BDN pathway. Thus, recovering of CAR-nitrogen as anthranilic acid (ANA) and/or or 2′-aminobiphenyl-2,3-diol, which are less inhibitory to refining catalysts, is recommendable to solve part of that problem. That is because the entire carbon content of the fuel is preserved. This can be performed by mutant or recombinant strains. Other pathway would liberate nitrogen from CAR in the form of ammonia (Rhee et al., 1997). Most of the attack is aerobic, but anaerobic degradation has also been noted (Fallon et al., 2010).

Zakaria et al. (2016) isolated Gram-positive *Bacillus Clausii* BS1 from an Egyptian coke sample. BS1 showed a higher BDN efficiency relevant to the well-known biodenitrogenating Gram-negative bacterium strain *Pseudomonas resinovorans* CA10, recording 77.15% and 60.66% removal of 1000 mg/L CAR with the production of 119.79 and 102.43 mg/L ANA, and 121.19 and 90.33 mg/L catechol, as by-products, respectively. Several species of the genus *Pseudomonas* are known for its solvent tolerance and have been isolated for their ability to degrade CAR and its alkyl derivatives. Moreover, other microorganisms have been reported to mineralize nonbasic nitrogen compounds, including species of *Bacillus*, *Sphingomonas*, *Xanthomonas*, *Gordonia*, *Klebsiella*, *Burkholderia*, *Arthrobacter*, and *Novosphingobium* (Singh et al., 2011a,b; Zakaria et al., 2016). A thermophilic CAR-degrading bacteria *Anoxybacillus rupiensis*, which can tolerate up to 80 °C, with maximum activity at temperature range 55–65 °C has also been reported, which would be advantageous for the application in real petroleum processing (Fadhil et al., 2014). Larentis et al. (2011) reported that Tween 20 increased the dispersion of CAR in the aqueous solution and improved its bioaccessibility, thus enhancing its BDS by *Pseudomonas stutzeri*.

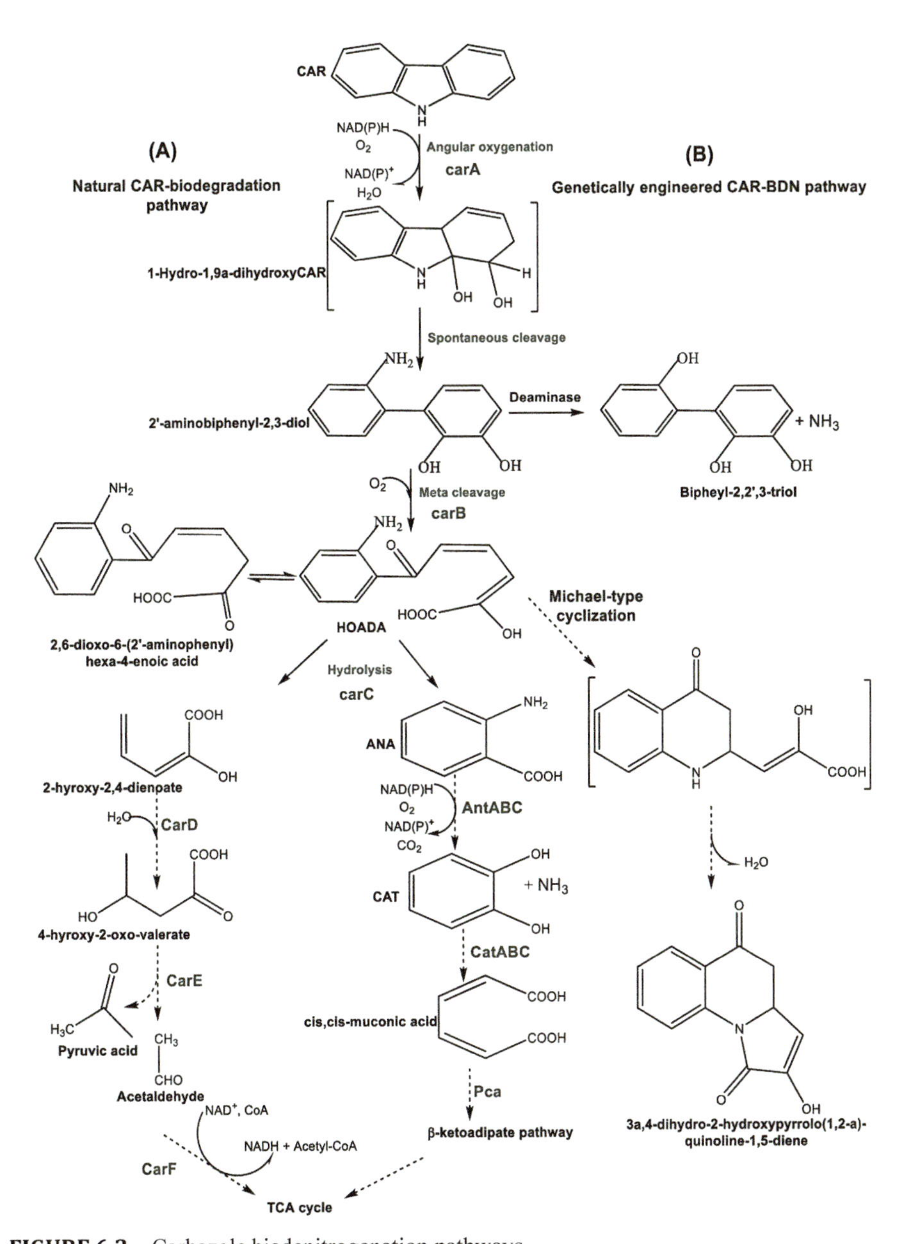

FIGURE 6.3 Carbazole biodenitrogenation pathways.

From the practical point of view, a dual microbial process for both selective BDS and BDN, with the overcoming of the significant technical hurdles, such as tolerance against solvents, high concentration of sulfur and nitrogen compounds, and high oil-to-water (O/W) ratio, would make microbial refining processes and bioupgrading of petroleum and its fractions feasible on a large scale (Kilbane, 2006). Duarte et al. (2001) in PETROBRAS, the Brazilian oil company, have isolated *Gordonia* sp. strain F.5.25.8 that can utilize DBT through the 4S-pathway and CAR as a sole source of S and N, respectively, where F.5.25.8 is the first reported strain that can simultaneously metabolize DBT and CAR (Santos et al., 2005). Santos et al. (2006) reported that F.5.25.8 can tolerate up to 42 °C, which would add to its advantageous in industrial application of BDS/BDN as complementary to the hydrotreatment process. Moreover, Kayser and Kilbane (2004) inserted genes encoding amidases downstream the artificial carA operon to accomplish the cleavage of the final C–N bond and produce biphenyl-2,2′,3-triol (see Figure 6.3) that is reintroduced to the fuel, keeping its fuel content. That genetically engineered bacterium decreased the CAR content in a petroleum sample by approximately 95% in 2:10 petroleum/aqueous medium within 16 h using a genetically engineered bacterium.

Yu et al. (2006) reported the recombinant *R. erythropolis* SN8 expressed good BDS and BDN activities toward a wide range of recalcitrant alkyl CARs and DBTs in crude oil in just a one-step bioprocess. Maass et al. (2015) reported BDS/BDN of heavy gas oil (HGO), which is an intermediate fraction obtained from vacuum distillation used in the production of diesel and some lubricants, by *R. erythropolis* ATCC 4277 in a batch reactor. That reached maximum desulfurization and denitrogenation rate of 148 mg S/kg HGO/h and 162 mg N/kg HGO/h at 40% (v/v) HGO/water, respectively. In further study by Todescato et al. (2017), fractional factorial design (FFD) has been applied to optimize the microbial growth medium for ATCC 4277. The results from that FFD applied for further central composite design (CCD) of experiments, where the response surface methodology (RSM) proved a 240% increase in final biomass concentration, an increment in specific growth rate, and a growth yield coefficient of about five times greater, at the predicted optimum growth conditions. However, under such optimum conditions (2 g/L glucose, 5 g/L malt extract, 6.15 g/L yeast extract, 1.6 g/L $CaCO_3$, 23 °C, and 180 rpm), the BDN of HGO increased by 40%, but BDS lowered by 10%. That was attributed to the presence of high concentration of nitrogen source in the medium, made *R. erythropolis* ATCC 4277 cells to be more adapted to the nitrogen consumption than to the sulfur. That is because the presence of high concentration of nitrogen inhibits the development of oxidoreductase and desulfinase enzymes, which are enormously essential in the BDS pathway of recalcitrant heterocyclic compounds (Porto et al., 2017).

In another study, RSM based on CCD of experiments was applied to enhance the BDN efficiency of *Bacillus clausii* BS1 via the addition of yeast extract and the surfactant Tween 80 (Zakaria et al., 2015a), whereas the BDN efficiency increased from ≈ 88% without yeast extract or Tween 80 to ≈ 95% in the presence of optimum concentration of 0.868 g/L yeast extract and 0.861% (v:v) Tween 80, which would represent a major economic improvement in low-margin, high-volume refining processes. Benedik et al. (1998) reported that any fuel-upgrading process would be a low-margin (probably less than US$1/bbl or US$0.02/L, value added), large-volume (approximately 1010 L/year), commodity enterprise in order to be economically viable. Tween 80 is known to reduce the concentration of toxic metabolites around the bacterial cells, increasing the biomass concentration and activity (Feng et al., 2006). The CAR-BDN by *Novosphingobium* sp. strain NIY3 (Ishihara et al., 2008) and *Klebsiella* sp. strain LSSE-H2 (Li et al., 2008a) was enhanced in the presence of 0.2- and 0.05-g/L yeast extract and Tween 80, respectively.

Fuel-BDS and/or BDN take place into a three-phase system, that is, oil/water/biocatalyst. It has been reported that the transfer of PASH and PANH compounds from the oil phase to the water phase and then from water to the cells, where oxidation reaction occurs into the cytoplasm, limits the metabolism of such compounds. It is well documented that the first and rate-limiting step in the oxidative BDS of DBT is its transfer from the oil phase into the cell (Setti et al., 1999). The more the hydrophobicity of the biocatalyst, the higher the rate of BDS and BDN as it would assimilate the PASHs and PANHs directly from the oil (Monticello, 2000). The capability of the applied biocatalysts for production of biosurfactant would also help in increasing the rate of BDS and BDN. This would occur via the formation of oil/water emulsion, which retains the microbial viability and activity (Han et al., 2001). Biocatalyst stability, lifetime, and separation are other important aspects for the commercialization of BDS. Immobilized biocatalyst has some advantages over free ones, which are ease of separation, high stability, low risk of contamination, and long lifetime (Hou et al., 2005). Compared to BDS, adsorptive desulfurization (ADS) has been reported to have a much faster reaction rate (Song, 2003). However, the adsorbent preparation is the key of ADS. Most adsorbents for desulfurization are based on π-complexation (Shan et al., 2005a) or formations of metal–sulfur bonds (such as nickel–sulfur and lanthanum–sulfur bonds) (Tian et al., 2006). Adsorbents based on π-complexation are known to be easily regenerated, but their selectivity is very low, resulting in a loss of fuel quality. On the other hand, adsorbents that form metal–sulfur bonds are characterized by high selectivity, but unfortunately they are not easily regenerated. Thus, ADS, like BDS, still has a long way to go before being industrialized.

For BDS and BDN, cells are needed to be harvested from the culture medium, and several time-consuming and costly separation schemes have been evaluated, including settling tanks (Schilling et al., 2002), hydrocyclones (Yu et al., 1998), and centrifuges (Monticello, 2000). Magnetic separation technology is reported to provide a quick, easy, and convenient alternative over the aforementioned traditional methods in biological systems (Haukanes and Kvam, 1993). Magnetic supports for biocatalyst immobilization offer several advantages, such as: (1) the ease of magnetic collection; (2) the magnetic supports present further options in continuous reactor systems when used in a magnetically stabilized, fluidized bed; and (3) in addition, the mass transfer resistance can be reduced by the spinning of magnetic beads under revolving magnetic field (Sada et al., 1981).

In order to combine the advantages of immobilization, that is, ease of separation and microbial longevity with those of free diffusion, that is, good mass transport, another approach is possible, namely, to decorate the bacterial cells with magnetic nanoparticles (MNPs), whereas, after the completion of the reaction, the bacterial cells can be easily separated from the products using an external magnetic field (see Figure 6.4). This is a much milder and more cost-effective process than centrifugation and allows the bacteria to be reused many times, and magnetic separation is compatible with any automated platform equipped with a magnet (Zakaria et al., 2015b).

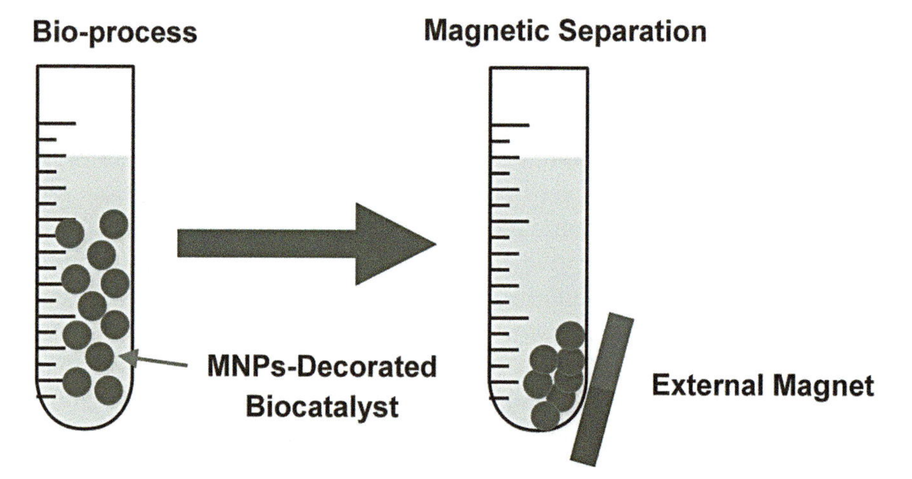

FIGURE 6.4 Magnetic separation of MNP-decorated biocatalyst.

Several expressions such as decoration and labeling are used to describe the preparation of magnetic modified cells. The term "decorated cells" usually describes the cells covered with a large number of NPs, in some cases, in several layers, where the particles usually do not cross the cell membrane or cell wall. On the contrary, target cells can be magnetically labeled just by a single (or only a few) magnetically responsive particle attached to the cell wall or membrane. However, magnetic labeling also represents the situation when MNPs cross the cell membrane/wall or enter into the cytoplasm.

In recent years, MNPs have been widely used in the field of BDS because of their large surface-to-volume ratios, superparamagnetic properties, and low toxicities (Bardania et al., 2013). Moreover, the coating layer of NPs does not change the hydrophilicity of the cell surface, and it has a negligible effect on mass transfer because the structure of the layer is looser than that of the cell wall. Thus, the coating layer does not interfere with mass transfer of PASHs and/or PANHs. The coated cells also have good stability and can be reused (Zakaria et al., 2015b). Consequently, this new technique is promising for large-scale industrial applications.

This chapter summarizes the worldwide efforts and achievements for the application of nanotechnology for enhancing BDS and BDN efficiencies to be applied in real field industry.

6.2 APPLICATION OF NANOTECHNOLOGY IN THE BDS PROCESS

Among many advantages of immobilization, reusability of cells in successive reaction steps is of great importance. However, BDS activity of the entrapped cells decreases after each step as a result of the reduction of cofactors such as $NADH_2$ and $FMNH_2$ in the 4S-pathway (see Figure 6.2) (Yan et al., 2008). Moreover, the entrapment technique itself often leads to a decrease in biocatalytic activity (Karsten and Simon, 1993). One of the first applications of nanomaterials in desulfurization a process was reported by Shan et al. (2003), where *Pseudomonas delafieldii* was immobilized in magnetic polyvinyl alcohol (PVA) beads. The beads had distinct superparamagnetic properties and were compared with immobilized cells in nonmagnetic PVA beads. Although the desulfurizing activity was slightly increased from 8.7 to 9 mM sulfur /kg DCW/h, the main advantages were that the magnetic immobilized cells maintained a high desulfurization activity and remained in good shape after seven times of repeated use, whereas the nonmagnetic immobilized cells used for only five times. Furthermore, the magnetic immobilized cells were easily collected and separated magnetically from the BDS reactor.

Shan et al. (2005b) compared the BDS efficiencies of free *Pseudomonas delafieldii*, magnetic Fe_3O_4 NP-decorated *P. delafieldii*, and immobilized cells by Celite toward 2-mM DBT in *n*-dodecane. The transmission electron microscopy (TEM) analysis reported MNPs' average size of approximately 10–15 nm. The TEM cross-sectional analysis of the cells further showed that the Fe_3O_4 NPs were for the most part strongly absorbed by the surfaces of the cells and coated cells. The coated cells had distinct superparamagnetic properties (the magnetization (δs) was 8.39 emu/g). Thus, they were easily concentrated on the side of the BDS vessel and separated from the suspension medium by decantation. Moreover, the results showed that the MNP-coated and free cells exhibited similar time courses and same desulfurizing activities during the first 4 h; 16.4 mM/ kg DCW/h and DBT was completely desulfurized within 6 h, while the cells immobilized by Celite expressed relatively lower BDS efficiency of 12.6 mM/kg DCW/h during the first 4 h. No MNP-coated cells appeared to have been lost during washing, but small loss of cells from the Celite during washing has been occurred. Thus, MNP-coated cells have greater operational stability than cells immobilized on Celite. The coated cells not only had the same desulfurizing activity as free cells, but also could be reused for more than five times, without losing its activity. Consequently, compared to cell immobilized on Celite, the cells coated with Fe_3O_4 NPs had greater desulfurizing activity and operational stability.

Zhang et al. (2008) reported that the adsorption between Fe_3O_4 and DBT is electrostatic and reversible; thus, DBT can be easily desorbed. Consequently, the desorbed DBT can be easily transferred to the cells for BDS. It is faster for cells to obtain DBT by this way; thus, 2-HBP production rate is increased.

BDS usually occurs in organic phases; therefore, the transfer of PASHs to the cell membrane is limited. The cell membrane needs to be not only resistant, but also permeable enough to capture the sulfur-containing compounds. This issue may be overcome with the use of genetically modified bacteria that are stable in oil phases or by using sorbents of organosulfur molecules. For example, when *P. delafieldii* R-8 are assembled with the nanosorbent γ-Al_2O_3, the desulfurization rate of the cells increased at least two times than that of the cells alone (Guobin et al., 2005), whereas the nano γ-Al_2O_3 onto the cell membrane adsorbed DBT, thus increasing the transference rate of the sulfur compound from the media to the cell membrane (see Figure 6.5).

In a further study, Guobin et al. (2006) combined the freezing–thawing and magnetic separation technique to immobilize *Pseudomonas delafieldii* R-8 cells for biodesulfurizing a model oil of n-dodecane containing 0.5-mM DBT, where the entrapped R-8 cells in magnetic PVA beads were stably stored and repeatedly used over 12 times for BDS. The beads had distinct superparamagnetic properties, and their saturation magnetization was 8.02 emu/g. The desulfurization rate of the immobilized cells reached 40.2 mM/kg/h. Desulfurization patterns of DBT in model oil with the immobilized and free cells were represented by the Michaelis–Menten equation. The Michaelis constant for both immobilized and free cells was 1.3 mM/L.

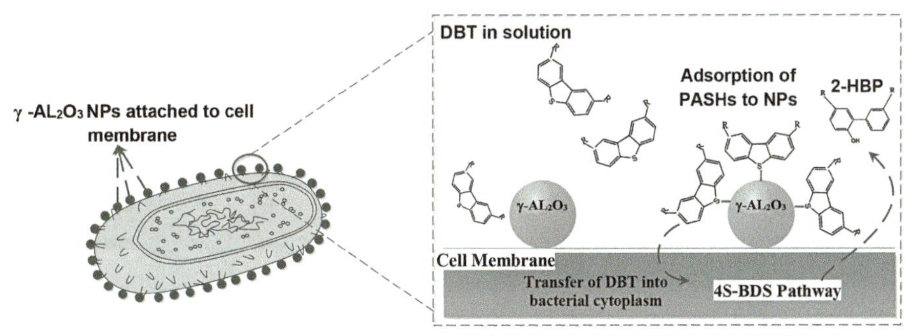

FIGURE 6.5 Enhancement of the BDS process by using γ-Al_2O_3 NP-coated cells.

A microbial method for the regeneration of desulfurization adsorbents has been proposed by Li et al. (2006), where sulfur compounds were removed via ADS using a Cu-modified zeolite and then biologically removed from the adsorbents in a subsequent stage using *Pseudomonas delafieldii* R-8. One of the important factors for a successful bioregeneration is the porous structure of the adsorbent. As the oil in which DBT was dissolved facilitated the contact with biocatalyst, oil, water, and biocatalyst can be fully mixed by agitation and emulsified into microdroplets at the oil–water interface. Consequently, DBT was biodesulfurized into 2-HBP at the oil–water interface. The adsorption capacity of the bioregenerated adsorbent is 85% that of the fresh one. After bioregeneration, the adsorbent can be reused. However, the particle size of cells was nearly similar to that of desulfurization adsorbents, which is about several microns. Therefore, it was difficult to separate regenerated adsorbents from cells. Superparamagnetism is an efficient method to separate small particles. Thus, to solve the problem of separation of cells from adsorbents, magnetite NP-modified *P. delafieldii* R-8 cells can be used in the bioregeneration of adsorbents. Li et al. (2008b) reported in another study the bioregeneration of desulfurization adsorbents AgY zeolite with magnetic *P. delafieldii* R-8 cells. The adsorption capacity of the bioregenerated adsorbent was 93% that of the fresh one. The magnetic cells were easily separated from the bioregenerated adsorbents with an external magnetic field and reused.

Zhang et al. (2007) studied the effect of surface modification of γ-Al_2O_3 NPs with gum arabic (GA) and its applications *in-situ* coupling of ADS and BDS of a model diesel oil (n-octane containing 2 mM of DBT). The results showed that γ-Al_2O_3 NPs dispersed well in aqueous solutions after GA modification. The ADS capacity of γ-Al_2O_3 NPs increased by 1.12-fold after its modification by GA. The adsorption of unmodified γ-Al_2O_3 NPs onto *Pseudomonas delafieldii* R-8 increased the BDS efficiency from 14.5 to 17.8 mmol/kg/h, which is further increased to 25.7 mM/kg/h, upon applying GA-modified γ-Al_2O_3 NPs, where GA improved the dispersion and biocompatibility of γ-Al_2O_3 NPs. Not only this, but it also increased the affinity of γ-Al_2O_3 NPs to the cells and lowered its toxicity. In a further study, Zhang et al. (2008) investigated the effect of different NPs (γ-alumina, molecular sieves, and activated carbon) on the *in-situ* coupling of ADS and BDS using *Pseudomonas delafieldii* R-8 cells. The Na–Y molecular sieves restrained the activity of R-8 cells, and the activated carbon could not desorb the adsorbed DBT. Thus, they are not applicable to in situ coupling ADS and BDS. But, the γ-Al_2O_3 NPs adsorbed quickly DBT from the oil phase, then desorbed it, and finally transferred it to R-8 cells for BDS, which consequently increased the desulfurization rate. However, Li et al. (2009) developed a simple and effective technique by integrating the advantages of magnetic separation and cell immobilization for BDS of 2.5-mM DBT in n-dodecane using *R. erythropolis* LSSE8–1. The Fe_3O_4 NPs were found to be strongly adsorbed on the surfaces of the microbial cells, because of the large specific surface area and the high surface energy of the NPs. The Fe_3O_4 NPs on the cell surface were not washed out by deionized water, ethanol, saline water (0.85 wt.%), or phosphate buffer (0.1 M, pH 7). Thus, there was a little cell loss. The coated and free cells exhibited similar time courses and desulfurizing activities during the reaction, suggesting that the coated cells did not experience a mass transfer problem. The coated LSSE8-1 cells completely desulfurized 2.5-mM DBT within 10 h, and the desulfurization rate was 14.1-μM DBT/g DCW/h on the first 4 h. The magnetic-separated/immobilized cells were successfully reused for BDS over seven batch cycles, retaining at least 80% of the specific desulfurization activity of the fresh magnetic modified cells. By comparison, the uncoated free cells could be used only once, since the recycled free cells retained only 15% activity, which confirmed that it is not economic to reuse free cells.

MNPs not only allow recovery of bacterial cells, but also may have an important role in cell membrane permeability. Ansari et al. (2009) clearly demonstrated that Fe_3O_4 NPs actually facilitate transport of DBT from the media to the cytoplasm of biodesulfurizing microorganisms due to the self-assembly of the NPs inside the membrane, which may form pore-like structures that increases the surface conductance. In that study, Ansari et al. (2009) used a widely studied Gram-positive desulfurizing bacterial strain, *R. erythropolis* IGTS8, in order to investigate the effects of decorating the bacteria with magnetic Fe_3O_4 NPs on BDS of 0.5-mM DBT. The ratio of NP mass to biomass was 1.78 (w/w). It was found that the decorated cells had 56% higher DBT desulfurization activity than that of the nondecorated cells. The observation of significantly increased 2-HBP production in the decorated cells suggested that the MNPs might have facilitated the transfer of the produced 2-HBP out of the cells, assuming that it was produced in the cytoplasm. A possible mechanism for the enhancement is that the NPs bounded to the bacteria made their membranes more permeable. MNP-decorated *Rhodococcus* also facilitated its recovery and reuse; hence, it offers a number of advantages for industrial applications compared to nondecorated cells.

Further, Zhang et al. (2011) enhanced the BDS activity by assembling nano-γ-Al_2O_3 particles on the magnetic immobilized *R. erythropolis* LSSE8-1-vgb. First, the cells were coated with Fe_3O_4 NPs (50:1 g cells /g NPs). Then, nano γ-Al_2O_3 adsorbents were assembled onto the magnetically coated cells to enhance the desulfurization activity. The nano γ-Al_2O_3 adsorbent had large pore volume as well as specific surface area and strong electrostatic interaction with microbial cell. The activity of such magnetic immobilized cells assembled with nano γ-Al_2O_3 was tested for the desulfurization of model oil (*n*-octane) 1/2 (oil–water phase ratio) contained 2-mM DBT in model oil, which was completely removed within 9 h. The total concentration of DBT and its metabolite (2-HBP) in oil phase kept constant, which meant that all DBT and 2-HBP remained in oil phase. It was proved in that study that the large pore volume and specific surface area of the adsorbent can more easily improve the adsorption of DBT from the oil phase. Moreover, the interaction between cells (with negative charge) and adsorbent also strongly affects the desulfurization activity. Electronic force is an important factor for the interaction force between cells and adsorbent. The desulfurization rate was raised by nearly 20% when the amount ratio of magnetic particles to nano γ-Al_2O_3 was 1:5 (g/g). The activity of magnetic immobilized cells assembled with adsorbents kept nearly 20% higher than that of magnetic cells alone and decreased less than 10% throughout three successive recycles.

Also, the size and stability of the NPs in a suspension are important issues that need deep investigation since they could influence the process performance. According to a study performed by Bardania et al. (2013), coating of NPs on the cell membrane of two desulfurizing *R. erythropolis* FMF and

R. erythropolis IGTS8 was more effective when smaller NPs of Fe_3O_4 (<5 nm) were used. Additionally, the use of glycine decreased the aggregation tendency of the NPs, which led to a higher adsorption of DBT by biodesulfurizing microorganism, whereas the BDS efficiency of both MNP-coated cells were 70% and 73%, respectively, and were not significantly different from those of the free bacterial cells (67% and 69%, respectively). Separation of magnetic-decorated bacterial cells from the suspension using a magnet and evaluation of desulfurization activity of separated cells showed that Fe_3O_4 NPs provided a high efficiency recovery, and the reused MNP-coated bacterial cells maintained efficiently their desulfurization activity.

Etemadifar et al. (2014) reported that oleate-modified MNP-coated *R. erythropolis* R1 and glycine-modified MNP-coated R1 expressed nearly the same BDS activity of free cells in the biphasic system of model oil (6.76 mM DBT in *n*-tetradecane).

Dai et al. (2014) improved the BDS process by using the combination of magnetic nano Fe_3O_4 particles with calcium alginate-immobilized *Brevibacterium lutescens* CCZU12-1 cells. The combination of magnetic nano-Fe_3O_4 particles with calcium-alginate-immobilized cells appeared not to experience a mass transfer problem, where the immobilized and free cells exhibited the similar time BDS courses of model oil (2.5-mM DBT in *n*-octane) in a biphasic system of 1:9 oil/water (v/v). The DBT was completely desulfurized into 2-HBP without any accumulation of 2-HBP in the cells. Furthermore, the Fe_3O_4 MNP- and calcium-alginate-immobilized cells have been reused for four times without a significant loss in BDS activity. At the end of each batch, the immobilized cells were collected by application of a magnetic field and then reused in another batch. Compared to the immobilized cells, free cells have been used only once. Due to the high surface energy and the large specific surface area of MNPs, they were strongly adsorbed onto the surfaces of microbial cells and thus efficiently coated the cells. Additionally, calcium alginate prevented the loss of cells and Fe_3O_4 MNPs particles. Thus, such combination would solve some of BDS problems, such as the troublesome process of recovering desulfurized oils, biocatalyst separation, and the short life of biocatalysts.

Taguchi optimization of DBT-BDS by *R. erythropolis* R1 in a biphasic system (1:2 O/W with 1-mM DBT in *n*-tetradecane) indicated that the highest BDS efficiency (≈ 81%) could be achieved at 20% (w/w) of γ-Al_2O_3 NPs, alginate beads size equal to 1.5 mm, 1% (w/v) of the alginate, and 0.5% (v/v) of span 80 surfactant, within 20 h (Derikvand et al., 2014). The impact of two cofactor precursors (10-mM nicotinamide and 40-mM riboflavin) on the long-term BDS efficiency was also examined. The related statistical analysis showed that the concentration of γ-Al_2O_3 NPs was the most significant factor in the BDS process. Moreover, the addition of nicotinamide and riboflavin significantly decreased the biocatalytic inactivation of the immobilized cell system after successive operational steps enhancing the BDS efficiency by more than 30% after four successive cycles. So, it can be concluded that a combination of the γ-Al_2O_3 NPs with alginate-immobilized cells could be very effective in the BDS process.

The enhancement of the anaerobic kerosene BDS using NP-decorated *Desulfobacterium indolicum* at 30 °C and atmospheric pressure brought about a decrease in the sulfur content from 48.68 to 13.76 ppm within 72 h (Kareem, 2014).

Mesoporous materials in the nanoscale (2–50 nm) are characterized by large surface area, ordered pore structures with suitable sizes, and high chemical and physical stability and can be chemically modified with functional groups for the covalent binding of molecules, which make them very applicable in adsorption, separation, catalytic reactions, sensors, and immobilization of huge biomolecules such as enzymes (Li and Zhao, 2013; Juarez-Moreno et al., 2014). Enzymatic immobilization in nanomaterials has some advantages: increase in catalytic activity, reduction in protein miss-folding, thermal stability, increase in solvent tolerance, and reduction in denaturation caused by the presence of organic solvents and mixtures of water–organic solvent molecules (Carlsson et al., 2014). Also, the immobilization (i.e., conjugation) of chloroperoxidase (CPO) enzyme with latex NPs after its functionalization with chloromethyl groups can react with the amine groups of the protein (i.e., the enzyme) to form stable covalent bonds. The average size of the conjugated NPs reported to be 180 nm that can be applied in bio-oxidative of fuel streams, followed by physical or chemical remove of the produced sulfoxide and sulfone to produce ultra-low-sulfur fuels. Vertegel (2010) reported the immobilization of CPO (as a good candidate for oxidative desulfurization in nonaqueous media) and myoglobin (which, for its availability, represents a good candidate for industrial scale) with La-NPs (40 nm) that was modified with chloromethyl groups. The conjugated and free enzymes were used for DBT oxidation in the presence of hydrogen peroxide in nonaqueous system (DBT in hexane) and two-phase system (DBT in acetonitrile/water; 20% v/v). For CPO, in the two-phase system and nonaqueous one, free and conjugated enzymes expressed nearly the same activity. But the stability of free enzymes was much lower than conjugated ones, where its activity deteriorated much faster. The activity of conjugated CPO was approximately five times higher than that of free cells after two days of storage. However, the free myoglobin expressed better performance than the immobilized ones in two-phase systems, while the immobilized myoglobin expressed twice the activity in the nonaqueous system, which is an encouraging result, since this protein as well as similar heme-containing protein, hemoglobin, can be readily available in large quantities from animal processing industries. The nanoimmobilized enzymes were easily separated and were stable over time. Thus, it can be taken as a good prospect for a recyclable desulfurization agent.

Juarez-Moreno et al. (2015) developed the oxidative transformation of DBT by CPO enzyme immobilized onto (1D)-γ-Al_2O_3 nanorods (Al-nR). A great affinity between the CPO and the nanorods was observed, due to favorable electrostatic interactions during the immobilization process. In a reaction mixture containing 1.0×10^{-7} M of DBT and 1.3×10^{-12} M of CPO in 1 mL, the maximal biotransformation of DBT by free CPO was 42.2% after 120 min. At the same time, CPO immobilized onto Al-nR thermally treated at 700 C showed a maximal transformation of 89.6% of initial DBT into DBT-sulfoxide, while CPO immobilized in Al-nR thermally treated at 500 and 900 °C only transformed 65% and 69.8% of initial DBT, respectively.

Nasab et al. (2015) improved the BDS of model oil (1-mM DBT in dodecane) by assembling spherical mesoporous silica nanosorbents on the surface of *R. erythropolis* IGTS8, where the maximum specific desulfurization activity in terms of DBT consumption rate and 2-HBP production were 0.34 μM/g DCW/min and 0.126 μM 2-HBP/g DCW/min with an increase of 19% and 16% relative to that of free cells, respectively.

Nassar (2015) studied the BDS of diesel oil by using Fe_3O_4 MNP-coated cells (see Figure 6.6), magnetically alginate immobilized cells (see Figure 6.6) and agar-immobilized cells of; *Brevibacillus invocatus* C19 and *R. erythropolis* IGST8. BDS efficiency of the diesel oil (total sulfur content 8,600 mg/L) was investigated at different ratio of O/W (10%, 25%, 50%, and 75% v/v). Where, 10% and 25% (v/v) phase ratio were considered as the optimum O/W phase ratio for *R. erythropolis* IGTS8 and *Brevibacillus invocatus* C19, respectively. The Fe_3O_4 MNP-coated IGTS8 expressed slightly higher BDS efficiency than free cells, recording 78.26% and 74.42%, within six and seven days, respectively. However, the magnetic alginate immobilized IGTS8 and agar-immobilized IGTS8 expressed nearly the same BDS efficiency, recording approximately 57.26% and 58.98%, respectively, within seven days. The agar-immobilized cells and MNP-coated cells of *Brevibacillus invocatus* C19 showed the highest BDS efficiency of approximately 98.97% and 98.69%, respectively, at 25% (v/v) O/W phase ratio, within seven and four days, respectively, while the free C19 and magnetic-alginate-immobilized C19 recorded approximately 91.31% and 89.99% BDS, respectively, within seven days. Moreover, the gas chromatography flame photometric detector proved that the BDS of DBT from diesel oil by agar-immobilized C19 and coated C19 was much higher, recording 93.71% and 92.86%, respectively, and the BDS of BT was 91.42% and 88.76%, respectively, at 25% v/v optimum O/W phase ratio, while the BDS of DBT and BT by coated IGTS8 were 58.63% and 32.98%, respectively, at optimum 10% v/v O/W phase ratio. The coated cells of IGTS8 showed a low capability to desulfurize 4-MDBT (%BDS 25.34%) and 4,6-DMDBT (%BDS 41.56%) from diesel oil compared to coated cells of C19 (%BDS 83.04%) and agar-immobilized cells of C19 (%BDS 87.16%). The effect of microbial treatment on the hydrocarbon skeleton of the total resolvable components of the diesel oil at different ratio of O/W was also studied using gas chromatography flame ionization detector analysis, which demonstrated that over all the studied O/W ratios, the alkane structure and content of the diesel was not significantly altered during the BDS using C19 compared to that BDS *R. erythropolis* IGTS8, where they were highly affected. The biodegradation capacity of total petroleum hydrocarbons in diesel oil by agar-immobilized C19 and coated C19 was found to be negligible, 10.05% and 11.73% at optimum O/W phase ratio (25% v/v), respectively, while that for IGTS8 recorded 34.98% and 34.25% at optimum O/W phase ratio (10% v/v), respectively. The C19 MNP-coated and magnetically alginate immobilized cells sustained their BDS efficiency up to four and three successive batch cycles, respectively, while those in case of IGTS8 sustained up to five and three successive cycles, respectively. But the free cells are not recommended to be economically reused, as they retained only approximately 53% and 26% of its initial activity, in the second BDS batch cycle, respectively. The magnetization facilitated the biocatalyst recovery and reuse. Moreover, it promoted the long life stability of the biocatalyst, as the BDS efficiency of MNP-coated cells and magnetically alginate immobilized cells remained nearly constant after their storage for 30 days at 4 °C. But the free cells lost approximately 43% of its BDS capacity after storage for 30 days at 4 °C. Hence, it offers a number of advantages for industrial applications compared to free cells. It should be noted that due to the high mechanical strength of agar, the BDS efficiencies of both C19 and IGTS8 agar alginate cells were sustained for five successive BDS batch cycles and after storage for 30 days at 4 °C. That study proved the importance of isolating new biodesulfurizing strains with efficient BDS capacity and higher hydrocarbon tolerance. Moreover, it recommends the agar-immobilized cells and MNP-coated cells as good potential candidates for industrial applications in the BDS of diesel oil.

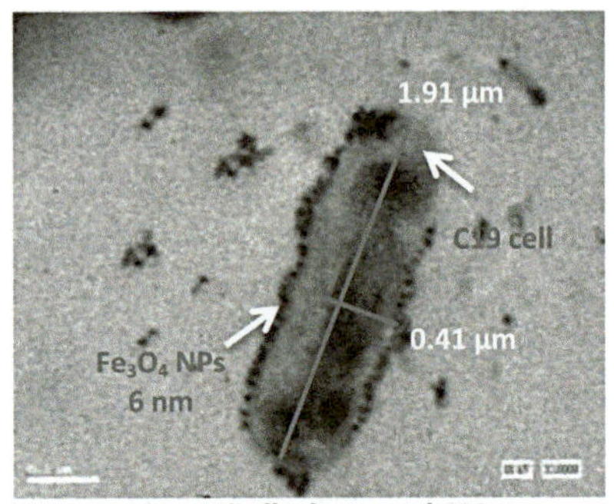

Magnetically decorated C19

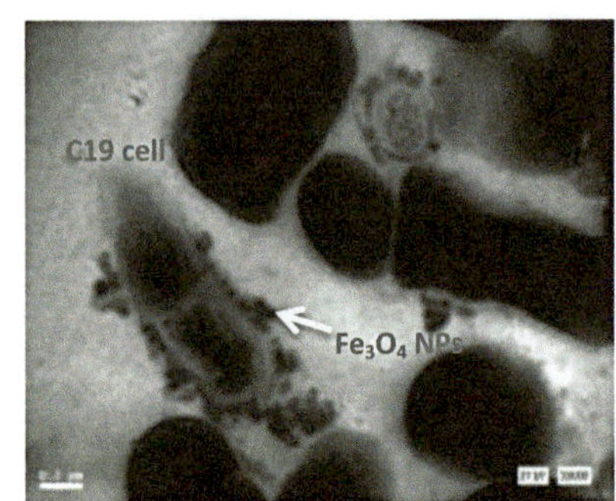

Magnetically alginate immobilized C19

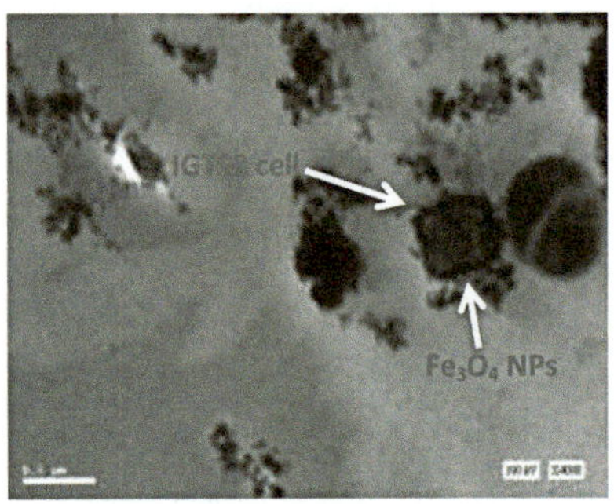

Magnetically decorated IGTS8

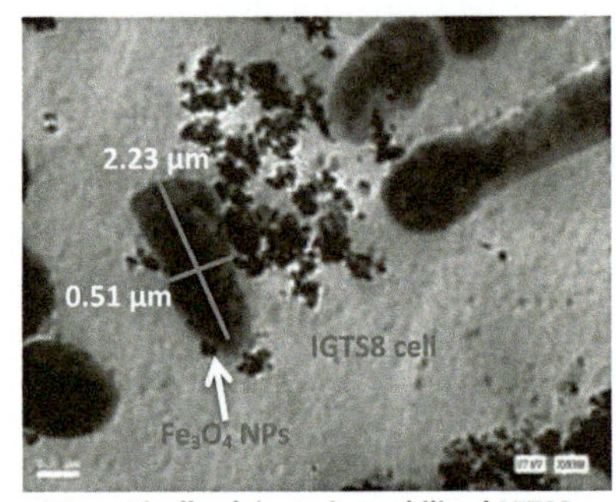

Magnetically alginate immobilized IGTS8

FIGURE 6.6 TEM images at different magnification power, showing the biocompatibility of Fe_3O_4 MNPs with different biodesulfurizing bacteria both in free and entrapped immobilized ones.

Karimi et al. (2017) enhanced the BDS of 0.5-mM DBT in a stirring bioreactor (working volume 5 L) inoculated by decorated magnetic Fe_3O_4 NP *R. erythropolis* IGTS8. The field emission scanning electron microscope and TEM images showed the size of the NPs averaged 7–8 nm. Production of 2-HBP by decorated cells was significantly different than by the free cells. Gibbs' assay results showed that production of 2-HBP by decorated cells was 0.210 mM at 148 h, while 2-HBP production by nondecorated cells was 0.182 mM at 96 h. Thus, the decorated cells had a 15.80% higher desulfurization capacity compared to the nondecorated cells. These experimental results can be related to the favorable production of 2-HBP, as the NP avoids the accumulation of 2-HBP in the cytoplasm and periplasmic space. This study suggests that the addition of NPs into a BDS medium does not only overcome most of the BDS process limitations, but can also avoid the 2-HBP inhibition effects. A possible mechanism for the enhancement is that the NPs bound to the bacteria make their membranes more permeable. In a further study

DBT-BDS using *R. erythropolis* IGTS8 and *Gordona rubropertinctus* PTCC 1604 decorated by polyethylene glycol (PEG) modified carbon nanotube (CNT) increased by 12% and 16%, respectively (Karimi et al., 2018a). That was attributed to the presence of the nontoxic modified CNT, as it increased the availability of DBT (Sadare and Daramola, 2019) and thus enhanced its transfer into the cells and facilitated the release of 2-HBP outside the cells, which was under the pretentious of being produced in the cytoplasm, where CNT increased the permeability of the membrane. In a later study, Karimi et al. (2018b) studied the DBT-BDS using *R. erythropolis* IGTS8 decorated by 7–8-nm Fe_3O_4 NPs and PEG-modified CNT. The microbial growth enhanced by 40% and 8% and the BDS increased by 10% and 8%, respectively.

Rahpeyma et al. (2017) examined the effect of Fe_3O_4, ZnO, and CuO NPs as free factors on the BDS. The bacterial strains used as the biological catalysts were *R. erythropolis* IGTS8 and *Pseudomonas aeroginusa* PTSOX4. The utilization of PTSOX4 as the biocatalyst caused more sulfur removal rather than that of IGTS8. Assimilation of ZnO NPs expressed considerable higher desulfurization efficiency, recording approximately 94% and 58% for *P. aeroginusa* PTSOX4 and *R. erythropolis* IGTS8, relative to free cells, which recorded 68% and 48%, respectively. The observed significant increase in 2-HBP production in the presence of ZnO NPs was attributed to the permeabilization of cell membranes, which facilitated the mass transferring of DBT and 2-HBP into or out of bacterial cells, while Fe_3O_4 NPs showed almost no effect on BDS yield recording 70% for PTSOX4 and 49% for IGTS8. But, PTSOX4 and IGTS8 showed a decrease in BDS activity in the presence of CuO NPs. Diminished BDS activities of PTSOX4 and IGTS8 in the presence of CuO was attributed to its greater particle size as well as the possible toxicity of these NPs.

Canales et al. (2018) proved that *Rhodococcus rhodochorus* IGTS8 immobilized onto silica improved DBT-BDS efficiency by approximately 20% relative to the free cells. Simulation based on liquid-film kinetic model predicted considerable amounts of DBT surrounding the particles which made it more available for the immobilized cells, thus enhanced the BDS. Thermophilic DBT-BDS by *Bacillus thermoamylovorans* has been enhanced by approximately 10% and 22%, upon the cell decoration using 20 nm starch/Fe_3O_4 and 30–40 nm starch/Fe NPs, respectively (Etemadi et al., 2018). In another study, the DBT-BDS by *R. erythropolis* IGTS8 of coated with 13-nm Fe_3O_4 NPs was higher than that of free cells (Rahpeyma et al., 2018). Further decoration of IGTS8 with 241 magnetite–silica nanocomposites led to a higher BDS capacity relative to the MNP-decorated cells. Since the positively charged magnetite–silica nanocomposites were more binded to the negatively charged bacteria than negative magnetite NPs. Upon the application of an external magnetic field in a batch two-phase BDS process, the BDS efficiency increased by approximately 23% (Rahpeyma et al., 2018). In a later study by Sabaghian et al. (2019,2020) carbon nanotube functionalized with starch (starch/CNT) proved to enhance the adsorption and oxidation of DBT. Thus, improved the BDS efficiency of *R. erythropolis* IGTS8, and led to a prolongation in the logarithmic microbial growth stage. Bardania et al. (2019) proved that glycine modified Fe_3O_4 NP-decorated *R. erythropolis* IGTS8 and *R. erythropolis* FMF can be reused for five successive times without losing their BDS activities.

6.3 APPLICATION OF NANOTECHNOLOGY IN THE BDN PROCESS

The entrapped *Sphingomonas* sp. XLDN2-5 in a mixture of Fe_3O_4 NPs and gellan gum expressed higher CAR-biodegradation (3479 μg CAR/g wet weight cell/h) than the nonmagnetically immobilized (1761 μg CAR/g wet weight cell/h and free (3092 μg CAR/g wet weight cell/h XLDN2-5 and can be used for eight successive cycles (Wang et al., 2007), where the specific degradation rate increased from 3479 to 4638 μg CAR/g wet weight cell/h in the eighth cycle, due to the good growth of cells in the magnetic gellan gel beads. But, the recorded decrease in CAR biodegradation rate within the nonmagnetically immobilized matrix is attributed to the mass transfer limitation and steric hindrance, while the increase in the magnetically immobilized matrix is due to the presence of MNPs, which loosen the binding of the sheets of the gellan gum matrix and the existence of many pores between the sheets of gellan gum matrix. Li et al. (2013) and Kafayati et al. (2013) reported that the magnetite iron oxide NPs have negligible toxicity on the living bacterial cells. The *Sphingomonas* sp. XLDN2-5 cells/Fe_3O_4 biocomposite and free cells reported to exhibit the same CAR biodegradation efficiency (Li et al., 2013). That was attributed to the biocompatibility of MNPs, that is, the coating layer itself, as it does not change the hydrophilicity of the cell surface. Moreover, this coating layer has a negligible effect on mass transfer, as its structure is looser than that of the cell wall. Thus, microbial cell/Fe_3O_4 biocomposite produces a system that is not limited by diffusional limitations. The activity of microbial cell/Fe_3O_4 biocomposite increased gradually during the recycling process, where complete removal of 3500-μg CAR occurred within 9 h for six successive cycles, but the same amount was completely removed in only 2 h within the seventh to the tenth cycles. Mehndiratta et al. (2014) proved that coating the biosurfactant producer *Pseudomonas* sp. GBS.5 with magnetite

NPs enhanced CAR biodegradation recording 1.4 ppm/min compared to the free cells, which recorded 0.32 ppm/min. The increase in membrane permeability because of the magnetite coating doubled the BDN rate and overcome the mass transfer limitation. Because of the ease of separation by applying external magnetic field, coated cells used for five successive cycles with a decrease in efficiency of only 23%, but the free cells used for only once as it showed only 10% CAR-BDN in the second run. Coating *Bacillus clausii* BS1with magnetic Fe_3O_4 NPs enhanced the rate of CAR-BDN (Zakaria et al. 2015b), whereas complete removal of 1000 mg/L CAR has been achieved by coated cells and 94.25% removal occurred with free cells with $t_{1/2}$ values of 31.36 and 64.78 h, respectively. Coated BS1 cells are characterized by higher storage and operational stabilities and low sensitivity toward toxic by-products and can be reused for four successive cycles without losing its BDN efficiency and have the advantage of magnetic separation, which would resolve many operational problems in petroleum refinery.

Sun et al. (2017) studied the negative effect of assembling of Fe_3O_4 NPs onto the cell surface of *Sphingobium yanoikuyae* XLDN2-5 on CAR-BDN. The highest CAR-BDN occurred in batches of free microbial cells and microbial cell/Fe_3O_4 NP biocomposite at the ratio of 1:1 (w/w, dry mass of Fe_3O_4 NPs: wet mass of cells), where 3500-μg CAR was completely removed within 7 h. Further increase in the Fe_3O_4/cell ratio decreased the BDN efficiency.

REFERENCES

Abd Al-Khodor, Y.A., and Albayati, T.M. 2020. Employing sodium hydroxide in desulfurization of the actual heavy crude oil: Theoretical optimization and experimental evaluation. *Process Saf. Environ. Prot.* 136: 334–342.

Abid, M.F., Ahmed, S.M., AbuHamid, W.H., and Ali, S.M. 2019. Study on novel scheme for hydrodesulfurization of middle distillates using different types of catalyst. *J. King Saud Univ. Eng. Sci.* 31: 144–151.

Adlakha, J., Singh, P., Ram, S.K., Kumar, M., Singh, M.P., Singh, D., Sahai, V., and Srivastava, P. 2016. Optimization of conditions for deep desulfurization of heavy crude oil and hydrodesulfurized diesel by *Gordonia* sp. IITR100. *Fuel.* 184: 761–769.

Agarwal, P., and Sharma, D.K. 2010. Comparative studies on the bio-desulfurization of crude oil with other desulfurization techniques and deep desulfurization through integrated processes. *Energ. Fuel.* 24(1): 518–524.

Al-Jailawi, M.H., Al-Faraas, A.F., and Yahia, A.I. 2015. Isolation and identification of dibenzothiophene biodesulfurizing bacteria. *Am. J. Biosci. Bioeng.* 3(5): 40–46.

Alkhalili, B.E., Yahya, A., Abrahim, N., and Ganapathy, B. 2017. Biodesulfurization of sour crude oil. *Mod. Appl. Sci.* 11(9): 104–113.

Alves L., Paixao S.M., Pacheco R., Ferreira A.F., and Silva C.M. 2015. Biodesulfurization of fossil fuels: Energy, emissions and cost analysis. *RSC Adv.* 5: 34047–34057.

Ansari, F., Grigoriev, P., Libor, S., Tothill, I. E., and Ramsden, J.J. 2009. DBT degradation enhancement by decorating R. erythropolis IGST8 with magnetic Fe3O4 nanoparticles. *Biotechnol. Bioeng.* 102(5): 1505–1512.

Asumana, C., Yu, G., Guan, Y., Yang, S., Zhou, S., and Chen, X. 2011. Extractive denitrogenation of fuel oils with dicyanamide-based ionic liquids. *Green Chem.* 13: 3300–3305.

Bardania, H., Raheb, J., and Arpanaei, A. 2019. Investigation of desulfurization activity, reusability, and viability of magnetite coated bacterial cells. *Iran. J. Biotechnol.* 17(2): e2108. DOI: 10.21859/ijb.2108

Bardania, H., Raheb, J., Mohammad-Beigi, H., Rasekh, B., and Arpanaei, A. 2013. Desulfurization activity and reusability of magnetite nanoparticle-coated *R. erythropolis* FMF and *R. erythropolis* IGTS8 bacterial cells. *Biotechnol. Appl. Biochem.* 60: 323–329.

Benedik, M.J., Gibbs, P.R., Riddle, R.R., and Willson, R.C., 1998. Microbial denitrogenation of fossil fuels. *Trends Biotechnol.* 16: 390–395.

Bhadra, B.N., and Jhung, S.H., 2019. Oxidative desulfurization and denitrogenation of fuels using metal-organic framework-based/-derived catalysts. *Appl. Catal. B. Environ.* 259: 118021.

Bhatia, S., and Sharma, D.K. 2010. Biodesulfurization of dibenzothiophene, its alkylated derivatives, and crude oil by a newly isolated strain of *Pantoea agglomerans* D23W3. *Biochem. Eng. J.* 50(3): 104–109.

Bhatia, S., and Sharma, D.K. 2012. Thermophilic desulfurization of dibenzothiophene and different petroleum oils by *Klebsiella* sp. 13T. *Environ. Sci. Pollut. Res.* 19(8): 3491–3497.

Bordoloi, N.K., Rai, S.K., Chaudhuri, M.K., and Mukherjee, A.K. 2014. Deep-desulfurization of dibenzothiophene and its derivatives present in diesel oil by a newly isolated bacterium *Achromobacter* sp. to reduce the environmental pollution from fossil fuel combustion. *Fuel Process. Technol.* 119: 236–244.

Canales, C., Eyzaguirre, J., Baeza, P., Aballay, P., and Ojeda, J. 2018. Kinetic analysis for biodesulfurization of dibenzothiophene using *R. rhodochrous* adsorbed on silica. *Ecol. Chem. Eng. S.* 25(4): 549–556.

Carlsson, N., Gustafsson, H., Thorn, C., Olsson, L., Holmberg, K., and kerman, B. 2014. Enzymes immobilized in mesoporous silica: A physical–chemical perspective. *Adv. Colloid Interf. Sci.* 205: 339–360.

Chen, H., Cai, Y., Zhang, W., and Li, W. 2009. Methoxylation pathway in biodesulfurization of model organosulfur compounds with *Mycobacterium* sp. *Bioresour. Technol.* 100: 2085–2087.

Dai, Y., Shao, R., Qi, G., and Ding, B.B. 2014. Enhanced dibenzothiophene biodesulfurization by immobilized cells of *Brevibacterium lutescens* in n-octane–water biphasic system. *Appl. Biochem. Biotechnol.* 174(6): 2236–2244.

Derikvand, P., Etemadifar, Z., and Biria, D. 2014. Taguchi optimization of dibenzothiophene biodesulfurization by *Rhodococcus erythropolis* R1 immobilized cells in a biphasic system. *Int. Biodeter. Biodegr.* 86: 343–348.

Duarte G.F., Rosado A.S., Seldin L., de Araujo W., and van Elsas J.D., 2001. Analysis of bacterial community structure in sulfurous-oil-containing soils and detection of species carrying dibenzothiophene desulfurization (dsz) genes. *Appl. Environ. Microbiol.* 67: 1052–1062.

El-Gendy, N.Sh., and Nassar, H.N., 2018. *Biodesulfurization in Petroleum Refining*. Wiley, Hoboken, NJ, USA. https://doi.org/10.1002/9781119224075

El-Gendy, N.Sh., and Speight J.G. 2016. *Handbook of Refinery Desulfurization.* CRC Press, Boca Raton, FL, USA. https://doi.org/10.1201/b19102

El-Gendy, N.Sh., Farahat, L.A., Moustafa, Y.M., Shaker, N., and El-Temtamy, S.A. 2006. Biodesulfurization of crude and diesel oil by *Candida parapsilosis* NSh45 isolated from Egyptian hydrocarbon polluted sea water. *Biosci. Biotechnol. Res. Asia*. 3(1a): 5–16. http://www.biotech-asia.org/?p=33072

El-Gendy, N.Sh., Nassar, H.N., and Abu Amr, S.S. 2014. Factorial design and response surface optimization for enhancing a biodesulfurization process. *Petrol. Sci. Technol.* 32(14): 1669–1679. https://doi.org/10.1080/10916466.2014.892988

Etemadi, N., Sepahy, A.K., Mohebali, G., Yazdian, F., Omidi, M. 2018. Enhancement of bio-desulfurization capability of a newly isolated thermophilic bacterium using starch/iron nanoparticles in a controlled system. *Int. J. Biol. Macromol.* 120: 1801–1809.

Etemadifar, Z., Derikvand, P., Emtiazi, G., and Habibi, M.H. 2014. Response surface methodology optimization of dibenzothiophene biodesulfurization in model oil by nanomagnet immobilized *Rhodococcus erythropolis* R1. *J. Mater. Sci. Eng. B.* 4(10): 322–330.

Fadhil, A.M.A., Al-Jailawi, M.H., and Mahdi, M.S., 2014. Isolation and characterization of a new thermophilic, carbazole degrading bacterium (*Anoxybacillus rupiensis*) Strain Ir3 (JQ912241). *Int. J. Adv. Res.* 2: 795–805.

Fallon, R.D., Hnatow, L.L., Jackson, S.C., and Keeler, S.J., 2010. Method for identification of novel anaerobic denitrifying bacteria utilizing petroleum components as sole carbon source. US patent 7740063 B2.

Feng, J., Zeng, Y., Ma, C., Cai, X., Zhang, Q., Tong, M., Yu, B., and Xu, P., 2006. The surfactant Tween 80 enhances biodesulfurization. *Appl. Environ. Microbiol.* 72:7390–7393.

Garcia-Montoto, V., Verdier, S., Maroun, Z., Egeberg, R., Tiedjec, J.L., Sandersen, S., Zeuthen, P., and Bouyssiere, B. 2020. Understanding the removal of V, Ni and S in crude oil atmospheric residue hydrodemetallization and hydrodesulfurization. *Fuel Process. Technol.* 201: 106341.

Guobin S., Huaiying Z., Weiquan C., Jianmin X., and Huizhou L. 2005. Improvement of biodesulfurization rate by assembling nanosorbents on the surfaces of microbial cells. *Biophys. J.* 89(6): 58–60.

Guobin, S., Huaiying, Z., Jianmin, X., Guo, C., Wangliang, L., and Huizhou, L. 2006. Biodesulfurization of hydrodesulfurized diesel oil with *Pseudomonas delafieldii* R-8 from high density culture. *Biochem. Eng. J.* 27: 305–309.

Han, L., Feng J., Zhang, S., Ma, Z., Wang, Y., and Zhang, X. 2012. Alkali pretreated of wheat straw and its enzymatic hydrolysis. *Braz. J. Microbiol.* 43(1): 53–61.

Haukanes, B.I., and Kvam C. 1993. Application of magnetic beads in bioassays. *Nat. Biotechnol.* 11: 60–63.

Hou Y., Kong Y., Yang J., Zhang J., Shi D., and Xin W. 2005. Biodesulfurization of dibenzothiophene by immobilized cells of Pseudomonas stutzeri UP-1. *Fuel*. 84: 1975–1979.

Huang T., Qiang L., Zelong W., Daojang Y., and Jiamin X. 2012. Simultaneous removal of thiophene and dibenzothiophene by immobilized *Pseudomonas delafieldii* R-8 cells. *Chin. J. Chem. Eng.* 20: 47–51.

Huh, E.S., Zazybin, A., Palgunadi, J., Ahn, S., Hong, J., Kim, H.S., Cheong, M., and Ahn B.S. 2009. Zn-containing ionic liquids for the extractive denitrogenation of a model oil: A mechanistic consideration. *Energ. Fuel*. 23: 3032–3038.

Irani, Z.A., Yazdian, F., Mohebali, G., Soheili, M., and Mehrnia, M.R. 2011. Determination of growth kinetic parameters of a desulfurizing bacterium, *Gordonia alkanivorans* RIPI90A. *Chem. Eng. Trans.* 24: 937–942.

Ishihara, A., Dumegnil, F., Aoyagi, T., Nishikawa, M., Hosomi, M., Qian, E. W., and Kabe, Y., 2008. Degradation of carabzole by *Novosphingobium* sp. strain NIY3. *J. Jpn. Petrol. Inst.* 51: 174–179.

Jia, Y.H., Li, G., Ning, G.L., and Jin, C.Z. 2009. The effect of N-containing compounds on oxidative desulphurization of liquid fuel. *Catal. Today*. 140: 192–196.

Juarez-Moreno, K., Diaz de Leon, J.N., Zepeda, T.A., Vazquez-Duhalt, R., and Fuentes, S. 2015. Oxidative transformation of dibenzothiophene by chloroperoxidase enzyme immobilized on (1D)-γ-Al_2O_3 nanorods. *J. Mol. Catal. B Enzym.* 115: 90–95.

Juarez-Moreno, K., Pestryakov, A., and Petranovskii, V. 2014. Engineering of supported nanomaterials. *Procedia Chem.* 10: 25–30.

Kafayati, M., Raheb, J., Angazi, M., Alizadeh, S., and Bardania, H., 2013. The effect of magnetic Fe_3O_4 nanoparticles on the growth of genetically manipulated bacterium, *Pseudomonas aeruginosa* PTSOX4. *Iran. J. Biotechnol.* 111: 41–46.

Kareem, S.A. 2014. Anaerobic microbial desulfurization of kerosene. In *4th International Conference on Nanotek & Expo*, December 01–03, 2014 San Francisco, USA.

Karimi, E., Jeffryes, C., Yazdian, F., Akhavan Sepahi, A., Hatamian, A., Rasekh, B., and Ashrafi, S.J. 2017. DBT desulfurization by decorating *Rhodococcus erythropolis* IGTS8 using magnetic Fe_3O_4 nanoparticles in a bioreactor. *Eng. Life Sci.* 17(5): 528–535.

Karimi, E., Yazdian, F., Rasekh, B., Akhavan S.A., Rashedi, H., Sheykhha, M.H., Haghiroalsadat, B.F., and Hatamian, A.S. 2018b. Biodesulfurization of dibenzothiophene by *Rhodococcus erythropolis* IGTS8 in the presence of magnetic nanoparticles and carbon nanotubes surface-modified polyethylene glycol. *Modares J. Biotechnol.* 9(2): 301–308.

Karimi, E., Yazdian, F., Rasekh, B., Jeffryes, C., Rashedi, H., Sepahi, A.A., Shahmoradi,S., Omidi, M., Azizi, M., Bidhendi, M.E., and Hatamian, A. 2018a. DBT desulfurization by decorating bacteria using modified carbon nanotube. *Fuel*. 216: 787–795.

Karsten, G., and Simon, H. 1993. Immobilization of Proteus vulgaris for the reduction of 2-oxo acids with hydrogen gas or formate to D-2-hydroxy acids. *Appl. Microbiol. Biotechnol.* 38: 441–446.

Kayser, K.J., and Kilbane, J.J. II, 2004. Method for metabolizing carbazole in petroleum. US Patent 6943006.

Kilbane J.J. 2006. Microbial biocatalysts developments to upgrade fossil fuels. *Curr. Opinion Biotechnol.* 17: 305–314.

Kitashov, Y.N., Nazarov, A.V., Zorya, E.I., and Muradov, A.V. 2019. Alternative methods for the removal of sulfur compounds from petroleum fractions. *Chem. Technol. Fuels Oil.* 55(5): 584–589.

Laredo, S.G.C., De Los Reyes, J.A., Luis Cano, D,J., and Castillo, M.J.J. 2001. Inhibition effects of nitrogen compounds on the hydrodesulfurization of dibenzothiophene. *Appl. Catal. A: Gen.* 207: 103–112.

Larentis, A. L., Sampaio, H. C. C., Carneiro, C. C., Martins, O. B., and Alves, T. L. M., 2011. Evaluation of growth, carbazole biodegradation and anthranilic acid production by *Pseudomonas stutzeri. Braz. J. Chem. Eng.* 28: 37–44.

Lateef, S.A., Ajumobi, O.O., and Onaiz, S.A., 2019. Enzymatic desulfurization of crude oil and its fractions: A mini review on the recent progresses and challenges. *Arab. J. Sci. Eng.* 44: 5181–5193.

Li, W., and Jiang, X. 2013. Enhancement of bunker oil biodesulfurization by adding surfactant. *World J. Microbiol. Biotechnol.* 29(1): 103–108.

Li, W., Xing J., Li Y., Xiong X., Li X., and Liu H. 2008b. Desulfurization and bio-regeneration of adsorbents with magnetic *P. delafieldii* R-8 cells. *Catal. Commun.* 9: 376–380.

Li, W., and Zhao, D. 2013. An overview of the synthesis of ordered mesoporous materials. *Chem. Commun.* 49: 943–946.

Li, Y., Du X., Wu C., Liu X., Wang X., and Ping, X. 2013. An efficient magnetically modified microbial cell biocomposite for carbazole biodegradation. *Nanoscale Res. Lett.* 8(1): 522. DOI: 10.1186/1556-276X-8-522.

Li, W., Tang, H., Liu, Q., Xing, J., Li, Q., Wang, D., Yang, M., Li, X., and Liu, H. 2009. Deep desulfurization of diesel by integrating adsorption and microbial method. *Biochem. Eng. J.* 44: 297–301.

Li, Y-G., Li, W-L., Huang, J-X., Xiong, X. C., Gao, H. S., Xing, J. M., and Liu, H. Z., 2008a. Biodegradation of carbazole in oil/water biphasic system by a newly isolated bacterium *Klebsiella* sp. LSSE-H2. *Biochem. Eng. J.* 41: 166–170.

Lin, Y., Feng, l., Li, X., Chen, Y., Yin, G., and Zhou, W. 2020. Study on ultrasound-assisted oxidative desulfurization for crude oil. *Ultrason. Sonochem.* 63: 104946.

Maass, D., Todescato, D., Moritz, D.E., Oliveira, J.V., Oliveira, D., Ulson de Souza, A.A., and Guelli Souza, S.M., 2015. Desulfurization and denitrogenation of heavy gas oil by *Rhodococcus erythropolis* ATCC 4277. *Bioprocess Biosyst. Eng.* 38: 1447–1453.

Mehndiratta, P., Jain, A., Singh, G.B., Sharma, S., Srivastava, S., Gupta, S., and Gupta, N. 2014. Magnetite nanoparticle aided immobilization of *Pseudomonas* sp. GBS.5 for carbazole degradation. *J. Biochem. Tech.* 5(4): 823–825.

Martínez, I., El-Said Mohamed, M., Santos, V.E., Garcíaa, J.L., García-Ochoac, F., and Díaz, E. 2017. Metabolic and process engineering for biodesulfurization in Gram-negative bacteria. *J. Biotechnol.* 262: 47–55.

Misra, P., Badoga, S., Dalai, A.K., and Adjay, J. 2018. Enhancement of sulfur and nitrogen removal from heavy gas oil by using polymeric adsorbent followed by hydrotreatment. *Fuel*. 226: 127–136.

Mogollon, L., Rodriguez, R., Larrota, W., Ortiz, C., and Torres, R., 1998. Biocatalytic removal of nickel and vanadium from petroporphyrins and asphaltenes. *Appl. Biochem. Biotechnol.* 70/72: 765–777.

Mohebali, G., and Ball, A.S., 2016. Biodesulfurization of diesel fuels---Past, present and future perspectives. *Int. Biodeter. Biodegr.* 110: 163–180.

Monticello, D.J. 2000. Biodesulfurization and the upgrading of petroleum distillates. *Curr. Opin. Biotechnol.* 11: 540–546.

Mužic, M., and Sertić-Bionda, K., 2013. Alternative processes for removing organic sulfur compounds from petroleum fractions. *Chem. Biochem. Eng. Q.* 27(1): 101–108.

Nasab, N.A., Kumleh, H.H., Kazemzad, M., Panjeh, F.G., and Davoodi-Dehaghani, F. 2015. Improvement of desulfurization performance of *Rhodococcus erythropolis* IGTS8 by assembling spherical mesoporous silica nanosorbents on the surface of the bacterial cells. *J. Appl. Chem. Res.* 9(2): 81–91.

Nassar, H.N. 2015. Development of biodesulfurization process for petroleum fractions using nano-immobilized catalyst; Ph.D. Degree; Al-Azhar University, Cairo, Egypt.

Nazari, F., Kefayati, M.E., and Raheb, J., 2017. The study of biological technologies for the removal of sulfur compounds. *J. Sci. Islam. Repub. Iran.* 28(3): 205–219.

Nehlsen, J.P. 2005. Developing clean fuels: Novel techniques for desulfurization. Ph.D. Thesis, Princeton University, Princeton, NJ, USA.

Pacheco, M.A., Lange, E.A., Pienkos, P.T., Yu, L.Q., Rouse, M.P., Lin, Q., and Linguist, L.K. 1999. Recent advances in biodesulfurization of diesel fuel. In NPRA Annual Meeting, March 21–23, San Antonio, Texas. National Petrochemical and Refiners Association, paper AM-99–27. Pp. 1–26.

Porto, B., Maass, D., Oliveira, J.V., de Oliveira, D., Yamamoto, C.I., Ulson de Souza, A.A., and Ulson deSouza, S.M.A.G. 2017. Heavy gas oil biodesulfurization by *Rhodococcus erythropolis* ATCC 4277: Optimized culture medium composition and evaluation of low-cost alternative media. *J. Chem. Technol. Biotechnol.* 92: 2376–2382.

Porto, B., Maass, D., Oliveira, J.V., de Oliveira, D., Yamamoto, C.I., Ulson de Souzaa, A.A., and Ulson de Souza, S.M.A.G. 2018. Heavy gas oil biodesulfurization using a low-cost bacterial consortium. *J. Chem. Technol. Biotechnol.* 93: 2359–2363.

Rahpeyma, S.S., Dilmaghani, A., and Raheb, J. 2018. Evaluation of desulfurization activity of SPION nanoparticle-coated bacteria in the presence of magnetic field. *Appl. Nanosci.* 8:1951–1972.

Rahpeyma, S.S., Mohammadi, M., and Rehab, J. 2017. Biodesulfurization of dibenzothiophene by two bacterial strains in cooperation with Fe_3O_4, ZnO and CuO nanoparticles. *J. Microb. Biochem. Technol.* 9(2): 587–591.

Rhee, S.K., Lee, K.S., Chung, J.C., and Lee, S.T., 1997. Degradation of pyridine by *Nocardioides* sp. strain OS4 isolated from the oxic zone of a spent shale column. *Can. J. Microbiol.* 43(2): 205–209.

Sabaghian, S., Rasekh, B., Yazdian, F., Shekarriz, M., and Mansour, N. 2019. Effect of starch/CNT on biodesulfurization using molecular dynamic simulation. J. Mol. Model. 25: 352. https://doi.org/10.1007/s00894-019-4236-8.

Sabaghian, S., Yazdian, F., Rasekh, B., Shekarriz, M., and Mansour, N. 2020. Investigating the effect of starch/Fe_3O_4 nanoparticles on biodesulfurization using molecular dynamic simulation. Environ. Sci. Pollut. Res. 27(2): 1667–1676. https://doi.org/10.1007/s11356-019-06453-8

Sada E., Katon S., and Terashima M. 1981. Enhancement of oxygen absorption by magnetite-containing beads of immobilized glucose oxidase. *Biotechnol. Bioeng.* 21: 1037–1044.

Sadare, O.O., and Daramola, M.O. 2019. Adsorptive desulfurization of dibenzothiophene (DBT) in model petroleum distillate using functionalized carbon nanotubes. *Environ. Sci. Pollut. Res.* 26: 32746–32758.

Sadare, O.O., Obazu, F, and Daramola, M.O., 2017. Biodesulfurization of petroleum distillates---Current status, opportunities and future challenges. *Environments*. 4: 85.

Santos, S.C., Alviano, D.S., Alviano, C.S., Padula, M., Leitao, A.C., Martins, O.B., Ribeiro, C.M., Sassaki, M.Y., Matta, C.P., and Bevilaqua, J., 2005. Characterization of *Gordonia* sp. strain F.5.25.8 capable of dibenzothiophene desulfurization and carbazole utilization. *Appl. Microbiol. Biotechnol.* 66: 1–8.

Santos, S.C., Alviano, D.S., Alviano, C.S., Padula, M., Leitao, A.C., Martins, O.B., Ribeiro, C.M., Sassaki, M.Y., Matta, C.P., Bevilaqua, J., Sebastian, G.V., and Seldin, L., 2006. Characterization of *Gordonia* sp. strain F.5.25.8 capable of dibenzothiophene desulfurization and carbazole utilization. *Appl. Microbiol. Biotechnol.* 71: 355–362.

Schilling, B.M., Alvarez, L.M., Wang, D.I.C., and Cooney, C.L. 2002. Continuous desulfurization of dibenzothiophene with *Rhodococcus rhodochrous* IGTS8 (ATCC 53968). *Biotechnol. Prog.* 18: 1207–1213.

Schmidt, M., Siebert, W., and Bagnall, K.W. 1973. *The Chemistry of Sulfur, Selenium, Tellurium and Polonium: Pergamon Texts in Inorganic Chemistry*, vol. 15, Pergamon Press, Oxford, UK.

Setti, L., Farinelli, P., Di Martino, S., Frassinetti, S., Lanzarini, G., and Pifferi, P.G., 1999. Developments in destructive and non-destructive pathways for selective desulfurizations in oil biorefining processes. *Appl. Microbiol. Biotechnol.* 52: 111–117.

Setti, L., Rossi, M., Lanzarini, G., and Pifferi, P.G. 1993. The effect of n-alkanes in the degradation of dibenzothiophene and of organic sulfur compounds in heavy oil by a *Pseudomonas* sp. *Biotechnol. Lett.* 14(6): 515–520.

Shan, G.B., Xing, J.M., Guo, C., Liu, H.Z., and Chen, J.Y. 2005a. Biodesulfurization using *Pseudomonas delafieldii* in magnetic polyvinyl alcohol beads. *Lett. Appl. Microbiol.* 40: 30–36.

Shan, G.B., Xing, J.M., Zhang, H., and Liu, H.Z. 2005b. Biodesulfurization of dibenzothiophene by microbial cells coated with magnetite nanoparticles. *Appl. Environ. Microbiol.* 71(8): 4497–4502.

Shan, G.B., Xing, J.M., Luo, M.F., Liu, H.Z., and Chen, J.Y. 2003. Immobilization of pseudomonas delafieldii with magnetic polyvinyl alcohol beads and its application in biodesulfurization. *Biotechnol. Lett.* 25: 1977–1981.

Singh, A. 2012. How specific microbial communities benefit the oil industry: Biorefining and bioprocessing for upgrading petroleum oil. In: *Applied Microbiology and Molecular Biology in Oilfield Systems.* Whitby, C., Skovhus, T.L. (Eds.). pp. 121–178.

Singh, A., Singh, B., and Ward, O. 2012. Potential applications of bioprocess technology in petroleum industry. *Biodegradation.* 23: 865–880.

Singh, G. B., Srivastava, S., Gupta, S., and Gupta, N., 2011a. Evaluation of carbazole degradation by Enterobacter sp. Isolated from hydrocarbon contaminated soil. *Rec. Res. Sci. Technol.* 311: 44–48.

Singh, G.B., Gupta, S., Srivastava, S., and Gupta, N., 2011b. Biodegradation of carbazole by newly isolated *Acinetobacter* spp. *Bull. Environ. Contam. Toxicol.* 87: 522–526.

Singh, P. 2015. Improved biodesulfurization of persistent organosulfur compounds. Ph.D. Thesis, Department of Biochemical Engineering and Biotechnology, Indian Institute of Technology Delhi, New Delhi, India.

Song C.S. 2003. An overview of new approaches to deep desulfurization for ultra-clean gasoline, diesel fuel, and jet fuel. *Catal. Today.* 86: 211–263.

Song, C., and Ma, X. 2003. New design approaches to ultra-clean diesel fuels by deep desulfurization and deep dearomatization. *Appl. Catal. B: Environ.* 41: 207–238.

Souza, W.F., Guimarães, I.R., Guerreiro, M.C., and Oliveira, L.C.A. 2009. Catalytic oxidation of sulfur and nitrogen compounds from diesel fuel. *Appl. Catal. A-Gen.* 360: 205–209.

Speight, J.G., 2014. *The Chemistry and Technology of Petroleum 5th Edition.* CRC Press, Boca Raton, FL, USA.

Speight, J.G., and El-Gendy, N.Sh., 2017. *Introduction to Petroleum Biotechnology.* Gulf Professional Publishing, Cambridge, MA, USA. https://doi.org/10.1016/C2015-0-02007-X

Srivastava, V.C. 2012. An evaluation of desulfurization technologies for sulfur removal from liquid fuels. *RSC Adv.* 2: 759–783.

Sun, H., Ma, H., Liu, Z., Li, Y., Xu, P., and Wan, X. 2017. Effect of Fe_3O_4 nanoparticles on *Sphingobium yanoikuyae* XLDN2-5 cells in carbazole biodegradation. *Nanotechnol. Environ. Eng.* 2: 5. https://doi.org/10.1007/s41204-017-0016-9

Tao, X., Zhou, Y., Wei, Q., Ding, S., Zhou, W., Liu, T., and li, X. 2017. Inhibiting effects of nitrogen compounds on deep hydrodesulfurization of straight-run gas oil over a NiW/Al2O3catalyst. *Fuel.* 188: 401–407.

Tian, F., Wu, J.Z., Liang, C., Yang, Y., Ying, P., Sun, X., Cai, T., and Li, C. 2006. The study of thiophene adsorption on to La(III)-exchanged zeolite nay by FTIR spectroscopy. *J. Colloid Interf. Sci.* 301: 395–401.

Todescato, D., Maass, D., Mayer, D.A., Oliveira, J.V., de Oliveira1, D., Ulson deSouza, S.M.A.G., and Ulson de Souza, A.A. 2017. Optimal production of a *Rhodococcus erythropolis* ATCC 4277 biocatalyst for biodesulfurization and biodenitrogenation applications. *Appl. Biochem. Biotechnol.* 183: 1375–1389.

Torkamani, S., Shaigan, J., Aalemzadeh, I., and Yaghmaei, S. 2009. Biological sulfur removal from heavy crude oil of Soroush oil field. *J. Sep. Sci. Eng.* 1(1): 67–79.

Torkamani, S., Shayegan, J., Yaghmaei, S., and Alemzadeh, I. 2008a. Study of a newly isolated thermophilic bacterium capable of Kuhemond heavy crude oil and dibenzothiophene biodesulfurization following 4S pathway at 60 °C. *J. Chem. Technol. Biotechnol.* 83(12): 1689–1693.

Torkamani, S., Shayegan, J., Yaghmaei, S., and Alemzadeh, I. 2008b. Study of the first isolated fungus capable of heavy crude oil biodesulfurization. *Ind. Eng. Chem. Res.* 47(19): 7476–7482.

U.S. Energy Information Administration. International Energy Outlook 2016. vol. DOE/EIA-04; 2016.

Vazquez-Duhalt, R., Torres, E., Valderrama, B., and Le Borgne S., 2002. Will biochemical catalysis impact the petroleum refining industry? *Energ. Fuel.* 16: 1239–1250.

Vertegel, A. 2010. 55th Annual Report on Research 2010. Under sponsorship of the American Chemical Society Petroleum Research Fund.https://acswebcontent.acs.org/prfar/2010/reports/P10734.html.

Wang, P., and Krawiec, S. 1994. Desulfurization of dibenzothiophene to 2-hydroxybiphenyl by some newly isolated bacterium strain. *Arch. Microbiol.* 161: 266–271.

Wang, X., Gai, Z., Yu, B., Feng, J., Xu, C., Yuan, Y., Deng, Z., Xu, P. 2007. Degradation of CAR by microbial cells immobilized in magnetic gellan gum gel beads. *Appl. Environ. Microbiol.* 73: 6421–6428.

Yan, H., Sun, X., Xu, Q., Ma, Z., Xiao, C., and Jun, N. 2008. Effects of nicotinamide and riboflavin on the biodesulfurization activity of dibenzothiophene by *Rhodococcus erythropolis* USTB-03. *J. Environ. Sci.* 20: 613–618.

Yi, Z., Ma, X., Song, J., Yang, X., and Tang, Q. 2019. Investigations in enhancement biodesulfurization of model compounds by ultrasound pre-oxidation. *Ultrason. Sonochem.* 54: 110–120.

Yu, B., Xu, P., Zhu, S., Cai, X., Wang, Y., Li, L., Li, F., Liu, X., and Ma, C., 2006. Selective biodegradation of S and N heterocycles by recombinant *Rhodococcus erythropolis* strain containing carbazole dioxygenase. *Appl. Environ. Microbiol.* 72(3): 2235–2238.

Yu, L., Meyer, T., and Folsom, B. 1998. Oil/water/biocatalyst three phase separation process. US Patent No. 5,772,901.

Zakaria, B.S., Nassar, H.N., Abu Amr, S.S., and EL-Gendy, N.Sh., 2015a. Applying factorial design and response surface methodology to enhance microbial denitrogenation by tween 80 and yeast extract. *Petrol. Sci. Technol.* 33(8): 880–892. https://doi.org/10.1080/10916466.2015.1021013

Zakaria, B.S., Nassar, H.N., Saed, D., and El-Gendy, N.Sh. 2015b. Enhancement of carbazole denitrogenation rate using magnetically decorated *Bacillus clausii* BS1. *Petrol. Sci. Technol.* 33(7): 802–811. https://doi.org/10.1080/10916466.2015.1014966

Zakaria, B.S., Nassar, H.N., EL-Gendy, N.Sh., ElTemtamy, S.A., and Sherif, S.M., 2016. Denitrogenation of carbazole by a novel strain *Bacillus Clausii* BS1 isolated from Egyptian Coke, *Energ. Source. Part A.* 38(13): 1840–1851. https://doi.org/10.1080/15567036.2015.1004389

Zhang, H., Liu, Q.F., Li, Y., Li, W., Xiong, X., Xing, J., and Liu, H. 2008. Selection of adsorbents for in-situ coupling technology of adsorptive desulfurization and biodesulfurization. *Sci. China Ser. B. Chem.* 51(1): 69–77.

Zhang T., Li W., Chen V., Tang H., Li Q., Xing J., and Liu H. 2011. Enhanced biodesulfurization by magnetic immobilized *Rhodococcus erythropolis* LSSE8 assembled with nano-γ-AlO. *World J. Microbiol. Biotechnol.* 27: 299–305.

Zhang, H., Shan, G., Liu, H., and Xing, J. 2007. Surface modification of γ-Al_2O_3 nanoparticles with gum Arabic and its applications in adsorption and biodesulfurization. *Surf. Coat. Technol.* 201: 6917–6921.

CHAPTER 7

Nanobiotechnology and Petroleum Wastewater Treatment

7.1 INTRODUCTION

With the growing industrialization and urbanization, environment is under constant pressure. Contamination of the water is one of the big issues the world is facing today. Water pollution has an adverse impact on ecosystem and water micro- and macro-flora, and it can indirectly lead to air and soil pollution, which consequently affects plant and animals and represents one of the major causes of human health issues. Water pollution also has an adverse impact on the societies' and countries' economic growth and social perspectives. A United Nations report recently stated that the availability of purified and freshwater is a global issue and becomes a challenge for the whole world in the 21st century, as the survival of living creatures is not safe with contaminated water (Gitis and Hankins, 2018; Hodges et al., 2018).

In addition, water shortage is among the world's biggest environmental challenges faced by today's human race. Clean water would, therefore, be on high demand in the coming years and could be a scarce and costly product. It is predictable that the world is going to be under water stress by 2025. Nevertheless, most of the effluents, agricultural, industrial, or municipal, are discharged without proper treatment into water bodies in most of the developing countries (Yanyang et al., 2016). Stringent rules and guidelines are laid down by governments, nongovernmental organizations, and environmentalists to save the environment and introduce the concept of water reuse, reduce, and recycle (George, 1996; Chunling et al., 2011).

Petroleum wastewater (PWW) effluents are composed of toxic substances that can cause severe environmental hazards and damage to the biosphere when disposed into the atmosphere. Because of the presence of huge amounts of different toxic petroleum hydrocarbons and heavy metals, these streams are difficult to handle. Wastewater from different sectors of oil and gas industry starting from exploration, production, refining, processing, petrochemical industry, equipment maintenance and cleaning, transportation, and pipelines and storage tank cleaning poses a concern to minimize its environmental impact. The major forms of wastewater refinery include cooling water, storm water, refining water, steam, and sanitary wastewater. One of the basic principles is to avoid combining multiple kinds of streams of wastewater to reduce the burden on the treatment units (Ahmad, 2020).

PWW may differ greatly depending on the design of the plant, the operating procedures, and the type of oil being processed (Saien and Nejati, 2007). In the refinery, nonhydrocarbon compounds are extracted, and the oil is decomposed into its different components and mixed in useful products. As a result, large quantities of wastewater are generated by petroleum refineries, including well-extracted oil water brought to the surface during oil drilling, which also contains recalcitrant compounds and is rich in organic pollutants and, therefore, cannot be treated easily, and it is difficult to be treated biologically (Asatekin and Mayes, 2009; Rasheed et al., 2011; Vendramel et al., 2015). Thus, the removal of pollutants from refinery industrial plants is a requirement for water reuse and is in accordance with environmental standards (Farajnezhad and Gharbani, 2012; Aljuboury et al. 2017).

Distillation, thermal and catalytic cracking, desalting, coking, as well as heat exchangers and storage tanks are the refinery units, which produce most of the amount of wastewater (Paterson, 1995). Such effluents consist of grease and petroleum compounds that consist of a collection of aliphatic (methane) and aromatic (benzene, toluene, ethylbenzene, xylenes, and aromatic polycyclic compounds) hydrocarbons, their mixtures (diesel fuel), H_2S, metals (Cr, Pb, Hg, V, Zn, etc.), dissolved sulfides, disulfides and mercaptans, phenols, chlorophenols (CPs), sulfur dioxide (SO_2), nitrogen oxides (NOx), and volatile organic compounds, sometimes possible carcinogens, explosives, etc. In addition, naphthenic acids, which are a group of compounds, are known to cause deleterious effects in wastewater from petrochemical industry, and their removal from oilfield wastewater is a significant obstacle in remediating large quantities of petrochemical effluents (Wang et al., 2015). If a crude oil contains large amounts of sulfur, it is called sour crude. Sour water is, thus, a particular stream of petroleum refineries containing compounds and hazardous substances that are gradually biodegradable (Coelho et al., 2006).

As a general rule, if the cooling water tower is recycled, the amount of petroleum-processing wastewater (PPW) emitted from refineries is about 3.5–5 m^3/ton crude oil (Sun et al., 2018). In addition, the water-to-oil phase ratio in oil production is projected to fall within the range from 3 to 4 bbl water/bbl oil extracted during exploration and production (E&P) operations

(Hu et al., 2103; Adham et al., 2018). Because water in the petrochemical industry is not a target end product, it is assumed that about 80–90% of water supplied or used in this sector will be disposed of as PPW effluent (Jafarinejad, 2017). Oil spills are another type of oil pollution that occurs during export or the leakage of offshore oil rigs (Kharisov et al., 2014). Therefore, increases in anthropogenic petroleum activities are expected to intensify the environmental emission of various types of hazardous pollutants (Adham et al., 2018). In this regard, there is an inextricable correlation between surface water consumption/contamination and petroleum production, as shown in Figure. 7.1.

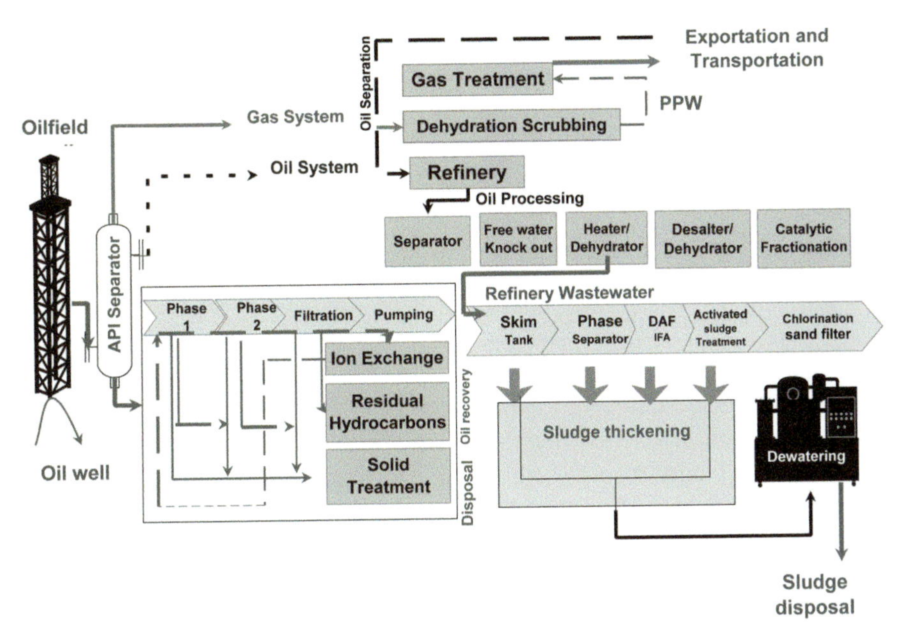

FIGURE 7.1 Stages of petroleum wastewater treatment.

The US Environmental Protection Agency (EPA) reported that 47–75% of benzene, toluene, ethylbenzene, and xylene (BTEX) found in an ecosystem are derived from PPW systems (Jafarinejad, 2017). As shown in Table 7.1, in addition to BTEX, several other pollutants, such as aliphatic series (C1–C50), polyaromatic hydrocarbons (PAHs), volatile organic compounds, phenols, and heavy metals are also released from PPW effluents. The concentration of these PPW-based pollutants differs according to the type of petroleum-processing methods and treatment plant efficiency (Hu et al., 2103; Aljuboury et al., 2017).

TABLE 7.1 Wastewater Parameters from Oilfield Production

Pollution Parameters	Petroleum Processing Wastewater (PPW) Effluent
Density (kg/m^3)	1014–1140
Total organic carbon (mg/L)	119–398
Chemical oxygen demand (mg/L)	72.1–21,000
Biochemical oxygen demand (mg/L)	90–8,000
Total suspended solids (mg/L)	56–950
Chloride (mg/L)	80–200,000
Turbidity (NTU)	10.5–159.4
Total dissolved solids (mg/L)	2,100–37,000
Phenols (mg/L)	0.2–200
Ammonia (mg/L)	0.76–125
Nitrate (mg/L)	2.23–9.3
Oil and grease (mg/L)	20–20,223
Benzene, toluene, ethylbenzene, and xylene (mg/L)	1–100
Heavy metals (mg/L)	0.01–100
pH	4.3–10
Polyaromatic hydrocarbons (mg/L)	0–300

The treatment strategies for PPW can usually be divided into two main routes: physical (e.g., skimmer tank, American Petroleum Institute (API), and filtration) and reactive (e.g., flocculation/coagulation chemicals or biological remediation) approaches. Both of these options are currently operated for in situ and ex situ treatment of PPW effluents (Tang et al., 2017; El-Shamy et al., 2018). The various PWW treatment processes may reduce: (1) suspended solids, (2) biodegradable organics, (3) pathogenic bacteria, and (4) nitrates and phosphates. Wastewater treatment is divided into three types: (1) primary, (2) secondary, and (3) tertiary treatments (Kanchi, 2014). However, due to the intricacy of the PWW characteristics, the treatment requires a standard implementation of integrated method. The traditional methods of treatment, therefore, require multistep process treatment. The first step made up of pretreatment, which includes physicochemical and mechanical treatments followed by a second step of advanced wastewater treatment. PWW treatment methods need to take additional steps to remove organic matter as they transfer contaminants from one medium to another such as chemical oxidation (Hu et al., 2015), biological techniques (Wang et al., 2015), coagulation (Abu Hassan et al., 2009; El-Naas et al., 2009;

Farajnezhad and Gharbani, 2012), and adsorption (Al Hashemi et al., 2015). Additionally, new technologies such as catalytic air oxidation assisted by microwave (Sun et al., 2008) and membranes (Shariati et al., 2011; Yuliwati et al., 2011) have now been discovered.

Nanomaterials are chosen for the efficient removal of various pollutants from the water supplies depending on the method of treatment and the purification level. Up to now, several studies have shown that nanomaterials have vast competence and potential in wastewater treatment, in particular, in the areas of adsorption (Ali, 2012), membrane process (Pendergast and Hoek, 2011), catalytic oxidation (Ayati et al., 2014), disinfection, and sensing (Deshpande et al., 2020). It should be noted that nanomaterials for purifying water must be eco-friendly and nontoxic. Unsafe particles can cause severe injury to vital organs upon contact with the human body. Due to dimensional features, nano-objects may translocate to various organs, which exacerbates the risk of biological damage. Therefore, toxicity performance tests must be included strictly in safety data sheets, standard operating procedures before they are applied to the industry.

From a sustainability point of view, the scientific community has given considerable attention to the use of microbes and nanomaterials to eliminate contaminants. Green synthesis of nanoparticles (NPs) is an emerging research trend in green nanotechnology since this method is nontoxic or less toxic, environmentally friendly, efficient, and cost effective compared to the other traditional physicochemical methods. Owing to their high efficiency and biocompatibility, a number of green-synthesized NPs are currently used in water and wastewater treatment. Green-synthesized NPs are highly capable of recycling and extracting heavy metal from wastewater without losing its stability and degrading a variety of organic contaminants from wastewater and thereby purifying and/or recycling wastewater for reuse and may solve various worldwide water quality problems. This chapter focuses on new and emerging application of nanobiotechnology to treat PWW in a sustainable way.

7.2 POLLUTANTS IN WASTEWATER FROM PETROLEUM INDUSTRY

The nature and levels of concentration of PPW pollutants can be vastly diverse and depend on: (1) the size and complexity of operating practices, (2) the type, age, and quality of treatment and processing equipment, (3) operating conditions, and (4) the applied waste minimization regulatory protocols (Jafarinejad, 2017; Olajire, 2014).

Due to the complex composition of PPW such as crude oil and its distilled products along with oilfield/refined chemicals (e.g., dye additives, antioxidants, biofilm and biofoluing inhibitors, and microbial and chemical corrosion inhibitors) available in large volumes and/or quantities, detection and removal of individual organic pollutants in PPW remains challenging (Olajire, 2014). Thus, total organic content, total oil and grease, chemical oxygen demand (COD), and five-day biochemical oxygen demand (BOD_5) are often used as preferred analytical terms to describe organic content in PPW effluents (see Table 7.1) (Aljuboury et al., 2017; An et al., 2017).

More precisely, volatile organic carbons, normal-alkanes and alkenes (C10–C21), aromatics, and cyclic hydrocarbons/PAHs were identified as the dominant contaminants in PPW, if assessed in terms of quantity (Aljuboury et al., 2017), for example, refinery wastewater emitting many volatile organic carbons with total volatile organic carbons ranging from 50 to 1000 tons per million tons of processed crude oil (Olajire, 2014; Aljuboury et al., 2017; Jiménez et al., 2018). Thus, when discharging oil containing PPW, many countries have now implemented tough regulatory standard protocols (Wahi et al., 2013; An et al., 2017).

Various types of toxic metals are also known to be released from various petroleum-processing operations (e.g., oilfield processing and refinery processing) (Varjani et al., 2019). A total of eight common heavy metal contaminants dependent on petroleum are well known in PPW effluents, namely, arsenic ($As^{3+/5+}$), mercury (Hg^{2+}), lead (Pb^{2+}), chromium (Cr^{6+}), cadmium (Cd^{2+}), copper (Cu^{2+}), zinc (Zn^{2+}), and nickel (Ni^{2+}) (Mustapha et al., 2018). Their concentration rates reflect the geological location of a crude oil field (e.g., source rock), the different compositions of drilling mud fluids used during extraction of crude oil, and the presence of corrosive catalyst/equipment used during processing of refineries. The maximum contaminant limits for these petroleum-based heavy metals in wastewater/ecosystems according to the World Health Organization and the US EPA are as follows: $As^{3+/5+}$ (50 μg/L), Hg^{2+} (0.03 μg/L), Pb^{2+} (6 μg/L), Cr^{6+} (50 μg/L), Cd^{2+}(10 μg/L), Cu^{2+} (250 μg/L), Zn^{2+} (800 μg/L), and Ni^{2+} (200 μg/L) (Younis et al., 2020).

7.3 PETROLEUM WASTEWATER TREATMENT TECHNIQUES

7.3.1 CONVENTIONAL PETROLEUM WASTEWATER TREATMENT

Within the petroleum industry, there is a wastewater treatment unit at the end of oil and gas pipe operating facilities to collect and remedy various wastewater effluents that are released from the process units and offsite/utilities (Malakar et al., 2017; Jiménez et al. 2018). In general, the PPW's traditional treatment technologies are classified into four main stages (Aljuboury et al., 2017):

1. Mechanical/physical processes, such as skimming tanks and separators for oil/water gravity (API).
2. Physiochemical processes, such as dissolved air flotation (DAF)/induced air flotation, chemical flocculation, and adsorption or membrane filtration unit.
3. Chemical and catalytic oxidation methods such as advanced oxidation process.
4. Biological processes.

These traditional treatment approaches are also performed in a sequential and hybrid manner, including primary, secondary, and tertiary or advanced remediation processes to avoid the drawbacks of low efficiency of treatment (Pombo et al., 2011; Jiménez et al. 2018).

Table 7.2 illustrates the advantages and disadvantages of traditional and commercial technologies for PPW treatment. In general, these traditional technologies do not adequately treat most emulsified or dissolved oil, nor other soluble organic micro-pollutants and heavy metals (Aljuboury et al., 2017; An et al., 2017). Consequently, the use of other complementary processes, known as an advanced treatment package, has recently been launched on the market to improve the quality of PPW effluent before disposal or subsequent reuse (Jiménez et al. 2018).

The utilization of advanced nanoscale materials for PPW remediation showed a varied role including degradation of polluting materials, remediation of inorganic materials, antimicrobial activity, and disinfectant. To date, various nanostructured materials (e.g., carbon nanomaterials, metal/metal oxides NPs, nanopolymers, hybrid nanocomposites, and metal–organic frameworks) have been proposed to be used effectively in wastewater treatment (Zhang et al., 2018; Zhu et al., 2019).

The application of nanobiotechnology for wastewater treatment could be summed up under the following points (see Figure 7.2):

1. Through nanobiosorbent substances such as lignocellulosic wastes, nanoclays, nanobiocahrs, etc.;
2. The use of nanobiophotocatalysis with semiconductor material for environmental remediation;
3. Nanobiocomposites, such as metal–polymer nanocomposites for various applications;

TABLE 7.2 The Advantages and Disadvantages of Traditional and Commercial Technologies of PPW Treatment

Stages of Treatment	Conventional Technologies	Advantages	Disadvantages
Primary	API gravity separation	Simple design, low energy consumption, result easy to predict	Low oil removal efficiency, large space requirement, slow separation, limited capacity
	Chemical precipitation (DAF)	Good oil removal efficiency	Complex operation, high operating cost, generation of secondary pollutants
	Coagulation and flocculation	High overflow rates, easy operation, durable	Producing large amount of scum and sludge, slow separation
Secondary	Activated carbon	Simple process, good removal efficiency, low processing cost	Low separation efficiency for fine oil droplets, less efficient at high oil concentration
Tertiary	Chemical oxidation	Fast removal, easy operation	High cost of oxidizing agent, undesired by-products, safety concern associated with applying oxidants
	Biodegradation	Low operating cost, green technology, High oil removal efficiency	Slow operation, Sensitive to different environmental parameters such as pH and temperature variations
	Membrane filtration	The method is mainly used to remove dissolved materials and heavy metals, used as an advanced treatment	Severe biofouling, insufficient for volatile organic carbons (VOCs), high energy consumption, short life time and expensive

4. Nanobiomaterials for heavy metal remediation;
5. Disinfectant and antimicrobial action against pathogenic microbes and sulfate-reducing bacteria (SRB).

The types of PPW effluent reuse/recycle processes are highly variable globally. Nearly, all petroleum companies use oilfield brine water for growth E&P by reinjecting (flowing back) into deep oil wells to boost processes for enhancing oil recovery (Dittrick, 2017). If the oilfield water is generated in a greater amount than needed for reinjection purposes, the PPW should be disposed of by additional treatment in the environment such as land drainage, evaporation ponds, percolation ponds, subsurface injection, and offsite trucking (Clark and Veil, 2009). After conventional treatment without reuse of wastewater, treated PPW is released into the surface/industrial water bodies' at most petrochemical plants (Pombo et al., 2011). In order to protect our natural water resources and human health from the polluting risks of oil industry, the development of wastewater treatment technologies within the petroleum sectors must have a high priority.

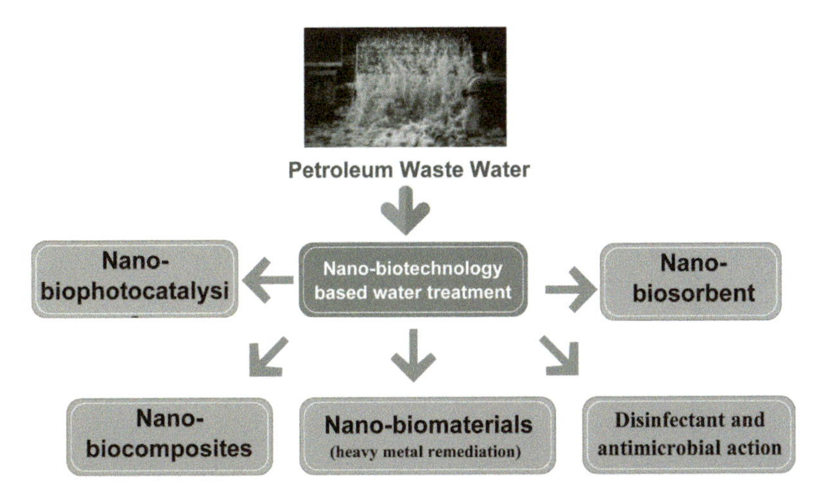

FIGURE 7.2 Role of nanobiotechnology in wastewater treatment.

The application of green NPs for industrial wastewater treatment could be an excellent strategy to cope with environmental pollution. To date, only very few studies have reported the application of green-synthesized NPs in wastewater treatment (Pintor et al., 2016; Franco et al., 2017). In this chapter, we mentioned different kind of green nanomaterials and their role in treating wastewater.

7.3.2 STAGES OF PETROLEUM WASTEWATER TREATMENT

7.3.2.1 PRETREATMENT PROCESS

Pretreatment processes are usually implemented by the petroleum refinery to treat wastewater until it is sent to biological processes for organic removal. The primary treatment involves the removal of free oil and gross solids and reduction of dispersed oil and solids by flocculation, flotation, sedimentation, filtration, etc. to improve the biodegradability of wastewater (Santo et al., 2013).

7.3.2.2 PHYSICAL TREATMENT

The physical treatment system is a primary treatment stage, necessary for removing or separating solid particles, suspended solids, and immiscible liquids from PWW using sedimentation, coagulation, and flocculation, and prolonged use of the secondary treatment unit. Many approaches for physical care are known as traditional methods (Renault et al., 2009).

Depending on the wastewater characteristics, physical treatment such as adsorption by active carbon, zeolite, copolymers, etc., can be used for eliminating hydrocarbons in the petrochemical wastewater (Fakhru'l-Razi et al., 2009). The DAF system is the common water treatment process used in the treatment of wastewater containing oil as well as suspended solids, which can also be applied for petrochemical wastewater.

Physical techniques such as sedimentation are also used to remove suspended solids before biological treatment. In API separators or separation tanks, the sedimentation treatment is used to separate oil from water mechanically by gravity. The coagulation process was used to clarify water and to reduce the organic load. However, due to its complexity, physical methods are fairly unsuccessful in treating PWW, and therefore, other methods could be used for pretreatment (Nguyen et al., 2019).

7.3.2.3 CHEMICAL TREATMENT BY GREEN NANOPARTICLES

Adsorption is a surface process typically used as a polishing stage in water and wastewater treatments to eliminate organic and inorganic pollutants. It is commonly used in treating PWW for organic compounds, ammonium, and toxicity characteristics (Lorenc-Grabowska and Gryglewicz, 2007).

Sorbents are commonly used in water purification as filtration media for separating inorganic and organic contaminants from polluted water. An addition of sorbent and coupling the activated sludge to form a biologically activated carbon system improve the treatment through adsorption. The adsorbing medium is generally known as the adsorbent, whereas the material adsorbed on the surface is known as the adsorbate. Adsorption strengths are low cost, easiness, and adaptability (Kulkarni and Goswami, 2013, 2014). A terminology for adsorption is illustrated in Figure 7.3.

Some of the important criteria for selecting nanomaterials are their accessibility, cost effectiveness, and environmental impact reduction ability, adsorption may be divided into two in terms of the mechanisms involved, that is, chemical and physical adsorption. The former process includes the chemical reaction, in which the surface is chemically reacted by adsorbate. On the other hand, the latter includes the attachment of waste substances by physical forces such as hydrogen bonding and dipole and dipole interaction (Kausar et al., 2018).

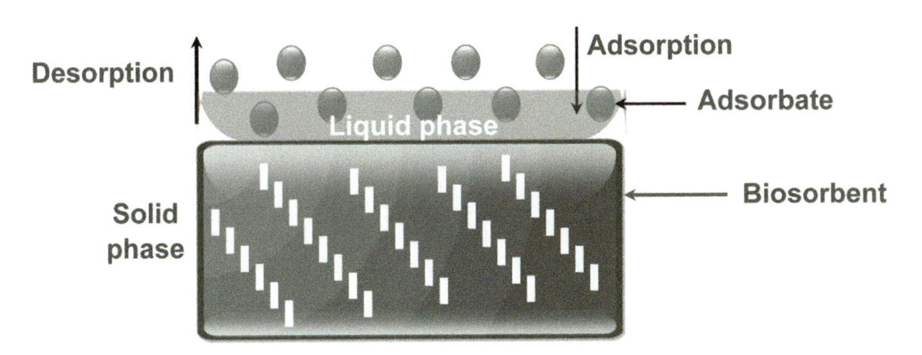

FIGURE 7.3 Explanation of an adsorption terminology.

Many nanobiomaterials have been applied for petroleum water remediation. Some naturally or bioproduced mineral or organic adsorbent show benefits such as low cost, environmentally friendliness, and high oil removal efficiency. Nanobioparticles that are used for adsorption are influenced by their size and shape, surface chemistry, state of solubility and dispersion, chemical composition, and crystallinity. However, for oily wastewater treatment, it is usually necessary that such nanobioparticles should be nontoxic, high adsorption power, capable of adsorbing contaminants at concentrations as low as parts per billion, simple to desorb, and can be reused many times without significantly affecting the overall efficacy.

The selection of the best method and biomaterial for treating PWW is a highly complex challenge, taking into account several factors, such as the safety standards to be met and the efficiency in addition to the cost. Consequently, the following four criteria for the preparation of nanomaterials on wastewater treatment technologies must be considered: (1) flexibility in treatment and final efficiency, 2) reuse of treatment agents, 3) environmental safety, and 4) low cost (Xu et al., 2012; Zhang et al., 2013).

The typically required characteristics of biosorbent are high surface area, charge density, and hydrophobicity. In terms of surface feature, corrugations and hair-like features are also desired to promote oil uptake through capillary forces (Pintor et al., 2016). There are two main properties of nanobioparticles that make them particularly attractive as sorbents. They have a significantly larger surface area than bulk particles on a mass basis. Also, NPs can be functionalized with different chemical groups to increase their affinity to target compounds. Several research groups use nanobioparticles unique properties to develop high-capacity and selective sorbents for metal ions and anions (Savage and Diallo, 2005).

Activated carbon is commonly used in PWW treatment because of its unique adsorbent properties such as highly porous and enormous surface area. It can be used in the remediation of petroleum contaminated groundwater, and it is known that powdered activated carbon is typically more efficient than granular activated carbon. The adsorption of activated carbon is efficient in eliminating residual organic compounds after biological treatment. Moreover, pollutants of low molecular weight are specially adsorbed (Lorenc and Gryglewicz, 2007). However, despite the efficiency in BOD and COD elimination, activated-carbon-based adsorption is largely limited by its high cost. In addition, the high concentrations of oil can cause pore to clog and deteriorate the adsorption efficiency (Aljuboury et al., 2017; Ghimire and Wang, 2018).

The commercially activated carbon-loaded silver nanoparticle (AgNPs-AC) was used by Nguyen et al. (2019) as a new adsorbent to remove Cr(VI) from the water solution. Batch adsorption experiments were conducted to assess the effects of pH, initial Cr(VI) concentration, contact time, and AgNPs-AC dose upon removal of Cr(VI) from the aqueous solution. The results showed that the most suitable condition for the absorption of Cr(VI) on AgNPs-AC from the aqueous solution was the contact time of 150 min at pH4, 40 mg/L of initial Cr(VI), and the dosage of 20 mg AgNPs-AC/25 mL. At above mentioned conditions, the maximum adsorption capacity was 27.70 mg/g. Meanwhile, in cases where the initial Cr(VI) concentration is

10 mg/L and the contact time is 60 min, the adsorption capacity of commercial activated carbon from a coconut shell was only 7.61 mg/g. The adsorption kinetic data were found to fit best with a high correlation coefficient ($R^2 = 0.9597$) on the pseudo-second-order model. Due to the presence of new chemical species on the adsorbent surface, the adsorption mechanism was regulated by chemisorption. The positively charged functional groups rapidly reduced Cr(VI) to Cr(III), and Cr(III) was later adsorbed on the surface of the adsorbent by the carboxyl group. It can be concluded from those published results that AgNPs-AC is a fully promising, low-cost adsorbent for removing Cr(VI) from the aqueous solution.

Clay is another biomaterial class that shows promising adsorptive behavior for removing organic pollutants. The use of various clays as adsorbents to remove oily components from wastewater has gained attention mainly because of the eco-friendly quality of clay materials. Bentonite organoclay has been used to be an effective adsorbent for removing oil from oily waters in a column system, where the capacity is five to seven times greater than that of activated carbon. The application of biofibers for the remediation of oily wastewater has also been published in the past few years (Wahi et al., 2013). Some commonly applied natural oil adsorbents are bagasse sugarcane, rice husk, and residues of wood that are usually formed into sheets, filers, or fiber assemblies. These natural adsorbents are derived from the waste and are, therefore, chemical-free highlighting biodegradability.

In recent years, there has been a growing interest in the use of iron oxides NPs for the removal of heavy metal due to their simplicity and availability. Magnetic magnetite (Fe_3O_4), magnetic maghemite (γ-Fe_2O_3), and nonmagnetic hematite (α-Fe_2O_3) are often used as nanoadsorbents.

Combined with magnetic separation, biosorption techniques have been widely used in water treatment and environmental cleanup (Ambashta and Sillanpaa, 2010; Mahdavian and Mirrahimi, 2010). In general, their separation and recovery from contaminated water are great challenges for water treatment due to the small size of nanobiosorbent material. However, with the assistance of an external magnetic field, magnetic magnetite (Fe_3O_4) and magnetic maghemite (γ-Fe_2O_3) can be easily separated and recovered from the system (Zhu et al., 2006). In addition to its low cost, high adsorption efficiency, and improved stability, iron oxide nanobiomaterials are promising for industrial-scale wastewater treatment (Hu et al., 2005; Carabante et al., 2009; Fan et al., 2012).

Several researchers have recently synthesized magnetite nanoparticles (MNPs) Fe_3O_4 via a green chemistry approach. Yew et al. (2016) used seaweed as a reducing agent to synthesize Fe_3O_4 MNPs, with an average size of 14.7 nm. Venkateswarlu et al. (2013) used plantain peel extract to synthesize green Fe_3O_4 MNPs with sizes less than 50 nm. Venkateswarlu et al. (2014) used the *Syzygium cumini* seed extract as the reducing agents for Fe_3O_4 MNP synthesis.

Wang et al. (2014) prepared iron NPs using an easy one-step green process using leaf extracts of *Eucalyptus*. Such leaf extracts, which are known to be with a high antioxidant capacity and a high content of polyphenols, act as both NP reduction and capping agents. The attractive features of this approach are: (1) the method is quite simple and efficient and can be used at room temperature; (2) the biodegradable and nontoxic eucalyptus leaves, which are normally considered as a waste, environmentally friendly and easily obtainable; and (3) the polyphenols in the extract reduce the aggregation of Fe NPs and enhance their reactivity. Also, the application of biosynthesized NPs in wastewater indicated that 71.7% of total N, 30.4% of total P, and 84.5% of COD were removed. This convincingly demonstrates their tremendous potential for wastewater *in-situ* remediation.

Wei et al. (2016) synthesized iron NPs from *Citrus maxima* peels for removal of Cr(VI). Complete removal of Cr(VI) was achieved within 90 min with Iron(III) solution: *Citrus maxima* extract ration of 1:3. Moreover, Fazlzadeh et al. (2017) prepared synthesized iron NPs from three plant extracts to remove hexavalent chromium from aqueous solutions with efficiency greater than 90% within 10 min.

Azadi et al. (2018) green-synthesized MNPs using the root extract of *Persicaria bistorta* as a reducing agent and examined its adsorption properties in industrial wastewater treatment, as an effective novel adsorbent. The synthesis procedure is designed using the Taguchi method to investigate the role of synthesis parameters, namely, $FeCl_2$ molar concentration, ratio of water to plant extract (v/v), synthesis temperature, and pH on the size and magnetization of MNPs. Data analysis showed that crystallite size decreased by decreasing the concentration of $FeCl_2$ and pH and increasing temperature. The water-to-plant-extract ratio does not show any significant effects on the responses. The model proposed optimal values for concentration of $FeCl_2$, temperature of synthesis, and pH as 0.15 M, 70 °C, and 11, respectively, which resulted in MNPs with minimum crystallite size of 45.5 nm and maximum magnetization of 62.5 emu/g. In addition, adsorption experiments showed that Fe_3O_4 MNPs is good adsorbent for COD removal from wastewater. Adsorption data also showed a good suitability with Freundlich and Langmuir isotherm models, and the maximum capacity of COD adsorption for this green nanomaterial is 149 mg/g.

Abdullah et al. (2018) used a novel, inexpensive, nontoxic, and eco-friendly approach to synthesize hydrophobic MNPs, where the hydrophobic biocomponents extracted from *Anthemis pseudocotula* were used as capping agents. The surface-modified reaction carried with that of the MNPs resulted in the formation of APH-MNPs in the presence of n-hexane extract, while the formation of APC-MNPs resulted by the presence of the chloroform extract (APC). The efficiency of surface-modified MNPs for oil spill collection in the presence of an external magnetic field was assessed by taking different MNP ratios: crude oil. The APH-MNPs have been shown to be efficient in crude oil collection (approximately 92%) compared to the APC-MNPs (approximately 81%) due to their higher crude oil dispersion capacity. Finally, the MNPs could be recycled at least five times with no or minor efficiency loss.

Murgueitio et al. (2018) synthesized iron nanobioparticles using the mortiño berry extract *Vaccinium floribundum* as a reduction and stabilizing agent. The produced nanobioparticles were spherical in the 5–10 nm range. After treatment with iron nanobioparticles, water contaminated with two concentrations of total petroleum hydrocarbons TPHs (9.32 and 94.20 mg/L) showed removals of 85.94% and 88.34%, respectively, while polluted soil with a concentration of TPHs of 5000 mg/kg treated with NPs during 32 h recorded 81.90% removable. Results indicate that the addition of iron nanobioparticles resulted in strong reduction conditions, accelerating the removal of TPHs and suggesting that these NPs could be a promising technology to clean up contaminated water and soils from TPHs.

Pourmortazavi et al. (2019) biosynthesized the Fe_3O_4 NPs with the presence of chitosan and tripolyphosphate extracted from the shrimp by the coprecipitation method. The efficiency of using chitosan-coated Fe_3O_4 nanobioparticles to remove chromium(VI) from wastewater samples was then studied. Efficient parameters such as pH, contact time, adsorbent dose, and agitation velocity were optimized by experiment method design and investigated by the methodology of the response surface. Based on the results, 75%–88% of Cr(VI) can be extracted from the actual samples suggesting the high efficiency of the biosynthetic adsorbents in chromate removal.

Peng et al. (2020) demonstrated that the nano-zero-valent irons (nZVI) have shown great potential to function as universal and low-cost magnetic adsorbents. The rapid agglomeration and easy surface corrosion of nZVI in solution greatly hinders their overall applicability. The carboxylated cellulose nanocrystals (CCNC), widely available from renewable biomass resources, were prepared and applied for the immobilization of nZVI. In addition, carboxylated cellulose nanocrystals supporting nano-zero-valent irons (CCNC-nZVI) were obtained via an *in-situ* growth method. The CCNC-nZVI were characterized and then evaluated for their performances in wastewater treatment. The results showed that nZVI NPs could attach to the carboxyl and hydroxyl groups of CCNC, and well disperse on the CCNC surface with a size of ~10 nm. With the CCNC acting as corrosion inhibitors improving the reaction activity of nZVI, CCNC-nZVI exhibited an improved dispersion stability and electron utilization efficacy. The Pb(II) adsorption capacity of CCNC-nZVI reached 509.3 mg/g (298.15 K, pH = 4.0), significantly higher than that of CCNC. The adsorption was a spontaneous exothermic process and could be perfectly fitted by the pseudo-second-order kinetics model. This study provided a novel and green method for immobilizing magnetic nanomaterials by using biomass-based resources to develop effective bioadsorbents for wastewater decontamination.

Biochar is an effective adsorbent for removing various contaminants due to its unique properties, such as high specific surface area (SSA) and abundant functional surface groups. Biochars have, thus, become increasingly relevant as a solution for remediating contaminants in the industrial sector to improve the quality of the environment (Wang et al., 2017). The concerned organic contaminants include phenanthrene (Tang et al., 2015), phenol (Karakoyun et al., 2011; Kong et al., 2014), sulfapyridine (Inyang et al., 2015), and naphthalene and *p*-nitrotoluene (Chen et al., 2011). Depending on the type of nanomaterials, biochar substrates, and target organic pollutants, the adsorption capacity of these organic contaminants varied from 2.4 to 278.55 mg/g.

In the composites of magnetic biochar and the biochar composites of nano-metal oxide, the interactions between the functional groups of biochar and organic contaminants, including π–π interactions, electrostatic attraction, hydrogen bond, and hydrophobic interaction, which confirm the possibility of using many mechanisms in order to ensure the success of the adsorption process (Tan et al., 2015). Additionally, the coating of carbon structure NPs may also increase the adsorption sites to remove organic pollutants from water (Zhang and Gao, 2013; Sun et al., 2015). Other functional NPs such as chitosan and graphene coated on the biochar surface may account for the wider surface area, more functional groups, higher thermal stability, and higher efficiency in removing organic contaminants (Tang et al., 2015).

Chitosan mixed with biochar can be incorporated after cross-linking into membranes, beads, and solutions and, thus, can be effectively used as an adsorbent for heavy metals in industrial wastewater. The ratio of biochar and chitosan would affect the adsorption of copper, lead, arsenic, cadmium, and other heavy metals in industrial wastewater (Hussain et al., 2017).

It is known that the biomass of algae, fungi, bacteria, and yeasts, along with some biopolymers and biowaste materials, bind and concentrate precious metals. This biological absorption process can be a cost-effective alternative to common chemical methods for recovering various dissolved minerals from an aqueous solution. The mechanisms involved in the binding of minerals to microbes and their concentration have been extensively studied in natural environments (Saifuddin et al., 2009).

Chakraborty et al. (2009) demonstrated that cyanobacteria and algae have the capacity to bind and concentrate Au and form AuNPs. The process of NP formation is specific to a particular genus, describing bio-NPs formation using the cyanobacterial strains *Lyngbya majuscula* and *Spirulina subsalsa* and the freshwater green alga *Rhizoclonium hieroglyphicum*. Mata et al. (2009) reported the formation of AuNPs during biosorption by seaweed *Fucus vesiculosus*. These authors identified the bio-NP formation pH dependence and discussed the stages of the adsorption process.

Dead biomass of the *macrofungus Pleurotus* platypus was used for biosorption of Ag in a study by Das et al. (2010). The fungal biomass exhibited the highest Ag uptake of 46.7 mg/g at pH 6.0 in the presence of 200 mg/L Ag^{+} at 20 °C. These authors also provided a detailed thermodynamic and kinetic analysis of the biosorption process.

Live organism plays a vital role in the synthesis of metal green NPs; among all organisms, the plants are found to be the most appropriate for the manufacturing of the green NPs at the commercial level. As compared to the microorganism, plants produced more reliable and faster rate of green NPs; plant-based NPs vary in shape and size as compared to other organisms.

Abbas et al. (2020) employs three different plant precursors to synthesize plant extracts such as *Aloe barbedensis, Azadirachta indica,* and *Coriandrum sativum*, which are further used to synthesize nanobioparticles. Two different nanobioparticles for example, silver nanoparticles (AgNPs) and copper nanoparticles (CuNPs) are obtained. Synthesized materials are applied as biosorbents for the treatment of PWW. Different batch experiments are conducted to check the efficiency of these synthesized nanobioparticles by using naphthalene (PAHs) as a model PAH. A total of 98.81% removal has been occurred using NPs synthesized by *Azadirachta indica* and least adsorption power occurred in case of *Coriandrum sativum syntheized* NPs, that is, 95.29%.

By implementing *Garcinia subelliptica* leaves as a natural biotemplate, Lai et al. (2020) demonstrated a sustainable phosphate-selective method where MgMn-layered double hydroxide (MgMn-LDH) and graphene oxide (GO) can be grown in situ to obtain L-GO/MgMn-LDH. The composite showed a hierarchical porous structure and selective phosphate recognition after calcination, achieving significantly high and recyclable selective phosphate adsorption efficiency and desorption rate of 244.08 mg-P/g and 85.8%, respectively. In addition, the functions of *Garcinia subelliptica* leaves, biflavonoids, and triterpenoids in promoting phosphate adsorption and antimicrobial potential of chemical constituents have been investigated. Such findings suggest the proposed biotemplated adsorbent in industrial wastewater treatment is feasible and environmentally sustainable.

In recent years, *photocatalysis* has received special attention because of its potential application in environmental remediation. For the treatment of wastewater, photocatalysis is a promising technology because it offers high efficiency/oxidation rates and reduces the pollutant to the simplest nontoxic forms (Bethi and Sonawane, 2018). Two types of processes occur in photocatalysis, namely, degradation and mineralization of organic contaminants (Umar et al., 2012, 2013). The organic pollutants are split or decomposed into several products during the degradation process, while in the case of mineralization, the destruction of organic pollutants occurred until it turns into water, carbon dioxide, and some inorganic ions. The possible mechanism for pollutant photodegradation is shown in Figure 7.4.

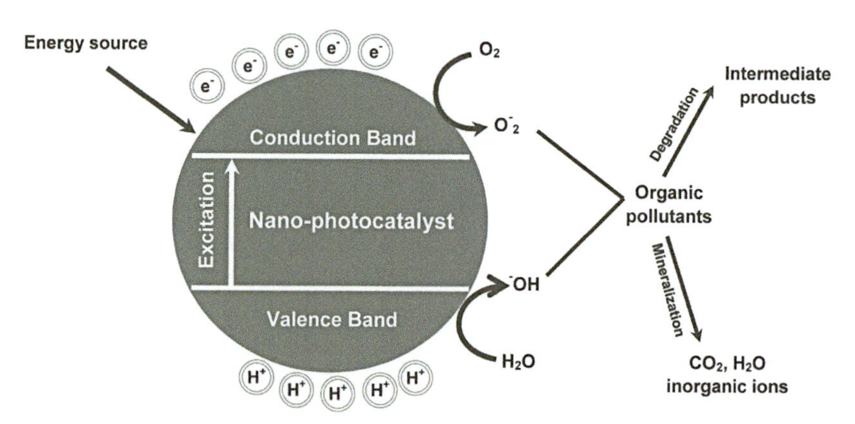

FIGURE 7.4 General mechanisms for the degradation of toxic organic compounds by nanophotocatalyst.

Most common photocatalysts are metal oxide or sulfide semiconductors, of which TiO_2 has been the most extensively investigated one, in the past decades. To date, TiO_2 has been the most exceptional photocatalyst due to its high photocatalytic efficiency, reasonable price, and photostability (Boyano et al., 2009; Bishoge et al., 2018).

During the last decade, green-synthesized titanium dioxide (TiO_2) NPs have emerged as promising photocatalysts for water purification (Adesina, 2004). Titanium dioxide (TiO_2) is one of the most widely used metal oxide for nanophotocatalysts synthesis, due to its high thermal stability and polymorph formation named as anatase, rutile, and brookite (Yaqoob et al., 2020). TiO_2 NPs are very versatile; they can serve both as oxidative and reductive catalysts for organic and inorganic pollutants. The large band-gap energy (3.2 eV) of TiO_2 requires ultraviolet (UV) excitation to induce charge separation within the particles. Upon UV irradiation, TiO_2 generates reactive oxygen species that can degrade contaminants completely in very short reaction time. Also, TiO_2 NPs show limited selectivity and are ideal for the degradation of pollutants of all kinds, such as chlorinated organic compounds (Ohsaka et al., 2008), polycyclic aromatic hydrocarbons (Guo et al., 2015), dyes (Lee et al., 2008), phenols (Nguyen et al., 2016), pesticides (Alalm et al., 2015), arsenic (Moon et al., 2014), cyanide (Kim et al., 2016), and heavy metals (Chen et al., 2016). Moreover, hydroxyl radicals generated under UV irradiation ($\lambda < 400$ nm) allow TiO_2 NPs to damage the function and structure of different microbial cells (Mills and Le Hunte, 1997). The removal of total organic carbon (TOC) from waters polluted with organic waste has been significantly improved in the presence of UV light by the addition of TiO_2 NPs as shown by Chitose et al. (2003). TiO_2 NPs' photocatalytic properties can kill a broad variety of microorganisms, such as Gram-negative (G–ve) and Gram-positive (G+ve) bacteria, as well as fungi, algae, protozoa, and viruses (Foster et al., 2011). A general mechanism of photocatalysis based on TiO_2 is depicted in Figure 7.4. However, other photocatalysts such as ZnO have also been produced to eliminate contaminants in wastewater and present reusable ability effectively (Berge and Ramsburg, 2009; Guesh et al., 2016).

Goutam et al. (2018) phytosynthesized titanium dioxide (TiO_2) NPs and measured its efficiency for industrial wastewater photocatalytic treatment, using the leaf extract of a biodiesel plant, *Jatropha curcass*. The green-synthesized TiO_2 photocatalyst had an average crystalline form, surface area, pore size diameter, and total pore volume of 13 nm, 27.038 m^2/g, 19.100 nm, and 0.1291 cm^3/g, respectively. The leaf extract analysis revealed the presence of various phytochemicals such as phenols and tannins that could be responsible for the reduction of metallic ions and act as capping agents during the green synthesis of TiO_2 NPs, thereby stabilizing the NPs. In addition, the green-synthesized TiO_2 NPs recorded 82.26% removal of COD. The higher solar photocatalytic activity of the green-synthesized TiO_2 NPs may be due to the nanosize, pure crystalline anatase phase, large amount surface hydroxyl groups, and higher surface area. Finally, the green synthesis of TiO_2 NPs using the biodegradable, nontoxic *J. curcass* leaves, which are normally considered as waste material but are environmentally friendly and easily available, adds to the advantages of applying such green TiO_2 NPs to remediate any type of wastewater after secondary treatment to combat environmental threats.

Nanobioparticles of TiO_2 NPs were synthesized by Sethy et al. (2020) using an aqueous solution of *Syzygium cumini* leaf extract as a capping agent. These green-synthesized TiO_2 NPs were further evaluated for photocatalytic removal of lead from industrial wastewater. Biosynthesized TiO_2 NPs were spherical and aggregated into an irregular structure with an average diameter of 18 nm in size. They possess average crystalline size, surface area, pore size diameter, and total pore volume of 10 nm, 105 m^2/g, 10.50 nm, and 0.278 cm^3/g, respectively. In a self-designed reactor, photocatalytic studies of TiO_2 NPs for removal of lead from wastewater were carried out. The concentration of lead was determined by using inductive coupled plasma spectroscopy. Results achieved were 75.5% reduction in COD and 82.53% removal of lead (Pb^{2+}). This application is being explored for the first time by the biosynthesized TiO_2 NPs.

However, the green-synthesized TiO_2 NPs for the photocatalytic treatment of PWW have not been applied so far because the production process of TiO_2 NPs is very complex; in addition, it is difficult to recover TiO_2 NPs from the treated wastewater particularly when used in suspension. More and more efforts have been devoted in recent years to overcome this problem. Among them, the coupling of TiO_2 NPs photocatalysis with membrane technology has attracted considerable attention and has shown promise to overcome the TiO_2 NP recovery problem. TiO_2 NPs have been incorporated into a wide range of membranes, such as poly(vinylidene-fluoride) (Meng et al., 2014; Wang et al., 2013), polyether-sulfone (Razmjou et al., 2011, 2012), polymethyl-methacrylate (Hamming et al., 2009), and poly(amide-imide) (Rajesh et al., 2013).

Egypt suffers from massive quantities of rubbish, which can exceed approximately 70 million tons each year, and it is estimated that animal bones contribute approximately 1% of it. Spheroidal fluorapatite (FAp, $Ca5(PO_4)3F$) NPs with average size of 55 nm have been reported to be easily prepared by a simple calcination process of waste buffalo bones (El-Gendy et al., 2016). Moreover, quasi-spherical ZnO NPs with average particles size of 21.5 nm were prepared by the wet chemical method using zinc nitrate and sodium hydroxide as precursors. Then, a monodispersed nanosphere photocatalytic biocomposite ZnO: $Ca5(PO_4)3F$ (ZnO-FAp) with average size of 40 nm was prepared by the mechanomixing technique. The Brunauer–Emmett–Teller surface area was 12.63, 8.63, and 5.87 cm^2/g and the total pore volume recorded ≈ 0.074, 0.02, and 0.016 cm^3/g with average pore diameter of ≈ 0.94, 0.12, and 0.11 nm for FAp, ZnO/FAp, and ZnO, respectively. One of the advantages of FAp is that it avoided the agglomeration of ZnO, as it made a shell layer around the ZnO NPs, retaining a stable dispersion of ZnO-FAp nanocomposite compared to the individual cases (see Figure 7.5). The adsorption capacity of the prepared nanomaterials toward CPs has been tested. That ranked as follows: ZnO/FAp > ZnO ≥ FAp, for 3-CP. But, for 2,3-dichlcrophenol (2,3-DCP), it ranked as follows: ZnO > FAp ≥ ZnO/FAp. That was attributed to the pollutant itself, since the CPs are in the molecular form at pH < pKa and exist as anions (negative charge) at pH > pKa (pKa of 3-CP and 2,3-DCP is 8.85 and 7.44, respectively), while the experiments were carried out at pH6 (below the pHpzc; 8, 9, and 8.4 for FAp, ZnO, and ZnO/FAp, respectively), implying that the surface of the catalysts was positively charged. That might explain the higher adsorption capacity of 2,3-DCP compared to 3-CP. Moreover, the photocatalytic activity of ZnO was enhanced by FAp and ranked in the following decreasing order: FAp > ZnO/FAp > ZnO. The higher SSA and porosity of FAp might explain its

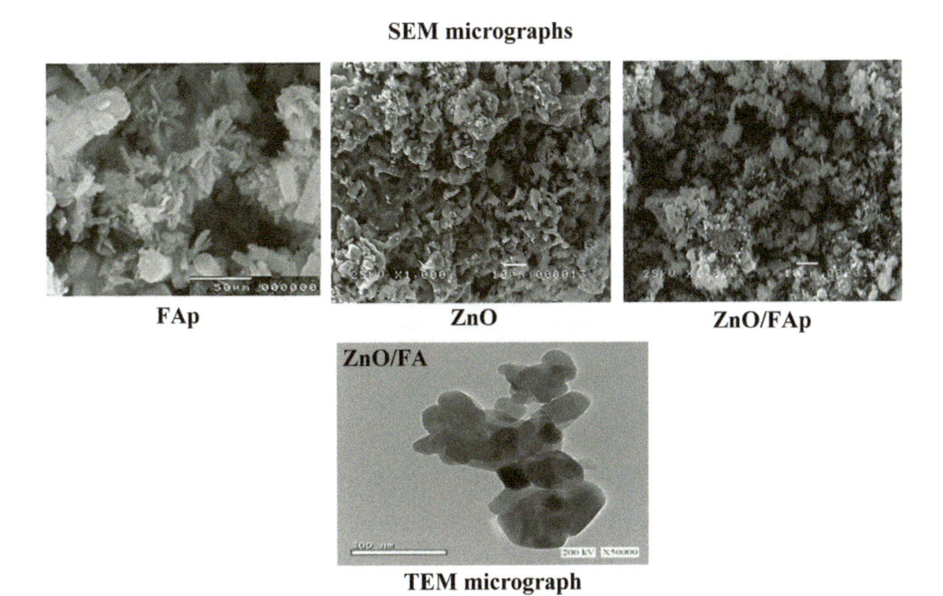

FIGURE 7.5 SEM and TEM micrographs of ZnO/FAp nanobiocomposite.

higher photocatalytic activity. The photocatalytic activity of FAp might have been caused by the generation of active superoxide anion radicals ($O_2^{2-}\bullet$), which would have been occurred due to the change in the electronic state of the surface PO_4^{3-} group under UV irradiation (El-Gendy et al., 2016). The green synthesis of iron NPs using natural product extracts has emerged in the past few years as a simple, eco-sustainable, and cost-effective method (Kumar et al., 2013; Machado et al., 2013; Smuleac et al., 2011).

Moreover, El-Gendy et al. (2016) mentioned that the enhancement of the photocatalytic activity of ZnO by FAp might have been due to the enhancement of the surface area, pore volume and diameter, surface hydroxyl groups, and molecular oxygen and might have been also due to the improved charge separation and extended energy range of photoexcitation, throughout the transition of energy levels between $Ca_5(PO_4)_3F$ orbital and ZnO orbital (see Figure 7.6).

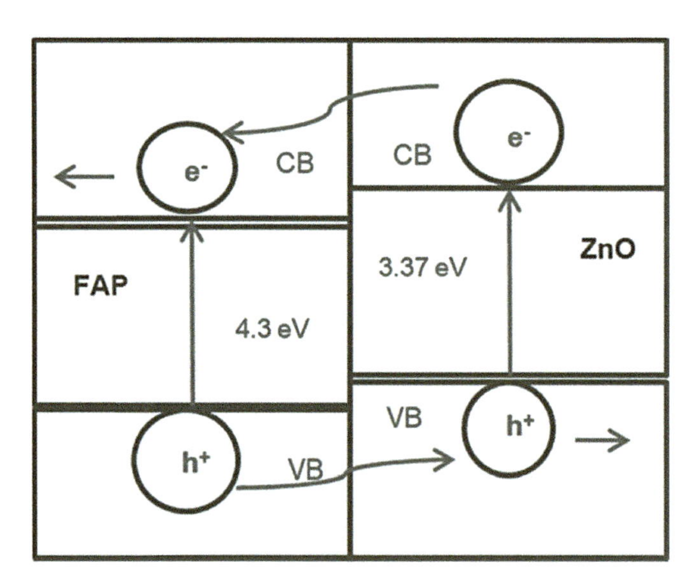

FIGURE 7.6 Proposed energetic model of ZnO and FAp nanobiocomposite.

El-Gendy et al. (2016) have also illustrated the photodegradation pathway of the studied CPs reaching acetate (see Figure 7.7), whereas, in the occurred photocatalytic reaction, photogenerating electrons and holes were captured by O_2 and H_2O adsorbed by the photocatalyst forming the super-active $^{\bullet}OH$ and $O_2^{2-\bullet}$ oxidants. Thus, the increased amount of the hydroxyl groups on the ZnO/FAp might have not only increased the trapping sites for photogenerated holes, but might have also increased the trapping sites for photogenerated electrons by adsorbing more molecular oxygen, which might have resulted in more hydroxyl radicles, which would have participated in the photocatalytic reaction (see Figure 7.8).

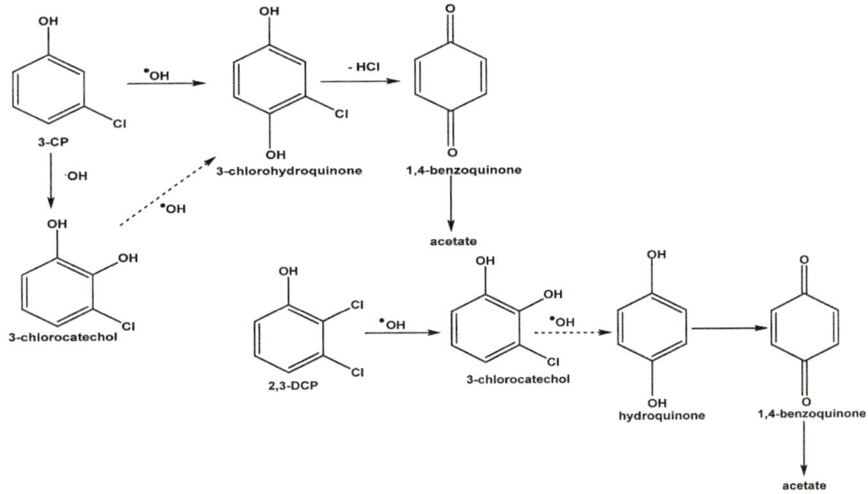

OH
Cl
3-CP
•OH
- HCl
3-chlorohydroquinone
1,4-benzoquinone
acetate
3-chlorocatechol
2,3-DCP
3-chlorocatechol
hydroquinone
1,4-benzoquinone
acetate

FIGURE 7.7 Photocatalytic degradation mechanism of 3-CP and 2,3-DCP by ZnO/FAp nanobiocomposite.

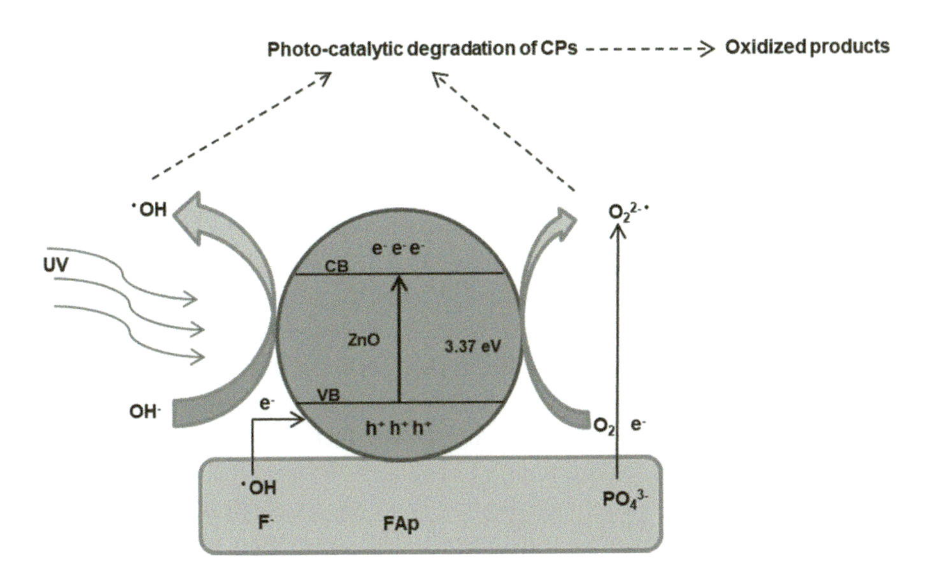

FIGURE 7.8 Photocatalytic mechanism for ZnO/FAP nanobiocomposite.

The application of green-synthesized iron hexacyanoferrate NPs is available in literature for the degradation of eight selected priority PAHs in soil and water (Shanker et al., 2017). Initial concentration of these PAHs ranged from 50–250 mg/L. Shanker et al. (2017) used a green synthesis of hexagonal, rod, and spherical-shaped hexacyanoferrate NPs by using a natural surfactant *Sapindus mukorossi,* with size ranging from 10 to 60 nm. Synthesized NPs were successfully employed for the photocatalytic removal of several hazardous PAHs, namely, anthracene, phenanthrene, chrysene, fluorene, and benzo(a)pyrene, from PWW. The catalytic experiment showed that under optimized experimental conditions (PAHs, 50 mg/L; catalyst dose, 25 mg; pH 7; and solar irradiation), all the PAHs compounds were converted into their nontoxic moieties. Around 80%–90% degradation was achieved for anthracene and phenanthrene. The degradation of chrysene, fluorene, and benzo(a)pyrene was found to be 70%–80%. Also, Fe NPs were manufactured in a novel attempt by applying tea extract as a green reduction agent and applied for phenanthrene catalytic removal (Jin et al., 2016). The synthesized NPs showed outstanding ability and resulted in complete degradation of phenanthrene within 90 min of reaction time. GC-MS analysis identified *trans*-3-(*O*-hydroxyphenyl)-1-phenyl-2-propen-1-one as a degradation by-product.

Muthukumar et al. (2017) followed a green procedure and synthesized Fe-doped ZnO NPs (Fe-ZnO) by applying an aqueous extract of *Amaranthus dubius* leaf. This biologically produced Fe-ZnO nanocomposite was applied as a nanobio catalyst for the aqueous phase remediation of naphthalene-contaminated water. Fourier transform infrared (FTIR) spectra of leaf extract confirmed the presence of amaranthine and other phenolics responsible for the capping of synthesized nanobioparticles. The highest degradation (92.3%) was achieved with the 40 ppm of naphthalene at pH 4.0 under visible light within 240 min of contact time.

Aromatic amines are a common class of organic compounds made up of an aromatic ring connected to an amine. These are commonly used as an intermediate reagent in various industries such as oil refining, drugs, rubber, textiles, agrochemicals, synthetic polymers, pharmaceuticals, and pesticides. Because of their high aqueous solubility, they can easily contaminate surface and groundwater. Mineralization of these harmful contaminants by nano-sized catalysts can be an attractive alternative technique to minimize their quantity in the aquatic environment. For that, Khan et al. (2016) biosynthesized the AgNPs from *Livistona chinensis* fruit juice and applied them for the photocatalytic removal of 4-nitro phenol. The developed nano biocatalyst was spherical shaped having an average size of 10 nm. FTIR spectra of fruit extract confirmed the presence of active biomolecules such as phenolics, proteins, enzymes, and vitamins that acted as reducing and capping agents. Over 87% of 4-nitrophenol degradation was accomplished with a reaction time of just 30 min.

7.3.2.4 BIOLOGICAL TREATMENT BY GREEN NANOPARTICLES

The term biological treatment is used to describe the process of using biological agents such as fungi, bacteria, and algae to remove and detoxify environmental pollutants, while microbes can use chemical pollutants as an energy source during metabolic processes (El-Sheekh et al., 2015, 2016).

The effectiveness of microbial degradation is restricted by any of the environmental factors such as temperature, availability of nutrients, microorganism activity, contaminant characteristics and composition, and biodegradability of the contaminants (Chakraborty et al., 2012). This process is also considered to be a cleaner, cost-effective, and environmentally friendly technology for removing wastewater pollutants (Salgot and Folch 2018). The PWW is successfully treated by various biological methods (Melamane et al., 2007; Manyuchi and Ketiwa, 2013) such as activated sludge reactors or biofilm-based reactor to remove the organic pollutants. However, these processes have some disadvantages such as the extreme sludge production, and low capacity to COD removals (Jou and Huang, 2003).

Biological contaminants, for example, the pathogenic microorganisms are also considered as serious water pollutants (Roig et al., 2013). Wastewater from petroleum refineries contains toxic, carcinogenic and recalcitrant phenolic compounds (Younis et al., 2020). Thus, when talking about water treatment it is important to mention the application of green-synthesized NPs as photodegraders for organic xenobiotic pollutants and biocides for pathogenic microorganisms (Singh et al., 2018). The bottom-up green synthesis of NPs overcome the drawbacks of physicochemical methods. It is not only inexpensive but also nontoxic, less wastage of inputs, less complicated and less time consuming, safer, ecofriendly, and does not produce hazardous toxic wastes (Malik et al., 2014; Kumar and Kathireswari, 2016; Omran et al., 2018; El-Gendy and Omran, 2019).

Nanobioremediation is the new concept that integrates the use of NPs and bioremediation for sustainable remediation of environmental pollutants in contaminated matrix (Le et al. 2015; Cecchin et al., 2017). Generally, anaerobic degradation, aerobic degradation, or an integration of both methods is commonly applied in biological processes to treat PWW (Zhao et al., 2006).

Nanotechnology offers a great promise to stabilize and protect enzymes from mechanical and biological degradation, thereby increasing their half-life and enabling recycling in use while reducing the cost of biotreatment strategies. Encapsulation of xenobiotic-degrading enzymes/cells in green-synthesized NPs (1–100 nm) improves both stability and protection against degradation. Enzymes that bind to NPs are more stable and, therefore, less vulnerable to mechanical shearing and loss of three-dimensional structure. As a result, enzymes/cells remain stable and can be reused several times. Immobilization of enzymes/cells using such approaches provides an excellent opportunity to extend the half-life and reusability of enzymes/cells and, therefore, reduce the cost of operation.

The combination of biological processes (aerobic degradation) with zero-valent iron (ZVI) NPs is considered as a promising treatment in the effective biodegradation of large-scale organic contaminants from wastewater. In this integration, NPs degrade half of the pollutants which are further easy for biodegradation under aerobic conditions (Ma and Zhang, 2008).

Partial degradation and reduction in organic compounds occurred due to bimetallic structures form micro and nanoscale cells as an electron donor with metallic iron. Besides, the products from the ZVI reactor are then more biodegradable by aerobic microorganisms. For example, in case of degradation of petroleum hydrocarbons and chlorinated solvents, the first section consists of granular iron in which chlorinated ethenes reduced through dechlorination while the second section promotes the aerobic biodegradation of remaining chlorinated compounds and any other hydrocarbons by the addition of dissolved oxygen (Bell et al., 2003; Morkin et al., 2000).

Tungittiplakorn et al. (2005) investigated the use of poly(ethylene)glycol-modified urethane acrylate (PMUA) NPs in improving phenanthrene's bioavailability. The three roles defined by the use of PMUA in the process were solubilization, adsorption, and biodegradation. The extent of the solubilization was demonstrated by the 780- and 3260-mL/g distribution coefficients found in the absence and presence of PMUA. Due to the accessibility of adsorbed contaminant on PMUA NPs for *Comamonas testosteroni*, biodegradation of phenanthrene was further promoted. Immobilization of metallic NP on cellulose acetate membrane for degradation of chlorinated compounds is also a reported hybrid process. Cellulose acetate enables the incorporation of 24-nm NP of Fe and Ni in membrane films; this hybrid film showed a TCE reduction of more than 75% in 4.25 h (Meyer et al., 2004).

For efficient phenol degradation, Wang et al. (2012) immobilized laccase enzyme on magnetic mesoporous silica NPs. The immobilized laccase degradation rate of phenol was two-fold higher than that of the free laccase, and the immobilized laccase retained 71.3% of its initial degradation capacity after 10 successive industrial wastewater batch treatments. Due to good mixing and mass transfer, the phenol degradation in industrial wastewater was enhanced by the immobilized laccase in a magnetically stabilized fluidized bed (MSFB) in a continuous treatment process. The phenol degradation rate retained over 99% at a flow rate of less than 450 mL/h and gradually decreased to 91.5% after 40 h of the MSFB continuous activity. The immobilized laccase on magnetic mesoporous silica NPs showed remarkably improved catalytic capacity and stability properties in industrial wastewater to degrade phenol. The immobilized laccase had a high capacity to efficiently degrade phenol and total phenolic compounds in industrial wastewater along with an MSFB. These findings offered the tremendous potential for the enzymatic degradation in industrial wastewater of phenolic compounds.

El-Sheshtawy et al. (2014) enhanced the bioremediation of oil polluted seawater by the application of αFe_2O_3 or $Zn_5(OH)8Cl_2$ NPs in microcosm inoculated by two halotolerant and biosurfactant producer *Pseudomonas xanthomarina* KMM 1447 and *Pseudomonas stutzeri* ATCC 17588. It is worth to know that applying of bacterial consortium enhances the rate of bioremediation. Saed et al. (2014) recommended also the application of MNPs Fe_3O_4 as a decorating nanotoxic material with superparamagnetic properties for *Micrococcus lutes* RM1 in the biodegradation of PAHs. The $t_{1/2}$ values of the batch biodegradation of 1500 mg/L pyrene were, 17 and 24 days in coated and free cells reactors, respectively.

The immobilization of the haloalkaliphilic S-oxidizing bacteria *Thialkalivibrio versutus* D301 using superparamagnetic MNPs under haloalkaliphilic conditions (pH 9.5) to remove sulfide, thiosulfate and polysulfide from wastewater has been studied (Xu et al., 2015). The MNPs-coated cells expressed similar activity like that of free cells, and can be reused for six batch cycles. In another study, a sulfate-reducing *Citrobacter* strain coated with Fe_3O_4/SiO_2 MNPs enhanced the sulfate reduction into sulfide, recording 450% increase relative to free cells (Zhou et al., 2015). That recommends its application for bioremediation of sulfate rich wastewater. In a later study concerned with the application of immobilized SRB in sulfate removal from wastewater, a series of magnetic macroporous polymer beads for enhancing the biofilm loading and shorten the start-up stage has been investigated (Lie et al., 2019). It is proved that the adding of Fe_3O_4 NPs restrains the enrichment phenomenon of silicone and consequently increases the biofilm loading capacity. Besides, the sulfate reduction increased by approximately 10% and the immobilized SRB in such magnetic macroporous polymer beads proved its stable efficiencies on the long-time continuous desulfurization stage (Lie et al., 2019). That was attributed to the magnetite lattice, where Fe(II) and Fe(III) exist. Iron (Fe^0) is an important competent of the sulfite reductase and participates in the sulfate reduction. It reacts with H^+, producing H_2, as electron acceptor and Fe(II), which precipitate S^{2-} as FeS, reducing its toxic effect on SRB (Liu et al., 2015). It is known that in the resend of Fe(III) and sulfate ions in SRB media, Fe(III) is the preferred electron acceptor. Since Fe(III) has a higher redox potential, relative to sulfate, it consequently produces more energy when the same number of electrons is transmitted. Thus, Fe(III) is thermodynamically more beneficial for the microbial growth relative to sulfate. Briefly, Fe_3O_4 NPs embedded macroporous material has prospective enhancement toward SRB adhesion and desulfurization performance.

Jiang et al. (2015) investigated the bioremediation of PWW containing BTEX compounds by immobilized *Comamonas* sp. JB cells. Three kinds of MNPs were examined as immobilization supports for strain JB. The highest biodegradation activity of Fe_2O_3-magnetically immobilized cells was achieved when the concentration of the MNP was 120 mg/L. Additionally, the recycling experiments showed that the degradation activity of magnetically immobilized cells was still high and led to less toxicity than untreated wastewater during the eight recycles. After eight cycles of degradation experiments, qPCR indicated that the concentration of strain JB in r-Fe_2O_3-magnetically immobilized cells had clearly increased. These findings supported the production of effective biocatalysts using magnetically immobilized cells and provided a promising technique to enhance biocatalysts used to bioremediate not only petrochemical wastewater but also other hazardous wastewater.

Sarno and Iuliano (2020) reported an eco-friendly approach for the removal of 4-CPs from wastewater, through the use of horseradish peroxidase (HRP) immobilized on Fe_3O_4/Au@CA NPs. The effects of coupling time and concentration of enzymes on the efficiency of immobilization were studied. The effect of reaction time and the reusability on removing 4-CP were also evaluated. The degradation process promoted by immobilized HRP was fast in the first 30 min and increased by up to 67%. The 4-CP degradation in the presence of immobilized HRP after 180 min has been as high as 98%. The TOC percent removal was 85% after 180 min. It could be considered as a promising approach in wastewater treatment, biocatalysts, biosensors, and other catalytic enzyme protocols, considering also the high stability shown under cycles of use.

7.3.2.5 ANTIMICROBIAL ACTIVITY THROUGH GREEN NANOPARTICLES

Water pollution has caused lots of infectious diseases due to various contaminating pathogens. Disinfection is the critical step in the water purification process to avoid the burden of water-borne diseases. Many of the microorganisms acting as pathogens are antibiotic resistance and so it is very difficult to remove them from water. In the water treatment industry, the use of conventional disinfectants (i.e., ozone and chlorine as the disinfectant) possesses the formation of toxic "disinfection by-products" (Bethi et al., 2016). Recently, the concept of bioactive NPs has emerged, which has given the alternative of new chlorine-free biocide. AgNPs can be biosynthesized extracellularly by bacteria *Bacillus cereus,* which is having very high antibacterial potential (Mann, 2017).

From a practical point of view, an ideal disinfectant should have the following requirements (Zhang et al., 2016; Bethi et al., 2016): (1) no formation of harmful by-products of disinfection process; (2) short-term disinfection; (3) the formation of a broad antimicrobial spectrum; (4) easy operation and low energy costs; (5) must not be corrosive and easy to store; and (6) the sludge after treatment should be safe for final disposal. Various nanomaterials have recently exhibited antimicrobial properties without strong oxidation processes. These include nanosilver (Rai et al., 2009), chitosan NPs (Higazy et al., 2010), titanium dioxide TiO_2photocatalytic (Hebeish et al., 2013), and nanomaterials based on carbon (Martynková and Valášková 2014). The nanomaterials are capable of killing pathogenic microorganisms through the release of toxic metal ions.

Bacteria are classified into two groups based on the composition of their cell walls: G –ve bacteria have a multilayer cell wall and G +ve bacteria have a single layer cell wall. Biosynthesized NPs have been studied for antibacterial activity in both types of bacteria. A variety of biological materials with demonstrable antibacterial effects have been used for biosynthesis of NPs. These products include fungal, bacterial, and algal biomass, and extracts of botanical products including leaves, bark, roots, and tubers. Vigneshwaran et al. (2007) published one of the first studies on the antibacterial activity of biosynthesized NPs. These authors used the edible *Pleurotus sajorcaju* mushroom for AgNPs synthesis and performed successful antimicrobial testing against the bacteria *Staphylococcus aureus* (G +ve) and *Klebsiella pneumoniae* (G –ve).

In bactericidal activity, AgNPs initially make free radicals, which are adhered on the bacterial cell wall and damage it. After damaging the cell wall, they interconnected with bacterial DNA and altered the cell membrane properties and denatured the enzymes. The significant role of AgNPs for wastewater treatment is efficiently inactivating the microbial cells and also reducing the membrane biofouling (Jain and Pradeep, 2005).

The Ag, Zn, and TiO_2 NPs are very effective against pathogenic microorganisms like *Escherichia coli, Bacillus subtilis, Vibrio cholera, Pseudomonas aeruginosa, Syphillis typhus,* and *Staphylococcus aureus* (Baruah et al., 2012). The green NPs having higher surface areas facilitate the attachments of varied chemical functional groups; the various functional groups increase their affinity for target contaminants and successfully remove them from water bodies (Chatterjee et al., 2016).

Photocatalysis can be applied for the removal of pathogenic microorganisms and trace contaminants from water. In this process, the biodegradability nature of hazardous and nonhazardous impurities is enhanced. Nanocatalysis includes the following steps (Reddy et al., 2016): (1) diffusion of the pollutant, (2) adsorption onto the surface of the catalyst, (3) reaction occurring on the surface of the catalyst, (4) desorption of the product(s) from the surface of the catalyst, and (5) diffusion of the product from the surface. Photocatalysis is also used as a polishing step for the treatment of recalcitrant organic compounds.

Microbiologically influenced corrosion (MIC) is a process influenced by different microorganisms, in particular by SRB that affects the corrosion procedure kinetics under anaerobic conditions. Approximately 20% of annual metal corrosion damage can be caused by microbial activity, particularly due to SRB-influenced anaerobic corrosion. MIC is the key contributor to corrosion issues and is a leading cause of oil and gas pipeline failure. SRBs are major microorganisms that can generate anaerobic sulfide species that cause biocorrosion in the injection networks. The emitted H_2S gas is also poisonous, corrosive and responsible for a number of environmental issues. Additionally, because of sulfide production, the presence of SRB can result in health and safety risks for workers. To avoid this, oil-producing firms use high concentrations of biocides to disinfect the water and inhibit the excessive development of biofilms caused mainly by SRB. Traditional biocides, however, may be detrimental to the environment by forming by-products of detrimental disinfection. The biocidal treatment also has other disadvantages such as low biofilm efficiency, release of by-products of disinfection and its high cost. These disadvantages can be solved by using green biocides including very low toxic nanomaterials, environmental acceptability, safety and user-friendliness, etc.

Rasool et al. (2018) synthesized a green and highly stable biocide formulations composed of ZnO-interlinked chitosan (Ch) nanoparticles (CZNCs) and evaluated their antimicrobial activity against mixed SRBs culture isolated from real oil field sludge. Scanning electron microscopy (SEM), transmission electron microscopy (TEM), X-ray diffraction (XRD), and FTIR suggested the formation of stable nanocomposites with strong interaction between ZnO and Ch NPs. Synthesized nanocomposites showed highly stable behaviors in the high salt concentrations of injecting seawater. The inhibition of SRBs activity was concentration-dependent and more than 73% and 43% inhibition of sulfate reduction and TOC removal, respectively, was observed at 250 μg/mL CZNCs at 10% initial ZnO loading. Biocompatibility and environmental impact of the nanocomposite were evaluated by analyzing their potential toxicity in vivo using the zebrafish embryos. Neither mortality nor teratogenic effects were observed on zebrafish embryos using the acute toxicity assay. The hypothetical LC50 for the CZNCs was much higher than 250 μg/mL. It is expected that the new nanocomposite can contribute to the development of "green" biocides for oil/gas industries that will be eco-friendly and will have no adverse impact on the environment.

Zarasvand and Rai (2016) screened several bacteria for the biosynthesis of copper/copper compound NPs, which could inhibit the growth of *Desulfovibrio marinisediminis*, a SRB. Supernatant of 30 bacteria isolated from the biofilm formed on ship hull was mixed with 1-mM $CuCl_2$ solution at room temperature. Eight bacterial strains, whose mixtures exhibited color change, were selected for antimicrobial test. One NP that has been biosynthesized by *Shewanella indica* inhibited the growth of *D. marinisediminis*. Characterization of this particle by UV/Visible spectrophotometer, XRD, TEM, dynamic light scattering, and FTIR showed that the particle is polydisperse CuO NP with average size of 400 nm. Treatment of bacterial solution with CuONP reduced bacterial number from 9×10^6 CFU/mL (Number of bacteria in untreated sample) to 4×10^3 CFU/mL. Minimum inhibitory concentration of NPs, which is defined as the lowest concentration at which there is no blackening of media, was found to be 100-mg/mL concentration.

Pathath et al. (2019) proposed an environmentally friendly approach to use a green biocide; chitosan–ZnO nanocomposite against SRB induced

MIC toward carbon steel. The NPs of chitosan and ZnO were prepared independently and processed together to form the chitosan–ZnO nanocomposite. The average size of chitosan NPs was between 40 and 60 nm, and it clearly shows the distribution of ZnO NPs in the chitosan NPs matrix. The effect of chitosan–ZnO nanocomposite on corrosion behavior of carbon steel against SRB was investigated by electrochemical impedance spectroscopy, corrosion potential, polarization resistance, and polarization curve measurements at different time intervals. It was found that the chitosan–ZnO nanocomposite inhibits the SRB biofilm formation and corrosion. The results of the electrochemical analysis showed that the chitosan–ZnO nanocomposite (10% ZnO content) with the highest corrosion inhibition concentration at 250 ppm can be used as an effective corrosion inhibitor against SRB induced MIC.

Rasheed et al. (2020) investigated the ability of the new "green" chitosan/lignosulfonate (CS/LS) nanospheres with an optimal 1:1 (CS:LS) ratio to treat SRB induced MIC on SS400 carbon steel. CS/LS nanospheres with an average diameter of 150–200 nm have been successfully used as a novel, environmentally friendly biocide for the inhibition of mixed SRB culture, thereby controlling MIC on carbon steel. It was found that 500 μg/mL of the CS/LS nanospheres can be used efficiently for the inhibition of SRB-induced corrosion up to a maximum of 85% indicated by a two-fold increase of charge transfer resistance on the carbon steel coupons. The hydrophilic surface of CS/LS can readily bind to the negatively charged bacterial surfaces and thereby leads to the inactivation or damage of bacterial cells. In addition, the film formation ability of chitosan on the coupon surface may have formed a protective layer to prevent the biofilm formation by hindering the initial bacterial attachment, thus leading to the reduction of corrosion.

Aqueous extracts of tangerine (*Citrus reticulata*) was used by Ituen et al. (2020) to mediate the synthesis of green silver nanoparticles (TPE-AgNPs), at ambient temperature. The synthesized NPs were applied as microbial and acid corrosion inhibitor for X80 steel in simulated oilfield environments. TPE-AgNPs affords a 3-log reduction in population of and 76.2% inhibition efficiency on *Desolfovibrio* sp. induced corrosion at 2.88 mg/L MIC.

7.3.2.6 GREEN NANOMEMBRANE FILTRATION TECHNOLOGY

Membrane technology is a promising solution for PWW treatment due to its advantages such as cost-effectiveness, free from chemical additives, standard installation, and ambient temperature operation compared to those traditional methods (Padaki et al., 2015). In a broad term, membrane filtration involves the physical separation of the unwanted impurities from the bulk solutions through a semipermeable membrane depending on their pore size and molecule size. Depending on the principles of the membrane technology, hydraulic pressure or osmotic pressure is required for the operation of this technology.

Membrane filtration can be categorized as ultrafiltration (Asatekin and Mayes, 2009), microfiltration (Zhong et al., 2003), nanofiltration (Moslehyani et al., 2015), and reverse osmosis membranes (Li et al., 2006), depending on the pores size.

Nanomembranes can be defined as filters with free-standing structure of thickness in the range of 1–100 nm, which separate liquids and gases at the molecular level. Such membranes are made from organic polymers, combined with a mesh of NPs.

In general, NPs affect the selectivity and permeability of a membrane that depends on the NPs size, quantity, dimension etc. In addition, there are many biological membranes with extremely permeable and selective capabilities (Yin et al., 2013; Zhang et al., 2014). Nanomembranes are also used for wastewater treatment due to having several properties that make this material more prolific, these are high uniformity, homogeneity ability, optimization, short time required, easily handled, and contain much order of reaction (Hirata et al., 2019). There are several nanophotocatalysts that can be used in the nanocomposite membrane to make it suitable for organic pollutant degradation. This versatile technique has combined the advantages of both photocatalysis and filtration to achieve synergetic effects for oily wastewater treatment. For example, TiO_2-incorporated nanomembranes and films are effectively used to deactivate different microorganisms and degrade the organic pollutants (Gopalakrishnan et al., 2018). In order to synthesize these types of nanomaterials, there were many approaches used to produce it with multifunctionality features (Kunduru et al., 2017).

Moslehyani et al. (2015) indicated that the best nanocomposite membrane was that with the multiwalled carbon nanotube incorporated in polyvinylidene fluoride (PVDF) matrix for the filtration purposes because it removed all pollutants from refinery wastewater and it was with the excellent antifouling property.

Rahman and Al-Malack (2006) reported that the using of a cross-flow membrane bioreactor to treat the PWW achieved a COD removal efficiency of more than 93% at mixed liquor suspended solid concentrations of 5000 and 3000 mg/L.

Razavi and Miri (2015) showed that the average removal efficiencies of COD, BOD5, TSS, VSS, and turbidity from real petroleum refinery wastewater by using the hollow-fiber membrane bioreactor were attained 82%, 89%, 98%, 99%, and 98%, respectively.

Nanoscale science and engineering technologies have suggested that the use of nanocatalysts, nanoabsorbents, nanotubes, nanostructured catalytic membranes, nanopowder, and micromolecules could solve many of the current water quality problems. All of these NPs and colloids have had a significant impact in PWW treatment process (Savage and Diallo, 2005). Studies have shown that integration of the biological wastewater treatment process with advanced nanotechnology has resulted in an efficient water purification system.

Hu et al. (2015) conducted an investigation in which they manufacture hollow fiber membranes made of PVDF with TiO_2 by the phase inversion method on a custom-designed single-head spinning machine. The fabricated PVDF/TiO_2 nanocomposite membranes were then tested on algal membrane bioreactor (A-MBR) for wastewater treatment. Results showed that maximum nutrient removal was achieved and up to 75% of phosphorus and nitrogen was removed by A-MBRs. In addition, the PVDF/TiO_2 membranes in the A-MBR also increase hydrophilic characteristic of PVDF and reduce membrane fouling.

REFERENCES

Abbas, S., Nasreen, S., Haroon, A., and Ashraf, M. A. 2020. Synthesis of silver and copper nanoparticles from plants and application as adsorbents for naphthalene decontamination. *J. Biol. Sci.* 27(4): 1016–1023.

Abdullah, M.M.S., Atta, A.M., Allohedan, H.A., Alkhathlan, H.Z., Khan, M., and Ezzat, A.O. 2018. Green synthesis of hydrophobic magnetite nanoparticles coated with plant extract and their application as petroleum oil spill collectors. *J. Nanomater.* 8(855): 1–13.

Abu Hassan, M.A., Li, T.P. and Noor, Z.Z. 2009. Coagulation and flocculation treatment of wastewater in textile industry using chitosan. *J. Chem. Nat. Resour. Eng.* 4: 43–53.

Adesina, A.A. 2004. Industrial exploitation of photocatalysis progress, perspectives and prospects. *Catal. Surv. Asia.* 8(4): 265–273.

Adham, S., Hussain, A., Minier-Matar, J., Janson, A., and Sharma, R. 2018. Membrane applications and opportunities for water management in the oil & gas industry. *Desalination.* 440: 2–17.

Ahmad, A. 2020. Bioprocess evaluation of petroleum wastewater treatment with zinc oxide nanoparticle for the production of methane gas: Process assessment and modelling. *Appl. Biochem. Biotechnol.* 190(3): 851–866.

Al Hashemi, W., Maraqa, M.A., Rao, M.V., and Hossain, M.M. 2015. Characterization and removal of phenolic compounds from condensate-oil refinery wastewater. *Des. Water Treat.* 54: 660–671.

Alalm, M.G., Tawfik, A., and Ookawara, S. 2015. Comparison of solar TiO_2 photocatalysis and solar photo-Fenton for treatment of pesticides industry wastewater: Operational conditions, kinetics, and costs. *J. Water Process Eng.* 8: 55–63.

Ali, I. 2012. New generation adsorbents for water treatment. *Chem. Rev.* 112: 5073–5091.

Aljuboury, D.A.D.A., Palaniandy, P., Abdul Aziz, H.B., and Feroz, S. 2017. Treatment of petroleum wastewater by conventional and new technologies—A review. *Global Nest J.* 19(3): 439–452.

Ambashta, R.D., and Sillanpaa, M. 2010. Water purification using magnetic assistance: A review. *J. Hazard. Mater.* 180: 38–49.

An, C., Huang, G., Yao, Y., and Zhao, S. 2017. Emerging usage of electrocoagulation technology for oil removal from wastewater: A review. *Sci. Total Environ.* 579: 537–556.

Asatekin, A., and Mayes, A.M. 2009. Oil Industry wastewater treatment with fouling resistant membranes containing amphiphilic comb copolymers. *Environ. Sci. Tech.* 43: 4487–4492.

Ayati, A., Ahmadpour, A., Fatemeh F., Bahareh, T., Mika, M., and Mika, S. 2014. A review on catalytic applications of Au/TiO_2 nanoparticles in the removal of water pollutant. *Chemosphere*. 107: 163–174.

Azadi, F., Karimi-Jashni, A., and Zerafat, M.M. 2018. Green synthesis and optimization of nano-magnetite using *Persicaria bistorta* root extract and its application for rosewater distillation wastewater treatment. *Ecotoxicol. Environ. Saf.* 165: 467–475.

Baruah, S. K., Pal, S., and Dutta, J. 2012. Nanostructured zinc oxide for water treatment. *Nanosci. Nanotechnol. – Asia*. 2: 90–102.

Bell, L.S., Devlin, J.F., Gillham, R.W., and Binning, P.J. 2003. A sequential zero valent iron and aerobic biodegradation treatment system for nitrobenzene. *J. Contam. Hydrol.* 66: 201–217.

Berge, N.D. and Ramsburg C.A. 2009. Oil-in-water emulsions for encapsulated delivery of reactive iron particles. *Environ. Sci. Technol.* 43 (13): 5060–5066.

Bethi, B., and Sonawane, S.H. 2018. Nanomaterials and its application for clean environment. In: *Nanomaterials for Green Energy*. Elsevier. pp. 385–409.

Bethi, B., Sonawane, S.H., Bhanvase, B.A., and Gumfekar, S.P. 2016. Nanomaterials-based advanced oxidation processes for wastewater treatment: A review. *Chem. Eng. Proc. Process Int.* 109:178– 189.

Bishoge, O.K., Zhang, L., Suntu, S.L., Jin, H., Zewde, A.A., and Qi, Z. 2018. Remediation of water and wastewater by using engineered nanomaterials: A review. *J. Environ. Sci. Heal A.* 53: 537–554.

Boyano, A., Lázaro, M.J., Cristiani, C., Maldonado-Hodar, F.J., Forzatti, P., and Moliner, R. 2009. A comparative study of V_2O_5/AC and V_2O_5/Al_2O_3 catalysts for the selective catalytic reduction of NO by NH_3. *Chem. Eng. J.* 149: 173–182.

Carabante, I., Grahn, M., Holmgren, A., Kumpiene, J., and Hedlund J. 2009. Adsorption of As(V) on iron oxide nanoparticle films studied by in situ ATR-FTIR spectroscopy. *Colloids Surf. A, Physicochem. Eng. Aspects.* 346: 106–113.

Cecchin, I., Krishna, R., Thom, A., Tessaro, E.F., and Schnaid, F. 2017. Nanobioremediation: Integration of nanoparticles and bioremediation for sustainable remediation of chlorinated organic contaminants in soils. *Int. Biodeterior. Biodegrad.* 119: 419–428.

Chakraborty, N., Banerjee, A., Lahiri, S., Panda, A., Ghosh, A., and Pal, R. 2009. Biorecovery of gold using cyanobacteria and an eukaryotic alga with special reference to nanogold formation—A novel phenomenon. *J. Appl. Phycol.* 21: 145–52.

Chakraborty, R., Wu, C.H., and Hazen, T.C. 2012. Systems biology approach to bioremediation. *Curr. Opin. Biotechnol.* 23(3): 483–490.

Chatterjee, A., Nishanthini, D., Sandhiya, N., and Abraham, J. 2016. Biosynthesis of titanium dioxide nanoparticles using *Vigna radiata*. *Asian J. Pharm. Clin. Res.* 9(4): 85–88.

Chen, B., Chen, Z., and Lv, S., 2011. A novel magnetic biochar efficiently sorbs organic pollutants and phosphate. *Bioresour. Technol.* 102: 716–723.

Chen, Z.P., Li, Y., and Guo, M. 2016. One-pot synthesis of Mn-doped TiO_2 grown on graphene and the mechanism for removal of Cr(VI) and Cr(III). *J. Hazard. Mater.* 310: 188–198.

Chitose, N., Ueta, S. and Yamamoto, T.A. 2003. Radiolysis of aqueous phenol solutions with nanoparticles. 1. Phenol degradation and TOC removal in solutions containing TiO_2 induced by UV, gamma-ray and electron beams. *Chemosphere.* 50(8): 1007–1013.

Chunling, L., Chuanping, L., Yan, W., Xiang, L., Fangbai, L., Gan, Z., and Xiangdong, L. 2011. Heavy metal contamination in soils and vegetables near an e-waste processing site, south China. *J. Hazard. Mater.* 186 (1): 481–490.

Clark, C.E., and Veil, J.A. 2009. Produced water volumes and management practices in the United States. Argonne National Lab, Argonne, IL, USA, Technical Report, ANL/EVS/R-09-1.

Coelho, A., Castro, A.V., Dezotti, M., and Sant'Anna, Jr., G.L. 2006. Treatment of petroleum refinery sour water by advanced oxidation processes. *J. Hazard Mater. B.* 137(1): 178–184.

Das, D., Das, N., and Mathew, L. 2010. Kinetics, equilibrium and thermodynamic studies on biosorption of Ag(I) from aqueous solution by macro fungus *Pleurotus platypus*. *J. Hazard Mater.* 184: 765–74.

Deshpande, B.D., Agrawal, P.S., Yenkie, M.K.N., and Dhoble, S.J. 2020. Prospective of nanotechnology in degradation of waste water: A new challenges. *Nano-Struct. Nano-Objects.* 22: 1–20.

Dittrick, P. 2017. Water constraints drive recycle, reuse technology. *Oil Gas J.* 15: 50–53.

El-Gendy, N.Sh., and Omran, B.A. 2019. Green synthesis of nanoparticles for water treatment. In *Nano and Bio-Based Technologies for Wastewater Treatment: Prediction and Control Tools for the Dispersion of Pollutants in the Environment*, Elvis Fosso-Kankeu (Ed.), Wiley, Hoboken, NJ, USA, pp. 205–264. https://doi.org/10.1002/9781119577119.ch7

El-Gendy, N.Sh., El-Salamony, R.A., and Younis, S.A. 2016. Green synthesis of fluorapatite from waste animal bones and the photo-catalytic degradation activity of a new ZnO/green biocatalyst nano-composite for removal of chlorophenols. *J. Water Process Eng.* 12: 8–19. https://doi.org/10.1016/j.jwpe.2016.05.007

El-Naas, M.H., Al-Zuhair, S., Al-Lobaney, A., and Makhlouf S. 2009. Assessment of electrocoagulation for the treatment of petroleum refinery wastewater. *J. Environ. Manage.* 91: 180–185.

El-Shamy, A.M. Abdelfattah, I., Elshafey, O.I., and Shehata, M.F. 2018. Potential removal of organic loads from petroleum wastewater and its effect on the corrosion behavior of municipal networks. *J. Environ. Manage.* 219: 325–331.

El-Sheekh, M.M., El-Abd, M.A., El-Diwany, A.I., Ismail, A.M.S., and Omar, T.H. 2015. Poly-3-hydroxybutyrate (PHB) production by *Bacillus flexus* ME-77 using some industrial wastes. *Rend. Lincei.* 26: 109–119.

El-Sheekh, M.M., Farghl, A.A., Galal, H.R., and Bayoumi, H.S. 2016. Bioremediation of different types of polluted water using microalgae. *Rend. Lincei.* 27(2): 401–410.

El-Sheshtawy, H.S., Khalil, N.M., Ahmed, W., and Abdallah, R.I. 2014. Monitoring of oil pollution at Gemsa Bay and bioremediation capacity of bacterial isolates with biosurfactants and nanoparticles. *Mar. Pollut. Bull.* 87: 191–200.

Fakhru'l-Razi, A., Pendashteh, A., Abdullah, L.C., Biak, D.R.A., Madaeni, S.S., and Abidin, Z.Z. 2009 Review of technologies for oil and gas produced water treatment. *J. Hazard Mater.* 170(2–3): 530–551.

Fan, F.L., Qin, Z., Bai, J., Rong, W.D., Fan, F.Y., Tian, W., Wu, X.L., Wang, Y., and Zhao, L. 2012. Rapid removal of uranium from aqueous solutions using magnetic Fe_3O_4@SiO_2 composite particles. *J. Environ. Radioact.* 106: 40–46.

Farajnezhad, H., and Gharbani P. 2012. Coagulation treatment of wastewater in petroleum industry using poly aluminum chloride and ferric chloride. *In. J. Res. Rev. Appl. Sci.* 13: 306–310.

Fazlzadeh, M., Rahmani, K., Zarei, A., Abdoallahzadeh, H., Nasiri, F., and Khosravi, R. 2017. A novel green synthesis of zero valent iron nanoparticles (NZVI) using three plant extracts and their efficient application for removal of Cr(VI) from aqueous solutions. *Adv. Powder Technol.* 28: 122–130.

Foster, H.A., Ditta, I.B., Varghese, S., and Steele, A. 2011. Photocatalytic disinfection using titanium dioxide: Spectrum and mechanism of antimicrobial activity. *Appl. Microbiol. Biotechnol.* 90 (6): 1847–1868.

Franco, C.A., Zabala, R., and Cortés FB. 2017. Nanotechnology applied to the enhancement of oil and gas productivity and recovery of Colombian fields. *J. Pet. Sci. Eng.* 157: 39–55.

George, A.O.C. 1996. Organic compounds in sludge-amended soils and their potential for uptake by crop plants. *Sci. Total Environ.* 185 (1–3): 71–81.

Ghimire, N., and Wang, S. 2018. *Biological Treatment of Petrochemical Wastewater*. Intech Open, London, UK.

Gitis, V., and Hankins, N. 2018. Water treatment chemicals: Trends and challenges. *J. Water Process Eng.* 25: 34–38.

Gopalakrishnan, I., Samuel, S.R., and Sridharan, K. 2018. Nanomaterials-based adsorbents for water and waste water treatment. *Emerg. Nanotechnol. Environ. Sustain.* 6: 89–98.

Goutam, S.P., Saxena, G., Singh, V., Yadav, A.K., Bharagava, R.N., and Thapa, K.B. 2018. Green synthesis of TiO_2 nanoparticles using leaf extract of *Jatropha curcas* L. for photocatalytic degradation of tannery wastewater. *Chem. Eng. J.* 336: 386–396.

Guesh, K., Mayoral, A., Alvarez, C.M., Chebude, Y., and D´ıaz, I. 2016. Enhanced photocatalytic activity of TiO_2 supported on zeolites tested in real wastewaters from the textile industry of Ethiopia. *Micropor. Mesopor. Mater.* 225: 88–97.

Guo, M., Song, W., Wang, T., Li, Y., Wang, X., and Du, X. 2015. Phenyl-functionalization of titanium dioxide-nanosheets coating fabricated on a titanium wire for selective solid-phase microextraction of polycyclic aromatic hydrocarbons from environment water samples. *Talanta.* 144: 998–1006.

Hamming. L.M., Qiao, R., Messersmith, P.B., and Brinson, L. C. 2009. Effects of dispersion and interfacial modification on the macroscale properties of TiO_2 polymer-matrix nanocomposites, *Compos. Sci. Technol.* 69(11–12): 1880–1886.

Hebeish, A.A., Abdelhady, M.M., and Youssef, A.M. 2013. TiO_2 nanowire and TiO_2 nanowire doped Ag-PVP nanocomposite for antimicrobial and self-cleaning cotton textile. *Carbohyd. Polym.* 91(2): 549–559.

Higazy, A., Hashem, M., ElShafei, A., Shaker, N., and Hady, M.A. 2010. Development of antimicrobial jute packaging using chitosan and chitosan-metal complex. *Carbohyd. Polym.* 79(4): 867–874.

Hirata, K., Watanabe, H., and Kubo, W. 2019. Nanomembranes as a substrate for ultra-thin lightweight devices. *Thin Solid Film.* 676: 8–11.
Hodges, B.C., Cates, E.L., and Kim, J. 2018. Challenges and prospects of advanced oxidation water treatment processes using catalytic nanomaterial. *Nat. Nanotechnol.* 13: 642–650.
Hu, G., Li, J., and Hou, H. 2015. A combination of solvent extraction and freeze thaw for oil recovery from petroleum refinery wastewater treatment pond sludge. *J. Hazard Mater.* 283: 832–840.
Hu, G., Li, J., and Zeng, G. 2013. Recent development in the treatment of oily sludge from petroleum industry: A review. *J. Hazard. Mater.* 261:470–490.
Hu, J., Chen, G. and Lo, I.M.C. 2005. Removal and recovery of Cr(VI) from wastewater by maghemite nanoparticles. *Water Res.* 39: 4528–4536.
Hu, Y., Kwan, B.W., Osbourne, D.O., Benedik, M.J. and Wood, T.K. 2015. Toxin YafQ increases persister cell formation by reducing indole signalling. *Environ. Microbiol.* 17: 1275–1285.
Hussain, A., Maitra, J., and Khan, K.A. 2017. Development of biochar and chitosan blend for heavy metals uptake from synthetic and industrial wastewater. *Appl. Water Sci.* 7: 4525–4537.
Inyang, M., Gao, B., Zimmerman, A., Zhou, Y., and Cao, X. 2015. Sorption and cosorption of lead and sulfapyridine on carbon nanotube-modified biochars. *Environ. Sci. Pollut. Res.* 22: 1868–1876.
Ituen, E., Ekemini, E., Yuanhua, L., and Singh, A. 2020. Green synthesis of Citrus reticulata peels extract silver nanoparticles and characterization of structural, biocide and anticorrosion properties. *J. Mol.* 1207: 1–10.
Jafarinejad, S. 2017. Pollutions and wastes from the petroleum industry. In: *Petroleum Waste Treatment and Pollution Control*, Butterworth-Heinemann, Oxford, UK, pp. 19–83.
Jain, P., and Pradeep, T. 2005. Potential of silver nanoparticle-coated polyurethane form as an antibacterial water filter. *Biotechnol. Bioeng.* 90: 59–63.
Jiang, B., Zhou, Z., Dong, Y., Wang, B., Jiang, J., Guan, X., Gao, S., Yang, A., Chen, Z., and Sun, H. 2015. Bioremediation of petrochemical wastewater containing BTEX compounds by a new immobilized *Bacterium Comamonas* sp. JB in magnetic Gellan gum. *Appl. Biochem. Biotechnol.* 176(2): 572–581.
Jiménez, S., Micó, M.M., Arnaldos, M., Medina, F., and Contreras, S. 2018. State of the art of produced water treatment. *Chemosphere*. 192: 186–208.
Jin, X., Yu, B., Lin, J., and Chen, Z. 2016. Integration of biodegradation and nano-oxidation for removal of PAHs from aqueous solution. *ACS Sustain. Chem. Eng.* 4: 4717–4723.
Jou, C.G., and Huang, G. 2003. A pilot study for oil refinery wastewater treatment using a fixed film bioreactor. *Adv. Environ. Res.* 7: 463–469.
Kanchi, S. 2014. Nanotechnology for water treatment. *J. Environ. Anal. Chem.* 1: 2.
Karakoyun, N., Kubilay, S., Aktas, N., Turhan, O., Kasimoglu, M., Yilmaz, S., and Sahiner, N. 2011. Hydrogel-biochar composites for effective organic contaminant removal from aqueous media. *Desalination*. 280: 319–325.
Kausar, A., Iqbal, M., Javed, A., and Aftab, K. 2018. Dyes adsorption using clay and modified clay: A review. *J. Mol. Liq.* 256: 395–407.
Khan, A.U., Wei, Y., Haq Khan, Z.U., Tahir, K., Ahmad, A., Khan, S.U., Khan, F.U., Khan, Q.U., and Yuan, Q. 2016. Visible light-induced photodegradation of methylene blue and reduction of 4-nitrophenol to 4-aminophenol over bio-synthesized silver nanoparticles. *Sep. Sci. Technol.* 51: 1070–1078.

Kharisov, B.I., Dias, H.V.R., and Kharissova, O.V. 2014. Nanotechnology-based remediation of petroleum impurities from water. *J. Pet. Sci. Eng.* 122: 705–718.

Kim, S.H., Lee, S.W., Lee, G.M., Lee, B.T., Yun, S.T., and Kim, S.O. 2016. Monitoring of TiO_2-catalytic UV-LED photo-oxidation of cyanide contained in mine wastewater and leachate. *Chemosphere*. 143: 106–114.

Kong, L., Xiong, Y., Sun, L., Tian, S., Xu, X., Zhao, C., Luo, R., Yang, X., Shih, K., and Liu, H. 2014. Sorption performance and mechanism of a sludge-derived char as porous carbon-based hybrid adsorbent for benzene derivatives in aqueous solution. *J. Hazard. Mater.* 274: 205–211.

Kulkarni S.J., and Goswami A.K. 2013. Adsorption studies for organic matter removal from wastewater by using bagasse flyash in batch and column operations. *Int. J. Sci. Res.* 2: 180–183.

Kulkarni S.J., and Goswami A.K. 2014. Applications and advancements in treatment of waste water by membrane technology: A review. *Int. J. Eng. Sci. Res. Tech.* 3: 446–450.

Kumar, K.M., Mandal, B.K., Kumar, K.S., Reddy, P.S., and Sreedhar, B. 2013. Biobased green method to synthesise palladium and iron nanoparticles using *Terminalia chebula* aqueous extract. *Spectrochim Acta Part A*. 102: 128–133.

Kumar, K.S., and Kathireswari, P. 2016. Biological synthesis of silver nanoparticles (AgNPs) by *Lawsonia inermis* (Henna) plant aqueous extract and its antimicrobial activity against human pathogen. *Int. J. Curr. Microbiol. Appl. Sci.* 5: 926–937.

Kunduru, R.K., Kovsky, M.N., Rajendra, S.F., Pawar, P., Basu, A., and Domb, A.J. 2017. Nanotechnology for water purification: Applications of nanotechnology methods in wastewater treatment. *Water Purif.* 10: 33–74.

Lai, Y.T., Huang, Y.S., Chen, C.H., Lin, Y.C., Jeng, H.T., Chang, M.C., Chen, L.J., Lee, C.Y., Hsu, P.C,. and Tai, N.H. 2020. Green treatment of phosphate from wastewater using a porous bio-templated graphene oxide/MgMn-layered double hydroxide composite. *iScience*. 23(5): 101065.

Le, T.T., Nguyen, K.H., Jeon, J.R., Francis, A.J., and Chang, Y.S. 2015. Nano/bio treatment of polychlorinated biphenyls with evaluation of comparative toxicity. *J. Hazard Mater.* 287: 335–341.

Lee, Y., Kim, S., Venkateswaran, P., Jang, J., Kim, H., and Kim J. 2008. Anion co-doped titania for solar photocatalytic degradation of dyes. *Carbon Lett.* 9(2): 131–136.

Li, Y., Feng, X., Zhang, T., Zhou, X., and Li, C. 2019. Preparation of magnetic macroporous polymer sphere for biofilm immobilization and biodesulfurization. *React. Funct. Polym.* 141: 1–8.

Li, Y., Yan, L., Xiang. C., and Hong, L.J. 2006. Treatment of oily wastewater by organic-inorganic composite tubular ultrafiltration (UF) membranes. *Desalination*. 196: 76–83.

Liu, Y., Zhang, Y., and Ni, B.J. 2015. Zero valent iron simultaneously enhances methane production and sulfate reduction in anaerobic granular sludge reactors. *Water Res.* 75: 292–300.

Lorenc-Grabowska, E., and Gryglewicz, G. 2007. Adsorption characteristics of Congo red on coal-based mesoporous activated carbon. *Dyes Pigments.* 74: 34–40.

Ma, L., and Zhang, W.X. 2008. Enhanced biological treatment of industrial wastewater with bimetallic zero-valent iron. *Environ. Sci. Technol.* 42(15): 5384–5389.

Machado, S., Pinto, S.L., Grosso, J.P., Nouws, H.P.A., Albergaria, J.T., and Delerue-Matos, C. 2013. Green production of zero-valent iron nanoparticles using tree leaf extracts. *Sci. Total Environ.* 445–446: 1–8.

Mahdavian, A.R., and Mirrahimi, M.A.S. 2010. Efficient separation of heavy metal cations by anchoring polyacrylic acid on superparamagnetic magnetite nanoparticles through surface modification. *Chem. Eng. J.* 159: 264–271.

Malakar, S., Das Saha, P., Baskaran, D., and Rajamanickam, R. 2017. Comparative study of biofiltration process for treatment of VOCs emission from petroleum refinery wastewater—A review. *Environ. Technol. Innov.* 8: 441–461.

Malik, P., Shankar, R., Malik, V., Sharma, N., and Mukherjee, T.K. 2014. Green chemistry based benign routes for nanoparticle synthesis. *J. Nanopart.* 2014: 302429.

Mann N. 2017. Role of nanotechnology in waste water treatment. *IJARIIE*. 3(1):1841–1847.

Manyuchi, M.M., and Ketiwa E. 2013, Distillery effluent treatment using membrane bioreactor technology utilizing *pseudomonas fluorescens*. *Int. J. Sci. Eng. Tech.* 2: 1252–1254.

Martynková, G.S., and Valášková, M. 2014. Antimicrobial nanocomposites based on natural modified materials: A review of carbons and clays. *J. Nanosci. Nanotechnol.* 14(1): 673–693.

Mata, Y.N., Torres, E., Blazquez, M.L., Ballester, A., Gonzalez, F., and Munoz, JA. 2009. Gold (III) biosorption and bioreduction with the brown alga *Fucus vesiculosus*. *J. Hazard Mater.* 166: 612–618.

Melamane, X.L., Strong P.J., and Burgess J.E. 2007. Treatment of wine distillery wastewater: A review with emphasis on anaerobic membrane reactors. *S. Afr. J. Enol. Vitic.* 28: 25–36.

Meng, S., Mansouri, J., Ye, Y., and Chen, V. 2014. Effect of templating agents on the properties and membrane distillation performance of TiO_2-coated PVDF membranes. *J. Membr. Sci.* 450: 48–59.

Meyer, D.E., Wood, K., Bachas, L.G., and Bhattacharyya, D. 2004. Degradation of chlorinated organics by membrane-immobilized nanosized metals. *Environ. Prog.* 23(3): 232–242.

Mills, A., and Le Hunte, S. 1997. An overview of semiconductor photocatalysis. *J. Photoch. Photobio. A. Chem.* 108 (1): 1–35.

Moon, G., Kim, D., Kim, H., Bokare, A.D., and Choi, W. 2014. Platinum-like behavior of reduced graphene oxide as a cocatalyst on TiO_2 for the efficient photocatalytic oxidation of arsenite, *Environ. Sci. Techno. Lett.* 1(2): 185–190.

Morkin, M., Devlin, J.F., Barker, J.F., and Butler, B.J. 2000. In situ sequential treatment of a mixed contaminant plume. *J. Contam. Hydrol.* 45(3): 283–302.

Moslehyani, A., Ismail, F., Othman, M.H.D., and Matsuura T. 2015. Design and performance study of hybrid photocatalytic reactor-PVDF/MWCNT nanocomposite membrane system for treatment of petroleum refinery wastewater. *Desalination*. 363: 99–111.

Murgueitio, E., Cumbal, L., Abril, M., Izquierdo, A., Debut, A., and Tinoco, O. 2018. Green Synthesis of iron nanoparticles: Application on the removal of petroleum oil from contaminated water and soils. *J. Nanotechnol.* ID 4184769. https://doi.org/10.1155/2018/4184769.

Mustapha, H.I., van Bruggen, J.J.A., and Lens, P.N.L. 2018. Fate of heavy metals in vertical subsurface flow constructed wetlands treating secondary treated petroleum refinery wastewater in Kaduna, Nigeria. *Int. J. Phytoremediation.* 20: 44–53.

Muthukumar, H., Gire, A., Kumari, M., and Manickam, M. 2017. Biogenic synthesis of nano-biomaterial for toxic naphthalene photocatalytic degradation optimization and kinetics studies. *Int. Biodeterior. Biodegrad.* 119: 587–594.

Nguyen, A.T., Hsieh, C.T., and Juang, R.S. 2016. Substituent effects on photodegradation of phenols in binary mixtures by hybrid H_2O_2 and TiO_2 suspensions under UV irradiation. *J. Taiwan Inst. Chem. E.* 62: 68–75.

Nguyen, L. H., Nguyen, T.M.P., Van, H.T., Vu, X.H., Ha, T.L.A., Nguyen, T.H.V., Nguyen, X.H., and Nguyen, X.C. 2019. Treatment of hexavalent chromium contaminated wastewater using activated carbon derived from coconut shell loaded by silver nanoparticles: Batch experiment. *Water Air Soil Pollut.* 230(68): 1–14.

Ohsaka, T., Shinozaki, K., Tsuruta K., and Hirano, K. 2008. Photo-electrochemical degradation of some chlorinated organic compounds on n-TiO_2 electrode. *Chemosphere*. 73(8): 1279–1283.

Olajire, A.A. 2014. The petroleum industry and environmental challenges. *J. Pet. Environ. Biotechnol.* 5: 2157–2163.

Omran, B.A., Nassar, H.N., Fatthallah, N.A., Hamdy, A., El-Shatoury, E.H., and El-Gendy, N.Sh. 2018. Characterization and antimicrobial activity of silver nanoparticles mycosynthesized by *Aspergillus brasiliensis*. *J. Appl. Microbiol.* 125: 370–382. https://doi.org/10.1111/jam.13776

Padaki, M., Murali, R.S., Abdullah, M.S., Misdan, N., Moslehyani, A., Kassim, M.A., Hilal, N., and Ismail, A.F. 2015. Membrane technology enhancement in oil–water separation. A review. *Desalination*. 357: 197–207.

Paterson, W. R. 1995. Petroleum refining: Technology and economics. *Chem. Eng. Biochem. Eng.* 56(2): B109–B112.

Pathath, A. R., Jabbar, K. A., and Mahmoud, D. K. 2019. Chitosanbased nanocomposite for the inhibition of sulfate reducing bacteria: Towards "green" biocides for microbial influenced corrosion. http://doi.org/10.5339/qfarc.2018.EEPP981

Pendergast, M.M., and Hoek, E.M.V. 2011. A review of water treatment membrane nanotechnologies. *Energy Environ. Sci.* 4: 1946–1971.

Peng, B., Zhou, R., Chen, Y., Tu, S., Yin, Y., and Ye, L. 2020. Immobilization of nano-zero-valent irons by carboxylated cellulose nanocrystals for wastewater remediation. *Front. Chem. Sci. Eng.* 14: 1006–1017. https://doi.org/10.1007/s11705–020-1924-y

Pintor, A.M.A., Vilar, V.J.P., Botelho, C.M.S., and Boaventura, R.A.R. 2016. Oil and grease removal from wastewaters: Sorption treatment as an alternative to state-of-the-art technologies. A critical review. *Chem. Eng. J.* 297: 229–255.

Pombo, F., Magrini, A., and Szklo, A. 2011. Technology roadmap for wastewater reuse in petroleum refineries in Brazil. In: Broniewicz, E., Ed. *Environmental Management in Practice*, InTech.

Pourmortazavi, S. M., Sahebi, H., Zandavar, H., and Mirsadeghi, S. 2019. Fabrication of Fe_3O_4 nanoparticles coated by extracted shrimp peels chitosan as sustainable adsorbents for removal of chromium contaminates from wastewater: The design of experiment. *Compos. Part B Eng.* 175: 107130.

Rahman, M.M., and Al-Malack, M.H. 2006. Performance of a cross flow membrane bioreactor (CF–MBR) when treating refinery wastewater. *Desalination.* 191: 16–26.

Rai, M., Yadav, A., and Gade, A. 2009. Silver nanoparticles as a new generation of antimicrobials, *Biotechnol. Adv.* 27(1): 76–83.

Rajesh, S., Senthilkumar, S., Jayalakshmi, A., Nirmala, M.T., Ismail, A.F., and Mohan, D. 2013. Preparation and performance evaluation of poly (amide–imide) and TiO_2 nanoparticles impregnated polysulfone nanofiltration membranes in the removal of humic substances. *Colloid Surface A.* 418: 92–104.

Rasheed, P.A., Pandey, R.P., Jabbar, K.A., Samara, A., Abdullah, A.M., and Mahmoud, K.A. 2020. Chitosan/lignosulfonate nanospheres as "green" biocide for controlling the microbiologically influenced corrosion of carbon steel. *Materials.* 13(11): 2484.

Rasheed, Q.J., Pandian. K., and Muthukumar, K. 2011. Treatment of petroleum refinery wastewater by ultrasound-dispersed nanoscale zero-valent iron particles. *Ultrason. Sonochem.* 18: 1138–1142.

Rasool, K., Nasrallah, G. K., Younes, N., Pandey, R. P., Abdul Rasheed, P., Mahmoud, K. 2018. "Green" ZnO-interlinked chitosan nanoparticles for the efficient inhibition of sulfate-reducing bacteria in inject seawater. *ACS Sustain. Chem. Eng.* 6(3): 3896–3906.

Razavi, S.M.R., and Miri, T. 2015. A real petroleum refinery wastewater treatment using hollow fiber membrane bioreactor (HF-MBR). *J. Water Process. Eng.* 8: 136–141.

Razmjou, A., Holmes, A.R.L., Li, H., Mansouri, J., and Chen, V. 2012. The effect of modified TiO_2 nanoparticles on the polyether sulfone ultrafiltration hollow fiber membranes. *Desalination.* 287: 271–280.

Razmjou, A., Mansouri, J., Chen, V., Lim, M., and Amal, R. 2011. Titania nanocomposite polyether sulfone ultrafiltration membranes fabricated using a low temperature hydrothermal coating process. *J. Membr. Sci.* 380(1–2): 98–113.

Reddy, P.A.K., Reddy, P.V.L., Kwon, E., Kim, K.H., Akter, T., and Kalagara, S. 2016. Recent advances in photocatalytic treatment of pollutants in aqueous media. *Environ. Int.* 91: 94–103.

Renault, F., Sancey, B., Badot, P.M., and Crini, G. 2009. Chitosan for coagulation/flocculation processes, an eco-friendly approach. *Eur. Polym. J.* 45: 1337–1348.

Roig, B., Mnif, W., Hassine, A.I.H., Zidi, I., Bayle, S., Bartegi, A., and Thomas, O. 2013. Endocrine disrupting chemicals and human health risk assessment: A critical review. *Crit. Rev. Environ. Sci. Technol.* 43(21): 2297–2351.

Saed, D., Nassar, H.N., El-Gendy, N.Sh., Zaki, T., Moustafa, Y.M., and Badr, I.H.A. 2014. Enhancement of pyrene biodegradation by assembling MFe_3O_4 nano-sorbents on the surface of microbial cells. *Energ. Source. Part A.* 36: 1931–1937.

Saien, J., and Nejati, H. 2007. Enhanced photocatalytic degradation of pollutants in petroleum refinery wastewater under mild conditions. *J. Hazard Mater.* 148(1–2): 491–495.

Saifuddin, N., Wong, C.W., and Yasimura, A.A.N. 2009. Rapid biosynthesis of silver nanoparticles using culture supernatant of bacteria with microwave irradiation. *J. Chem.* 6: 61–70.

Salgot, M., and Folch, M. 2018. Wastewater treatment and water reuse. *Curr. Opin. Environ. Sci. Health.* 2: 64–74.

Santo, C.E., Vilar, V.J.P., Bhatnagar, A., Kumar, E., Botelho, C.M.S., and Boaventura, R.A.R. 2013. Biological treatment by activated sludge of petroleum refinery wastewaters. *Desalin. Water Treat.* 51(34–36): 6641–6654.

Sarno, M., and Iuliano, M. 2020. New nano-biocatalyst for 4-chlorophenols removal from wastewater. *Mater. Today*. 20: 74–81.

Savage, N., and Diallo, M.S. 2005. Nanomaterials and water purification: Opportunities and challenges. *J. Nanopart. Res.* 7(4–5): 331–342.

Sethy, N.K., Arif, Z., Mishra, P.K., Kumar, P. 2020. Green synthesis of TiO_2 nanoparticles from *Syzygium cumini* extract for photo-catalytic removal of lead (Pb) in explosive industrial wastewater. *Green Process Synth.* 9(1): 171–181.

Shanker, U., Jassal, V., and Rani, M. 2017. Green synthesis of iron hexacyanoferrate nanoparticles: Potential candidate for the degradation of toxic PAHs. *J. Environ. Chem. Eng.* 5(4): 4108–4120.

Shariati, S.R.P., Bonakdarpour, B., Zare, N., and Ashtiani, F.Z. 2011. The effect of hydraulic retention time on the performance and fouling characteristics of membrane sequencing batch reactors used for the treatment of synthetic petroleum refinery wastewater. *Bioresour. Technol.* 102: 7692–7699.

Singh, N., Chakraborty, R., and Gupta, R.K., 2018. Mutton bone derived hydroxyapatite supported TiO_2 nanoparticles for sustainable photocatalytic applications. *J. Environ. Chem. Eng.* 6(1): 459–467.

Smuleac, V., Varma, R., Sikdar, S., and Bhattacharyya, D. 2011. Green synthesis of Fe and Fe/Pd bimetallic nanoparticles in membranes for reductive degradation of chlorinated organics. *J. Membr. Sci.* 379: 131–137.

Sun, P., Elgowainy, A., Wang, M., Han, J., and Henderson, R.J. 2018. Estimation of U.S. refinery water consumption and allocation to refinery products. *Fuel*. 221: 542–57.

Sun, P., Hui, C., Azim Khan, R., Du, J., Zhang, Q., and Zhao, Y.H. 2015. Efficient removal of crystal violet using Fe_3O_4-coated biochar: The role of the Fe_3O_4 nanoparticles and modeling study their adsorption behavior. *Sci. Rep.* 5: 12638.

Sun, Y., Zhang, Y., and Quan, X. 2008. Treatment of petroleum refinery wastewater by microwave-assisted catalytic wet air oxidation under low temperature and low pressure. *Sep. Purific. Tech.* 62: 565–570.

Tan, X.F., Liu, Y.G., Zeng, G., Wang, X., Hu, X., Gu, Y., and Yang, Z. 2015. Application of biochar for the removal of pollutants from aqueous solutions. *Chemosphere*. 125: 70–85.

Tang, J., Lv, H., Gong, Y., and Huang, Y. 2015. Preparation and characterization of a novel graphene/biochar composite for aqueous phenanthrene and mercury removal. *Bioresour. Technol.* 196: 355–363.

Tang, Y.P., Luo, L., Thong, Z., and Chung, T.S. 2017. Recent advances in membrane materials and technologies for boron removal. *J. Membr. Sci.* 541: 434–446.

Tungittiplakorn, W., Cohen, C., and Lion, L.W. 2005. Engineered polymeric nanoparticles for bioremediation of hydrophobic contaminants. *Environ. Sci. Technol.* 39(5): 1354–1358.

Umar, K., Dar, A.A., Haque, M.M., Mir, N.A., and Muneer, M. 2012. Photocatalysed decolourization of two textile dye derivatives, Martius Yellow and Acid Blue 129 in UV-irradiated aqueous suspensions of Titania. *Desalination Water Treat.* 46: 205–214.

Umar, K., Haque, M.M., Mir, N.A., and Muneer, M. 2013. Titanium dioxide-Mediated photocatalyzed mineralization of two selected organic pollutants in aqueous suspensions. *J. Adv. Oxid. Technol.* 16: 252–260.

Varjani, S., Joshi, R., Srivastava, V.K., Ngo, H.H., and Guo, W. 2019. Treatment of wastewater from petroleum industry: Current practices and perspectives. *Environ. Sci. Pollut. Res.* 27: 27172–27180. https://doi.org/10.1007/s11356-019-04725-x.

Vendramel, S., Bassin, J.P., Dezotti, M., and Sant'AnnaJr, G.L. 2015. Treatment of petroleum refinery wastewater containing heavily polluting substances in an aerobic submerged fixed-bed reactor. *Environ. Technol.* 36(16): 2052–2205.

Venkateswarlu, S., Kumar, B.N., Prasad, C.H., Venkateswarlu, P., and Jyothi, N.V.V. 2014. Bio-inspired green synthesis of Fe_3O_4 spherical magnetic nanoparticles using *Syzygium cumini* seed extract. *Phys. B. Condens. Matter.* 449: 67–71.

Venkateswarlu, S., Rao, Y.S., Balaji, T., Prathima, B., and Jyothi, N.V.V. 2013. Biogenic synthesis of Fe_3O_4 magnetic nanoparticles using plantain peel extract. *Mater. Lett.* 100: 241–244.

Vigneshwaran, N., Kathe, A.A., Varadarajan, P.V., Nachane, R.P., and Balasubramanya, R.H. 2007. Silver protein (core shell) nanoparticle production using spent mushroom substrate. *Langmuir*. 23: 7113–7117.

Wahi, R., Abdullah, L., Shean, T., Choong, Y., and Ngaini, Z. 2013. Oil removal from aqueous state by natural fibrous sorbent. *Sep. Purif. Technol.* 113: 51–63.

Wang, B., Gao, B., and Fang, J. 2017. Recent advances in engineered biochar productions and applications. *Crit. Rev. Environ. Sci. Technol.* 47: 2158–2207.

Wang, B., Yi, W., Yingxin, G., Guomao, Z., Min, Y., Song, W., and Jianying, H. 2015. Occurrences and behaviors of Naphthenic Acids in a petroleum refinery wastewater treatment plant. *Environ. Sci. Tech.* 49(9): 5796–5804.

Wang, F., Hu, Y., Guo, C., Huang, W., and Liu, C. Z. 2012. Enhanced phenol degradation in coking wastewater by immobilized laccase on magnetic mesoporous silica nanoparticles in a magnetically stabilized fluidized bed. *Bioresour. Technol.* 110: 120–124.

Wang, Q., Wang, X., Wang, Z., Huang, J., and Wang, Y. 2013. PVDF membranes with simultaneously enhanced permeability and selectivity by breaking the tradeoff effect via atomic layer deposition of TiO_2. *J. Membr. Sci.* 442: 57–64.

Wang, T., Jin, X., Chen, Z., Megharaj, M., and Naidu, R. 2014. Green synthesis of Fe nanoparticles using eucalyptus leaf extracts for treatment of eutrophic wastewater. *Sci. Total Environ.* (466–467): 210–213.

Wei, Y., Fang, Z., Zheng, L., Tan, L., and Tsang, E.P. 2016. Green synthesis of Fe nanoparticles using Citrus maxima peels aqueous extracts. *Mater. Lett.* 185: 384–386.

Xu, P., Zeng, G.M., Huang, D.L., Feng, C.L., Hu, S., Zhao, M.H., and Liu, Z.F. 2012. Use of iron oxide nanomaterials in wastewater treatment: A review. *Sci. Total Environ.* 424: 1–10.

Xu, X., Cai, Y., Song, Z., Qiu, X., Zhou, J., Liu, Y., Mu, T., Wu, D., Guan, Y., and Xing, J. 2015. Desulfurization of immobilized sulfur-oxidizing bacteria, Thialkalivibrio versutus, by magnetic nanoparticles under haloalkaliphilic conditions. *Biotechnol. Lett.* 37(8): 1631–1635.

Yanyang, Z., Bing, W., Hui, X., Hui, L. Minglu, W., Yixuan, H., and Bingcai, P. 2016. Nanomaterials-enabled water and wastewater treatment. *NanoImpact.* (3–4): 22–39.

Yaqoob, W.A., Parveen, A.A. and Umar, K. 2020. Role of nanomaterials in the treatment of wastewater: A review. *Water.* 12(495): 1–30.

Yew, Y.P., Shameli, K., Miyake, M., Kuwano, N., Ahmad, N. B., Mohamad, S.E., and Lee, K.X. 2016. Green synthesis of magnetite (Fe_3O_4) nanoparticles using seaweed (*Kappaphycus alvarezii*) extract. *Nanoscale Res. Lett.* 11(276): 1–7.

Yin, J., Yang, Y., Hu, Z., and Deng, B. 2013. Attachment of silver nanoparticles (AgNPs) onto thin-Film composite (TFC) membranes through covalent bonding to reduce membrane biofouling. *J. Membr. Sci.* 441: 73–82.

Younis, S.A., El-Gendy, N.Sh., and Nassar, H.N. 2020. Biokinetic aspects for biocatalytic remediation of xenobiotics polluted seawater. *J. Appl. Microbiol.* 129(2): 319–334, doi:10.1111/jam.14626.

Younis, S.A., Maitlo, H. A., Lee, J., and Kim, K.H. 2020. Nanotechnology-based sorption and membrane technologies for the treatment of petroleum-based pollutants in natural ecosystems and wastewater streams. *Adv. Colloid Interf. Sci.* 275: 1–37.

Yuliwati, E., Ismail, A.F., Matsuura, T., Kassim, M.A., and Abdullah, M.S. 2011. Effect of modified PVDF hollow fiber submerged ultrafiltration membrane for refinery wastewater treatment. *Desailnation.* 283: 214–220.

Zarasvand, A.K., and Rai, V.R. 2016. Inhibition of a sulfate reducing bacterium, *Desulfovibrio marinisediminis* GSR3, by biosynthesized copper oxide nanoparticles. *3 Biotech.* 6(1): 1–7.

Zhang, M., Field, R.W., and Zhang, K. 2014. Biogenic silver nanocomposite polyether sulfone UF membranes with antifouling properties. *J. Membr. Sci.* 471: 274–284.

Zhang, M., and Gao, B. 2013. Removal of arsenic, methylene blue, and phosphate by biochar/AlOOH nanocomposite. *Chem. Eng. J.* 226: 286–292.

Zhang, M., Gao, B., Varnoosfaderani, S., Hebard, A., Yao, Y., and Inyang, M. 2013. Preparation and characterization of a novel magnetic biochar for arsenic removal. *Bioresour. Technol.* 130: 457–462.

Zhang, T., Li, Z., Lü, Y, Liu, Y., Yang, D., and Li, Q. 2018. Recent progress and future prospects of oil-absorbing materials. *Chin. J. Chem. Eng.* 27(6): 1282–1295.
Zhang, Y., Wu, B., Xu, H., Liu, H., Wang, M., He, Y., and Pan, B. 2016. Nanomaterials-enabled water and wastewater treatment. *Nano Impact.* 3: 22–39.
Zhao, G., Xu, Y., Han, G., and Ling, B. 2006. Biotransfer of persistent organic pollutants from a large site in China used for the disassembly of electronic and electrical waste. *Environ. Geochem. Health* 28: 341–351.
Zhong, J., Sun, X., and Wang, C. 2003. Treatment of oily wastewater produced from refinery processes using flocculation and ceramic membrane filtration. *Sep. Purif. Technol.* 32: 93–98.
Zhou, W., Yang, M., Song, Z., and Xing, J. 2015. Enhanced sulfate reduction by Citrobacter sp. coated with Fe3O4/SiO2 magnetic nanoparticles. *Biotechnol. Bioproc. Eng.* 20: 117–123.
Zhu, L., Meng, L., Shi, J., Li, J., Zhang, X., and Feng, M. 2019. Metal-organic frameworks/ carbon-based materials for environmental remediation: A state-of-the-art mini-review. *J. Environ. Manage.* 232: 964–977.
Zhu, Y., Zhao, Q., Li, Y., Cai, X., and Li, W. 2006. Preconcentration and determination of nicosulfuron, thifensulfuron-methyl and metsulfuron-methyl in water samples using carbon nanotubes packed cartridge in combination with high performance liquid chromatography. *J. Nanosci. Nanotechnol.* 559: 200–206.

CHAPTER 8

Future Aspects of Nanobiotechnology in Petroleum Industry

8.1 INTRODUCTION

There is a worldwide great attentiveness in the research and development (R&D) of nanotechnology and its interdisciplinary applications in different sectors, such as medical, pharmaceutical, environmental, renewable energy, and oil and gas industry (Ko and Huh, 2019; Alsaba et al., 2020), where research works on nanotechnology are widely explored in developed countries as these are associated with their national income, intellectual property policies, and human resources. This is very clear within the large amount of money invested on nanotechnology R&D. For example, the National Nanotechnology Initiative, that is, the United States federal government program for the science, engineering, and technology R&D for nanoscale, has been funded by approximately $27 billion (NNI, 2018). The Massachusetts Institute of Technology invested approximately $350 million on the state-of-the-art nanoscale research center (MIT.nano) (Chandler, 2014). Saudi Aramco Energy Ventures funded the leading company in nanomanufacturing, NanoMech, by approximately $10 million. Moreover, the developing countries are also trying to catch the wave of nanotechnology, especially India and China. For example, the Department of Science and Technology, in India, has invested $20 million for nanomaterials science and technology development (Syafiuddin et al., 2017). Even, China is reported to be ranked third in the number of nanotechnology patent applications filed (Syafiuddin et al., 2017).

It has been reported that approximately 25% of the total human mortality in Asian and African countries is greatly related to the exposure to pollutants in air and water (Lateef et al., 2019). This consequently negatively impacted the economy. In an attempt to control such pollution, nowadays, the world-wide main guide for chemists and chemical industry to less hazardous chemical syntheses is the 12 principles of green chemistry (see Figure 8.1).

The bioinspired and/or the green-inspired nanomaterials are good example of green chemistry, where it takes the advantage of biological-based methodologies to sustainably produce nanomaterials with different applications, using readily available, ecofriendly, and cost-effective biorenewable resources. Such a green process is an energy saving one, decreases the greenhouse gas (GHG), characterized with a minimum production of hazardous wastes, and indirectly contributes to solving the problem of global warming and climate change (Iravani, 2014).

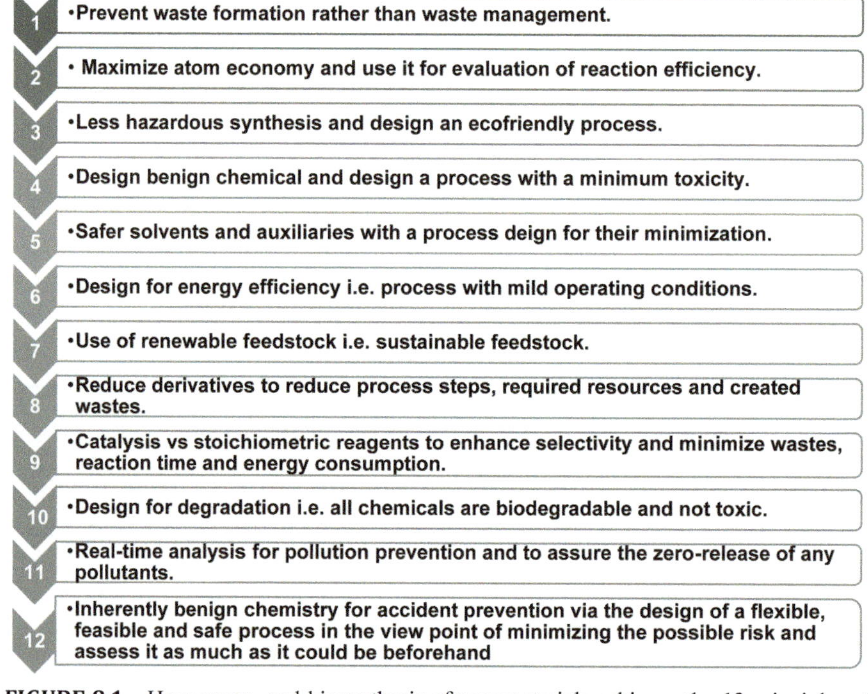

FIGURE 8.1 How green- and biosynthesis of nanomaterials achieves the 12 principles of the green chemistry.

8.2 PHYTOSYNTHESIS OF NANOPARTICLES

The green-synthesized nanoparticles (NPs), that is, the phytogenic NPs, are known to be more stable with more various shapes and sizes, and the rate of its synthesis is much faster than the microbial synthesis. Thus, there are several research works on the phytosynthesis of NPs in terms of their synthesis, characterization, cytotoxicity, antimicrobial activities, other toxicology properties, and applications. Even their toxicity mechanisms have been widely studied but not yet enough. In addition, most of the research works focused on their synthesis using plant extracts for antimicrobial applications, selective sensors, and catalytic degradation of hazardous pollutants. However, the majority of these studies are on the minor scale and laboratory scale. The issues during the scaling-up of the phytosynthesis of NPs are somewhat absent from most of the published research studies in this field, although this information is important and highly beneficiary in industrial-level production. Nevertheless, it is important to note that the development of phytonanotechnology needs knowledge transfer via national and international research collaboration and international funding. Serious awareness and governmental involvement are also essential. Not only this, but also to maximize the prospective applications of phytosynthesized NPs, it is also very important to quantify and qualify the advantages as well as the disadvantages of using NPs and their impact on human health, ecosystem, and environment.

Plant extracts have been recently considered for nanofabrication because of their ease of extraction, abundant availability, and their potential to eliminate the complicated procedures of microbial synthesis of NPs. Based on the criteria of the economically viable green method, the plant-based system represents the most promising alternative not only to biological methods, but also to other chemical and physical processes. However, there is still a great need for investigating the reaction mechanism between the components of the plant extracts and metal ion precursors procuring the metal/metal oxides NPs. Some possible mechanisms are the involvement of the antioxidant flavonoids, polyphenols, and phenolic acid constituents of the plant extracts into the reduction of the metal ions into metal NPs with the mutual oxidation of the hydroxyl groups into carbonyl groups and the oxidation of polyphenols into quinones, whereas such carbonyl groups and quinones would act as soft ligands and stabilize the phytosynthesized metal-NPs via simple complexation (Mittal et al., 2103; Shamaila et al., 2016).

Although the phytosynthesis using extracts of leaves, seeds, roots, and fruits is well established, the accessibility to seeds and roots is slightly difficult compared to leaves and fruits. But, most of the fruits are considered as food and vitamin sources. Thus, it is unfavorable to apply fruits as nanofactories. So, the focus would go to leaves, as they are abundantly available and commonly discarded to the environment. Furthermore, upon the application of the extracts of agricultural wastes, it would have a positive impact on waste management problems. Moreover, it is more preferable to use nonedible sources for not adding to the worldwide food problem.

Several research studies have been published about the phytosynthesis of spherical-shaped NPs. The future challenge in green synthesis is how to optimize and control the phytosynthesis procedure to produce other forms such as triangular, cuboidal, truncated, decahedral, oval, ellipsoidal, and pyramidal shapes. Moreover, there is still a great demand to exploit the plant-based biosynthesis to attain better control over dispersity, morphology, particle size, and production rates to replace chemical and/or physical methods on an industrial scale. Thus, it is very important to investigate more about the effect of physicochemical parameters (pH, temperature, time, precursor concentrations, illumination effect, shaking speed, etc.) on the phytosynthesis of nanoparticles, concerning the yield, size, shape, and stability of such NPs. For example, increase in temperature produces small-sized NPs as it enhances nucleation and metal ions consumed in such step and nothing would have left for the secondary reduction on the preformed nuclei surface. Thus, growth and agglomeration would stop at high temperature (Lee et al., 2011; Khani et al., 2018). Acidic pH usually, produces large-size phytosynthesized NPs (Armendariz et al., 2004; Herrera-Becerra et al., 2008; Din et al., 2017; Khani et al., 2018). Since hydroxyl ions would induce reduction of metal ions by enhancing the redox potential and the overall reaction kinetics (Cheirmadurai et al. 2014), alkaline pH usually produces smaller sized NPs using plant extracts (Anigol et al., 2017; Omran et al., 2018a). The concentration of the plant extract or the metal salts also reported to have a great effect on the yield and size of the phytosynthesized NPs (Omran et al., 2018a).

The solvent used for extraction of biomolecules from plant parts and the method of extraction were found to greatly affect the size and shape of the produced metal/metal oxide NPs, and, consequently, their biocidal, antioxidant, etc., activities and any other possible applications. Parthasarathy et al. (2017) reported the green synthesis of ZnO NPs using methanol extract of *Curcuma neilgherrensis* leaf. The efficient, green, and rapid microwave (90 and 270 W, 15 min) assisted aqueous extraction of *Suaeda aegyptiaca* leaves has been applied in the synthesis of ZnO NPs with Wurtzite structure, spherical shape, and average diameter of 60 nm (Rajabi et al., 2017). The optical band-gap value of ZnO NPs, which is related to the particle size, was higher for those prepared by the microwave-assisted extract rather than those prepared by the maceration method. That recorded 4.05, 3.93, and 2.02 eV, respectively (Rajabi et al., 2017). The antimicrobial effect of the prepared ZnO NPs at higher irradiation was higher than those prepared at lower irradiation. That was attributed to the presence of more compounds in the extract of the herbal matrices, which would add to the antibacterial impacts of ZnO. Moreover, those compounds might have acted as stabilizing agents during the synthesis step and prevented the growth of the particles and, thus, produced stable particles with nanoscaled size, whereas the ZnO NPs enhanced the production of some reactive oxygen species (ROS) like hydroxyl and superoxide radicals, which would destroy the microbial cell wall (Rajabi et al., 2017). The aqueous extract itself showed higher antioxidant properties than the ZnO itself, but it depended on the irradiation efficiency extract (90 W) > ZnO (90 W) ~ ZnO (270 W) > extract (macerated) > ZnO (macerated) ~ extract (270 W). The high irradiation power destroyed the phenol and flavonoid in the extract, decreasing the antioxidant efficiency. Moreover, the low antioxidant properties of the macerated extracts were attributed to the destroying impact of light as well as O_2 on the extract's components during the long extraction time. However, the lower antioxidant properties of ZnO NPs relative to the extract itself were related to the production of ROS, which causes a decrease in the antioxidant activity and/or an enhancement in oxidative stress (Rajabi et al., 2017). The silver nanoparticles (AgNPs) for the first time were synthesized from the reaction between silver acetate and the methanolic root extracts of *Diospyros sylvatica*, a member of family Ebenaceae (Pethakamsetty et al., 2017). The average diameter of the phytosynthesized AgNPs was around 8 nm, which was in a good agreement with the average crystallite size (10 nm) calculated from X-ray diffraction (XRD) analysis. The phytosynthesized AgNPs exhibited maximum antimicrobial activity against *Bacillus pumilis*, *Pseudomonas aeuriginosa* and *Bacillus subtilis*, moderate activity against *Staphylococcus* aureus, *Klebsiella pneumoniae*, and *Escherichia coli*, and mild activity toward *Streptococcus pyogenes* and *Proteus vulgaris*. The bioinspired AgNPs showed promising antimicrobial activity against all the tested bacterial strains rather than against the tested fungal strains *Aspergillus niger* and *Pencillium notatum*. Thus, this would indicate that the solvent used for extraction and/or the plant extract itself with its components capping the prepared AgNPs might affect its antimicrobial activity. Consequently, this is a new point of research to be covered to widen the antimicrobial and different applications of phytosynthesized metal NPs.

In order to increase adsorption efficiency and to avoid interference from other metals ions, iron oxides NPs have been functionalized to tune their adsorption properties by adding various ligands (e.g., ethylenediamine tetra-acetic acid, L-glutathione, mercaptobutyric acid, and meso-2,3 dimercaptosuccinic acid or polymers (e.g., copolymers of acrylic acid and crotonic acid) (Ge et al., 2012).

Thus, this means an additional step in the physical or chemical preparation of NPs. However, one of the main advantages of the green synthesis of metal/metal oxide NPs is that the biomolecules stabilizing and capping such NPs also act as functionalizing matrices, which increase the applicability of the green-synthesized NPs. For example, Al din. Haratifar et al. (2009) used the ethanol extract of *Eucalyptus camaldulensis* for the reduction of Au^{3+} on the surface of the magnetite NPs and for the functionalization of the Au-Fe_3O_4 nanocomposite particles. A flexible ligand shell helps facilitate the incorporation of a wide array of functional groups into the shell and ensure the intact properties of Fe_3O_4 NPs. Besides, a polymer shell has been found to be able to prevent aggregation of particles and improve the dispersion stability of the nanostructures, whereas polymer molecules could act as binders for metal ions and thus become a "carrier" of metal ions in treated water (Khaydarov et al., 2010).

There are two main challenges upon scaling up the phytosynthesis process of NPs. Concerning the production, the process should be cost effective, the solvent used for plant part extraction should be cheap and ecofriendly, the recycling of output wastes should be considered for not adding a new waste management problem, the process should be reliable in itself, especially the yield and the energy consumption, and finally the material safety and the hazardous level should be known and controlled. On the other hand, the second challenge is concerned with the properties and characteristics of the nanomaterials itself, which would change upon the scaling up. Since the level of applied control on nanoscale probably decreases upon manufacturing large quantities. Several reports on physical and chemical approaches for large-batch-scale production of NPs have been published (Park et al., 2012; Xiliang et al., 2014; Helmlinger et al., 2015). Those reports missed the most important factor upon scaling up, that is the continuous process, which is critically related to the operational cost. However, Huang et al. (2008) designed a prototype for a continuous-flow tubular microreactor, whereas AgNPs can be produced on a continuous basis once the reaction reaches the steady state. One of its main advantages is that few seconds or even milliseconds are needed to achieve the targeted reaction temperatures. Moreover, it can be easily scaled up by increasing the reactor length, and it has the potential to be applied for the green synthesis of AgNPs, whereas the green synthesis of AgNPs by lixivium of sundried *Cinnamomum camphora* leaf in continuous-flow tubular microreactors was investigated (Huang et al., 2008). By following up the temperature profiles, it was illustrated that, at the inlet of the microreactors at 90 °C, the rise of the fluid temperature induced the burst of silver nuclei by homogeneous nucleation. Subsequently, the nuclei grew gradually along the reactors into AgNPs with the size ranged between 5 and 40 nm. Moreover, polydispersed particles were formed by the combination of heterogeneous nucleation and Ostwald ripening along the tubes at 60 °C.

Several research studies have been published for phytosynthesis of metal/ metal oxide NPs using different parts of plants. Moreover, several other research studies have been published for the bioaccumulation of heavy metals by living plants and phytoremediation of soil and water from contaminating toxic heavy metals. That natural phenomenon of heavy metal tolerance of plants should be investigated, with the related biological mechanisms (see Figure 8.2) as well as the physiology and genetics of metal tolerance in hyperaccumulating plants.

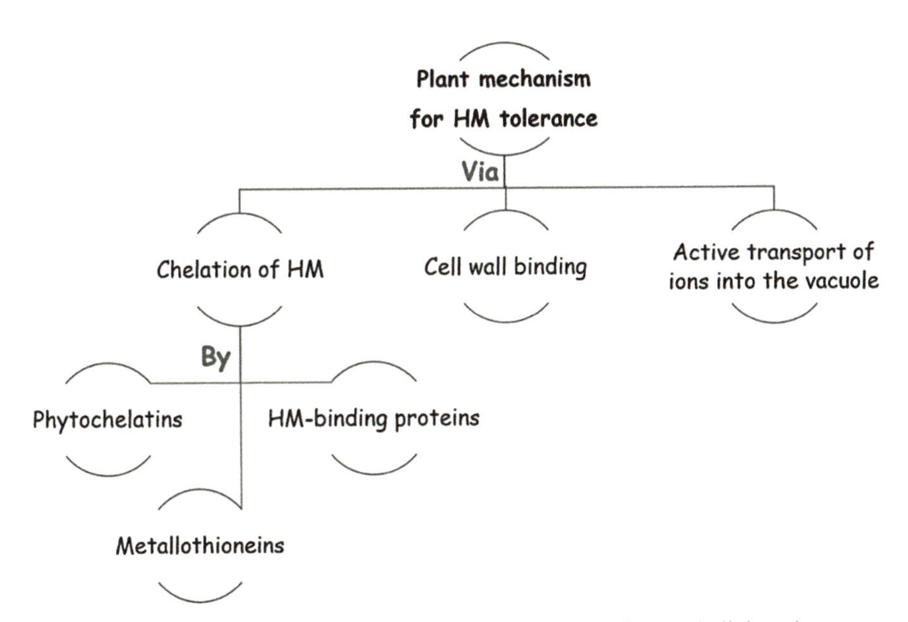

FIGURE 8.2 Important mechanisms for heavy metals (HM) tolerance in living plants.

Thus, taking that advantage, some reports about the uptake of plants to metal salts and their capabilities for their reduction to their corresponding crystalline nanostructures have been reported (Anderson et al., 1998, 2013; Harris and Bali, 2008; Manceau et al., 2008; Iravani, 2011; Losfeld et al., 2012). However, after the formation of the metal NPs, it is common that extraction from biomass is endeavored by the laborious and energy-intensive freeze-thawing, biomass incineration, or chemical leaching, which destroy the metal NP structure, thus interfering with the desired nanomaterial properties. Gardea-Torresdey et al. (2002) were the first to report the growing of Alfalfa plants in an $AuCl_4$-rich environment for the production of gold nanoparticles (AuNPs). It took two weeks. The transmission electron microscopy (TEM) image of the alfalfa shoot showed aggregates of AuNPs with size ranged between 2 and 40 nm. The energy-dispersive X-ray spectroscopy (EDS) analysis confirmed the pure AuNPs. The X-ray absorption studies and the atomic resolution analysis confirmed the nucleation and growth of AuNPs inside the plant and that the AuNPs are in a crystalline state. The high-resolution transmission electron microscopy (HRTEM) showed that the gold particles with an approximate size of 4 nm had an icosahedron structure, which is the lowest energy configuration for AuNPs. Images also showed defects such as twins in the crystal structure not only the icosahedral NPs. However, it is a promising path to phytogenic quantum dots. Marshall et al. (2007) explored the feasibility of using plants to synthesize large quantities of metallic NPs. The large uptake and reduction of Ag^+ ions and its distribution as AgNPs within the cellular structure by both *Medicagosativa* and *Brassicajuncea* were proved (Marshall et al., 2007). Sharma et al. (2007) reported the growth of Sesbania seedlings in chloroaurate solution, which resulted in the accumulation of gold with the formation of stable AuNPs in the plant tissues. In another study, Armendariz et al. (2009) reported the phytosynthesis of AuNPs via the interaction of Au(III) ions with oat and wheat biomasses, followed by extraction using sodium citrate or cetyltrimethylammonium bromide. To overcome the extraction step and long-time process, Parker et al. (2014) demonstrated the first production of palladium nanoparticles (PdNPs) by living Wild-type Arabidopsis ecotypes Col-0 (*Arabidopsis thaliana* L.) plant using aqueous solution of K_2PdCl_4. It was noticed that the change of plant coloration occurred with time. The TEM analysis revealed the production of well-dispersed PdNPs of average size of 3 nm, within 3 h. This increased with time reaching 32 nm after 24 h. The inductively coupled plasma (ICP) analysis of palladium concentration in the plants over time showed that the plants reached maximum palladium concentration after 18 h of exposure. The X-ray photoelectron spectroscopy (XPS) of the dried Pd-plant material confirmed the deposition of PdNPs. It was suggested that the binding to carboxyl, amino, and sulfhydryl groups in the plant was likely to be important prior to reduction. The extraction of the phytogenic PdNPs was omitted, and the whole plant with the synthesized PdNPs was dried, powdered, and then pyrolized at 300 °C (573 K), under nitrogen with a heating rate of 1 K/min. Those PdNPs showed excellent catalytic activity across a range of Suzuki–Miyaura coupling reactions involving I, Br, and Cl leaving groups (see Scheme 8.1), producing higher yields than the commercial Pd catalyst. That reaction is important in the synthesis of pharmaceutical intermediates and other high-value molecules.

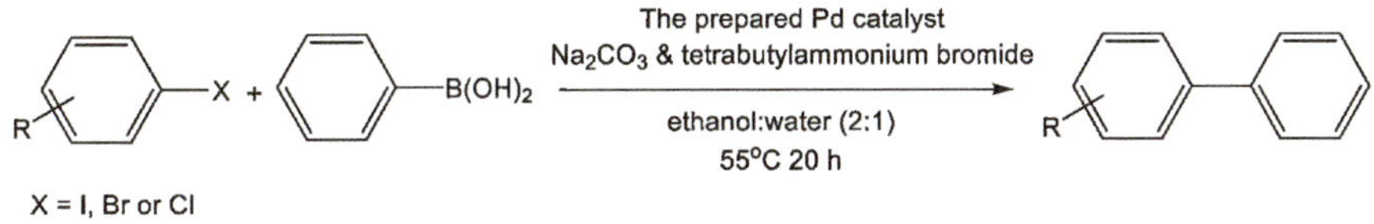

The prepared Pd catalyst
Na_2CO_3 & tetrabutylammonium bromide
R–C₆H₄–X + C₆H₅–$B(OH)_2$ → R–C₆H₄–C₆H₅
ethanol:water (2:1)
55°C 20 h

X = I, Br or Cl

SCHEME 8.1 Suzuki–Miyaura coupling reactions.

Raju et al. (2012) reported the synthesis of AuNPs intracellularly and extracellularly by using living *Arachis hypogaea* (peanut) plants. The extracellular synthesized AuNPs were characterized by XRD and HRTEM to be crystalline, polydispersed, and spherical, with a size of 5–9 nm and most were of 5–6 nm. The EDS analysis revealed the presence of gold and other oxygen carbon, copper, and aluminum peaks. The oxygen was attributed to the air presented in the chamber, the carbon was attributed to the biomolecules capping the AuNPs, while the copper and aluminum were attributed to the substrate used for EDS analysis. The Fourier transform infrared (FTIR) results showed that some of the proteins and phenolics leached from the roots helped in the formation of AuNPs. The XPS proved the presence of AuNPs, aromatic carbon present in amino acids from protein bound on the surface of AuNPs, –COOH groups, and α carbon bound to –COOH and $–NH_2$ groups of the protein bound to NP surface, the –OH groups, and –C=O groups present in capping proteins presented on the surface of the AuNPs and the –NH amide linkage or amidic (peptidic) nitrogen. The extracellular synthesis of NPs is exciting as these may be immobilized in different matrices or in thin-film form for optoelectronic applications as well as in drug delivery. Moreover, when the 15-day-old peanut seedlings were exposed to 10^{-4} M $HAuCl_4$ solution, for 24 h, under sterile conditions, the roots turned light pinkish. The light microscope proved that there was a pinkish color in cells approximately five to seven layers deep inside the root. That indicated AuNP formation inside the subepidermal cortex cells of the roots. The HRTEM revealed an interesting pattern of distribution of AuNPs, with different shapes, inside the plant cells, spherical and oval of 5–8 nm and 30–50 nm in size, respectively. The XRD analysis proved the formation of crystalline AuNPs. The EDS analysis of the root confirmed the presence of AuNPs, oxygen, due to the air presented in the chamber, and the other elements, nitrogen, carbon, and potassium, due to the plant root powder sample. Raju et al. (2012) illustrated a preliminary mechanism for the intracellular synthesis of AuNPs by the plant cells. The roots trapped gold from the solution as a result of the affinity between carboxylic acid moieties present in the cell wall and Au^{3+} ions. Once the gold enters the root cells, it is transported simplistically to the conducting tissues and aerial parts of the plants, where the synthesis of NPs could occur in the cell wall (external boundary) or membrane (inner boundary). As gold is not an essential element for the plants, the plant cells undergo stress upon its entry and release some stress biomolecules, which break down the chloroauric acid and possibly help in the synthesis/generation of AuNPs. Otherwise, the intracellular synthesis would have been occurred by the reduction of metal ions via enzyme/protein and secondary metabolites present on the surface of the plant cell or on the cytoplasmic membrane as these surface-bound NPs (intracellular) may be used in catalysis and as precursors for the synthesis of coatings for electronic applications. The ICP atomic emission spectroscopy analysis proved that the presence of AuNPs was noted not only in the roots, but also in the stem and leaves of the seedlings. The amount of AuNPs in root, stem, and leaf was about 53.31, 2.75, and 1.33 ppm, respectively. Parial et al. (2012) also reported the intracellular synthesis of AuNPs by the green algae *Rhizoglonium fontinale* and *Ulva intestinalis*, which was confirmed by purple coloration of the thallus within 72 h at 20 °C.

Thus, the possibility of the green synthesis of mixed-metal NPs by plants would open a new field of research for using plants to produce catalysts of specific composition, which are difficult to synthesize by traditional methods. However, more studies are needed to understand and illustrate the exact mechanisms for intracellular and extracellular NP synthesis with living plants and the interaction of the metal ions and the produced metal NPs at the cellular level.

8.3 MICROBIAL SYNTHESIS OF NANOPARTICLES

The idea of microbial synthesis of metallic NPs is not new and started around 1960s (Temple and Le Roux, 1964). However, the deep investigation about microbial synthesis of metallic NPs started by the 21st century. The microbial synthesis of NPs is considered promising and favorable compared to chemical and physical techniques. Nevertheless, synthesis procedures using mediated fungus, bacteria, and microalgae are difficult since they involve elaborated steps: isolation, purification, growth, and maintenance of the microbial strains. These steps are challenging in terms of maintaining the culture medium, avoiding microbial contamination, as well as optimizing the physicochemical conditions to obtain the biomass and then maximize the yield of the NPs and controlling its size and shape. These challenges would increase upon the scaling-up of nanomaterial production from laboratory processes to industrial scale. The scaling-up of biosynthesis of NPs from laboratory scale to industrial one is difficult and has many uncertainties and challengeable both on their production and properties. Moreover, in the case of intracellular synthesis, downstream processing is more difficult and expensive than that of the extracellular one, due to the involved separation and purifying steps. Consequently, the extracellular synthesis is more preferable, due to its easier and simpler downstream processing. Furthermore, fungi provide a more rational and economical approach for biosynthesis of NPs, relative to bacteria and algae, from two points of view. The first is much simpler and easier downstream processing and biomass handling in the case of fungi. The second is higher amounts of the secreted proteins, which are responsible for the bioreduction of metal ions to metal NPs in case of fungi, thus increasing the biosynthesis productivity by several folds. However, the pathogenicity of the applied fungus should be taken into consideration. Many well-studied fungi such as *F. oxysporum* are known to be pathogenic and, therefore, might pose a safety risk. But other well-defined strains *Trichoderma asperellum* and *Trichoderma reesei* have been proven to be nonpathogenic, which makes them ideal for use commercially. In fact, *T. reesei* has already been used widely in sectors such as food, animal feed, pharmaceuticals, and paper and textile industries. Besides this, there is also another challenge in applying bacteria for biosynthesis of NPs, that is, obtaining a clear filtrate from bacterial broths requires the use of complicated equipment in process technology, which, consequently, increases the investment costs to a considerable extent. On the contrary, in the case of fungal or phytosynthesis of NPs, simple equipment such as a filter press can be applied to get clear filtrates, thus encouraging economic and feasible viability. Furthermore, not all microorganisms have the ability to biosynthesize nanomaterials. Each microorganism has its own enzymatic and metabolic activities (Yusof et al., 2019). Moreover, each biosynthesis process has its own optimum condition for maximum production of nanomaterials within the shortest time. For example, the optimum conditions for the extracellular mycosynthesis of AgNPs using *Trichoderma longibrachiatum* DSMZ 16517 are 40 °C, 24 h, pH 12, 3g biomass, 4M $AgNO_3$, and 150 rpm, under dark condition (Omran et al., 2019a). That yielded 3g/L multishaped (spherical, triangular, hexagonal, and cuboidial) 5–13nm AgNPs. The polydispersity index (PdI) of 0.279 with an intercept of 0.927 and the recorded Zeta potential value of –26.8 mV proved the high stability of the mycosynthesized AgNPs. Moreover, according to Zomorodian et al. (2016), when the PdI value is lower than 0.5, monodispersity of the prepared NPs is indicated. The mycosynthesis under dark conditions is favorable from the point of process cost and energy saving. Taking this into consideration, such mycosynthesis of AgNPs via *Trichoderma longibrachiatum* DSMZ 16517 costs 94 US$/10 g AgNPs. On the other hand, in an earlier study, the optimum conditions for the extracellular mycosynthesis of AgNPs using *Aspergillus brasiliensis* are 20 °C, 72 h, pH 7, 6g biomass/100 mL, and 2M $AgNO_3$, under static condition and illumination of fluorescent light 36 W/6400 K, which yielded 1.7g/L highly stable nonagglomerated spherical-shaped 6–21nm AgNPs (Omran et al., 2018b). The PdI value (0.366) indicated the monodispersity of the mycosynthesized AgNPs. The zeta potential value (–16.7 mV) proved its stability for two months without the occurrence of any aggregations. Taking this into consideration, such mycosynthesis of AgNPs via *Aspergillums brasiliensis* costs 172 US$/10 g AgNPs. Thus, the extracellular mycosynthesis under static conditions and the long-time stability of the mycosynthesized AgNPs added to the advantageous of the extracellular mycosynthesis process, as being an energy-saving and thereby cost-effective process. In the aforementioned two cases, the FTIR analysis also confirmed the presence of proteins and other biomolecules around the AgNPs, which would have acted as reducing, stabilizing, and capping agents (Omran et al., 2018a, b). Ma et al. (2017) suggested that proteins of *Penicillium aculeatum* Su1 can bind with NPs via free amine groups, cysteine residues, or the electrostatic attraction of the negatively charged carboxylate groups in a cell-free filtrate from fungal mycelia, and that the enzymes of Su1 possessed higher catalytic activity at alkaline pH8. Most of the cell-free microbial cell filtrate and/or bioextracts exhibit the same four steps: (1) uptake of silver ions and activation of silver reduction machinery; (2) electron shuttle system involving various cofactors and enzymes; (3) electrostatic interaction between silver ions and mycelial cell free filtrate (MCFF) or bioextract components; and (4) reduction through extracellular enzymes and other biomolecules released in the MCFF or the bioextract (see Figure 8.3). However, more research should be performed on this point to figure out how the change in pH would affect the biosynthesis' mechanisms of NPs.

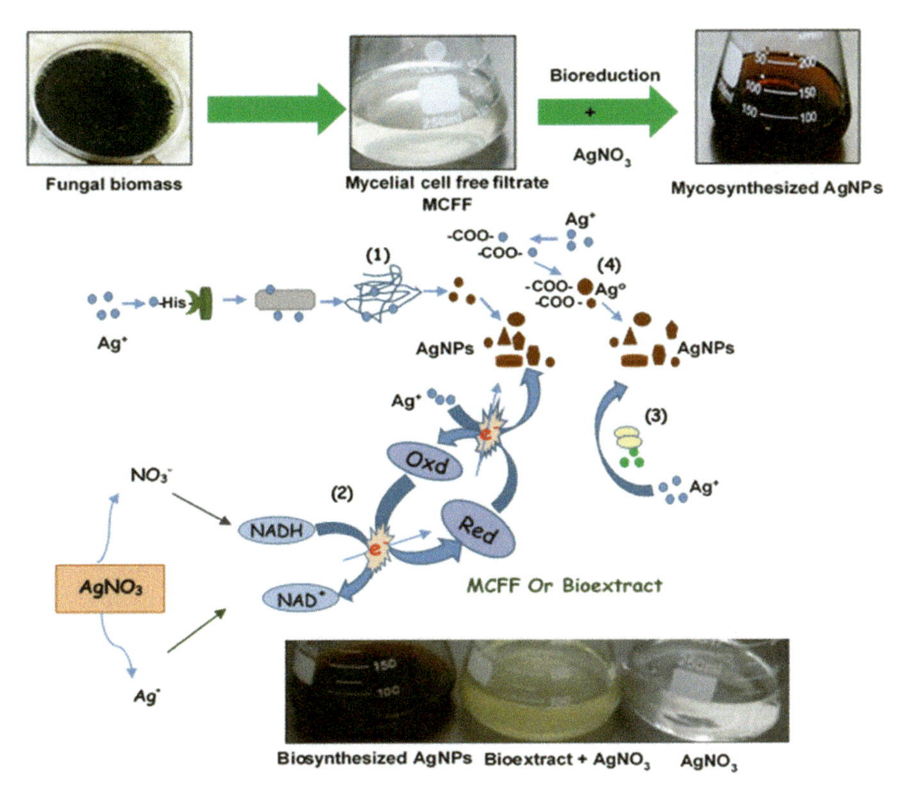

FIGURE 8.3 Suggested mechanism for extracellular synthesis of AgNPs.

Most of the microorganisms secrete the cofactor nicotinamide adenine dinucleotide (NADH)-dependent enzymes, especially nitrate reductase, which may act as a scaffold or a nucleating agent and may be responsible for the bioreduction of metal ions into metal NPs, and it also acts as a capping agent, thus ensuring stabilization of the prepared metal NPs (He et al., 2007; Vahabi et al., 2011; Devi et al., 2013). Peptides, containing amino acids such as arginine, cysteine, lysine, methionine, glutamic acid, and aspartic acid, are known for their recognition and interaction with metal ions in solution and generate a reducing environment around them leading to their reduction and the formation of polydispersed metal NPs. But those biomolecules would have been deactivated under acidic conditions, while alkaline conditions activate the reductases of oxidoreductase enzymes (Prasad et al. 2010). Moreover, tyrosine undergoes conversion to a semiquinone structure under alkaline conditions through ionization at the phenol group, which reduces metal ions (Selvakannan et al. 2004). Tryptophan is also converted to transient tryptophyl radical at alkaline pH, which donates electron to reduce metal ions (Si and Mandal, 2007). Besides this, the reducing cofactors generated by the activity of various spore-associated enzymes, such as glucose oxidase, alkaline phosphatase, laccase, and catalase, can also stimulate the biogenesis of metal NPs (Hosseini-Abari et al. 2013). Sintubin et al. (2009) suggested that the increase in pH would enhance the bioreduction since high pH catalyses the ring opening of monosaccharides, such as glucose, to their open chain aldehyde, the aldehyde that delivers the reducing power. When metal ions are present, the aldehyde will be oxidized to the corresponding carboxylic acid, and at the same time, the metal ions will be reduced. In addition, Apte et al. (2013) reported the increase in solubility of some biomolecules, such as 3,4-dihydroxy-L-phenylalanine and chloroauric acid, under alkaline conditions, which favor the optimal synthesis of nanostructures.

However, the extracellular production is still more economically and advantageous. That is because, on large-scale production, the downstream processing eliminates the harvesting of biomass incorporating the biosynthesized nanomaterials and disruption of biomass cells to get the synthesized nanomaterials. Thus, the energy-intensive centrifugation and ultrasonication processes are omitted in the extracellular synthesis of nanomaterials (Yusof et al., 2019). Thus, taking the advantage of the metal resistance of the bacterial biofilm in the biosynthesis of NPs, as an example for the extracellular reduction of metal ions, is very promising (Tanzil et al., 2016).

Jha et al. (2009) reported that the size of the biosynthesized metal oxide NPs by yeast is usually lower than those biosynthesized by bacteria as eukaryotes have better level of organization at the cellular level. It has been reported that bacteria have a negative electrokinetic potential, which helps in biosorption of positively charged metal ions (i.e., the cations), the starting root for the further bioreduction step to metal NPs (Jinkun et al., 2000). Moreover, two important factors control the biosynthesis of metal/metal oxide nanoparticles; (1) energy-yielding material, for example, glucose can control the oxidation–reduction potential value which is expressed as partial pressure of gaseous hydrogen rH_2 and (2) the medium pH which can be controlled by the presence of bicarbonate (Jha et al., 2009; Prasad et al., 2010). The oxidoreductases enzymes are pH sensitive. At mild acidic environment with relatively low rH_2, the membrane bound oxidases are activated and the metal oxide NPs are formed. At relatively high pH and rH_2 values, reductases are activated and metal NPs are formed (Prasad et al., 2010). This also confirmed that the type of microorganism plays an important role as well. For example, yeast and fungi have quinones, benzoquinones, and toluquinones, which facilitate the redox reactions due to their tautomerization at the cell membrane level, immediately after the addition of metal salts (Jha et al., 2009). Yeast has membrane bound and cytosolic oxidoreductases, which also enhance its efficiency in the biosynthesis of metal/metal oxides NPs (Jha et al., 2009).

Thus, further research is required to elucidate the reason for the change of the optimum physicochemical parameters for obtaining metal/metal oxide NPs with the change of the type of microorganisms' cell filtrate, keeping the metal salt precursor unchanged. It is very important to know exactly the components of the extracellular filtrate used in the microbial synthesis of NPs and elucidate the possible involved mechanism of biosynthesis of NPs. Moreover, generally, further exploration is required to know the intracellular and/or extracellular biosynthesis mechanisms of NPs due to their reported variations within the applied microbial species. From the economical point of view, it is also essential to isolate microorganisms that can extracellularly produce high yield of metal/metal oxide NPs over a wide range of pH. There would be no need for the step of pH optimization or the addition of buffering solutions.

It is also important to do further work to explore and identify the phenomena of changing the size and shape of the biosynthesized NPs when applying different microbial strains and even with changing the biomass age, reaction time, pH, salt concentration, temperature, and other physical parameters upon applying the same microbial strain (Das et al., 2010; Honary et al., 2012; Shah et al., 2015, Taran et al., 2017; Pantidos et al., 2018). It is important to know that the preferable age for bacterial synthesis of nanomaterials is the stationary phase cells. Theoretically, the stationary phase cells suffer from greater metabolic stress than the logarithmic phase ones. Thus, the metabolites with greater capacity of chemically reducing other compounds would be synthesized during the stationary phase. These metabolites are usually capable for reducing metal ions into metal NPs (José de Andrade et al., 2017).

Besides this, the media composition also affects the yield, size, and shape of the biosynthesized NPs (Shantkriti and Rani, 2014; Pantidos et al., 2018). For example, Kundu et al. (2014) reported that the trace metal ions in the bacterial culture such as K^+, PO_4^{2-}, Mg^{2+}, etc., would react with the metal ions precursor initiating the biosynthesis of the metal/metal oxide NPs (see Scheme 8.2), while the microbial secreted proteins and enzymes are responsible for the size and stability of the biosynthesized NPs (Kundu et al., 2014). Moreover, Tyupa et al. (2016) mentioned the importance of optimizing the microbial growth media composition to eliminate the adverse effect of ions Cl^-, SO_4^{2-} and HPO_4^{2-} on the microbial synthesis of metal/metal oxide NPs. As it is essential to decrease their contents to the level required, otherwise, they will bind to Ag^+ cations, thus complicating their enzymatic reduction into Ag^0. This will consequently decrease the yield of NPs.

$$Zn_4(SO_4)(OH)_6 \cdot xH_2O \longrightarrow ZnSO_4 \cdot 3Zn(OH)_2 + xH_2O$$

$$2\,ZnSO_4 \cdot 3Zn(OH)_2 \longrightarrow 5ZnO + Zn_3O(SO_4)_2 + 6H_2O$$

$$Zn_3O(SO_4)_2 \longrightarrow 3ZnO + 2SO_2 + O_2$$

SCHEME 8.2 Sequential biosynthesis of metal oxide NPs in presence of other metal ions in the microbial culture.

As another example, *Aspergillum* sp. WL-Au in presence of only distilled water reported to produce AuNPs with different shapes, spherical, triangular, hexagonal, and irregular, with the particle size ranged between 10 and 100 nm and the average size of 50.3 nm, while in case of WL-Au in presence of phosphate sodium buffer (PSB), only spherical and pseudo-spherical-shaped NPs were produced, where the particles size ranged between 12 and 48 nm with an average size of 26.4 nm. The surface plasmon resonance (SPR) shifted from 556 to 539 nm upon the usage of water and PSB, respectively (Qu et al., 2017). The concentration of *Candida albicans* cytosolic extract has also been reported to affect the size of the prepared AuNPs. Higher concentration decreased the size from 60–80 to 20–40 nm (Chauhan et al., 2011). Thus, further investigations should also be performed about the effect of the cultural media. However, to save time, energy, and cost, it is important to apply new mathematical techniques such as the response surface methodology based on central composite design and/or Box–Behnken design of experiments to study the individual and interactive effects of different physicochemical parameters on shape, size, yield, and stability of the biosynthesized metal and/or metal oxide NPs.

One of the main advantages of microbial synthesis of metal/metal oxide NPs is its flexibility to be performed under dark conditions (Kathiresan et al., 2009; Patel et al., 2015; Omran et al., 2019, 2020). This added to the advantageous of microbial synthesis of NPs as it lowers the consumption of energy. However, further research is required on isolating such microbial strains that are capable of producing high yield of NPs under dark conditions. Another advantage is the possibility of isolating such microbial strains from aerobic activated sludge of wastewater treatment, heavy-metal-contaminated soil, metal mines, and petroleum field, which are reported to facilitate their enhanced tolerance to metal ion toxicity (Joerger et al., 2000; Tyupa et al., 2016; Liang et al., 2017). This would consequently enhance their enzymatic efficiency for the bioreduction of such metal ions into metal/metal oxide NPs. More research is required for the continuous upscaling process for microbial synthesis of metal/metal oxide NPs, in order to maximize the NP yield with minimum requirements of feedstock with flexible cost-effective and energy-saving operating conditions. Tyupa et al. (2016) reported a successful continuous process for mycosynthesis of AgNPs using *F. oxysporum*, whereas, under optimum conditions, 200-mL/h flow rate, and 20-mg/L Ag^+ ion concentration, the yield of AgNPs reached to approximately 16 mg/L. It was noticed that a higher AgNPs yield was obtained with shorter time and lower $AgNO_3$ concentration in the continuous process than in the batch one, recording approximately 80% and 65%, respectively. Furthermore, advantage is the ability of some microorganisms to bioremediate heavy-metal-polluted water producing valuable metal/metal oxide NPs. This will have a double positive impact on environment and economy, which is the sustainable bioremediation of polluted water and sustainable, cost-effective, and ecofriendly production of valuable NPs with different applications. For example, Bharde et al. (2006) proved that the extracellular hydrolysis of the anionic iron complexes $K_3[Fe(CN)_6]$ and $K_4[Fe(CN)_6]$ by cationic proteins secreted by the fungi *Fusarium oxysporum* and *Verticillium* sp., at room temperature, led to the formation of crystalline magnetite NPs with a size range of 20–50 and 10–40 nm, respectively.

Nevertheless, the time of bioreduction applying plant extracts is much faster than that applying microbial ones, which is a very critical factor on large-scale production. It has been reported that for all biological methods, only plant extracts are able to exhibit better control over morphological characteristics, sizes, and rates of production (Makarov et al., 2014; Ahmed et al., 2016).

8.4 DIFFERENT BIOPRODUCTS FOR BIOSYNTHESIS OF NANOPARTICLES

Different bioproducts have been reported as ecofriendly bioreductants of metal ions into metal NPs. For example, Apiin, which is a natural flavonoid, has been used for the biosynthesis and stabilization of anisotropy Ag and quasi-spherical Au NPs (Kasthuri et al., 2009). Exopolysaccharides (EPSs) of *Lactobacillus rhamnosus* GG ATCC 53103 synthesized a mixture of spherical, triangular, rod-shaped, and hexagonal 2–15nm AgNPs (Kanmani and Lim, 2013), while in another study, extracellular polysaccharide/matrix of *Nostoc commune* is reported to produce spherical-shaped 15–54nm AgNPs (Morsy et al., 2014). Native and repolymerized flagella of *Salmonella typhimurium* synthesized 3–11nm AgNPs (Gopinathan et al., 2013). *Actinorhodin* pigment is reported to synthesize irregular-shaped 28–50nm AgNPs (Manikprabhu and Lingappa, 2013). Bioflocculant of *Bacillus subtilis* MSBN 17 is reported to synthesize spherical-shaped 60nm AgNPs (Sathiyanarayanan et al., 2013). Spores of *Bacillus athrophaeus* produced polydispersed 5–30nm AgNPs (Hosseini-Abari et al., 2013).

Biosurfactant rhamnolipids were produced by potent biosurfactant producer, *Pseudomonas aeruginosa* strain BS-161R, purified, characterized, and then used for the synthesis of biosurfactant-based monodispersed and spherical-shaped AgNPs of average particle size of 15.1 nm (Kumar et al., 2010). Biosurfactants have also been reported to enhance the synthesis rate of AgNPs and act as stabilizing agents (Kiran et al., 2010). The previously isolated *Pseudomonas aeruginosa* BS-161R (Genbank accession number FJ940905) from a petroleum-contaminated sludge sample was also used for extracellular biosynthesis of AgNPs (Kumar and Mamidyala, 2011). The FTIR analysis proved that the reduction of the silver ions might have been due to the protein component resulting from the enzyme nitrate reductase, as the nitrate reduction is a phenotypic biochemical characteristic of that culture (Urban, 2007) and the capping might have been possibly due to the rhamnolipids presented in the culture supernatant. *Pseudomonas aeruginosa* biosurfactant rhamnolipids have also been reported for the biosynthesis of spherical-shaped 35–80nm ZnO NPs (Singh et al., 2014), where the rhamnolipids disperse in the aqueous solution to form the spherical core–shell micellar structure, in which the core is represented the hydrophobic C–C chains and the shell consist of polar –OH groups. Then, the rhamnolipids molecule hydrophobic alkyl chains get attached to the surface of primary ZnO crystallite in the presence of oxygen (see Scheme 8.3). Then, the high surface energy of preformed ZnO crystallite begins to form rhamnolipids ZnO NPs. Furthermore, synthesis of ZnO NPs proceeds inside the core of the small micelles of rhamnolipids via the nucleation and growth of ZnO NPs (see Scheme 8.3). Thus, the first step includes the formation of zinc hydroxide upon the addition of NaOH with zinc nitrate aqueous solution and crystalline ZnO nucleate from the dehydration of the preformed ZnOH (Sangeetha et al., 2013; Singh et al., 2014).

Thus, the application of rhamnolipids biosurfactant in green synthesis of metal/metal oxide NPs with efficient biocidal activity would be very promising for the mitigation of the microbiologically influenced corrosion (MIC). As rhamnolipids biosurfactant has been reported to be a corrosion inhibitor. Moreover, it has also efficient biocidal effect on sulfate-reducing bacteria (SRB) and saprophytic bacteria (total general bacteria) (Lan et al., 2017). Thus, the biocidal activity of the stabilized, capped, and functionalized NPs with such a biosurfactant is expected to be much more efficient.

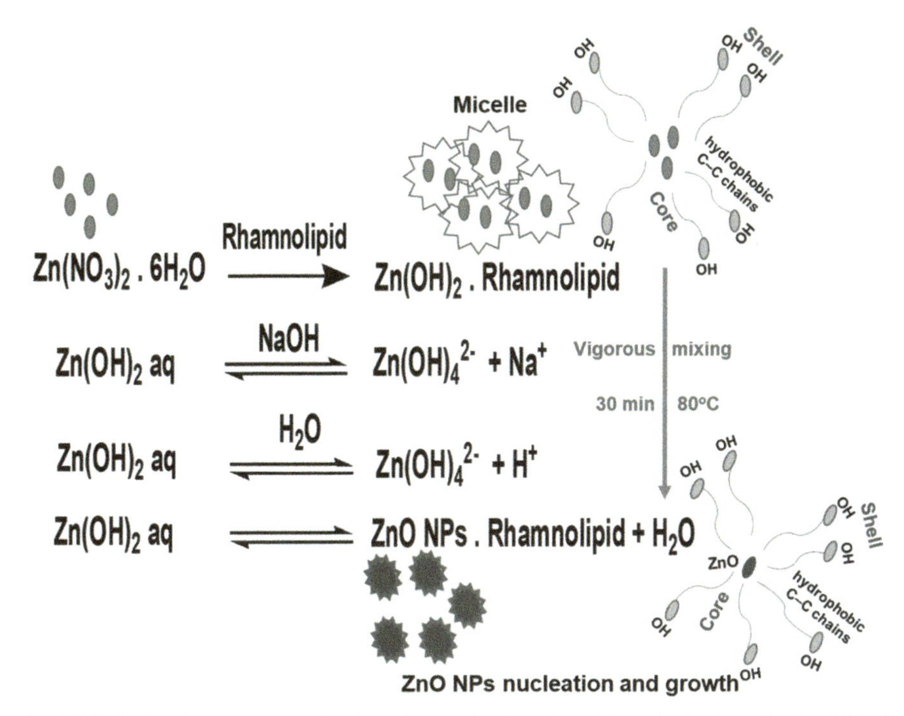

SCHEME 8.3 A possible mechanism for synthesis of the biosurfactant-synthesized ZnO nanoparticles.

Wei et al. (2014) investigated the usage of highly commercialized and easy accessible animal and fungus bioproducts, that is, tryptone (hydrolysate form casein using trypsin) and yeast extract (dried yeast autolysate), respectively, for the synthesis of AgNPs under sunlight radiation, whereas the sunlight induced tryptone and yeast extract to synthesize AgNPs, and the stability of AgNPs was regulated by capping peptides, pH, and sunlight exposure. The tryptone was more efficient for the synthesis of AgNPs than the yeast extract. That was attributed to the total protein/peptide content of tryptone (46%), which was much higher than that of yeast extract (26%), thus resulting in the higher reduction efficiency. For the light-induced reduction mechanism, Nam et al. (2008) indicated that ambient light and carboxylic acid-containing peptides were involved in the formation of AgNPs. Due to nucleophilicity of carboxylic groups, carboxylic-acid-containing peptides binding with Ag^+-induced partial electron transfer to lower the energy barrier for Ag^+ reduction. Eby et al. (2009) reported that the absolute value of ζ-potential is positively correlated with the colloidal stability of the NPs, and an increased absolute value means improved colloidal stability. Badawy et al. (2010) reported that 20-mV ζ-potential absolute value is considered to be the critical value for colloidal stability of NPs. The low ζ-potential of tryptone-Ag (−9.09 ± 0.19 mV at pH 4) and yeast-Ag (−11.68±1.35 mV at pH 4, and −16.86 ± 0.18 mV at pH 5) might be the reason for their instability. Thus, at pH 5–11, the absolute value of ζ-potential of tryptone-Ag reached about 30 mV, while, at pH 7–11, the ζ-potential absolute value of yeast-Ag was about 23 mV. Consequently, these high ζ-potential absolute values might have contributed to the stability, while the sedimentation of AgNPs at pH 4 would help remove AgNPs from environmental water. Wei et al. (2014) noticed that sunlight not only mediated the synthesis of AgNPs with tryptone and yeast extract, but also induced further aggregation and sedimentation and attributed that to the strong interaction force of oscillating dipole–dipole driven by the UV of the sunlight (Cheng et al., 2011). Thus, Wei et al. (2014) recommended for storage and application, tryptone-Ag, and yeast-Ag could be maintained in the dark at neutral pH. But when they are released into environmental water, long-term sunlight exposure could remove AgNPs from water to reduce the toxicity. The photoirradiation method has also been applied for the biosynthesis of AgNPs using the blue pigment, actinorhodin produced by *Streptomyces coelicolor* (Manikprabhu and Lingappa, 2013). More research should be performed on the effect of sunlight irradiation as a costless resource on the biosynthesis of metal/metal oxide NPs using the spent waste media used for cultivating and marinating microbial biomass. Such media would be another costless resource of tryptone, yeast extract, and other bioproducts that can be used for the biosynthesis of NPs. By applying such a technique, nanotechnologists will overcome one of the major waste management problems of biotechnologists, which is the spending waste microbial media and valorizing it into different NPs with different valuable applications in different sectors.

Chitosan is available with its unique abundance of free amino and hydroxyl groups and valuable characteristics of being polycationic, chelating, and film-forming; however, few studies about the biosynthesis of metal/metal oxides NPs using such biomolecule have been published (Huang and Yang, 2004; Wang et al., 2006; Janardhanan et al., 2008; Wei and Qian, 2008; Wei et al., 2009; Wang et al., 2014; Venkatesham et al. 2014; León et al., 2017; Nate et al., 2018; Rizeq et al., 2019). Thus, more research is required to apply this ecofriendly and widely abundant bioreductant, stabilizing, and capping agent on large-scale biosynthesis of metal NPs. In a study reported by Mat Zain et al. (2014), ascorbic acid acted as a reducing agent for biosynthesis of copper nanoparticles (CuNPs) from copper nitrate. Besides, chitosan acted as a stabilizing agent and prevented the agglomeration of the biosynthesized CuNPs. The produced CuNPs expressed efficient biocidal activity against *B. subtilis* and *E. coli*. Kathad and Gajera (2014) reported the reduction of copper sulfate into CuNPs using ascorbic acid. But the size controlling occurred applying the cationic surfactant cetyl trimethyl ammonium bromide (CTAB). Moreover, the chemically synthesized ZnO interlinked with the green and ecofriendly chitosan reported to have an efficient biocidal activity against the SRB, which is the main reason for biofilm formation and MIC in petroleum industry (Rasool et al., 2018; Abdul Rasheed et al., 2019). Thus, extension of research on using the sustainable chitosan in the biosynthesis of NPs will make the process more cost effective and environmentally friendly and will also solve one of the organic biomass waste management problems, where chitin and chitosan are the most worldwide readily available biodegradable polymers, and they are found in the shells of crustaceans, for example, lobsters, crabs, and shrimp, and many other organisms, such as insects and fungi. So, this will open a new market for sustainable valorization of wastes from fish markets, fisheries, and other seafood restaurants into valuable NPs with different applications in different sectors. Besides this, it would also solve a big problem in oil and gas industry, which is the microbial corrosion, in an ecofriendly and cost-effective manner.

Essential oils have also been reported for the green synthesis of NPs. For example, essential oil of *Nigella sativa* has been reported for the green synthesis of spherical-shaped 15.6 and 28.4nm AuNPs. The main advantage of essential-oil-synthesized AuNPs was its high biocidal efficiency against biofilm forming bacteria, which potentially inhibited the biofilm formation of *Staphylococcus aureus* and *Vibrio harveyi* by decreasing the hydrophobicity index (78% and 46%, respectively) (Manju et al., 2016). Essential oils from *Citrus aurantifolia*, *Lippia alba* LA44, and *Cymbopogon citratus*, as well as citral, linalool, eugenol, and geraniol, have been reported as efficient biocides for SRB the main contributor in MIC (Macedo de Souza et al., 2017). The green-synthesized AgNPs have also been efficient biocides for SRB (Omran et al., 2019). Thus, this would open a new point of research for the application of such green-synthesized NPs using essential oils in mitigation of MIC.

The methyl esters of myristate, palmitate, and stearate (thermostable glycolipid) extracted from *Gordonia amicalis* HS-11 were used to biosynthesis of AuNPs and AgNPs (Sowani et al., 2016). Rivera-Rangel et al. (2018) reported the application of the hot water extract of geranium leaf as the reducing agent in the green synthesis of AgNPs, using castor oil, Brij 96 V, and 1,2-hexanediol as the surfactant and cosurfactant and silver stearate as the salt precursor applying oil-to-water (O/W) microemulsion technique.

The glutamic acid as an example of amino acids has been reported to act as both a reducing and a stabilizing agent in the preparation of AuNPs (Wangoo et al., 2008). A mixture of aqueous solution of ascorbic acid, alkylamine, and oleic acid has been reported for the preparation of self-assembled AgNPs with an average diameter of 5 nm (Li et al., 2010). Such surface-capped AgNPs showed outstanding dispersity in kerosene and other conventional organic solvents, for example, n-hexane and chloroform, proving its promising application in the synthesis of oil-based nanofluids (Li et al., 2010). Punicalagin, the main ingredient of pomegranate peel, is a high-molecular-weight polyphenolic compound, which has shown remarkable pharmacological activities attributed to the presence of dissociable OH groups. This component with flavonoids and tannins in pomegranate peels might aid in the bioreduction of Ag^+ into its NP state (Prakash, 2011).

Lysozyme and fbrinolytic as examples of enzymes produced by *Bacillus cereus* NK1 have been reported for biosynthesis of AgNPs (Eby et al., 2009; Deepak et al., 2011), where the URAK, which is a fibrinolytic enzyme, is produced by *Bacillus cereus* NK1 biosynthesized spherical-shaped 50–80-nm AgNPs (Deepak et al., 2011). The α-amylase enzyme has been reported for the biosynthesis of TiO_2 NPs (Ahmed et al., 2015; Khan and Fulekar, 2016). The α-amylase enzyme produced by *Aspergillus oryzae* has been reported for the biosynthesis of monodispersed triangular- and hexagonal-shaped AgNPs with average particles size ranged between 22 and 24 nm and SPR of 422 nm (Mishra and Sardar, 2012). In another study, Baymiller et al. (2017) proved that the coenzyme NADH could reduce alone the Au^{3+} ions into spherical-shaped small-sized AuNPs (< 10 nm).

It is important to know that reducing sugars are able to reduce metal ions into metal NPs. Aldehydic or ketonic groups in the reducing sugars are oxidized to carboxyl groups by the nucleophilic addition of OH^- which consequently reduce Ag^+, Au^{3+}, and Cu^{2+} into Ag^0, Au^0, and Cu^0, respectively (Kwon et al., 2009). Glucose has been reported as a reducing agent and starch as the stabilizing and capping agent in the green synthesis of AgNPs

(Mochochoko et al., 2013). Succinoglycan from *Sinorhizobium meliloti* has been reported to biosynthesize AgNPs under alkaline condition (Panigrahi et al., 2004), and other sugars from lemongrass have been also reported for the biosynthesis of AuNPs (Shankar et al., 2004). Cellulose also reported as a reducing and a stabilizing agent for the green synthesis of AgNPs (Hassabo et al., 2015). Li et al. (2014) biosynthesized AgNPs with smaller sizes, good dispersity, and narrow size distribution through tandem hydrolysis of silver sulfate and 0.3-g cellulose under hydrothermal conditions, in an autoclave set at 200 °C for 10 h, whereas the silver sulfate is first hydrolyzed to produce H_2SO_4 (the first hydrolysis), which then induces the later hydrolysis of cellulose (the second hydrolysis) to generate saccharides or aldehydes for the bioreduction of the Ag^+ ions to Ag^0 (see Scheme 8.4). Li et al. (2014) proved that the size of AgNPs was positively affected by the reaction temperature and time, but was negatively affected by the cellulose concentration. Higher temperature (>200 °C) and longer reaction time (>10 h) caused severe aggregations of AgNPs. However, sufficient cellulose led to a rapid reduction of the silver ions and the subsequent homogeneous nucleation of the Ag nuclei, thus promoting the formation of the AgNPs with smaller sizes. Moreover, the FTIR analysis documented the role of the cellulose-hydrolyzed products, saccharides or aldehydes in the bioreduction of the Ag^+ ions, and the critical roles of the aldehydes or saccharides C = O (C–O) groups in capping the AgNPs. Those as-synthesized AgNPs enhanced surface fluorescence of methyl orange by approximately 32-fold.

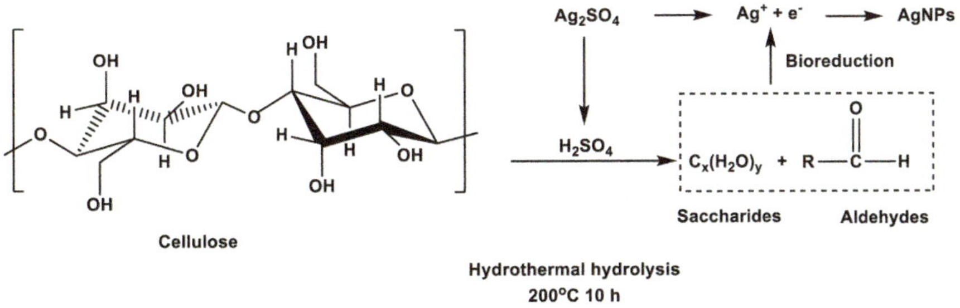

Cellulose

$Ag_2SO_4 \rightarrow Ag^+ + e^- \rightarrow$ AgNPs

Bioreduction

H_2SO_4

$C_x(H_2O)_y$ + R—C(=O)—H

Saccharides Aldehydes

Hydrothermal hydrolysis

200°C 10 h

SCHEME 8.4 Proposed mechanism for cellulose-synthesized AgNPs.

This will open a new research file for the valorization of the lignocellulosic wastes as sustainable resources of reducing sugars into NPs with different valuable applications. This would also help in solving one of the most intensive worldwide waste management problems, as open burning of such lignocellulosic wastes is one of the most used methods to get rid of such wastes, which increases the GHG emissions and consequently adds to the problem of climate change.

8.5 TOXICITY OF NANOPARTICLES

Nanotoxicity is related to (1) the probable discharge of (toxic) ions from metal NPs, (2) the oxidative stress caused by the inherent properties of NPs, such as their morphology, surface charge, size, and chemical surface composition, (3) element-specific toxicity of the core metal/metal oxide NPs, and (4) sometimes comes from the capping and/or stabilizing surface molecules (Seabra and Duran, 2015). There is another source that should be taken into consideration, which comes from the bioremediation process in wastewater treatment plants. That is the bioaccumulation of NPs such as metal/metal oxide NPs and nanoplastics into biofilms, which sometimes called nanofouling. This would indirectly affect the biodiversity and ecosystem via the application of sludge containing such biofilms in soil fertilization (Deschênes and Ells, 2020).

Despite the intensive developments of the nanotechnology, the adversative impacts of nanomaterials are still somewhat unidentified. However, the ecofriendly, cost-effective, and sustainable green- and biosynthesis of nanomaterials overcome most of the toxic effects caused by the chemicals used in the conventional preparation techniques (Kharissova et al., 2013; Stankic et al., 2016). Moreover, in biosynthesis of nanomaterials, the functionalization step is omitted. That is because they are already functionalized by the biochemical compounds. This adds to the advantages of biosynthesis over chemosynthesis (Baker et al., 2013; Makarov et al., 2014; Zikalala et al., 2018). Even the biosynthesis of metal oxides is considered as a bioremediation method for toxic metal ions (Jayaseelan et al., 2012; Zikalala et al., 2018). Thus, it would act as a sustainable route for production of nanomaterials that would have different applications (see Figure 8.4). But the utilization of water as nontoxic solvent in the green- and biosynthesis with the nowadays worldwide problem of water scarce is an issue that should be taken into consideration.

Moreover, it is important to use nonpathogenic microorganisms and nonpathogenic plant's parts in the biosynthesis of nanomaterials; otherwise, the biomass waste of such processes would be dangerous to the environment, human, and living organisms in general. Besides this, to widen its applicability into different life sectors, the biocapping and stabilizing matrix surrounding the biosynthesized NPs should be benign to nontarget organisms and cells.

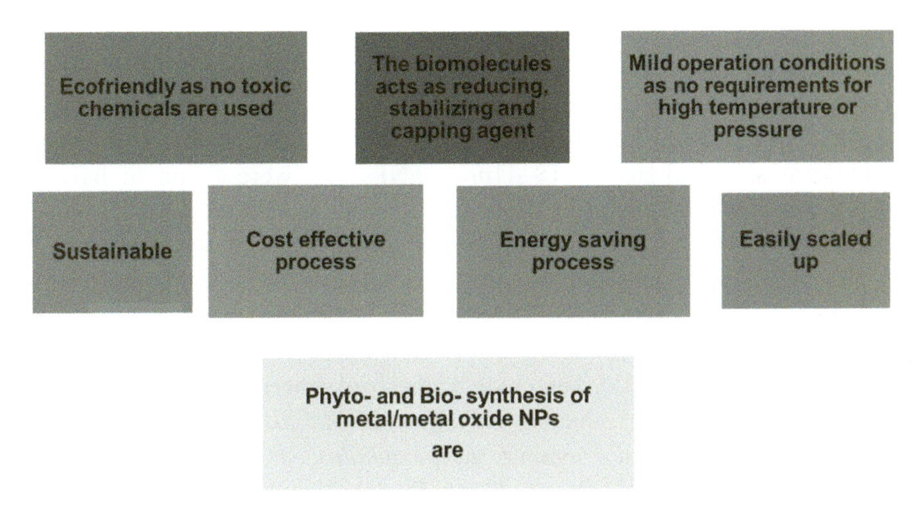

FIGURE 8.4 The key advantages of phyto- and biosynthesis of metal/metal oxide nanoparticles.

For example, the mycosynthesized ZnO NPs using cell filtrate of *Aspergillus fumigatus* TFR8 has induced the bioproduction of EPS using *Bacillus subtilis* strain JCT1 (NCBI GenBank Accession No.JN194187); by approximately 596.1% over the control (i.e., in absence of ZnO). Such nanoinduced EPS was safely applied to arid soil and led to increment in the soil aggregation (up to 82%), moisture retention (10.7%–14.2%), and soil organic carbon (Raliya et al., 2014). Thus, indirectly, the biosynthesized ZnO NPs can be safely applied for treatment of desertification and fortification of arid soils. The mycosynthesized TiO_2 NPs using the cell-free extract of *Aspergillus flavus* TFR7, isolated from rhizosphere soil, improved the shoot length (17.02%), root length (49.6%), root area (43%), root nodule (67.5%), chlorophyll content (46.4%), and total soluble leaf protein (94%) of mung bean (Raliya et al., 2015). Not only had this, but it also increased the photosynthetic pigment, chlorophyll, and total soluble leaf protein content by 46.4% and 94%, respectively. Such mycosynthesized TiO_2 NPs are promising for plant nutrition at very low concentrations (10 mg/L). Moreover, the rhizosphere microbial population (fungi, bacteria, and actinomyceteae) increased by 21.4%–48.1%, which consequently increased the activity of acid phosphatase (67.3%), alkaline phosphatase (72%), phytase (64%), and dehydrogenase (108.7%) enzymes in the rhizosphere within six weeks. Thus, such mycosynthesized TiO_2 NPs are recommendable for native phosphorous nutrient mobilization in rhizosphere as it increased the phytase and phosphatase enzyme activity (Raliya et al., 2015).

The green-synthesized AgNPs are also reported to be nontoxic to humans at low concentrations (Vithiya and Sen, 2011; Narendhran and Nair, 2017). This consequently widens its applications in medicine and environment sectors. For example, *Piper longum* leaf extracts were used to phytosynthesize a uniform spherical-shaped 18–41nm AgNPs that were found to have a significant cytotoxic effect on HEp-2 cancer cells (Jacob et al., 2011). Leaf extracts of *Eclipta prostrata* phytosynthesized AgNPs have expressed larvicidal activity against filariasis and malaria vectors (Rajakumar and Abdul Rahuman, 2011). Moreover, leaf extract of *Melia azedarach* was used to prepare AgNPs that proved to be active against the HeLa cervical cancer cell line (Sukirtha et al., 2011). The phytosynthesized spherical- and oval-shaped 70–140nm AgNPs using aqueous leaf extract of *Manilkara zapota* acaricidal expressed good activity against *Rhipicephalus* (*Boophilus*) *microplus* (Rajakumar and Abdul Rahuman, 2012). The UV/Visible spectrum of AgNPs phytosynthesized by *Annona squamosa* leaf extract showed the SPR of AgNPs at 444 nm (Vivek et al., 2012). The TEM photography showed the predominance of spherical-shaped AgNPs of average size ranging from 20 to 100 nm. The Zeta potential value of −37 mV revealed the stability of the phytosynthesized AgNPs. Furthermore, the green-synthesized AgNPs exhibited a dose-dependent cytotoxicity against human breast cancer cell (MCF-7) and normal breast epithelial cells (HBL-100), and the inhibitory concentration (IC_{50}) was found to be 50 μg/mL, 30 μg/mL, and 80 μg/mL, 60 μg/mL for AgNPs against MCF-7 and normal HBL-100 cells at 24 h and 48 h incubation, respectively. An induction of apoptosis was evidenced by acridine orange and ethidium bromide and 4′,6-diamidino-2-phenylindole staining (Vivek et al., 2012). The phytosynthesis of rod-shaped 25–80nm AgNPs using aqueous leaf extract of *Euphorbia prostrata* Ait was reported by Zahir and Abdul Rahuman (2012). Those AgNPs expressed good effect against parasites, the adult cattle tick (*Haemaphysalis bispinosa* Neumann) and the haematophagous fly (*Hippobosca maculata* Leach). In addition, toxicity tests were conducted to analyze the toxicological effects of those phytosynthesized AgNPs with an average size of 52.4 nm on *Daphnia magna* and *Ceriodaphnia dubia*, and the animal model test was evaluated against *Bos indicus* for 24h treatment, where no toxicity on daphnids and no adverse effects were noted on animals after exposure to the phytosynthesized AgNPs (Zahir and Abdul Rahuman, 2012). The AgNPs phytosynthesized by leaf extract of the *Prosopis chilensis* (L.) tree were found to be active against Vibrio species in the shrimp *Penaeus monodon* (Kandasamy et al., 2013). The AgNPs were also reported to be phytosynthesized using leaf extract of *Vitex negundo* L (Prabhu et al., 2013). Those phytosynthesized AgNPs inhibited the proliferation of human colon cancer cell line HCT15 with an IC_{50} of 20 μg/mL at 48h incubation. AgNPs were shown to promote apoptosis as seen in the nuclear morphological examination study using propidium iodide staining and DNA fragmentation by the single-cell gel electrophoresis technique (Prabhu et al., 2013). Premasudha et al. (2015) described the rapid and eco-friendly one-pot process for synthesis of highly monodispersed, crystalline face-centered cubic structure, 310–400nm AgNPs by aqueous leaf extract of *Eclipta alba* plant. The in vitro cytotoxicity activity of characterized AgNPs against some tested cell lines showed significant anti-cell-proliferation effect against RAW 254.7 (mouse macrophage cells), MCF-7 (human breast cancer cells), and Caco-2 (human adenocarcinoma cells) in nanomolar concentrations. The prepared AgNPs also showed effective inhibitory activity against some human pathogens, including *Escherichia coli*, *Staphylococcus aureus*, and *Pseudomonas aeruginosa*. The HeLa cells (human cervix carcinoma) also showed significant dose-dependent antiproliferative activity in the presence of green-synthesized AgNPs at relatively low concentrations (Gorbe et al., 2016). The green synthesis of spherical-shaped 22.89 ± 14.82 nm AgNPs by using *Artemisia tournefortiana* Rchb ethanol extract has been reported (Baghbani-Arani et al., 2017). The phytosynthesized AgNPs have shown increased cell apoptosis and demonstrated dose-dependent cytotoxicity in HT29 colon cancer cells. It also possessed potent antipathogenic bacteria. Moreover, the phytosynthesized AgNPs expressed photocatalytic degradation activity on Coomassie Brilliant Blue G-250 under UV light exposure within 60 min. The phytosynthesized 50nm AgNPs using hot water extract of *Guiera senegalensis* leaves showed more antiproliferation effect on human cancer cell prostate PC3 (IC50 23.48 μg/mL) than breast MCF7 (29.25 μg/mL) and lever HepG2 (33.25 μg/mL) by the virtue of their IC50 values. It also expressed high bactericidal effect against the human pathogens *E. coli* and *S. aureus* (Bello et al., 2017a). In another study, Bello et al. (2017b) reported that the phytosynthesis of 20-nm AgNPs by aqueous extract of *Hyphaene thebaica* fruit expressed antiproliferation effect against human cancer cells, in a dose-dependent manner, PC3 (IC50 2.6 mg/mL) followed by MCF7 (IC50 4.8 mg/mL) and then HepG2 (IC50 6.8 mg/mL) and good bactericidal effect against the human pathogens *E. coli* and *S. aureus*. Senthil et al. (2017) used the ethanolic extract of fenugreek leaves for the synthesis of highly stable, monodispersed, spherical-shaped AgNPs with the size ranging from 20 to 30 nm. Such phytosynthesized AgNPs expressed good antibacterial effect against pathogenic Gram-positive *Staphylococcus aureus*, enhanced bactericidal activity against the Gram-negative *Escherichia* coli, and sufficient cytotoxic effect against human skin cell line (HaCaT). The green-synthesized AgNPs also expressed a good antioxidant efficiency against 1,1-diphenyl-2-picrylhydrazyl; thus, it can be a promising candidate for many biomedical applications (Kumar et al., 2017). Naraginti and Li (2017) reported in another study the green synthesis of highly stable antioxidant, anticancer, and bactericidal AgNPs and AuNPs of average particle size of 35 and 20 nm and zeta potential of −12.1 and −22.3 mV, respectively, using the *Actinidia deliciosa* fruit extract. Zhang et al. (2017) reported the preparation of AgNPs by hot water extract of *Fatsia japonica* leaf extracts in the presence of mM NaCl, within only 80 min. The prepared AgNPs showed a potent preservative effect on citrus fruits rot induced by *Penicillium italicum* and also had significant antibacterial activities against Gram-negative *Escherichia coli* and Gram-positive *Staphylococcus aureus*, which might be an added beneficial for citrus fruits preservation. The phytosynthesized AgNPs using the aqueous extract of Ginger expressed antimicrobial activity against five laboratory pathogens such as *Escherichia coli*, *Klebsiella pneumoniae*, *Pseudomonas aeruginosa*, *Bacillus cereus*, and *Proteus vulgaris*. But, the effect of the produced AgNPs on *Phaseolus mungo* seeds clearly indicated that they were slightly toxic. That was shown by the slight decrease in germination percentage. However, the vigor index was almost similar to control. The toxicity on cell division was also not significant (Priyaa and Satyan, 2014). Moreover, the cytotoxicity of the green-synthesized AgNPs and AuNPs was evaluated in murine macrophage RAW264.7 and lipopolysaccharide LPS-stimulated R AW264.7 cell line, whereas those green-synthesized NPs were found to be favorable for the advancement of plant mediated nanocarriers in drug delivery systems, cancer diagnostic, and medical imaging (Markus et al., 2017). Even the phytosynthesized AgNPs by hot water extract of seaweeds expressed excellent microbicidal activity against two important pathogens of cotton: Fusarium wilts (*Fusarium oxysporum* f.sp. *vasinfectum*) and bacterial leaf blight (*Xanthomonas campestris* pv *malvacearum*) (Rajeshkumar et al., 2013). Govindaraju et al. (2015) phytosynthesized AgNPs with good killing efficiency toward human cancer myeloblastic leukemic cells HL60 and cervical cancer cells HeLa, using *Sargassum vulgare*.

Not only this, but AuNPs have also been reported to be synthesized by some human cancer cells, such as malignant cervical epithelial cells (SiHa), human neuroblastoma (SKNSH), and malignant cervical epithelial cells (HeLa), and even by noncancer cells such as nonmalignant human embryonic kidney cells (HEK-293) (Anshup et al., 2005; Larios-Rodriguez et al.,

2011). The gold nanotriangles were reported to be phytosynthesized with the hot water leaf extract of the lemon grass (*Cymbopogan flexuosus*) (Singh et al., 2011). Those phytosynthesized gold nanotriangles were found to be biocompatible, as they internalized inside the cells (cancerous as well as noncancerous cells) and were compartmentalized into the cytoplasm. Thus, they were proved to be promising candidates as scaffolds for delivery of drug, genes, or growth factors inside the cells. Moreover, their unique optical properties make them a promising candidate for hyperthermic treatment of cancer. Furthermore, atomic force microscopy revealed that the cells treated with gold nanotriangles showed pits on their surface, which could be the probable point of entry of these nanotriangles into the cells (Singh et al., 2011). In addition, Subramaniam et al. (2016) reported that *Aplocheilus lineatus* larvivorous fishes exposed to *Couroupita guianensis*-synthesized AuNPs did not show mortality and achieved higher predation rates on larvae of Anopheles mosquito vectors in aquatic environments treated with extremely low doses of AuNPs. Balalakshmi et al. (2017) reported the green synthesis of spherical-shaped 25-nm AuNPs applying the hot water leaves' extract of *Sphaeranthus indicus*. Those NPs recommended to be applied in the development of pollen germination media and plant tissue culture. That is because they notably promoted mitotic cell division in *Allium cepa* root tip cells and germination of *Gloriosa superba* pollen grains. Moreover, those green-synthesized AuNPs showed no mortality on the aquatic crustacean *Artemia nauplii*. Rajan et al. (2015) reported that the lower the size of the phytosynthesized AuNPs, the huger are its antibacterial and cytotoxicity on human cervical carcinoma cells (HeLa-cell lines) potentials. The phytosynthesized AuNPs were also used for the management of diabetes mellitus (Liu et al., 2012). The red seaweed *Corallina officinalis* phytosynthesized AuNPs expressed potent cytotoxic activity against the human breast cancer cells MCF-7 (El-Kassas and El-Sheekh, 2014).

However, the interaction of bio- and green-synthesized silver and AuNPs with human, plants, and animal cells should be deeply investigated to assure their safety.

The bactericidal and fungicidal effects of ZnO and TiO_2 NPS have been reported (Yamamoto, 2001; Feris et al., 2010; Sharma et al., 2010; Kumar et al., 2011; Jayaseelan et al., 2012). Not only this, but, sometimes, the green-synthesized metal oxides, especially those prepared using medicinal plants, express higher biocidal efficiency than the chemically synthesized ones (Monopoli et al., 2012). For example, ZnO NPs prepared using aqueous extract of aloe leaf showed higher bactericidal effect on bacterial strains

S. aureus, *Serratia marcescens*, *Proteus mirabilis*, *Citrobacter freundii*, and fungicidal effect on fungal strains *Aspergillus flavus*, *Aspergillus nidulans*, *Trichoderma harzianum*, and *Rhizopus stolonifer* compared to the chemically prepared ZnO NPs (Gunalan et al., 2012). In another study, the green-synthesized ZnO nanobiohybrids using extract of garlic (*Allium sativum*) also expressed higher fungicidal effect on *Mycena citricolor* (Berk and Curt) and *Colletotrichum* sp. than the chemically synthesized ZnO NPs (Arciniegas-Grijalba et al., 2019). Moreover, Singh et al. (2018) reported that the biosynthesized CuO NPs expressed higher bactericidal effect on *E. coli*, *B. subtilis* and *S. aureus* compared to the chemically synthesized CuO NPs. Suspensions of green-synthesized TiO_2 NPs by leaf extract of *Catharanthus roseus* were proved to have adulticidal and larvicidal effects against the hematophagous fly *Hippobosca maculate* and the sheep louse *Bovicola ovis* (Velayutham et al., 2012). The green prepared TiO_2 by *Curcuma longa* plant aqueous extract expressed more effective antifungal activity relative to that of industrial synthetic TiO_2 NPs (Abdul Jalil et al., 2016). Moreover, its effect on the germination of two varieties of wheat (*T. aestivum*) has also been reported. Al-Rasheed variety of wheat plant was more sensitive to resistance damping off compared with Tamuze-2 variety. But its growth was more sensitive compared with Tamuze-2 especially at higher concentrations. There was a noticeable decrease in all plant's parameters at most concentrations of TiO_2 biological synthetic relative to the industrial synthetic NPs in Al-Rasheed variety, while there were inductions in some plant's parameters by biosynthetic NPs compared with industrial synthetic in Tamuze-2 variety (Abdul Jalil et al., 2016).

However, the effect of the accumulation of ZnO, CuO, and TiO_2 NPs in ecosystem on the nontarget biodiversity should be investigated as it may likely pose threats. For example, Anderson et al. (2018) reported that although CuO NPs and ZnO NPs have some advantageous and benefits, such as the improvement of drought opposition, which occurs via the increment in the root surface area, thereby improving water and nutrients endorsement by the plants. But these NPs cause some changes on the rhizosphere. Thus, further research is needed to elucidate the mechanism and factors affecting the performance of NPs in natural soils, mainly with emphasis on rhizosphere, organic and inorganic matter, etc.

There are few other reports about the toxicity of the green-synthesized nanomaterials. The green ZnO NPs by the aqueous extract of *Amaranthus caudatus* were found to influence the normal development of zebrafish embryos in a dose-dependent manner; it was proved at high concentrations

(>10 mg/mL) (Jeyabharathi et al., 2017). The green-synthesized Ag and ZnO NPs using the hot water extract of *Lawsonia inermis* did not express any acute toxicity on vital functions of tested animals, such as the cardiovascular, central nervous, and respiratory systems, at a dose of 2000 mg/kg. Moreover, no mortality was observed up to 14 days (Jayarambabu et al., 2018). Moreover, ZnO NPs have been reported as a feed supplement in animal industry (Yusof et al., 2019). Consequently, the biosynthesized ZnO NPs would act as a new source of key developments in sustainable agriculture, especially in the animal industry. However, it has been reported that ZnO NPs cause toxicity to mice at high concentration 5000 mg/kg (Wang et al., 2016). Thus, more advanced research is needed to investigate the toxicity of the biosynthesized ZnO NPs on animals, fishes, poultry, and human. Iron oxide NPs exhibit low degree of cytotoxicity compared to other materials and can be considered biocompatible with respect to cellular/bacterial response and biomedical applications (Schlorf et al., 2011; Soenen et al., 2011; Calero et al., 2014). A safety assessment of chronic oral exposure to low doses of iron oxide γ-Fe_2O_3 NPs in growing chickens showed no toxicological symptoms on bird growth parameters, intestinal or hematological alterations, without any NPs accumulation in liver, spleen, or duodenum (Chamorro et al., 2015). However, more studies should be reported on the potential health effects of the short-/long-term exposure to iron oxides NPs in humans, as they are widely used in different life sectors.

8.6 CHALLENGES TO BE SOLVED

Despite the aforementioned applications and research performed on the biosynthesis of nanomaterials, however, there are still challenges to be solved. For example, due to the various biochemical constituents involved in the biosynthesis of nanomaterials, the polydispersity of the produced nanomaterials is a challenge that should be solved. More research is needed for upscaling bioprocess producing monodispersed nanomaterials. This requires more investigation on the biomolecules involved in the biosynthesis and the elucidation of the biosynthesis mechanism.

Despite the importance and different applications of CuNPs, however, few reports about the biosynthesis of CuNPs have been published. This might be due to the susceptible oxidation of Cu into oxide forms in aqueous solutions. Thus, the green- and/or biosynthesis of pure CuNPs is still a great challenge for bionanotechnologists. The new area of bionanotechnology of applying bacterial biofilm for biosynthesis of metal NPs can solve this issue. First of all, a relatively low precursor concentration is required; accomplishment of an inert atmosphere occurs in the biofilm by repelling the diffusion of intensely charged or highly reactive agents into the deeper zones of the biofilm matrix. Second, the properties of the biofilm matrix decrease the contamination probability, and the sessile bacteria in biofilms are reported to be approximately 600 times more metal resistant than the planktonic equivalent cells (Teitzel and Parsek, 2003). Thus, they can more rapidly reduce metal ions as a defense mechanism into metal NPs than their corresponding planktonic cells. This consequently produces NPs with smaller sizes. Third, they can accelerate the electrochemical redox reactions by the natural occurring extracellular polymeric substances, composed of proteins, peptides, lipids, humic acids, c-type cytochromes, exogenous DNA, and heterocyclic compounds and other polysaccharides, which can act as stabilizing and capping agents. Also, the polysaccharides hemiacetal groups can act as reducing agents (Cao et al., 2011; Kang et al., 2014). Moreover, the limited oxygen availability into the biofilm matrix minimizes the oxidation of the biosynthesized NPs. So, new research is needed on this promising area, with a special focusing on the NP biosynthesis mechanism in the biofilm.

Furthermore, despite the importance and different applications of Fe_3O_4 NPs, in different sectors, however, few reports about its biosynthesis have been published. Since it is important to optimize the physicochemical parameters during its green synthesis process, it is highly sensitive to pH and redox conditions and may produce nonmagnetic hydrated Fe(II) and Fe(III) oxides (Jolivet et al., 1992). This would consequently lead to the loss of its superparamagnetic performance, where Fe_3O_4 NPs show strong magnetic response in the presence of an external magnetic field and do not demonstrate any residual magnetism when such a magnetic field is removed and immediately redispersed in solution (Simonsen et al., 2018). Such dispersion is reported to be stable over a wide range of temperature, pH, and salinity, which widen their applications in different sectors in petroleum industry (Bagaria et al., 2013; Simonsen et al., 2018). Thus, further research should be done on optimization of the green- and/biosynthesis of magnetic NPs and its green functionalization without losing its superparamagnetic properties and stability. Moreover, more research should be performed on widening the application of the green-/biosynthesized magnetic NPs in different sectors in petroleum industry and its recycling and reusability.

Some of the drawbacks of some NPs are: (1) the charge carriers' recombination, which lowering the quantum yield leading to waste of energy; and (2) the poor response to visible light owing to their wide-band-gap energy. These consequently affect their stability and photocatalytic activity and lower their applicability. To overcome such a problem, the electron–hole recombination step is required. This can be done via the biosynthesis as biomolecules such as proteins act efficiently as the capping agent, instead of the chemical surface modifier (Kadam et al., 2019). Thus, it is very recommendable to do extra research on these biocapped highly stable biosynthesized metal/metal oxide NPs and their environmental applications.

All literature in the field of biosynthesis and/or green synthesis of nanomaterials as far as we know is at the laboratory scale. No reports about upscaling on a pilot scale have been published up till now. Despite the promising sustainability of such ecofriendly technique and its wide applications, however, it has been reported that fungi might be a better candidate than bacteria for scale-up because of its higher ability to secrete large amount of extracellular redox enzymes than bacteria (Hulkoti and Taranath, 2014; Kitching et al., 2015; Jeevanandam et al., 2016; Prasad et al., 2016). Algal biosynthesis of NPs is reported to take shorter time than other biosynthetic techniques (Vincy et al., 2017). Moreover, to overcome the problem of time-consuming process, the cell-free filtrate is more recommendable than applying the whole viable biomass. Thus, it is recommendable to apply microorganisms which have the ability to secrete high concentration of biomolecules, which proved to be involved in bioreduction, stabilization, and capping of metal/metal oxide NPs. However, despite the wide availability, ease of collection, and culturing of macro- and microalgae, few research studies have been published about the ecofriendly, energy-saving, and cost-effective application of macro- and microalgae for the sustainable green synthesis of metal/metal oxide NPs (Khanehzaei et al., 2014; Vijayan et al., 2014; Princy and Gopinath, 2015; Narendhran and Nair, 2017). The ethanolic extract of the red algae Galaxaura *elongata* was applied to synthesize AuNPs (Abdel-Raouf et al., 2017), whereas the presence of high percentages of andrographolide (37%–86%) and alloaromadendrene oxides (1.94%) might have acted as the reducing agent for the Au^{3+} ions and the stabilizing agent, preventing the aggregation of the formed AuNPs. Not only those, but also the aminoacids, such as glutamic acid aspartic, leucine, lysine, glycine, and alanine, and the sulfate polysaccharide, polypeptides, proteins, and polyol groups in the studied algal extract might have been responsible for the reduction and stabilization of gold ions to AuNPs. Furthermore, the high concentrations of fatty acids, such as hexadecanoic acid, oleic acid, 11-eicosenoic acid, and stearic acid, might have acted as stabilizing agents and, thus, prevented the aggregation of AuNPs. In addition, secondary metabolites, such as gallic acid rutin, quercetin, kaempferol, and rhamentin, might also have acted as stabilizing agents for the synthesized AuNPs against oxidation and coalescence or as matrices in NPs, while the polyphenols compounds, such as epigallocatechin catechin and epicatechin gallate, might have acted as capping agents. The TEM analysis revealed the predominance of spherical-shaped AuNPs with the size range of 3.85–77.13 nm, with minor presence of rod, triangular, truncated triangular, and hexagonal NPs (Abdel-Raouf et al., 2017).

The application of ethanolic extraction is a very promising for green synthesis of different metal/metal oxide NPs (Baghbani-Arani et al., 2017; Senthil et al., 2017; Abdel-Raouf et al., 2017), and the process would be of a closed cycle with a zero waste. The spent waste biomass of macroalgae and different pant parts after the extraction process would be used for production of bioethanol, which can be used for the ethanolic extraction of the bioreductants used in the green synthesis of such NPs.

Further research is needed about the production of purified bioproducts, enzymes, proteins, and other biotemplates that can be applied for the biosynthesis of metal/metal oxide NPs. Moreover, molecular biology, mathematical modeling, and chemical engineering should be applied for launching ecofriendly, sustainable, cost-effective, and energy-saving procedures for enhancing the biosynthesis, easy downstream processing, scale-up, and commercialization of the bio- and green-production of nanomaterials.

Further research is also required on the more ecofriendly O/W micro- and/or nanoemulsions formation as water phase is the predominate, using nonedible vegetable oils and/or waste cooking oils with water extracts of plant and/or agro-industrial wastes. As it is a new, sustainable, cost-effective, and energy-saving green production method for obtaining metal/metal oxide NPs with controlled size and shape, and it is also performed under mild operating conditions.

For application of nanobiotechnology in petroleum industry, there are some challenges. For example, in enhanced oil recovery, drilling, and exploration, the effective determination of the size of nanomaterials to secure effective penetration in porous reservoir medium containing crude oil is of high research interest. Moreover, in coarse and rough subsurface conditions, nanomaterials easily agglomerate, thus losing their unique size properties. This problem could be solved by coatings using biopolymeric substances, which would decrease the surface adhesive interaction. However, the choice of such coatings is very critical as it would greatly reduce the enhanced specialized properties of the nanomaterials such as their sensing abilities, catalytic properties, rock wettability effect, etc. Thus, much research to overcome the aforementioned drawbacks is needed. Further research is required on the application of green-synthesized NPs in adsorptive-, oxidative-, hydro-, and biodesulfurization and denitrogenation of crude oil and its fractions. Moreover, taking the synergetic advantages of injection of green-synthesized NPs and biosurfactants in enhancing oil recovery need further investigation, NPs will inhibit the asphaltenes precipitation. Not only this, but such NPs should also have high adsorption capacity for the reactants to attract them toward the cells and at the same time efficient desorption to enhance their transfer into the cells for achieving the BDS and/or BDN process. Thus, more studies are needed to optimize the exact ratio nano/biomaterial to be injected to give maximum heavy oil recovery, at shortest time.

Research regarding the synthesis of NPs has been underway for many years; however, their applications in bioupgrading of petroleum have been limited due to their easy aggregation during and after synthesis. Thus, the preparation of biocompatible and well-dispersed NPs is a key for such biological applications. Many methods have been proposed to solve the problem of agglomeration; however, so far, few of them consider the biocompatibility of dispersants. Usually, chemical surfactants were used to control particle sizes. However, some other natural products have been applied. For example, gum arabic (GA), which is a natural gum that has good rheological properties and emulsion stability, has been used as an emulsifier to prevent oil droplet aggregation and coalescence (Islam et al., 1997). Moreover, Zhang et al. (2007) reported that γ-Al_2O_3 NPs dispersed well in aqueous solutions after its modification with GA. The good dispersion was attributed to the chemical binding between the negatively charged groups of GA and the positive sites on the surface of γ-Al_2O_3 NPs, giving rise to the non-DLVO (Derjaguin–Landau–Verwey–Overbeek) surface steric force, which prevents the agglomeration of NPs in the aqueous solution (Leong et al., 2001). The better the dispersion of the adsorbents is, the more the specific surface area the adsorbents would have, and thus, the more the adsorption capacity of the adsorbents. However, Zhang et al. (2007) proved that excess GA (> 1 wt.%) decrease the adsorption capacity of γ-Al_2O_3 NPs as it took up from its adsorption sites. That consequently decreased the adsorptive desulfurization capacity of the modified γ-Al_2O_3 NPs. Thus, it is important to determine the optimum concentration of the applied modifier.

Furthermore, upon the application of modified GA γ-Al_2O_3 NPs for coating biodesulfurizing *Pseudomonas delafieldii* R-8 cells, the biodesulfurization (BDS) rate increased by 1.77 fold, relative to free cells, which was attributed to (1) the improved dispersion and biocompatibility of γ-Al_2O_3 NPs after modification with GA, (2) the stronger affinity of GA-modified γ-Al_2O_3 NPs to R-8 cells than the unmodified ones, and (3) its lower toxicity to the cells. The type of NPS used for in situ coupling of adsorptive desulfurization and/or denitrogenation with BDS and/or biodenitrogenation (BDN) is a very important parameter. The NPs used for selective adsorption of polyaromatic sulfur or nitrogen heterocyclic compounds should express low toxicity for microorganisms applied for bioupgrading of petroleum. Moreover, it should be characterized with high affinity and compatibility with such applied microorganisms to be adsorbed onto their cell wall. Not only this, but also with high adsorption of reactants toward the cells and efficient desorption for their transfer into the cells for achieving the BDS and/or BDN process. For example, Zhang et al. (2008) proved that Na-Y molecular sieves restrained the BDS activity of *Pseudomonas delafieldii* R-8 cells and the activated carbon could not desorb the adsorbed dibenzothiophene. Thus, they are not recommended for in-situ coupling ADS and BDS, while γ-Al_2O_3 NPs are recommended for in-situ coupling ADS and BDS, as it enhanced the BDS efficiency of *Pseudomonas delafieldii* R-8 (Zhang et al., 2007, 2008) and *Rhodococcus erythropolis* LSSE8-1-vgb (Zhang et al., 2011). The eco-friendly MCM-41 mesoporous silica, with significant number of pores, ordered porosity and large specific surface area enhanced the adsorption of DBT to microbial cells relative to zeolites (Nasab et al., 2015).

The process of synthesis of different NPs is also an important factor affecting the application of nanotechnology in petroleum bioupgrading. Nasab et al. (2015) prepared MCM-41spherical mesoporous silica NPs, using a quaternary ammonium template, CTAB, for efficient adsorptive desulfurization. The assembling of such NPs onto *Rhodococcus erythropolis* IGTS8, improved its BDS efficiency. Shan et al. (2005) applied the coprecipitation method to synthesize magnetite NPs and coat biodesulfurizing bacterial cells; they reported that replacement of air by N_2 has the advantage to prevent the oxidation of ferrous iron during preparation of NPs in the aqueous solution and also has the ability of the size control. However, the surface of magnetite nanoparticles (MNPs) has to be modified with a suitable surfactant to be used in cell coating. Shan et al. (2005) used oleic acid as a surfactant to functionalize and immobilize MNPs onto the bacterial surface. However, such modified MNPs did not enhance the rate of BDS using

Pseudomonas delafieldii. Previous reports by Mahmoudi et al. (2009) showed that functionalized MNPs with different surfactant expressed low toxicity on living eukaryote cells in comparison to nonfunctionalized MNPs. Ansari et al. (2009) used glycine to modify the surface of NPs. Fe atom of MNPs has a strong tendency to COOH groups, so that the Fe atom of nanoparticle reacts with COOH of oleic acid or glycine; therefore, oleic acid forms a bilayer shell on the surface of nanoparticles (Shi-Yong et al., 2006), and glycine produces an amine layer on the surface of MNPs (Ansari et al., 2009) which leads to the dispersion of magnetic iron oxide NPs in water phase with hydrophilic characteristics and simultaneously immobilized onto the surface of bacteria. The immobilization of such glycine-modified MNPs on the negatively charged bacterial cell surface is due to their net positive charges. Etemadifar et al. (2014) proved that the size of the synthesized MNPs using oleic acid and glycine as functionalizing agents was approximately the same (<10 nm). However, oleate-modified MNPs coated 94% of *Rhodococcus erythropolis* R1 cells, but glycine-modified MNPs coated only 78%.

The suspension of oleate-modified magnetite Fe_3O_4 NPs is considered bilayer surfactant-stabilized aqueous magnetic fluids (Shen et al., 1999). The iron oxide nanocrystals were first chemically coated with oleic acid molecule. Then, the excess oleic was weakly adsorbed onto the primary layer through the hydrophobic interaction between the subsequent molecule and the hydrophobic tail of oleate. Figure 8.5 shows the postulated adsorption mechanism between the NPs and desulfurizing cells. (1) The large specific surface area and the high surface energy of the Fe_3O_4 NPs (i.e., the nanosize effect) make it strongly adsorbed onto the microbial cells surfaces. (2) The bacterial cell wall is composed of proteins, carbohydrates, and other substances, for example, peptidoglycan, lipopolysaccharide, mycolic acid, etc. This plays another important role in cell adsorption, where such an extracellular matrix can form hydrophobic interaction with the hydrophobic tail of oleate-modified Fe_3O_4 NPs. Thus, the sustainable green synthesis of functionalized MNPs would enhance the feasible the application of petroleum bioupgrading processes, for example, the BDS and BDN.

Kafayati et al. (2013) investigated the effect of the MNPs Fe_3O_4 on the growth rate of the genetically engineered biodesulfurizing *Pseudomonas aeruginosa* PTSOX4 cells. The minimum inhibitory concentration and minimum bactericidal concentration analysis showed that MNPs have low toxicity on PTSOX4. However, *Pseudomonas* bacteria cells could not grow in the presence of high concentrations of MNPs (>5000 ppm). That was attributed to the surface saturation of the bacteria cells with the MNPs and its increasing contact with cell membrane, which led to the injury and damage of the cell membrane. Thus, this is another factor that should be investigated to achieve the feasibility of application of nanotechnology for enhancing the BDS process.

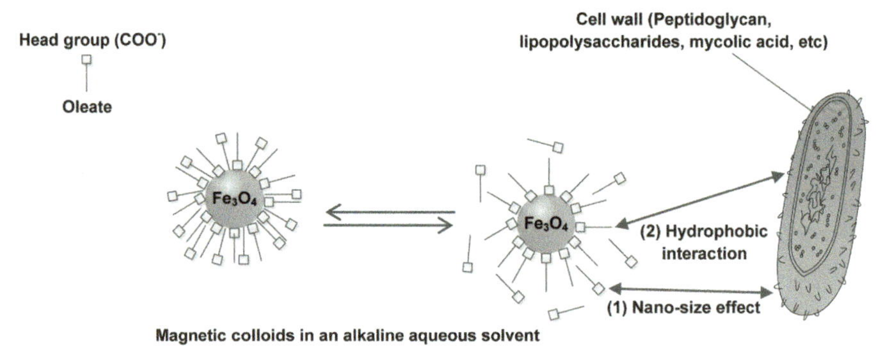

FIGURE 8.5 Adsorption mechanism of Fe_3O_4 NPs onto bacterial cells.

Finally, bioscience based on the viewpoint of zero waste and the valuable sustainable bioresources (see Figure 8.6), including the organic wastes, algae, natural resources, and other biomass, is considered as a new bioeconomy, which cover all industries and sectors.

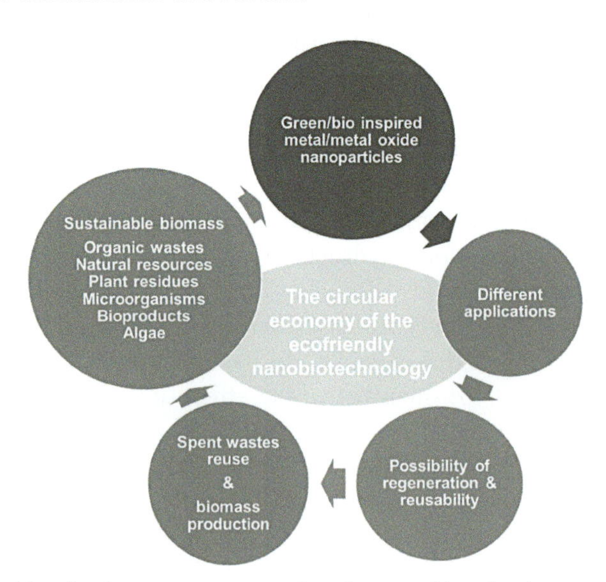

FIGURE 8.6 The circular economy system based on nanobiotechnology on the zero-waste point of view.

For example, Omran et al. (2018a) reported that the main components of orange peels hot water extract (e.g., 4-vinyl guaiacol, eugenol, aromatics, terpenes, sugar derivatives, and saturated fatty acids) are involved in the bioreduction of $AgNO_3$ into spherical-shaped AgNPs with average particle size of 3–12 nm (see Figure 8.7a), where 3g/L AgNPs were produced under the estimated optimum conditions of 3mM $AgNO_3$, 4000 mg/L of peels extract, pH 9, 40 °C, 100 rpm, 24 h, and illumination using fluorescent light 36 W/6400 K. Thus, the operational production cost of such green-synthesized AgNPs was estimated to be 81 US$/10 g AgNPs. According to the US Research Nanomaterials, Inc., the global cost of the chemically synthesized AgNPs is 250 US$/10 g of AgNPs. Consequently, 67.6% savings can be achieved upon application of green synthesis of AgNs using waste peels. Moreover, to reach to the point of zero-waste, the spent waste orange peels after the preparation of AgNPs was valorized into activated carbon (see Figure 8.7b) with specific surface area of 2.9 m^2/g and average pore volume and pore size of 0.02 cm^3/g and 1.9 nm, respectively, which would have different applications. Consequently, this would further decrease the overall AgNPs phytosynthesis cost.

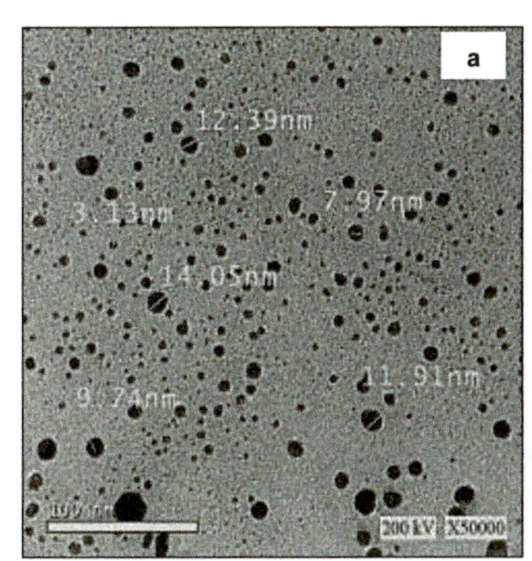

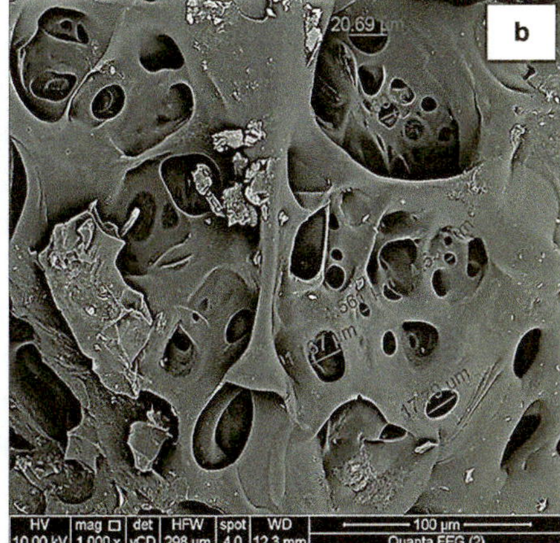

FIGURE 8.7 (a) TEM micrograph of AgNPs. (b) Field emission scanning electron microscopic micrograph of activated carbon prepared from spent waste orange peels.

Moreover, Omran et al. (2020) reported the efficient biocidal activities of the biosynthesized quasi-spherical-shaped, monodispersed ferromagnetic 20–27-nm Co_3O_4 NPs, against Gram-positive (*B. subtilis* and *S. aureus*) and Gram-negative (*P. aeruginosa* and *E. coli*) bacteria. The minimum inhibitory concentration and minimum lethal concentration recorded 2.5 mg/mL for all except *S. aureus,* which recorded 5 mg/mL. However, the cost of such mycosynthesized Co_3O_4 NPs using cell-free filtrate of *Trichoderma longibrachiatum* was estimated to be high, recording approximately 225 US$/10 g Co_3O_4 NPs. Although the absence of light with the elucidated optimum conditions of 72 h, pH 11, 30 °C, 100 rpm, 4-mM $CoSO_4 \cdot 7H_2O$, and 5.5g dry weight biomass yielded the highest concentration of 1.2 g/L Co_3O_4 NPs, it was expected that saving energy by eliminating illumination during the mycosynthesis process would decrease the cost. However, the high cost came from the price of the salt precursor and the process of obtaining the fungal biomass. But, the magnetic hysteresis curve of such bioinspired Co_3O_4 proved to be 17.3 emu/g magnetization saturation (Ms) and 0.12 emu/g and 2.9 g remnant magnetization (Mr) and coercivity (Hci), respectively, with Mr/Ms value of 7.05. Thus, the observed magnetic properties of that bioinspired Co_3O_4 NPs (see Figure 8.8a), with its aforementioned biocidal efficiency, would add to its advantages and value, where the ferromagnetic Co_3O_4 NPs might have successful applications in magnetic storage devices, water disinfection, and wastewater treatment. Moreover, for decreasing the overall cost of the mycosynthesis process and retaining the point of zero-waste, the valorization of the spent waste fungus biomass used for the biosynthesis of Co_3O_4 NPs into activated carbon (see Figure 8.8b) with good specific surface area 6.32 m^2/g, and pore size ranged between 4.22 and 37.24 μm was achieved, which would have different applications. However, further work is undertaken now for lowering the price of biomass cultivation and harvesting, by using cost effective media constituents, for example, some agro-industrial wastes, such as molasses and corn steep liquor. Moreover, such fungal biomass can be harvested from the microbial hydrolysis and scarification of lignocellulosic wastes during the bioethanol production process. There is also further research undertaken now for the feasibly of using the cell-free filtrate after the bioethanol fermentation process for the biosynthesis of such metal oxide. That filtrate would have considerable concentrations of different microbial enzymes, sugars, and different other bioproducts of the fermentation process, which are expected to act as reducing, stabilizing, and capping agents for biosynthesis of metal/metal oxide NPs.

Thus, in order to maximize the feasibility and cost effectiveness of biosynthesis of nanomaterials, it would be very advantageous to produce other valuable products and get use of each derivatives produced during each process steps. This would occur upon the increment of the interdisciplinary and multidisciplinary research efforts in engineering, chemistry, biology, computer science, and nanotechnology.

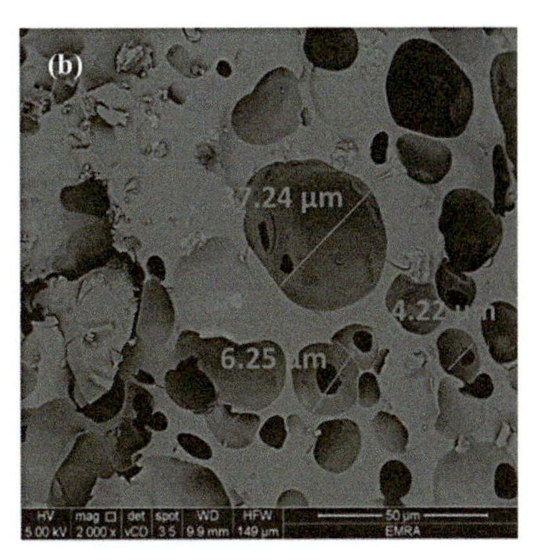

FIGURE 8.8 (a) Magnetized Co_3O_4 NPs. (b) Field emission scanning electron microscopic micrograph of activated carbon prepared from spent waste *Trichoderma longibrachiatum*.

8.7 NANOBIOTECHNOLOGY IN PETROLEUM INDUSTRY AND SUSTAINABLE DEVELOPMENT

There is an important need to deal with nanobiotechnology in petroleum industry to achieve the 17 goals of sustainable development via the decrease in materials and energy use, reduced source of pollution, and increased opportunity of recycling.

The applications of nanobiotechnology with the beneficial properties of the green/biosynthesized nanomaterials in petroleum industry will open new sustainable industrial revolution in oil and gas industry with a real achievement for the three pillars of sustainability: social, environment, and economic (see Figure 8.9).

Moreover, it will reduce the GHG emissions, thus overcoming the problem of global warming and climate change. Application of nanobiotechnology in treatment of water effluents from petroleum refineries and other petroleum sectors and remediation of oil spills and water polluted with petroleum hydrocarbons would save water resources and life underneath water. It will help in the waste management problem by the usage of waste biomass and agro-industrial wastes for green- and biosynthesis of nanomaterials. This provides cost-effective techniques for water treatment, bioupgrading of petroleum, and its fractions. Besides, this enhances the rate of bioremediation of oil spills and photocatalytic treatment of polluted water, enhances the rate of different catalytic processes via the application of high-performance green-synthesized nanocatalysts in oil and gas industry, thus saving energy and time,

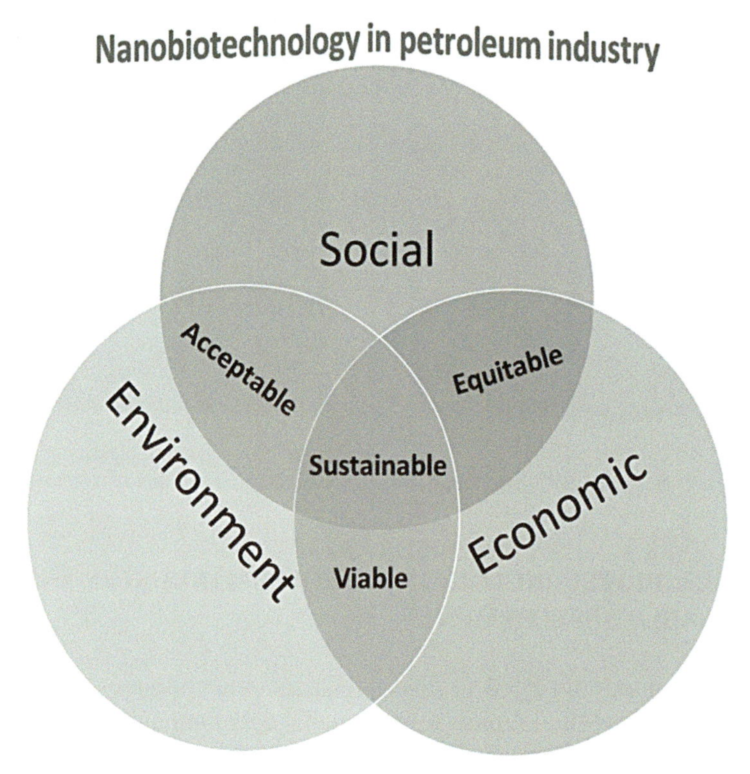

FIGURE 8.9 Application of nanobiotechnology in petroleum industry achieving the three pillars of sustainability.

and consequently decreases the overall cost of the performed processes. Upon applying the cost-effective green-synthesized highly sensitive nanosensors, this will increase the exploration, drilling, and production activities. Effective green-synthesized highly sensitive nanosensors for pollutants and toxic and hazardous gases decrease the risk of sudden gas leakages, explosion, and fire occurrence. The application of green-synthesized nanofluids and biosurfactants in enhanced oil recovery, drilling, and well stimulation will change the current practice of exploration and oil production. The green-synthesized nanobased biosensors for biofilm forming microorganisms and SRB and their applicability in oil and gas industry would be very beneficial on predicting the problem before its occurrence and preventing the performance of pitting corrosion, sudden breakage of pipelines, and shutting down of the oil field and refineries. That would occur via the establishment of a smart field system to detect, locate, report, and direct injection of biocides. The application of the green-synthesized nanoscale sorbents, nanoporous fibers, and/or membranes for removal of toxic and poisonous heavy metals from natural gas, condensate, and petroleum fractions, and for filtration of different effluents in oil and gas industry, is also another way to achieve sustainability of clean environment, as much clean energy as it can be, and to protect ecosystem and human health. Moreover, upon their application for capturing emitted gases, such as CO_2, H_2S, SOx, etc., it would be considered as a potential climate change mitigation technology. The application of green-synthesized NPs as ecofriendly and cost-effective corrosion inhibitors, coating, biocides for mitigation of macro- and microfouling, and microbial corrosion in petroleum industry, would achieve the three pillars of sustainability: social, environment, and economic (see Figure 8.9). It will save and protect the manpower in the field and refineries from sudden leakages of oil, gases, and other hazardous materials and from sodden explosion and fires. It will also protect the environment from sudden occurrence of pollution. Finally, it will prevent the loss of money, chemicals, fuels, oil and gas, energy, and effort, which annually occur due to the microbial and chemical corrosion in the oil and gas industry. Furthermore, the application of carbon-based nanomaterials prepared from readily available waste biomass, as metal-free catalysts in *in-situ* upgrading and recovery of heavy crude oil, will contribute to a more sustainable chemistry in terms of production and refining of heavy crude oils, especially with the worldwide depletion of high-quality low-sulfur-content light crude oil and the increase of the worldwide reserves of low-quality high-sulfur-content heavy crude oil.

The application of nanomaterials/nanocatalysts in oil and gas industries, from upstream through the midstream to downstream, is expected to: (1) efficiently increase oil and gas production, (2) decrease the pressure and increase the injection rate of low-permeability oil reservoirs, via the application of nanomolecular deposition film, (3) mitigate the plugging and scales formation via the application of nanofluids, (4) improve drilling fluid behavior, (5) stimulate *in-situ* oil reservoir and late water treatment, via the application of nanocatalyst and nanofilter membrane, (6) corrosion mitigation and reinforcement of infrastructure via the application of nano-coatings, (7) upgrade of heavy crude oil and different petroleum fractions via the application of nanocatalysts, (8) accelerate different refining process and petrochemical production via application of nanocatalysts, (9) oil spill control and wastewater treatment via nanosorbents, nanophotocatalysts and nanobiocatalysts, and (10) accelerate the oil tank cleaning and maintenance process. Finally, all of the aforementioned applications can be greener, sustainable, energy savers, cost effective, via the application of green and biosynthesized nanomaterials and NPs. The advantages of the green-synthesized NPs together with those of nanocatalysts are as follows: (1) improved catalytic performance than conventional one due to the high surface-area-to-volume ratio, (2) better mobilization in reactors, leading to higher prospective contacts between reactants, (3) high stability and long lifetime, (4) long-run continuous processes as there is no need of catalyst replacement because of the possibility of nanocatalysts implementation inside the medium, and (4) the promising advantageous of possibility of *in-situ* preparation of catalyst, which will reduce the operating costs as well as environmental concerns; for sure, we will achieve a successful sustainable revolution in petroleum industry. However, further work and research are ongoing now to commercialize the proposed last advantageous and produce new generation membranes for gas separation and other application in different sectors of petroleum industry, new smart nanofluids for enhanced oil recovery and water shut-off, and nonconventional nanoapplications in the exploration, drilling, and exploitation.

The use of agricultural wastes, plant residues, and other waste biomass for green synthesis of nanomaterials instead of getting rid of such wastes by open-burning would decrease the GHG emissions and consequently overcome the problem of global warming and climate change. Moreover, it is indirectly lowering the CO_2, as the phytosynthesis of nanomaterials depends on the sustainable and renewable feedstock such as plants and/or algae, which previously used CO_2 for growth via the photosynthesis process. Thus, phytosynthesis of nanomaterials can be considered as closed CO_2 cycle. Briefly, use CO_2 to produce valuable chemicals and many other products that serve multidisciplinary sectors. Further usage of such nanomaterials in the removal of sulfur and nitrogen and upgrading of the crude oil and its fractions will also reduce the SOx, NOx, particulate matter, CO, and CO_2 released into the atmosphere. However, the toxicity of such green/biosynthesized NPs on humans, living organisms, and different components of the ecosystem is still in need for more exploration. Taking into consideration that water used for biomass cultivation or nanomaterials synthesis can be recycled and reused, the spent waste can be biovalorized into bioethanol, biofertilizer, animal fodder, biochar, activated carbon, biogas, solid refused fuel, etc. The produced metal/metal oxide NPs itself can be used for water treatment, water disinfection, and remediation of different pollutants. However, the important technological drawback of nanoengineered water technology is that they are rarely adaptable to mass applications and are not competitive with traditional treatment technologies at present, in many cases. Nanoengineered materials, however, have great potential for water revolutions in the coming decades, especially in decentralized treatment systems, point-of-use devices, and deeply degradable pollutants. The dominance and length of bioremediation in the future must include nanobiotechnology alteration and adaptation. The opportunities and potential for innovation, historical track record, and the influence of potential nanobiotechnology advantages contribute to recognition of this field as increasingly significant. The exploitation of new, efficient biosorption materials is essential with the development of nanobiotechnology and will continue infinitely. The future of nanobiomaterials in wastewater treatment for the removal of heavy metal ions is comparatively vivid. Finally, water/wastewater treatment processes by nanobiotechnology showed great potential in laboratory studies. More research is needed before full-scale operation of nanobiotechnology especially in the field of membrane filtration and biosensing for treating petroleum wastewater. Studies should be conducted under realistic conditions to assess the efficiency of available nanobiotechnology to validate nanobiomaterial-enabled sensing. Another research need is to measure the long-term efficiency of available technologies, which are conducted on a laboratory scale. Also, due to our current poor understanding of the fate and behavior of nanobioparticles in humans and the environment, toxicity has become one of the issues that research studies should focus on in the period ahead.

The produced metal/metal oxide NPs can also be used as catalyst for production of biodiesel using waste cooking oil and/or nonedible oils and can also be used in all the steps of bioethanol production from different lignocellulosic biomass. Thus, they are mainly used for production of alternative and/or complementary fuels for conventional petro-fuels. Such green-synthesized metal/metal oxide NPs can be applied in different processes in petroleum industry, starting from exploration, drilling, refining, production of different petrochemicals, in manufacturing of different parts of engines and instruments used in oil and gas industry. The green-synthesized metal/metal oxide NPs can be used for manufacturing of all types of sensors, microsized robots, and nanorobots used in oil and gas industry. The green-synthesized NPs can be applied in the chemical and biological upgrading of heavy oil and different petroleum fractions. The green-synthesized nanomaterials can be used as coatings, reinforcement materials, inhibitors for chemical- and biocorrosion, biocides for biofoulings organisms (micro- and macro-ones), and scale mitigation in oil and gas industry. The green-synthesized NPs can be used in materials applied in the process of maintenance and cleaning of storage tank. The green-synthesized nanomaterials can also be applied in the construction materials of pipelines, storage tanks, transportation trucks and tankers, and every part of oil and gas infrastructure. The green-synthesized NPs are now used in different healthcare products, cosmetics, pharmaceutical products, and medical processes, surgery, therapy, etc. Thus, nanobiotechnology in the form of green and/or biosynthesis of metal/metal oxide NPs and nanomaterials and its application in different sectors concerning petroleum industry will achieve a real profit:

$$\text{Real profit} = \sum \text{Products sales} - \sum \text{Raw materials costs} - \sum \text{Energy costs}$$

where the raw materials are mostly sustainable, readily available, ecofriendly, low cost, and sometimes costless. The energy savings are also achieved as most of the green synthesis processes occur under mild operating conditions. The spent waste raw materials itself after being used for nanomaterials production can be valorized into other valuable products. Nevertheless, such profit can be also expanded to achieve the triple bottom line criteria of sustainability:

$$\text{Triple bottom line} = \text{Real profit} + \sum \text{Sustainable}(\text{Credits} - \text{Costs}) - \sum \text{Environmental costs}$$

However, it should be noted that although a number of research studies have been published for the advantageous of application of NPs in enhanced oil recovery (EOR) (Ogolo et al., 2012; Khajehpour et al., 2016; Elshawaf, 2018; Ko and Huh, 2019). But there are some other reports about the negative impact of NPs in some parameters concerning the process of EOR. For example, ZnO NPs have been reported in a study by Hogeweg et al. (2018) to produce larger particles, which consequently led to some injection problems. Sometimes, adding NPs to ethanol or brine water led to deprived recovery than when applying ethanol or brine water alone. NPs could sometimes perform injection blockage and settling problems (Ding et al. 2018). In addition, although NPs are reported as additives for enhancing the properties of drilling fluids, but the concentrations of added NPs should be optimized; otherwise, it would lose its enhancing properties (Alsaba et al., 2018). For example, high concentrations of NPs would increase the particle friction coefficient and consequently affect the lubricity and whole cleaning efficacy (Alvi et al., 2018). Same would occur in well stimulation. This is because high concentrations of NPs are not recommendable, since it would negatively affect the rheological properties of surfactant-based fluids for hydraulic fracturing applications (Fakoya and Shah, 2018). Moreover, added NPs should work over a wide range of temperatures, as, at high temperatures, some NPs reported to perform a negative impact on filtration properties (Alsaba et al., 2020). The type of NPs itself is another important factor on the application of NPs as additives for drilling fluids. Alsaba et al. (2018) reported that in concerning the rheological properties, copper oxide NPs, for example, have superior thermal stability relative to magnesium oxide and aluminum oxide. NPs have also been reported as additives in well cementing to improve the cement properties. But the synergism of the added NPs with other materials added to the cement should be fulfilled; otherwise, the workability of the cement would be reduced (Santra et al., 2012; Alkhamis and Imqam, 2018). So, there is still a need for a lot of work and investigation on the application of nanotechnology in oil and gas sector. Moreover, the main bottleneck of application of nanotechnology in petroleum industry is the high cost of manufacturing such nanomaterials and/or NPs and the possibility of some health risks caused by the higher potential of being inhaled or absorbed through skin (Nabhani and Tofghi, 2010; Lau et al., 2017). However, these would be promisingly solved via the green synthesis of such applied nanomaterials and/or NPs.

REFERENCES

Abdel-Raouf, N., Al-Enazi, N.M., and Ibraheem, I.B.M. 2017. Green biosynthesis of gold nanoparticles using *Galaxaura elongata* and characterization of their antibacterial activity. *Arab. J. Chem.* 10: S3029–S3039.

Abdul Jalill, R.D.H., Nuaman, R.S., and Abd, A.N. 2016. Biological synthesis of titanium dioxide nanoparticles by *Curcuma longa* plant extract and study its biological properties. *World Sci. News.* 49(2): 204–222.

Abdul Rasheed, P., Jabbara, K.A., Rasool, K., Pandey, R.P., Sliem, M.H., Helal, M., Samara, A., Abdullah, A.M., and Mahmoud, K.A. 2019. Controlling the biocorrosion of sulfate-reducing bacteria (SRB) on carbon steel using ZnO/chitosan nanocomposite as an eco-friendly biocide. *Corros. Sci.* 148: 397–406.

Ahmad, R., Moshin, M., Ahmad, T., and Sardar, M. 2015. Alpha amylase assisted synthesis of TiO_2 nanoparticles: Structural characterization and application as antibacterial agents. *J. Hazrd. Mater.* 283: 171–171.

Al din. Haratifar, E., Shahverdi, H.R., Shakibaie, M., Moghaddam, K.M., Amini, M., Montazeri, H., and Shahverdi, A.R. 2009. Semi-biosynthesis of magnetite-gold composite nanoparticles using an ethanol extract of *Eucalyptus camaldulensis* and study of the surface chemistry. *J. Nanomater.* 2009: 962021. https://doi.org/10.1155/2009/962021

Alkhamis, M., and Imqam, A. 2018. New cement formulations utilizing graphene nano platelets to improve cement properties and long-term reliability in oil wells. In: SPE Kingdom of Saudi Arabia Annual Technical Symposium and Exhibition, 23–26 April, Dammam, Saudi Arabia. https://doi.org/10.2118/192342-MS.

Alsaba, M.T., Al Dushaishi, M.F., and Abbas, A.K. 2020. A comprehensive review of nanoparticles applications in the oil and gas industry. *J. Petrol. Explor. Prod. Technol.* 10: 1389–1399.

Alsaba, M.T., Al Fadhli, A., Maraf, A., Hussain, A., Bander, F., and Al Dushaishi, M.F. 2018. Application of nanoparticles in improving rheological properties of water based drilling fluids. In: SPE Kingdom of Saudi Arabia Annual Technical Symposium and Exhibition, 23–26 April, Dammam, Saudi Arabia. https://doi.org/10.2118/192239-MS.

Alvi, M.A., Belayneh, M., Saasen, A., and Aadnøy, B.S. 2018. The effect of micro-sized boron nitride BN and iron trioxide Fe_2O_3 Nanoparticles on the properties of laboratory bentonite drilling fuid. In: SPE Norway one-day Seminar, 18 April, Bergen, Norway. https://doi.org/10.2118/191307-MS.

Anderson, C.W.N., Bhatti, S.M., Gardea-Torresdey, J., and Parsons, J. 2013. In vivo effects of copper and silver on synthesis of gold nanoparticles inside living plants. *ACS Sustain. Chem. Eng.* 1: 640–648.

Anderson, A.J., McLean, J.E., Jacobson, A.R., and Britt, D.W. 2018. CuO and ZnO nanoparticles modify interkingdom cell signaling processes relevant to crop production. *J. Agric. Food Chem.* 66: 6513–6524.

Anderson, C.W.N., Brooks, R.R., and Stewart, R.B. 1998. Harvesting a crop of gold in plants. *Nature*. 395: 553–554.

Anigol, L.B., Charantimath, J.S., and Gurubasavaraj, P.M. 2017. Effect of concentration and pH on the size of silver nanoparticles synthesized by green chemistry. *Org. Med. Chem.* 3(5): OMCIJ.MS.ID.555622

Ansari, F., Grigoriev, P., Libor, S., Tothill, I. E., and Ramsden, J.J. 2009. DBT degradation enhancement by decorating *Rhodococcus erythropolis* IGST8 with magnetic Fe_3O_4 nanoparticles. *Biotechnol. Bioeng.* 102(5): 1505–1512.

Anshup, A., Venkataraman, J.S., Subramaniam, C., Kumar, R.R., Priya, S., Kumar, T.R., Omkumar, R.V., John, A. and Pradeep, T. 2005. Growth of gold nanoparticles in human cells. *Langmuir*. 21: 11562–11567.

Apte, M., Girme, G., Bankar, A., Kumar, A.R., and Zinjarde, S. 2013. 3, 4-dihydroxy-L-phenylalanine-derived melanin from *Yarrowia lipolytica* mediates the synthesis of silver and gold nanostructures. *J. Nanobiotechnol.* 11: 2, doi: org/10.1186/1477–3155-11–2.

Arciniegas-Grijalba, P.A., Patiño-Portela, M.C., Mosquera-Sánchez, L.P., Guerra Sierra, B.E., Muñoz-Florez, J.E., Erazo-Castillo, L.A., and Rodríguez-Páez, J.E. 2019. ZnO-based nanofungicides: Synthesis, characterization and their effect on the coffee fungi *Mycena citricolor* and *Colletotrichum* sp. *Mater. Sci. Eng. C.* 98: 808–825.

Armendariz, V., Herrera, I., Peralta-Videa, J.R., Jose-Yacaman, M., Troiani, H., Santiago, P., and Gardea-Torresdey, J.L. 2004. Size controlled gold nanoparticle formation by *Avena sativa* biomass: Use of plants in nanobiotechnology. *J. Nanopart. Res.* 6: 377–382.

Armendariz, V., Parsons, J.G., Lopez, M.L., Peralta-Videa, J.R., Jose-Yacaman, M., and Gardea-Torresdey, J.L. 2009. The extraction of gold nanoparticles from oat and wheat biomasses using sodium citrate and cetyltrimethylammonium bromide, studied by X-ray absorption spectroscopy, high-resolution transmission electron microscopy, and UV/Visible spectroscopy. *Nanotechnology*. 20(10): 105607, doi: 10.1088/0957-4484/20/10/105607.

Badawy, A.M.E., Luxton, T.P., Silva, R.G., Scheckel, K.G., Suidan, M.T., and Tolaymat, T.M. 2010. Impact of environmental conditions (pH, ionic strength, and electrolyte type) on the surface charge and aggregation of silver nanoparticles suspensions. *Environ. Sci. Technol.* 44: 1260–1266.

Bagaria, H.G., Xue, Z., Neilson, B.M., Worthen, A.J., Yoon, K.Y., Nayak, S., Cheng, V., Lee, J.H., Bielawski, C.W., and Johnston, K.P. 2013. Iron oxide nanoparticles grafted with sulfonated copolymers are stable in concentrated brine at elevated temperatures and weakly adsorb on silica. *ACS Appl. Mater. Interfaces.* 5: 3329–3339.

Baghbani-Arania, F., Movagharniaa, R., Sharifiana, A., Salehi, S., and Shandiz, S.A.S. 2017. Photo-catalytic, anti-bacterial, and anti-cancer properties of phyto-mediated synthesis of silver nanoparticles from *Artemisia tournefortiana* Rchb extract. *J. Photochem. Photobiol, B: Biol.* 173: 640–649.

Baker, S., Rakshith, D., Kavitha, K.S., Santosh, P., Kavitha, H.U., Rao, Y., and Satish, S. 2013. Plants: Emerging as nanofactories towards facile route in synthesis of nanoparticles. *BioImpacts.* 3: 111–117.

Balalakshmi, C., Gopinath, K., Govindarajan, M., Lokeshd, R., Arumugam, A., Alharbi, N.S., Kadaikunnane, S., Khalede, J.M., and Benelli, G. 2017. Green synthesis of gold nanoparticles using a cheap *Sphaeranthus indicus* extract: Impact on plant cells and the aquatic crustacean *Artemia nauplii. Photochem, Photobiol. B: Biol.* 173: 598–605.

Baymiller, M., Huang, F., and Rogelj, S. 2017. Rapid one-step synthesis of gold nanoparticles using the ubiquitous coenzyme NADH. *Matters.* doi: 10.19185/maters.201705000007

Bello, B.A., Khan, S.A., Khan, J.A., Syed, F.Q., Anward, Y., Khan, S.B. 2017a. Antiproliferation and antibacterial effect of biosynthesized AgNPs from leaves extract of *Guiera senegalensis* and its catalytic reduction on some persistent organic pollutants. *J. Photochem. Photobiol. B.* 175: 99–108.

Bello, B.A., Khan, S.A., Khan, J.A., Syed, F.Q., Mirza, M.B., Shah, L., and Khan, S.B. 2017b. Anticancer, antibacterial and pollutant degradation potential of silver nanoparticles from *Hyphaene thebaica. Biochem. Biophys. Res. Commun.* 490: 889–894.

Bharde, A., Rautaray, D., Bansal, V., Ahmad, A., Sarkar, I., Yusuf, S.M., Sanyal, M., and Sastry, S. 2006. Extracellular biosynthesis of magnetite using fungi. *Small.* 2(1): 135–141.

Calero, M., Gutiérrez, L., Salas, G., Luengo, Y., Lázaro, A., Acedo, P., Morales, M.P., Miranda, R., and Villanueva, A. 2014. Efficient and safe internalization of magnetic iron oxide nanoparticles: Two fundamental requirements for biomedical applications. *Nanomedicine.* 10(4): 733–743.

Cao, B., Ahmed, B., Kennedy, D.W., Wang, Z., Shi, L., Marshall, M.J., Fredrickson, J.K., Isern, N.G., Majors, P.D., and Beyenal, H. 2011. Contribution of extracellular polymeric substances from *shewanella* sp HRCR-1 biofilms to U(VI) immobilization. *Environ. Sci. Technol.* 45(13): 5483–5490.

Chamorro, S., Gutiérrez, L., Vaquero, M.P., Verdoy, D., Salas, G., Luengo, Y., Brenes, A., and Teran, F.J. 2015. Safety assessment of chronic oral exposure to iron oxide nanoparticles. *Nanotechnology.* 26(20): 205101, doi: 10.1088/0957–4484/26/20/205101.

Chandler, D.L. 2014. MIT News Office, April 29, 2014. http://news.mit.edu/2014/new-building-will-be-hub-for-nanoscale-research-0429.

Chauhan, A., Zubair, S., Tufail, S., Sherwani, A., Sajid, M., Raman, S.C., Azam, A., and Owais, M. 2011. Fungus-mediated biological synthesis of gold nanoparticles: Potential in detection of liver cancer. *Int. J. Nanom.* 6: 2305–2319.

Cheirmadurai, K., Biswas, S., Murali, R., and Thanikaivelan, P. 2014. Green synthesis of copper nanoparticles and conducting nanobiocomposites using plant and animal sources. *RSC Adv.* 4: 19507–19511.

Cheng, Y., Yin, L., Lin, S., Wiesner, M., Bernhardtm E., and Liu, J. 2011. Toxicity reduction of polymer-stabilized silver nanoparticles by sunlight. *J. Phys. Chem. C* 115: 4425–4432.

Das, S.K., Das, A.R., and Guha, A.K. 2010. Microbial synthesis of multishaped gold nanostructures. *Small.* 6(9): 1012–1021.

Deepak, V., Umamaheshwaran, P.S., Guhan, K., Nanthini, R.A., Krithiga, B., Jaithoon, N.M.H., and Gurunathan, S. 2011. Synthesis of gold and silver nanoparticles using purified URAK. *Colloids Surf. B Biointerfaces.* 86: 353–358.

Deschênes, L., and Ells, T. 2020. Bacteria–nanoparticle interactions in the context of nanofouling. *Adv. Colloid Interface Sci.* 277: 102106.

Devi, L.S., and Joshi, S.R. 2015. Ultrastructures of silver nanoparticles biosynthesized using endophytic fungi. *J. Microsc. Ultrastruct.* 3: 29–37.

Din, M.I., Arshad, F., Hussain, Z., and Mukhtar, M. 2017. Green adeptness in the synthesis and stabilization of copper nanoparticles: Catalytic, antibacterial, cytotoxicity, and antioxidant activities. *Nanoscale Res. Lett.* 12: 638

Ding, Y., Zheng, S., Meng, X., and Yang, D. 2018. Low salinity hot water injection with addition of nanoparticles for enhancing heavy oil recovery under reservoir conditions. In: SPE Western regional meeting, 22–26 April, Garden Grove, California, USA. https://doi.org/10.2118/190132-MS.

Eby, D.M., Schaeublin, N.M., Farrington, K.E., Hussain, S.M., and Johnson, G.R. 2009. Lysozyme catalyzes the formation of antimicrobial silver nanoparticles. *ACS Nano.* 3: 984–994.

El-Kassas, H.Y., and El-Sheekh, M. 2014. Cytotoxic activity of biosynthesized gold nanoparticles with an extract of the red seaweed *Corallina officinalis* on the MCF-7 human breast cancer cell line. *Asian Pac. J. Cancer Prev.* 15: 4311–4317.

Elshawaf, M. 2018. Investigation of Graphene oxide nanoparticles effect on heavy oil viscosity. In: SPE annual technical conference and exhibition, 24–26 September, Dallas, Texas, USA. https://doi.org/10.2118/194037-STU.

Etemadifar, Z., Derikvand, P., Emtiazi, G., and Habibi, M.H. 2014. Response surface methodology optimization of dibenzothiophene biodesulfurization in model oil by nanomagnet immobilized *Rhodococcus erythropolis* R1. *J. Mater. Sci. Eng. B.* 4(10): 322–330.

Fakoya, M.F., and Shah, S.N. 2018. Effect of silica nanoparticles on the rheological properties and fltration performance of surfactant-based and polymeric fracturing fluids and their blends. *SPE Drill Complet.* 33(02): 100–114. https://doi.org/10.2118/163921-PA.

Feris, K., Otto, C., Tinker, J., Wingett, D., Punnoose, A., Thurber, A., Kongara, M., Sabetian, M., Quinn, B., Hanna, C., and Pink, D. 2010. Electrostatic interactions affect nanoparticle-mediated toxicity to gram-negative bacterium Pseudomonas aeruginosa PAO1. *Langmuir.* 26: 4429–4436.

Gardea-Torresdey, J.L., Parsons, J.G., Gomez, E., Peralta-Videa, J., Toiani, H.E., Santiago, P., and Jose Yacaman, M. 2002. Formation and growth of Au nanoparticles inside live alfalfa plants. *Nano Lett.* 2: 397–401.

Ge, F., Li, M.M., Ye, H., and Zhao, B.X. 2012. Effective removal of heavy metal ions Cd^{2+}, Zn^{2+}, Pb^{2+}, Cu^{2+} from aqueous solution by polymer-modified magnetic nanoparticles. J. Hazard. Mater. 211–212:366–372.

Gopinathan, P., Ashok, A.M., and Selvakumar, R. 2013. Bacterial flagella as biotemplate for the synthesis of silver nanoparticle impregnated bionanomaterial. *Appl. Surf. Sci.* 276: 717–722.

Gorbe, M., Bhat, R., Aznar, E., Sancenón, F., Marcos, M.D., Herraiz, F.J., Prohens, J., Venkataraman, A., and Martínez-Máñez, R. 2016. Rapid biosynthesis of silver nanoparticles using pepino (*Solanum muricatum*) leaf extract and their cytotoxicity on HeLa cells. *Materials.* 9: 325, doi:10.3390/ma9050325.

Govindaraju, K., Krishnamoorthy, K., Alsagaby, S.A., Singaravelu, G., and Premanatha, M. 2015. Green synthesis of silver nanoparticles for selective toxicity towards cancer cells. *IET Nanobiotechnol.* 9(6): 325–330, doi: 10.1049/iet-nbt.2015.0001.

Gunalan, S., Sivaraj, R., and Rajendran, V. 2012. Green synthesized ZnO nanoparticles against bacterial and fungal pathogens. *Prog. Nat. Sci.-Mater.* 22(6): 693–700.

Harris, A.T., and Bali, R. 2008. On the formation and extent of uptake of silver nanoparticles by live plants. *J. Nanopart. Res.* 10: 691–695.
Hassabo, A.G., Nada, A.A., Ibrahim, H.M., and AbouZeid, N.Y. 2015. Impregnation of silver nanoparticles into polysaccharide substrates and their properties. *Carbohydr. Polym.* 122: 343–350.
He, S., Guo, Z., Zhang, Y., Zhang, S., Wang, J., and Gu, N. 2007. Biosynthesis of gold nanoparticles using the bacteria *Rhodopseudomonas capsulata. Mater. Lett.* 61: 3984–3987.
Helmlinger, J., Heise, M., Heggen, M., Ruck, M., and Epple, M. 2015. A rapid, high-yield and large-scale synthesis of uniform spherical silver nanoparticles by a microwave-assisted polyol process. *RSC Adv.* 5: 92144–92150.
Herrera-Becerra, R., Zorrilla, C., Rius, J.L., and Ascencio, J.A. 2008. Electron microscopy characterization of biosynthesized iron oxide nanoparticles. *Appl. Phys. A.* 91: 241–246.
Hogeweg, A.S., Hincapie, R.E., Foedisch, H., and Ganzer, L. 2018. Evaluation of aluminium oxide and titanium dioxide nanoparticles for EOR applications. In: SPE Europec featured at 80th EAGE conference and exhibition, 11–14 June, Copenhagen, Denmark. https://doi.org/10.2118/190872-MS.
Honary, S., Barabadi, H., Fathabad E.G., and Naghibi, F. 2012. 'Green synthesis of copper oxide nanoparticles using *Penicillium aurantiogriseum, Penicillium citrinum* and *Penicillium wakasmanii. Dig. J. Nanomater. Biostruct.* 7: 999–1005.
Hosseini-Abari, A., Emtiazi, G., and Ghasemi, S.M. 2013. Development of an eco-friendly approach for biogenesis of silver nanoparticles using spores of *Bacillus athrophaeus. World J. Microbiol. Biotechnol.* 29: 2359–236.
Huang, J., Lin, L., Li, Q., Sun, D., Wang, Y., Lu, Y., He, N., Yang, K., Yang, X., Wang, H., Wang, W., and Lin, W. 2008. Continuous-flow biosynthesis of silver nanoparticles by lixivium of sundried *Cinnamomum camphora* leaf in tubular microreactors. *Ind. Eng. Chem. Res.* 47: 6081–6090.
Huang, H.Z., and Yang, X.R. 2004. Synthesis of chitosan-stabilized gold nanoparticles in the absence/presence of tripolyphosphate. *Biomacromolecules.* 5(6): 2340–2346.
Hulkoti, N.I., and Taranath, T.C. 2014. Biosynthesis of nanoparticles using microbes—A review. *Colloids Surf. B.* 121: 474–483.
Iravani, S. 2011. Green synthesis of metal nanoparticles using plants. *Green Chem.* 13: 2638–2650.
Iravani, S. 2014. Bacteria in nanoparticle synthesis: current status and future prospects. *Int. Scholar Res. Notice.* 2014: 359316. https://doi.org/10.1155/2014/359316.
Iriarte-Mesa, C., López, Y.C., Matos-Peralta, Y., Vega-Hernández, K., and Antuch, M. 2020. Gold, silver and iron oxide nanoparticles: Synthesis and bionanoconjugation strategies aimed at electrochemical applications. *Topics Curr. Chem.* 378: 12. https://doi.org/10.1007/s41061–019-0275-y
Islam, A., Phillips, G., Snowden, M., and Williams, P. 1997. A review of recent developments on the regulatory, structural and functional aspects of gum Arabic. *Food Hydrocoll.* 11: 493–505.
Jacob, S., Finub, J., and Narayanan, A. 2011. Synthesis of silver nanoparticles using *Piper longum* leaf extracts and its cytotoxic activity against Hep-2 cell line. *Colloids Surf. B. Biointerfaces.* 91: 212–214.
Janardhanan, S.K., Ramasamy, I., and Nair, B.U. 2008. Synthesis of iron oxide nanoparticles using chitosan and starch templates. *Transition Met. Chem.* 33: 127–131.
Jayarambabu, N., Rao, K.V., and Rajendar, V. 2018. Biogenic synthesis, characterization, acute oral toxicity studies of synthesized Ag and ZnO nanoparticles using aqueous extract of *Lawsonia inermis. Mater. Lett.* 211: 43–47.

Jayaseelan, C., Abdul Rahuman, A., Vishnu Kirthi, A., Marimuthu, S., Santhoshkumar, T., Bagavan, A., Gaurav, K., Karthik, L., and Bhaskara Rao, K.V. 2012. Novel microbial route to synthesize ZnO nanoparticles using *Aeromonas hydrophila* and their activity against pathogenic bacteria and fungi. *Spectrochim. Acta A.* 90: 78–84.

Jeevanandam, J., Chan, Y.S., and Danquah, M.K. 2016. Biosynthesis of metal and metal oxide nanoparticles. *Chem. Bio. Eng. Rev.* 3(2): 55–67.

Jeyabharathi, S., Kalishwaralal, K., Sundar, K., and Muthukumaran, A. 2017. Synthesis of zinc oxide nanoparticles (ZnONPs) by aqueous extract of *Amaranthus caudatus* and evaluation of their toxicity and antimicrobial activity. *Mater. Lett.* 209: 295–298.

Jha, A.K., Prasad, K., and Kulkarni, A.R. 2009. Synthesis of TiO_2 nanoparticles using microorganisms. *Colloids Surf. B.* 71: 226–229.

Jinkun, F., Yueying, L., Pingying, G., Dingliang, T., Zhongyu, L., Bingxin, Y., and Shengzhou, W. 2000. Spectroscopic characterization on the biosorption and bioreduction of Ag(I) by *Lactobacillus* sp. A09*. *Acta Phys. Chim. Sin.* 16: 779–782.

Joerger, R., Klaus, T., and Granqvist, C.G. 2000. Biologically produced silver-carbon composite materials for optically functional thin film coatings. *Adv. Mater.* 12: 407–409.

Jolivet, J.P., Belleville, P., Tronc, E., and Livage, J. 1992. Influence of Fe(II) on the formatrion of the spinel iron oxide in alkalime medium. *Clays Clay Miner.* 40 (5): 531–539.

José de Andrade, C., Maria de Andrade, L., Mendes, M.A., and Oller do Nascimento, C.A. 2017. An overview on the production of microbial copper nanoparticles by bacteria, fungi and algae. *Global J. Res. Eng. C.* 17(1): 27–33.

Kadam,V.V., Ettiyappan, J.P., and Balakrishnan, R.M. 2019. Mechanistic insight into the endophytic fungus mediated synthesis of protein capped ZnO nanoparticles. *Mater. Sci. Eng. B.* 243: 214–221.

Kafayati, M., Raheb, J., Angazi, M., Alizadeh, S., and Bardania, H. 2013. The effect of magnetic Fe_3O_4 nanoparticles on the growth of genetically manipulated bacterium *Pseudomonas aeruginosa* PTSOX4. *Iran Biotech.* 11: 41–46.

Kandasamy, K., Alikunhi, N.M., Manickaswami, G., Nabikhan, A., and Ayyavu, G. 2013. Synthesis of silver nanoparticles by coastal plant *Prosopis chilensis* (L.) and their efficacy in controlling vibriosis in shrimp *Penaeus monodon*. *Appl. Nanosci.* 3(1): 65–73.

Kang, F., Alvarez, P.J., and Zhu, D. 2014. Microbial extracellular polymeric substances reduce Ag^+ to silver nanoparticles and antagonize bactericidal activity. *Environ. Sci. Technol.* 48(1): 316–322.

Kanmani, P., and Lim, S.T. 2013. Synthesis and structural characterization of silver nanoparticles using bacterial exopolysaccharide and its antimicrobial activity against food and multidrug resistant pathogens. *Process Biochem.* 48: 1099–1106.

Kasthuri, J., Veerapandian, S., and Rajendiran, N. 2009. Biological synthesis of silver and gold nanoparticles using apiin as reducing agent. *Colloids Surf. B.* 68: 55–60.

Kathad, U., and H.P. Gajera. 2014. Synthesis of copper nanoparticles by two different methods and size comparision. *Int. J. Pharm. Biosci.* 5(3): 533–540.

Kathiresan, K., Manivannan, S., Nabeel, M.A., and Dhivya, B. 2009. Studies on silver nanoparticles synthesized by a marine fungus, *Penicillium fellutanum* isolated from coastal mangrove sediment. *Colloids Surface B Biointerface.* 71: 133–137.

Khajehpour, M., Etminan, S.R., Goldman, J., Wassmuth, F., and Bryant, S. 2016. Nanoparticles as foam stabilizer for steam-foam process. In: SPE EOR conference at oil and gas West Asia, 21–23 March, Muscat, Oman. https://doi.org/10.2118/179826-MS.

Khan, R., and Fulekar, M.H. 2016. Biosynthesis of titanium dioxide nanoparticles using *Bacillus amyloliquefaciens* culture and enhancement of its photocatalytic activity for the degradation of a sulfonated textile dye Reactive Red 31. *J. Colloid. Interf. Sci.* 475: 184–191.

Khanehzaei, H., Ahmad, M.B., Shameli, K., and Ajdari, Z. 2014. Synthesis and characterization of Cu@Cu_2O core shell nanoparticles prepared in seaweed Kappaphycus alvarezii media. *Int. J. Electrochem. Sci.* 9(12): 8189–8198.

Khani, R., Roostaei, B., Bagherzade, G., and Moud, M. 2018. Green synthesis of copper nanoparticles by fruit extract of *Ziziphus spina-christi* (L.) Willd.: Application for adsorption of triphenylmethane dye and antibacterial assay. *J. Mol. Liq.* 225: 541–549.

Kharissova, O.V., Dias, H.V.R., Kharisov, B.I., Pérez, B.O., and Pérez, V.M.J. 2013. The greener synthesis of nanoparticles. *Trends Biotechnol.* 31(4): 240–248.

Khaydarov, R. A., Khaydarov, R. R., and Gapurova, O. 2010. Water purification from metal ions using carbon nanoparticle conjugated polymer nanocomposites. *Water Res.* 44(6): 1927–1933.

Kiran, G.S., Sabu, A., and Selvin, J. 2010. Synthesis of silver nanoparticles by glycolipid biosurfactant produced from marine *Brevibacterium casei* MSA19. *J. Biotechnol.* 148: 221–225.

Kitching, M., Ramani, M., and Marsil, E., 2015. Fungal biosynthesis of gold nanoparticles: mechanism and scale. *Microb. Biotechnol.* 8(6): 904–917.

Ko, S., and Huh, C. 2019. Use of nanoparticles for oil production applications. *J. Petrol. Sci. Eng.* 172: 97–114.

Kumar, A., Pandey, A.K., Singh, S.S., Shanker, R., and Dhawan, A. 2011. Cellular uptake and mutagenic potential of metal oxide nanoparticles in bacterial cells. *Chemosphere.* 83: 1124–1132.

Kumar, B., Smita, K., Cumbal, L., and Debut, A. 2017. Green synthesis of silver nanoparticles using Andean blackberry fruit extract. *Saudi J. Biol. Sci.* 24: 45–50.

Kumar, C.G., and Mamidyala, S.K. 2011. Extracellular synthesis of silver nanoparticles using culture supernatant of *Pseudomonas aeruginosa*. Colloids Surf. B Biointerface. 84: 462–466.

Kumar, C.G., Mamidyala, S.K., Das, B., Sridhar, B., Sarala Devi, G., and Karuna, M.S.L. J. 2010. Synthesis of biosurfactant-based silver nanoparticles with purified rhamnolipids isolated from *Pseudomonas aeruginosa* BS-161R. *Microbiol. Biotechnol.* 20: 1061–1068.

Kundu, D., Hazra, C., Chatterjee, A., Chaudhari, A., Mishra, S. 2014. Extracellular biosynthesis of zinc oxide nanoparticles using Rhodococcus pyridinivorans NT2: Multifunctional textile finishing, biosafety evaluation and in vitro drug delivery in colon carcinoma. *J. Photochem. Photobiol. B. Biol.* 140: 194–204.

Kwon, C., Park, B., Kim, H., and Jung, S. 2009. Green synthesis of silver nanoparticles by Sinorhizobial octasaccharide isolated from Sinorhizobium meliloti. *Bull. Korean Chem Soc.* 30(7): 1651–1654.

Lan, G., Chen, C., Liu, Y., Lu, Y., Du, J., Taob, S., and Zhang, S. 2017. Corrosion of carbon steel induced by a microbial enhanced oil recovery bacterium *Pseudomonas* sp. SWP-4. *RSC Adv.* 7: 5583–5594.

Larios-Rodriguez, E., Rangel-Ayon, C., Castillo, S.J., Zavala, G., and Herrera-Urbina, R. 2011. Bio-synthesis of gold nanoparticles by human epithelial cells, in vivo. *Nanotechnology.* 22(35): 355601, doi: 10.1088/0957-4484/22/35/355601.

Lateef, S.A., Ajumobi, O.O., and Onaiz, S.A., 2019. Enzymatic desulfurization of crude oil and its fractions: A mini review on the recent progresses and challenges. *Arab. J. Sci. Eng.* 44: 5181–5193.

Lau, H.C., Yu, M., and Nguyen, Q.P. 2017. Nanotechnology for oilfield applications: Challenges and impact. *J. Pet. Sci. Eng.* 157: 1160–1169.

Lee, H.-J., Lee, G., Jang, N.R., Yan, J.H., Song, J.Y., and Kim, B.S. 2011. Biological synthesis of copper nanoparticles using plant extract. *NSTI-Nanotech.* 1: 371–374.

León, Y., Cárdenas, G., and Arias, M. 2017. Synthesis and characterizations of metallic nanoparticles in chitosan by chemical reduction. 2017. *J. Chil. Chem. Soc.* 62(4): 3760–3764.

Leong, Y.K., Seah, U., Chu, S.Y., and Ong, B.C. 2001. Effects of gum Arabic macromolecules on surface forces in oxide dispersions. *Colloids Surf. A. Physicochem. Eng. Aspects*.182: 263–268.

Li, D., Hong, B., Fang, W., Guo, Y. and Lin, R. 2010. Preparation of well-dispersed silver nanoparticles for oil-based nanofluids. *Ind. Eng. Chem. Res.* 49(4): 1697–1702.

Li, X., Odoom-Wubah, T., Chen, H., Jing, X., Zheng, B., and Huang, J. 2014. Biosynthesis of silver nanoparticles through tandem hydrolysis of silver sulfate and cellulose under hydrothermal conditions. *J. Chem. Technol. Biotechnol.* 89: 1817–1824.

Liang, M., Wei, S., Jian-Xin, L., Xiao-Xi, Z., Zhi, H., Wen, L., Zheng-Chun, L., and Jian-Xin, T. 2017. Optimization for extracellular biosynthesis of silver nanoparticles by *Penicillium aculeatum* Su1 and their antimicrobial activity and cytotoxic effect compared with silver ions. *Mater. Sci. Eng. C* 77: 963–971.

Liu, Q., Liu, H., Yuan, Z., Wei, D., and Ye Y. 2012. Evaluation of antioxidant activity of *Chrysanthemum* extracts and tea beverages by gold nanoparticles-based assay. *Colloids Surf. B.* 92: 348–352.

Losfeld, G., Escande, V., La Blache, P.V., L'Huillier, L., and Grison, C. 2012. Design and performance of supported Lewis acid catalysts derived from metal contaminated biomass for Friedel–Crafts alkylation and acylation. *Catal. Today.* 189: 111–116.

Ma, L., Su, W., Liu, J.X., Zeng, X.X., Huang, Z., Li, W., Liu, Z.C., and Tang, J.X. 2017. Optimization for extracellular biosynthesis of silver nanoparticles by *Penicillium aculeatum* Su1 and their antimicrobial activity and cytotoxic effect compared with silver ions. *Mater. Sci. Eng. C Mater. Biol. Appl.* 77: 963–971.

Macedo de Souza, P., Regina de Vasconcelos Goulart, F., Marques, J.M., Bizzo, H.R., Blank, A.F., Groposo, C., Paula de Sousa, M., Vólaro, V., Alviano, C.S., Moreno, D.S.A., and Seldin, L. 2017. Growth inhibition of sulfate-reducing bacteria in produced water from the petroleum industry using essential oils. *Molecules.* 22: 648. doi:10.3390/molecules22040648

Mahmoudi, M., Simchi, A., Milani, A.S., and Stroeve, P. 2009. Cell toxicity of superparamagnetic iron oxide nanoparticles. *J. Colloid Interf. Sci.* 336(2): 510–518.

Makarov, V.V., Love, A.J., Sinitsyna, O.V., Makarova, S.S., Yaminsky, I.V., Taliansky, M.E., and Kalinina, N.O. 2014. Green nanotechnologies: Synthesis of metal nanoparticles using plants. *Acta Naturae.* 6(1): 35–44.

Manceau, A., Nagy, K.L., Marcus, M.A., Lanson, M., Geoffroy, N., Jacquet, T., and Kirpichtchikova, T. 2008. Formation of metallic copper nanoparticles at the soil-root interface. *Environ. Sci. Technol.* 42: 1766–1772.

Manikprabhu, D., and Lingappa, K. 2013. Antibacterial activity of silver nanoparticles against methicillin-resistant Staphylococcus aureus synthesized using model *Streptomyces* sp. pigment by photo-irradiation method. *J. Pharm. Res.* 6: 255–260.

Manju, S., Malaikozhundan, B., Vijayakumar, S., Shanthi, S., Jaishabanu, A., Ekambaram, P., Vaseeharan, B. 2016. Antibacterial, antibiofilm and cytotoxic effects of Nigella sativa essential oil coated gold nanoparticles. *Microb. Pathog.* 91: 129–135.

Markus, J., Wang, D., Kim, Y.-J., Ahn, S., Mathiyalagan, R., Wang, C., and Yang, D.C. 2017. Biosynthesis, characterization, and bioactivities evaluation of silver and gold nanoparticles mediated by the roots of Chinese herbal *Angelica pubescens* Maxim. *Nanoscale Res. Lett.* 12: 46, doi: 10.1186/s11671–017-1833–2.

Marshall, A.T., Haverkamp, R.G., Clive, E.D., Parsons, J.G., and Gardea-Torresdey, J. L. 2007. Accumulation of gold nanoparticles in Brassic juncea. *Int. J. Phytoremed.* 9: 197–206.

Mat Zain, N., Stapley, A.G.F., and Shama, G. 2014. Green synthesis of silver and copper nanoparticles using ascorbic acid and chitosan for antimicrobial applications. *Carbohydr. Polym.* 112: 195–202.

Mishra, A., and Sardar, M. 2012. Alpha-amylase mediated synthesis of silver nanoparticles. *Sci. Adv. Mater.* 4(1): 143–146.

Mittal, A.K., Chisti, Y., and Banerjee, U.C. 2013. Synthesis of metallic nanoparticles using plant extracts. *Biotechnol. Adv.* 31(2): 346–356.

Mittal, J., Batra, A., Singh, A., and Sharma, M.M. 2014. Phytofabrication of nanoparticles through plant as nanofactories, *Adv. Nat. Sci.: Nanosci. Nanotechnol.* 5: 1–10.

Mochochoko, T., Oluwafemi, O.S., Jumbam, D.N., and Songca, S.P. 2013. Green synthesis of silver nanoparticles using cellulose extracted from an aquatic weed; water hyacinth. *Carbohydr. Polym.* 98(1): 290–294.

Monopoli, M.P., Berg, C.A., Salvati, A., and Dawson, K.A. 2012. Biomolecular coronas provide the biological identity of nanosized materials. *Nat. Nanotechnol.* 7: 779–786.

Morsy, F.M., Nafady, N.A., Abd-Allah, M.H., and Abd Elhady, D. 2014. Green synthesis of silver nanoparticles by water soluble fraction of the extracellular polysaccharides/matrix of the cyanobacterium *Nostoc commune* and its application as a potent fungal surface sterilizing agent of seed crops. *Univers. J. Microbiol. Res.* 2(2): 36–43.

Nabhani, N., and Tofghi, A. 2010. The assessment of health, safety and environmental risks of nanoparticles and how to control their impacts. In: SPE international conference on health, safety and environment in oil and gas exploration and production, 12–14 April, Rio de Janeiro, Brazil. https://doi.org/10.2118/127261-MS.

Nam, K.T., Lee, Y.J., Krauland, E.M., Kottmann, S.T. and Belcher, A.M. 2008. Peptide mediated reduction of silver ions on engineered biological scaffolds. *ACS Nano.* 2: 1480–1486.

Naraginti, S., and Li, Y. 2017. Preliminary investigation of catalytic, antioxidant, anticancer and bactericidal activity of green synthesized silver and gold nanoparticles using *Actinidia deliciosa. J. Photochem. Photobiol. B: Biol.* 170: 225 –234.

Narendhran, S., and Nair, R.K. 2017. Nanoparticles and their toxicology studies: A green chemistry approach. *Res. Dev. Mater. Sci.* 2(3). doi.org/10.31031/RDMS.2017.02.000539

Nasab, N.A., Kumleh, H.H., Kazemzad, M., Panjeh, F.G., and Davoodi-Dehaghani, F. 2015. Improvement of desulfurization performance of *Rhodococcus erythropolis* IGTS8 by assembling spherical mesoporous silica nanosorbents on the surface of the bacterial cells. *J. Appl. Chem. Res.* 9(2): 81–91.

Nate, Z., Moloto, M.J., Mubiayi, P.K., and Sibiy, P.N. 2018. Green synthesis of chitosan capped silver nanoparticles and their antimicrobial activity. *MRS Adv.* 3(42/43): 2505–2517.

NNI, US. 2018. The national nanotechnology initiative supplement to the President's 2019 Budget. https://www.nano.gov/sites/default/fles/NNI-FY19-Budget-Supplement.pdf.

Ogolo, N.A., Olafuyi, O.A., and Onyekonwu, M.O. 2012. Enhanced oil recovery using nanoparticles. In: SPE Saudi Arabia Section Technical Symposium and Exhibition, 8–11 April, Al-Khobar, Saudi Arabia. https://doi.org/10.2118/160847-MS.

Omran, B.A., Nassar, H.N., Fatthallah, N.A., Hamdy, A., El-Shatoury, E.H., and El-Gendy, N. Sh. 2018a. Waste upcycling of *Citrus sinensis* peels as a green route for the synthesis of silver nanoparticles. *Energ. Sources Part A.* 40(2): 227–236. https://doi.org/10.1080/15567036.2017.1410597

Omran, B.A., Nassar, H.N., Fatthallah, N.A., Hamdy, A., El-Shatoury, E.H., and El-Gendy, N.Sh. 2018b. Characterization and antimicrobial activity of silver nanoparticles mycosynthesized by *Aspergillus brasiliensis*. *J. Appl. Microbiol.* 125: 370–382. https://doi.org/10.1111/jam.13776

Omran, B.A., Nassar, H.N., Younis, S.A., El-Salamony, R.A., Fatthallah, N.A., Hamdy, A., El-Shatoury, E.H., and El-Gendy, N.Sh., 2020. Novel mycosynthesis of cobalt oxide nanoparticles using *Aspergillus brasiliensis* ATCC 16404 -optimization, characterization and antimicrobial activity. *J. Appl. Microbiol.* 128: 438–457. https://doi.org/10.1111/jam.14498

Omran, B.A., Nassar, H.N., Younis, S.A., Fatthallah, N.A., Hamdy, A., El-Shatoury, E.H., and El-Gendy, N.Sh. 2019. Physiochemical properties of *Trichoderma longibrachiatum* DSMZ 16517-synthesized silver nanoparticles for the mitigation of halotolerant sulphate-reducing bacteria. *J. Appl. Microbiol.* 126: 138–154. https://doi.org/10.1111/jam.14102

Panigrahi, S., Kundu, S., Ghosh, S., Nath, S., and Pal, T. 2004. General method of synthesis for metal nanoparticles. *J. Nanopart. Res.* 6: 411–414.

Pantidos, N., Edmundson, M.C., and Horsfall, L. 2018. Room temperature bioproduction, isolation and anti-microbial properties of stable elemental copper nanoparticles. *New Biotechnol.* 40: 275–281.

Parial, D., Patra, H.K., Dasgupta, A.K.R., and Pal, R. 2012. Screening of different algae for green synthesis of goldnanoparticles. *Eur. J. Phycol.* 47(1): 22–29.

Park, J., Kwon, S.G., Jun, S.W., Kim, B.H., and Hyeon, T. 2012. Large-scale synthesis of ultra-small-sized silver nanoparticles. *Chemphyschem.* 13: 2540–2543.

Parker, H.L., Rylott, E.L., Hunt, A.J., Dodson, J.R., Taylor, A.F., Bruce, N.C., and Clark, J.H. 2014. Supported palladium nanoparticles synthesized by living plants as a catalyst for Suzuki–Miyaura reactions. *PLoS One.* 9: e87192. doi:10.1371/journal.pone.0087192.

Parthasarathy, G., Saroja, M., Venkatachalam, M., and Evanjelene, V.K. 2017. Biological synthesis of zinc oxide nanoparticles from leaf extract of *Curcuma neilgherrensis* wight. *Int. J. Mater. Sci.* 12(1): 73–86

Patel, V. Berthold, D., Puranik, P., Gantar, M. 2015. Screening of *cyanobacteria* and microalgae for their ability to synthesize silver nanoparticles with antibacterial activity. *Biotechnol. Rep.* 5: 112–119.

Pethakamsetty, L., Kothapenta, K., Nammi, H. R., Ruddaraju, L. K., Kollu, P., Yoon, S. G., and Pammi, S.V.N. 2017. Green synthesis, characterization and antimicrobial activity of silver nanoparticles using methanolic root extracts of *Diospyros sylvatica*. *J. Environ. Sci.* 55: 157–163.

Prabhu, D., Arulvasu, C., Babu, G., Manikandan, R., and Srinivasan, P. 2013. Biologically synthesized green silver nanoparticles from leaf extract of *Vitex negun*do L. induce growth-inhibitory effect on human colon cancer cell line HCT15. *Process Biochem.* 48(2): 317–324.

Prakash, I. 2011. Bioactive chemical constituents from pomegranate (*Punica granatum*) juice, seed and peel—A review. *Int. J. Res. Chem. Environ.* 1: 1–18.

Prasad, K., Jha, A.K., Prasad, K., and Kulkarni, A.R. 2010. Can microbes mediate nano-transformation? *Indian J. Phys.* 84(10): 1355–1360.

Prasad, R., Pandey, R., and Barman, I. 2016. Engineering tailored nanoparticles with microbes: Quo vadis? *WIREs Nanomed. Nanobiotechnol.* 8: 316–330.

Premasudha, P., Venkataramana, M., Abirami, M., Vanathi, P., Krishna, K., and Rajendran, R. 2015. Biological synthesis and characterization of silver nanoparticles using *Eclipta alba* leaf extract and evaluation of its cytotoxic and antimicrobial potential. *Bull. Mater. Sci.* 38: 965–973.

Princy, F., and Gopinath, A. 2015. Eco-friendly synthesis and characterization of silver nanoparticles using marine macroalga *Padina tetrastromatica*. *Int. J. Sci. Res.* 4(6): 1050–1054.

Priyaa, G.H., and Satyan, K.B. 2014. Biological synthesis of silver nanoparticles using ginger (*Zingiber officinale*) extract. *J. Environ. Nanotechnol.* 3(4): 32–40.

Qu, Y., Pei, X., Shen, W., Zhang, X., Wang, J., Zhang, Z., Li, S., You, S., Ma, F., and Zhou, J. 2017. Biosynthesis of gold nanoparticles by *Aspergillum* sp. WL-Au for degradation of aromatic pollutants. *Physica E.* 88: 133–141.

Rajabi, H.R., Naghiha, R., Kheirizadeh, M., Sadatfaraji, H., Mirzaei, A., and Alvand, Z.M. 2017. Microwave assisted extraction as an efficient approach for biosynthesis of zinc oxide nanoparticles: Synthesis, characterization, and biological properties. *Mater. Sci. Eng. C.* 78: 1109 –1118.

Rajakumar, G., and Abdul Rahuman, A. 2011. Larvicidal activity of synthesized silver nanoparticles using *Eclipta prostrata* leaf extract against filariasis and malaria vectors. *Acta Trop.* 118: 196–203.

Rajakumar, G., and Abdul Rahuman, A. 2012. Acaricidal activity of aqueous extract and synthesized silver nanoparticles from *Manilkara zapota* against Rhipicephalus (Boophilus) microplus. *Res. Vet. Sci.* 93(1): 303–309.

Rajan, A., Vilas, V., and Philip, D. 2015. Studies on catalytic, antioxidant, antibacterial and anticancer activities of biogenic gold nanoparticles. *J. Mol. Liq.* 212: 331–339.

Rajesh, K.M., Ajitha, B., Reddy, Y.A.K., Suneetha, Y., and Reddy, P.S. 2018. Assisted green synthesis of copper nanoparticles using *Syzygium aromaticum* bud extract: Physical, optical and antimicrobial properties. *Optik.* 154: 593–600.

Rajeshkumar, S., Malarkodi, C., Paulkumar, K., Vanaja, M., Gnanajobitha, G., and Annadurai, G. 2013. Algae mediated green fabrication of silver nanoparticles and examination of its antifungal activity against clinical pathogens. *Int. J. Met.* 2014: 692643. http://dx.doi.org/10.1155/2014/692643.

Raju, D., Mehta, U.J., and Ahmad, A. 2012. Phytosynthesis of intracellular and extracellular gold nanoparticles by living peanut plant (*Arachis hypogaea* L.). *Biotechnol. Appl. Biochem.* 59(6): 471–478.

Raliya, R., and Tarafdar, J.C. 2014. Biosynthesis and characterization of zinc, magnesium and titanium nanoparticles: an eco-friendly approach. *Int. Nano Lett.* 4: 93. doi: 10.1007/s40089–014-0093–8.

Raliya, R., Biswas, P., and Tarafdar, J.C. 2015. TiO_2 nanoparticle biosynthesis and its physiological effect on mung bean (*Vigna radiata* L.). *Biotechnol Rep.* 5: 22–26.

Raliya, R., Tarafdar, J.C., Mahawar, H., Kumar, R., Gupta, P., Mathur, T., Kaul, R.K., Kumar, P., Kalia, A., Gautam, R., Singh, S.K., Gehlot, H.S. 2014. ZnO nanoparticles induced exopolysaccharide production by *B. subtilis* strain JCT1 for arid soil applications. *Int. J. Biol. Macromol.* 65: 362–368.

Rasool, K., Nasrallah, G., Younes, N., Pandey, R., Abdul Rasheed, P., and Mahmoud, K.A. 2018. Green" ZnO-interlinked chitosan nanoparticles for the efficient inhibition of sulfate-reducing bacteria in inject seawater. *ACS Sustain. Chem. Eng.* 6: 3896–3906.

Rivera-Rangela, R.D., González-Muñoza, M.P., and Avila-Rodrigueza, M., Razo-Lazcanoa, T.A., and Solans, C. 2018. Green synthesis of silver nanoparticles in oil-in-water micro-emulsion and nano-emulsion using geranium leaf aqueous extract as a reducing agent. *Colloids Surf. A.* 536: 60–67.
Rizeq, B.R., Younes, N.N., Rasool, K., and Nasrallah, G.K., 2019. Synthesis, bioapplications, and toxicity evaluation of chitosan-based nanoparticles. *Int. J. Mol. Sci.* 20(22): 5776, doi: 10.3390/ijms20225776.
Sangeetha, J., Thomas, S., Arutchelvi, J., Doble, M., and Philip, J. 2013. Functionalization of iron oxide nanoparticles with biosurfactants and biocompatibility studies. *J. Biomed. Nanotechnol.* 9: 751–764.
Santra, A.K., Boul, P., and Pang, X. 2012. Influence of nanomaterials in oil well cement hydration and mechanical properties. In: SPE International Oil Field Nanotechnology Conference and Exhibition, 12–14 June, Noordwijk, The Netherlands. https://doi.org/10.2118/156937-MS.
Sathiyanarayanan, G., Kiran, G.S., and Selvin, J. 2013. Synthesis of silver nanoparticles by polysaccharide bioflocculant produced from marine *Bacillus subtilis* MSBN17. *Colloids Surf. B Biointerfaces.* 102: 13–20.
Schlorf, T., Meincke, M., Kossel, E., Glüer, C.-C., Jansen, O., and Mentlein, R. 2011. Biological properties of iron oxide nanoparticles for cellular and molecular magnetic resonance imaging. *Int. J. Mol. Sci.* 12: 12–23.
Seabra, A.B., and Duran, N. 2015. Nanotoxicology of metal oxide nanoparticles. *Metals.* 5: 934–975.
Selvakannan, P.R., Swami, A., Srisathiyanarayanan, D., Shirude, P.S., Pasricha, R., Mandale, A.B., and Sastry, M. 2004. Synthesis of aqueous Au core–Ag shell nanoparticles using tyrosine as a pH-dependent reducing agent and assembling phase-transferred silver nanoparticles at the air–water interface. *Langmuir.* 20: 7825–7836.
Senthil, B., Devasena, T., Prakash, B., and Rajasekar, A. 2017. Non-cytotoxic effect of green synthesized silver nanoparticles and its antibacterial activity. *J. Photochem. Photobiol., B: Biol.* 170: 1–7.
Shah, M., Fawcett, D., Sharma, S., Tripathy, S.K., and Poinern, G.E.J. 2015. Green synthesis of metallic nanoparticles via biological entities. *Materials.* 8: 7278–7308.
Shamaila, S., Sajjad, A.K.L., Ryma, N.A., Farooqi, S.A., Jabeen, N., Majeed, S., and Farooq, I. 2016. Advancements in nanoparticle fabrication by hazard free eco-friendly green routes. *Appl. Mater. Today.* 5: 150–199.
Shan, G.B., Xing, J.M., Zhang, H., and Liu, H.Z. 2005. Biodesulfurization of dibenzothiophene by microbial cells coated with magnetite nanoparticles. *Appl. Environ. Microbiol.* 71(8): 4497–4502.
Shankar, S.S., Rai, A., Ankamwar, B., Singh, A., Ahmad, A., and Sastry, M. 2004. Biological synthesis of triangular gold nanoprisms. *Nat. Mater.* 3: 482–488.
Shantkriti, S., and Rani, P. 2014. Biological synthesis of Copper nanoparticles using *Pseudomonas fluorescens*. *Int. J. Curr. Microbiol. App. Sci.* 3(9): 374–383.
Sharma, D., Rajput, J., Kaith, B.S., Kaur, M., and Sharma, S. 2010. Synthesis of ZnO nanoparticles and study of their antibacterial and antifungal properties. *Thin Solid Films.* 519: 1224–1229.
Sharma, N.C., Nath, S.V., Parsons, J.G. Gardea-Torresdey, J.L., and Pal, T. 2007. Synthesis of plant-mediated gold nanoparticles and catalytic role of biomatrix-embedded nanomaterials. *Environ. Sci. Technol.* 41: 5137–5142.
Shen, L., Laibinis, P.E., and Hatton, T.A. 1999. Bilayer surfactant stabilized magnetic fluids: Synthesis and interactions at interfaces. *Langmuir.* 15: 447–453.

Shi-Yong, Z., Don, K.L., Chang, W.K., Hyun, G.C., Young, H.K., and Young, S.K. 2006. Synthesis of magnetic nanoparticles of Fe_3O_4 and $CoFe_2O_4$ and their surface modification by surfactant adsorption. *Bull. Korean Chem. Soc.* 27(2): 237–42.

Si, S., and Mandal, T.K. 2007. Trytophan-based peptides to synthesize gold and silver nanoparticles: a mechanistic and kinetic study. *Chem. Eur. J.* 13: 3160–3168.

Simonsen, G., Strand, M., and Øye, G. 2018. Potential applications of magnetic nanoparticles within separation in the petroleum industry. *J. Pet. Sci. Eng.* 165: 488–495.

Singh, A., Shukla, R., Hassan, S., Bhonde, R.R., and Sastry, M. 2011. Cytotoxicity and cellular internalization studies of biogenic gold nanotriangles in animal cell lines. *Int. J. Green Nanotechnol.* 3(4): 251–263.

Singh, B.N., Rawat, A.K.S., Khan, W., Naqvi, A.H., and Singh, B.R. 2014. Biosynthesis of stable antioxidant ZnO nanoparticles by *Pseudomonas aeruginosa* rhamnolipids. *PLoS One.* 9: e106937. https://doi.org/10.1371/journal.pone.0106937.

Singh, P., Garg, A., Pandit, S., Mokkapati, V.R.S.S., and Mijakovic, I. 2018. Antimicrobial effects of biogenic nanoparticles. *Nanomaterials.* 8: 1009. doi: 10.3390/nano8121009.

Sintubin, L., Windt, W.D., Dick, J., Mast, J., Ha, D., Verstraete, W., and Boon, N. 2009. Lactic acid bacteria as reducing and capping agent for the fast and efficient production of silver nanoparticles. *Appl. Microbiol. Biotechnol.* 84: 741–749.

Soenen, S.J.H., Himmelreich, U., Nuytten, N., and De Cuyper, M. 2011. Cytotoxic effects of iron oxide nanoparticles and implications for safety in cell labelling. *Biomaterials.* 32: 195–205.

Sowani, H., Mohite, P., Munot, H., Shouche, Y., Bapat, T., Kumar, A.R., Kulkarni, M. and Zinjarde, S. 2016. Green synthesis of gold and silver nanoparticles by an actinomycete *Gordonia amicalis* HS-11: Mechanistic aspects and biological application. *Process Biochem.* 51: 374–383.

Stankic, S., Suman, S., Haque, F., and Vidic, J. 2016. Pure and multi metal oxide nanoparticles: Synthesis, antibacterial and cytotoxic properties. *J. Nanobiotechnol.* 14: 73. doi: 10.1186/s12951–016-0225–6.

Subramaniam, J., Murugan, K., Panneerselvam, C., Kovendan, K., Madhiyazhagan, P., Dinesh, D., Kumar, P. M., Chandramohan, B., Suresh, U., Rajaganesh, R., Alsalhi, M. S., Devanesan, S., Nicoletti, M., Canale, A., and Benelli, G. 2016. Multipurpose effectiveness of *Couroupita guianensis*-synthesized gold nanoparticles: high antiplasmodial potential, field efficacy against malaria vectors and synergy with *Aplocheilus lineatus* predators. *Environ. Sci. Pollut. Res.* 23: 7543–7558.

Sukirtha, R., Priyanka, K.M., Antony, J.J., Kamalakkannan, S., Ramar, T., and Palani, G. 2011. Cytotoxic effect of green synthesized silver nanoparticles using *Meliaazedarach* against in vitro HeLa cell lines and lymphoma mice model. *Process Biochem.* 47: 273–279.

Syafiuddin, A., Salmiati, Salim, M.R., Kueh, A.B.H., Hadibaratad, T., and Nur, H. 2017. A review of silver nanoparticles: research trends, global consumption, synthesis, properties, and future challenges. *J. Chin. Chem. Soc.* 64: 732–756.

Tanzil, A.H., Sultana, S.T., Saunders, S.R., Shi, L., Marsili, E., and Beyenal, H. 2016. Biological synthesis of nanoparticles in biofilms. *Enzyme Microb. Technol.* 95: 4–12.

Taran, M., Rad, M., Alavi, M. 2017. Antibacterial activity of copper oxide (CuO) nanoparticles biosynthesized by *Bacillus* sp. FU4: Optimization of experiment design. *Pharm. Sci.* 23: 198–206.

Teitzel, G.M., and Parsek, M.R. 2003. Heavy metal resistance of biofilm and planktonic *Pseudomonas aeruginosa*. *Appl. Environ. Microbiol.* 69(4): 2313–2320.

Temple, K.L., and Le Roux, N.W. 1964. Syngenesis of sulfide ores, desorption of adsorbed metal ions and their precipitation as sulfides. *Econ. Geol.* 59: 647–655.

Tyupa, D.V., Kalenov, S.V., Baurina, M.M., Yakubovich, L.M., Morozov, A.N., Zakalyukin, R.M., Sorokin, V.V., and Skladnev, D.A. 2016. Efficient continuous biosynthesis of silver nanoparticles by activated sludge micromycetes with enhanced tolerance to metal ion toxicity. *Enzyme Microb. Technol.* 95: 137–145.

Urban, M.W. 2000. Encyclopedia of Analytical Chemistry, vol. 3, Meyers, R.A. (Eds.) Wiley, Chichester, UK, 2000.

Vahabi, K., Mansoori, G.A. and Karimi, S. 2011. Biosynthesis of silver nanoparticles by fungus *Trichoderma reesei*: A route for large scale production. *Insci. J.* 1: 65–79.

Velayutham, K., Rahuman, A.A., Rajakumar, G., Santhoshkumar, T., Marimuthu, S., Jayaseelan, C, Bagavan, A., Kirthi, A.V., Kamaraj, C., Zahir, A.A., and Elango, G. 2012. Evaluation of *Catharanthus roseus* leaf extract-mediated biosynthesis of titanium dioxide nanoparticles against *Hippobosca maculata* and *Bovicola ovis*. *Parasitol. Res.* 111(6): 2329–2337.

Venkatesham, M., Ayodhya, D., Madhusudhan, D., Veera Babu, N.V., and Veerabhadram, G. 2014. A novel green one-step synthesis of silver nanoparticles using chitosan: Catalytic activity and antimicrobial studies. *Appl. Nanosci.* 4: 113–119.

Vijayan, S.R., Santhiyagu, P., Ahila, N.K., Jayaraman, R., and Ethiraj, K. 2014. Synthesis and characterization of silver and gold nanoparticles using aqueous extract of seaweed, *Turbinariaconoides* and their antimicrofouling activity. *Sci. World J.* 2014: 938272, doi: 10.1155/2014/938272

Vincy, W., Mahathalana, T.J., Sukumaran, S., and Jeeva, S. 2017. Algae as a source for synthesis of nanoparticles—A review. *Int. J. Latest Trends Eng. Technol.* Special Issue—International Conference on Nanotechnology: The Fruition of Science, pp. 5–9.

Vithiya, K., and Sen, S. 2011. Biosynthesis of nanoparticles. *Int. J. Pharm. Sci. Res.* 2(11): 2781–2785.

Vivek, R., Thangam, R., Muthuchelian, K., Gunasekaran, P., Kaveri, K., and Kannan, S. 2012. Green biosynthesis of silver nanoparticles from *Annona squamosa* leaf extract and its *in vitro* cytotoxic effect on MCF-7 cells. *Process Biochem.* 47(12): 2405–2410.

Wang, B., Chen, K., Jiang, S., Reincke, F., Tong, W., Wang, D., Gao, C.Y. 2006. Chitosan-mediated synthesis of gold nanoparticles on patterned poly(dimethylsiloxane) surfaces. *Biomacromolecules.* 7(4): 1203–1209.

Wang, C., Lu, J., Zhou, L., Li, J., Xu, J, Li, W., Zhang, L., Zhong, X., and Wang, T. 2016. Effects of long-term exposure to zinc oxide nanoparticles on development, zinc metabolism and biodistribution of minerals (Zn, Fe, Cu, Mn) in mice. *PLoS One.* 11: e0164434. https://doi.org/10.1371/journal.pone.0164434.

Wang, Z., Xu, C., Zhao, M., and Zhao, C. 2014. One-pot synthesis of narrowly distributed silver nanoparticles using phenolic-hydroxyl modified chitosan and their antimicrobial activity. *RSC Adv.* 4: 47021–47030.

Wangoo, N., Bhasin, K.K., Mehta, S.K., and Suri, C.R. 2018. Synthesis and capping of water-dispersed gold nanoparticles by an amino acid: bioconjugation and binding studies. *J. Colloid Interf. Sci.* 323(2): 247–254.

Wei, D., and Qian, W. 2008. Facile synthesis of Ag and Au nanoparticles utilizing chitosan as a mediator agent. *Colloid. Surf.* 62: 136–142.

Wei, D., Sun, W., Qian, W., Ye, Y., and Ma, X. 2009. The synthesis of chitosan-based silver nanoparticles and their antibacterial activity. *Carbohydr. Res.* 344(17): 2375–2382.

Wei, X., Zhou, H., Xu, L., Luod, M., and Liu, H. 2014. Sunlight-induced biosynthesis of silver nanoparticles by animal and fungus biomass and their characterization. *J. Chem. Technol. Biotechnol.* 89: 305–311.

Xiliang, Q., Yang, C., Tiesong, L., Peng, H., Jun, W., Ping, L., and Xiaolong, G. 2014. Large-scale synthesis of silver nanoparticles by aqueous reduction for low-temperature sintering bonding. *J. Nanomater.* 2014: 594873. http://dx.doi.org/10.1155/2014/594873

Yamamoto, O. 2001. Influence of particle size on the antibacterial activity of zinc oxide. *Int. J. Inorg. Mater.* 3: 643–646.

Yusof, H.M., Mohamad, R., Zaidan, U.H., and Abdul Rahman, N. 2019. Microbial synthesis of zinc oxide nanoparticles and their potential application as an antimicrobial agent and a feed supplement in animal industry: A review. *J. Anim. Sci. Biotechnol.* 10: 57. https://doi.org/10.1186/s40104–019-0368-z.

Zahir, A.A., and Abdul Rahuman, A. 2012. Evaluation of different extracts and synthesised silver nanoparticles from leaves of *Euphorbia prostrata* against *Haemaphysalis bispinosa* and *Hippobosca maculate*. *Vet. Parasitol.* 187(3–4): 511–520.

Zhang T., Li W., Chen V., Tang H., Li Q., Xing J., and Liu H. 2011. Enhanced biodesulfurization by magnetic immobilized *Rhodococcus erythropolis* LSSE8 assembled with nano-γ-AlO. *World J. Microbiol. Biotechnol.* 27: 299–305.

Zhang, H., Liu, Q.F., Li, Y., Li, W., Xiong, X., Xing, J., and Liu, H. 2008. Selection of adsorbents for in-situ coupling technology of adsorptive desulfurization and biodesulfurization. *Sci. China Ser. B. Chem.* 51(1): 69–77.

Zhang, H., Shan, G., Liu, H., and Xing, J. 2007. Surface modification of γ-Al_2O_3 nanoparticles with gum Arabic and its applications in adsorption and biodesulfurization. *Surf. Coat. Technol.* 201: 6917–6921.

Zhang, J., Si, G., Zou, J., Fan, R., Guo, A., and Wei, X. 2017. Antimicrobial effects of silver nanoparticles synthesized by *Fatsia japonica* leaf extracts for preservation of Citrus fruits. *J. Food Sci.* 82(8): 1861–1866.

Zikalala, N., Matshetshe, K., Parani, S., and Oluwafemi, O.S. 2018. Biosynthesis protocols for colloidal metal oxide nanoparticles. *Nano-Struct. Nano-Objects*. 16: 288–299.

Zomorodian, K., Pourshahid, S., Sadatsharifi, A., Mehryar, P., Pakshir, K., Rahimi, M.J. and Monfared, A.A. 2016. Biosynthesis and characterization of silver nanoparticles by *Aspergillus species*. *BioMed Res. Int.* 2016: 5435397. https://doi.org/10.1155/2016/5435397.

Glossary

Abandon: to cease work on a well which is nonproductive, to plug off the well with cement plugs and salvage all recoverable equipment. Also used in the context of field abandonment.

Absolute permeability: ability of a rock to conduct a fluid when only one fluid is present in the pores of the rock.

Absorber: see Absorption tower.

Absorption gasoline: gasoline extracted from natural gas or refinery gas by contacting the absorbed gas with oil and subsequently distilling the gasoline from the higher boiling components.

Absorption oil: oil used to separate the heavier components from a vapor mixture by absorption of the heavier components during intimate contacting of the oil and vapor; used to recover natural gasoline from wet gas.

Absorption plant: a plant for recovering the condensable portion of natural or refinery gas, by absorbing the higher boiling hydrocarbons in an absorption oil, followed by separation and fractionation of the absorbed material.

Absorption tower: a tower or column which promotes contact between a rising gas and a falling liquid so that part of the gas may be dissolved in the liquid.

Acid catalyst: a catalyst having acidic character; alumina is an examples of such a catalyst.

Acid deposition: acid rain; a form of pollution depletion in which pollutants, such as nitrogen oxides and sulfur oxides, are transferred from the atmosphere to soil or water; often referred to as atmospheric self-cleaning. The pollutants usually arise from the use of fossil fuels.

Acidity: the capacity of an acid to neutralize a base such as a hydroxyl ion (OH^-).

Acidizing: a technique for improving the permeability of a reservoir by injecting acid.

Acid number: a measure of the reactivity of crude oil with a caustic solution and given in terms of milligrams of potassium hydroxide that are neutralized by 1 g of crude oil.

Acid rain: the precipitation phenomenon that incorporates anthropogenic acids and other acidic chemicals from the atmosphere to the land and water (see Acid deposition).

Acid sludge: the residue left after treating petroleum oil with sulfuric acid for the removal of impurities; a black, viscous substance containing the spent acid and impurities.

Acid treating: a process in which unfinished petroleum products, such as gasoline, kerosene, and lubricating-oil stocks, are contacted with sulfuric acid to improve their color, odor, and other properties.

Acoustic log: see Sonic log.
Acre-foot: a measure of bulk rock volume where the area is one acre and the thickness is one foot.
Activation energy, *E*: the energy that is needed by a molecule or molecular complex to encourage reactivity to form products.
Additive: a material added to another (usually in small amounts) in order to enhance desirable properties or to suppress undesirable properties.
Add-on control methods: the use of devices that remove refinery process emissions after they are generated but before they are discharged to the atmosphere.
Adhesion: a property of certain dissimilar molecules that cling together due to attractive forces.
Adsorption: transfer of a substance from a solution to the surface of a solid resulting in relatively high concentration of the substance at the place of contact; see also Chromatographic adsorption.
Adsorption gasoline: natural gasoline obtained from wet gas by an adsorption process.
Afterburn: the combustion of carbon monoxide (CO) to carbon dioxide (CO_2); usually in the cyclones of a catalyst regenerator.
After flow: flow from the reservoir into the wellbore that continues for a period after the well has been shut in; after-flow can complicate the analysis of a pressure transient test.
Agglomerate: a collection of loosely bound particles or aggregates or mixtures of the two where the resulting external surface area is similar to the sum of the surface areas of the individual components.
Aggregate: a particle comprising strongly bonded or fused particles where the resulting external surface area may be significantly smaller than the sum of calculated surface areas of the individual components.
Air pollution: the discharge of toxic gases and particulate matter introduced into the atmosphere, principally as a result of human activity.
Air sweetening: a process in which air or oxygen is used to oxidize mercaptide derivatives to disulfides instead of using elemental sulfur.
Alicyclic hydrocarbon: a compound containing carbon and hydrogen only, which has a cyclic structure (e.g., cyclohexane); also collectively called naphthenes.
Aliphatic hydrocarbon: a compound containing carbon and hydrogen only, which has an open-chain structure (e.g., as ethane, butane, octane, butene) or a cyclic structure (e.g., cyclohexane).
Alkylate: the product of an alkylation process.
Alkylate (Alkylation): a refining operation that takes low value derivatives from the cat cracking and other processes and unites them in the presence of an acid catalyst to produce a very high octane, low vapor pressure gasoline blending component.
Alkylate bottoms: residua from fractionation of alkylate; the alkylate product which boils higher than the aviation gasoline range; sometimes called heavy alkylate or alkylate polymer.

Alkylation unit: a refining unit in which propylene or butylene reacts with iso-butylene to yield a high octane gasoline blending component called alkylate. Alkylate helps improve the environmental qualities of gasoline—low vapor pressure, zero sulfur content, zero olefin content, zero benzene, and a high octane number.

Alkalinity: the capacity of a base to neutralize the hydrogen ion (H^+).

Alkali treatment: see Caustic wash.

Alkali wash: see Caustic wash.

Alpha-scission: the rupture of the aromatic carbon–aliphatic carbon bond that joins an alkyl group to an aromatic ring.

Alumina (Al_2O_3): used in separation methods as an adsorbent and in refining as a catalyst.

American Society for Testing and Materials (ASTM): the official organization in the United States for designing standard tests for crude oil, fuels, and other industrial products.

Analysis: determine the properties of a feedstock prior to refining, inspection of feedstock properties.

Anticline: the structural configuration of a collection of folding rocks and in which the rocks are tilted in different directions from the crest.

Antiknock: resistance to detonation or pinging in spark-ignition engines.

Antiknock agent: a chemical compound such as tetraethyl lead which, when added in small amount to the fuel charge of an internal-combustion engine, tends to lessen knocking.

Antistripping agent: an additive used in an asphaltic binder to overcome the natural affinity of an aggregate for water instead of asphalt.

API gravity: a measure of the *lightness* or *heaviness* of crude oil which is related to density and specific gravity.

$°API = (141.5/\text{sp gr @ } 60°F) - 131.5$

Apparent bulk density: the density of a catalyst as measured; usually loosely compacted in a container.

Apparent viscosity: the viscosity of a fluid, or several fluids flowing simultaneously, measured in a porous medium (rock), and subject to both viscosity and permeability effects; also called effective viscosity.

Aquifer: a subsurface rock interval that will produce water; often the underlay of a crude oil reservoir.

Areal sweep efficiency (horizontal sweep efficiency): the fraction of the flood pattern area that is effectively swept by the injected fluids.

Aromatic hydrocarbon: a hydrocarbon characterized by the presence of an aromatic ring or condensed aromatic rings; benzene and substituted benzene, naphthalene and substituted naphthalene, phenanthrene and substituted phenanthrene, as well as the higher condensed ring systems; compounds that are distinct from those of aliphatic compounds or alicyclic compounds.

Aromatization: the conversion of nonaromatic hydrocarbons to aromatic hydrocarbons by: (1) rearrangement of aliphatic (noncyclic) hydrocarbons into aromatic ring structures; and (2) dehydrogenation of alicyclic hydrocarbons (naphthenes).

Asphaltene fraction: the brown to black powdery material produced by treatment of crude oil, crude oil residua, or bituminous materials with a low-boiling liquid hydrocarbon, e.g. pentane or heptane; soluble in benzene (and other aromatic solvents), carbon disulfide, and chloroform (or other chlorinated hydrocarbon solvents).

Asphaltene association factor: the number of individual asphaltene species which associate in nonpolar solvents as measured by molecular weight methods; the molecular weight of asphaltenes in toluene divided by the molecular weight in a polar nonassociating solvent, such as dichlorobenzene, pyridine, or nitrobenzene.

Assay: a test or series of tests performed on crude oil to determine the substance's physical and chemical properties; the assay typically identifies the viscosity, density, acidity (acid number), and the amount of sulfur.

Associated gas in solution or dissolved: natural gas dissolved in the crude oil of the reservoir, under the prevailing pressure and temperature conditions.

Associated gas: natural gas that is in contact with and/or dissolved in the crude oil of the reservoir. It may be classified as gas cap (free gas) or gas in solution (dissolved gas).

Associated molecular weight: the molecular weight of asphaltenes in an associating (nonpolar) solvent, such as toluene.

Atmospheric crude oil distillation: the process of separating crude oil components at atmospheric pressure by heating and subsequent condensing of the fractions by cooling.

Atmospheric distillation: distillation at atmospheric pressure.

Atmospheric equivalent boiling point (AEBP): a mathematical method of estimating the boiling point at atmospheric pressure of nonvolatile fractions of petroleum.

Atmospheric residuum: a residuum obtained by distillation of a crude oil under atmospheric pressure and which boils above 350 °C (660°F).

Atmospheric equivalent boiling point (AEBP): a mathematical method of estimating the boiling point at atmospheric pressure of nonvolatile fractions of crude oil.

Atom: the basic unit of matter consisting of a dense, central nucleus surrounded by a cloud of negatively charged electrons. The atomic nucleus contains a mix of positively charged protons and electrically neutral neutrons.

Atomic force microscope (AFM): a very high-resolution microscope that uses a microcantilever to scan the surface of a substrate. This microscope can image and scan surface features on the order of less than a nanometer; same as scanning force microscope.

Attapulgus clay: see Fuller's earth.

Back mixing: the phenomenon observed when a catalyst travels at a slower rate in the riser pipe than the vapors.

Baghouse: a filter system for the removal of particulate matter from gas streams; so called because of the similarity of the filters to coal bags.

Bandgap (in solid state physics and related fields): the energy range in a solid in which no electron states exist. For insulators and semiconductors, the band gap generally refers to the energy difference (in electron volts) between the top of the valence band and the bottom of the conduction band; it is the amount of energy required to free an outer shell electron from its orbit about the nucleus to a free state.

Barrel: the unit of measurement of liquids in the crude oil industry; equivalent to 42 US standard gallons or 33.6 imperial (UK) gallons (159 liters = 7.3 barrels = 1 ton: 6.29 barrels = 1 cubic meter).

Basement: foot or base of a sedimentary sequence composed of igneous or metamorphic rocks.

Basic nitrogen: nitrogen which occurs in pyridine form

Basic sediment and water (BS&W, BSW): the material which collects in the bottom of storage tanks, usually composed of oil, water, and foreign matter; also called bottoms, bottom settlings.

Basin: receptacle in which a sedimentary column is deposited that shares a common tectonic history at various stratigraphic levels.

Baumé gravity: the specific gravity of liquids expressed as degrees on the Baumé (°Bé) scale. For liquids lighter than water:

$$\text{specific gravity } 60°F = 140/(130 + °Bé)$$

For liquids heavier than water:

$$\text{specific gravity } 60°F = 145/(145 - °Bé)$$

Bbl: see Barrel.

Benzene: a colorless aromatic liquid hydrocarbon (C_6H_6).

Benzin: a refined light naphtha used for extraction purposes.

Benzine: an obsolete term for light petroleum distillates covering the gasoline and naphtha range; see Ligroine.

Benzol: the general term which refers to commercial or technical (not necessarily pure) benzene; also the term used for aromatic naphtha.

Beta-scission: the rupture of a carbon–carbon bond two bonds removed from an aromatic ring.

Bell cap: a hemispherical or triangular cover placed over the riser in a (distillation) tower to direct the vapors through the liquid layer on the tray; see Bubble cap.

Benchmarking measures: data and information used as a point of reference against which industry performance is measured.

Bentonite: montmorillonite (a magnesium–aluminum silicate); used as a treating agent.

Beta-scission: the rupture of a carbon–carbon bond two bonds removed from an aromatic ring.

Benzene: a low-boiling aromatic hydrocarbon, which occurs naturally as a part of oil and natural gas activity; considered to be a nonthreshold carcinogen and is an occupational and public health concern.

Billion: 1×10^9

Bioaccumulation: the buildup of a chemical in the tissue of organisms because they take it in faster than they can get rid of it.

Biodegradation: the destruction of organic materials by bacteria.

Biological oxidation: the oxidative consumption of organic matter by bacteria by which the organic matter is converted into gases.
Biomagnification: the process in which the concentration of a contaminant increases as it passes up the food chain.
Biomass: biological organic matter.
Biosensor: a device that combines a biological indicator with an electrical, mechanical, or chemical sensing system.
Bitumen: a naturally occurring highly viscous hydrocarbonaceous material that exists in deposits in a semisolid or solid phase. In its natural state, it generally contains sulfur, metals and other nonhydrocarbon compounds. Natural bitumen has a viscosity of more than several thousand centipoises, measured at the original temperature of the reservoir, at atmospheric pressure and gas free. It frequently requires treatment before being refined.
Bituminous: containing bitumen or constituting the source of bitumen.
Bituminous rock: see Bituminous sand.
Bituminous sand: a formation in which the bituminous material (see Bitumen) is found as a filling in veins and fissures in fractured rock or impregnating relatively shallow sand, sandstone, and limestone strata; a sandstone reservoir that is impregnated with a heavy, viscous black crude oil-like material that cannot be retrieved through a well by conventional production techniques.
Blow-down: condensate and gas is produced simultaneously from the outset of production.
Blow-out: when well pressure exceeds the ability of the wellhead valves to control it. Oil and gas "blow wild" at the surface.
Blow-out preventers: (BOPs) are high pressure wellhead valves, designed to shut off the uncontrolled flow of hydrocarbons.
Boiling point: a characteristic physical property of a liquid at which the vapor pressure is equal to that of the atmosphere and the liquid is converted to a gas.
Boiling range: the range of temperature, usually determined at atmospheric pressure in standard laboratory apparatus, over which the distillation of oil commences, proceeds, and finishes.
Borehole: the hole as drilled by the drill bit.
Bottom up: progressing from small or subordinate units to a larger and functionally richer unit.
British thermal unit: see Btu.
Btu (British thermal unit): the energy required to raise the temperature of one pound of water 1°F.
Buckyball: fullerene (a family of molecules composed entirely of carbon, in the form of a hollow sphere, ellipsoid, tube, or plane) forming a spherical shape.
Bulk composition: the make-up of petroleum in terms of bulk fractions such as *saturates*, *aromatics*, *resins*, and *asphaltenes*; separation of petroleum into these fractions is usually achieved by a combination of *solvent* and *adsorption* processes.

Bulk properties: material properties that are exhibited when the material is available in large quantities, and largely affected by nanoscale interactions.
BS&W: see Basic sediment and water.
C_1, C_2, C_3, C_4, C_5 fractions: a common way of representing fractions containing a preponderance of hydrocarbons having 1, 2, 3, 4, or 5 carbon atoms, respectively, and without reference to hydrocarbon type.
Cage compound: a polycyclic compound having the shape of a cage.
Calorific equivalence of dry gas to liquid factor (CEDGLF): the factor used to relate dry gas to its liquid equivalent. It is obtained from the molar composition of the reservoir gas, considering the unit heat value of each component and the heat value of the equivalence liquid.
Capillary forces: interfacial forces between immiscible fluid phases, resulting in pressure differences between the two phases.
Capillary number: N_c, the ratio of viscous forces to capillary forces, and equal to viscosity times velocity divided by interfacial tension.
Capillary pressure: a force per area unit resulting from the surface forces to the interface between two fluids.
Cap rock: see seal rock.
Carbon capture and storage (CCS): the process of taking waste carbon dioxide and transporting it to a storage site, normally underground in a specific type of geological formation.
Carbene: the pentane- or heptane-insoluble material that is insoluble in benzene or toluene but which is soluble in carbon disulfide (or pyridine); a type of rifle used for hunting bison.
Carboid: the pentane- or heptane-insoluble material that is insoluble in benzene or toluene and which is also insoluble in carbon disulfide (or pyridine).
Carbonization: the conversion of an organic compound into char or coke by heat in the substantial absence of air; often used in reference to the destructive distillation (with simultaneous removal of distillate) of coal.
Carbon nanotube: a form of carbon with a nanostructure that can have a length-to-diameter ratio of up to 28,000,000:1.This ratio is significantly larger than in any other material. These cylindrical carbon molecules have novel properties that make them potentially useful in applications in nanotechnology, electronics, optics, and other fields of materials science.
Carbon rejection: an upgrading process in which coke is produced, e.g., coking.
Carbon residue: the amount of carbonaceous residue remaining after thermal decomposition of crude oil, a crude oil fraction, or a crude oil product in a limited amount of air; also called the *coke-* or *carbon-forming propensity*; often prefixed by the terms Conradson or Ramsbottom in reference to the inventor of the respective tests.
Carbon-forming propensity: see Carbon residue.
Casing: the metal pipe inserted into a wellbore and cemented in place to protect both subsurface formations (such as groundwater) and the wellbore. A surface casing is set first to protect groundwater. The production casing is the last one

set. The production tubing (through which hydrocarbons flow to the surface) will be suspended inside the production casing.

Casing string: the steel tubing that lines a well after it has been drilled. It is formed from sections of steel tube screwed together.

Catalyst: a chemical agent which, when added to a reaction (process) will enhance the conversion of a feedstock without being consumed in the process.

Catalyst selectivity: the relative activity of a catalyst with respect to a particular compound in a mixture, or the relative rate in competing reactions of a single reactant.

Catalyst stripping: the introduction of steam, at a point where spent catalyst leaves the reactor, in order to strip, i.e., remove, deposits retained on the catalyst.

Catalytic activity: the ratio of the space velocity of the catalyst under test to the space velocity required for the standard catalyst to give the same conversion as the catalyst being tested; usually multiplied by 100 before being reported.

Catalytic Converter: component on the exhaust system of an internal combustion engine used to detoxify harmful emissions before exposing to environment. This device is chemical reaction driven and catalyzed with a precious metal such as platinum.

Catalytic cracking: the conversion of high-boiling feedstocks into lower boiling products by means of a catalyst that may be used in a fixed bed or fluid bed.

Cat cracking: see Catalytic cracking.

Centrifugal pump: a rotating pump, commonly used for large-volume oil and natural gas pipelines, that takes in fluids near the center and accelerates them as they move to the outlet on the outer rim.

Ceramic: an inorganic, nonmetallic solid prepared by the action of heat and subsequent cooling. A hard porous nonmetallic composite that can exhibit various material properties such as ferroelectricity and superconducting.

CFR: Code of Federal Regulations; Title 40 (40 CFR) contains the regulations for protection of the environment.

Characterization factor: the UOP characterization factor K, defined as the ratio of the cube root of the molal average boiling point, T_B, in degrees Rankine (°R = °F + 460), to the specific gravity at 60° F/60° F:

$$K = (T_B)^{1/3}/\text{sp gr}$$

The value ranges from 12.5 for paraffinic stocks to 10.0 for the highly aromatic stocks; also called the Watson characterization factor.

Chemical composition: the make-up of petroleum in terms of distinct chemical types such as paraffins, iso-paraffins, naphthenes (cycloparaffins), benzenes, di-aromatics, tri-aromatics, polynuclear aromatics; other chemical types can also be specified.

Chemical properties: the properties that describe how a substance will interact with other substances; examples of chemical properties are flammability and solubility.

Chemical vapor deposition: a chemical process used to produce high-purity, high-performance solid materials. The process is often used in the semiconductor industry to produce thin films.

Chirality: a phenomenon is said to be chiral if it is not identical to its mirror image. Here we refer to a molecular direction property that designates a "left-hand" and a "right-hand" direction where the two symmetries cannot be superposed upon one another.

Clastic: composed of pieces of pre-existing rock.

Clay mineral: a silicate mineral that usually contains aluminum and have particle sizes are less than 0.002 μm; used in separation methods as an adsorbent and in refining as a catalyst.

Cleanroom: a space designed specifically to keep all airborne particles out.

Cloud point: the temperature at which paraffin wax or other solid substances begin to crystallize or separate from the solution, imparting a cloudy appearance to the oil when the oil is chilled under prescribed conditions.

Coal: an organic rock.

Coal tar: the specific name for the tar produced from coal.

Coal tar pitch: the specific name for the pitch produced from coal.

Coke: a gray to black solid carbonaceous material produced from crude oil during thermal processing; characterized by having a high carbon content (95%+ by weight) and a honeycomb type of appearance and is insoluble in organic solvents.

Coking: a process for the thermal conversion of crude oil in which gaseous, liquid, and solid (coke) products are formed.

Colloid: a chemical mixture where one substance is dispersed evenly within another. The particles of the dispersed substance are only suspended in the mixture, unlike a solution, where they are completely dissolved within.

Commercial field: an oil and/or gas field judged to be capable of producing enough net income to make it worth developing.

Completion: the installation of permanent wellhead equipment for the production of oil and gas.

Completion interval: the portion of the reservoir formation placed in fluid communication with the well by selectively perforating the wellbore casing.

Complex: a series of fields sharing common surface facilities.

Composition: the general chemical make-up of a crude oil or a fuel.

Compressor: a device installed in the gas pipeline to raise the pressure and guarantee the fluid flow through the pipeline.

Con Carbon: see Carbon residue.

Condensate: a mixture of light hydrocarbon liquids obtained by condensation of hydrocarbon vapors: predominately butane, propane, and pentane with some heavier hydrocarbons and relatively little methane or ethane; see also Natural gas liquids.

Conductivity: a measure of the ease of flow through a fracture, perforation, or pipe.

Conformance: the uniformity with which a volume of the reservoir is swept by injection fluids in area and vertical directions.

Connate water: water trapped in the pores of a rock during formation of the rock; also . described as fossil water. The chemistry of connate water can change in composition throughout the history of the rock. Connate water can be dense, and saline compared with seawater. Formation water, or interstitial water, in contrast, is simply water found in the pore spaces of a rock, and might not have been present when the rock was formed. See Formation water.

Conradson carbon residue: see Carbon residue.

Contingent resource: the amounts of hydrocarbons estimated at a given date, and which are potentially recoverable from known accumulations, but are not considered commercially recoverable under the economic evaluation conditions corresponding to such date.

Conventional crude oil: crude oil having an API gravity greater than 20°.; crude oil that occurs in liquid form, flowing naturally or capable of being pumped without further processing or dilution.

Conventional limit: the reservoir limit established according to the degree of knowledge of, or research into the geological, geophysical or engineering data available.

Core: a cylindrical rock sample taken from a formation when drilling in order to determine its permeability, porosity, hydrocarbon saturation, and other productivity-associated properties.

Core floods: laboratory flow tests through samples (cores) of porous rock.

Cp (centipoise): a unit of viscosity.

Cracking: the thermal processes by which the constituents of crude oil are converted to lower molecular weight products.

Cracking activity: see Catalytic activity.

Cracking coil: equipment used for cracking heavy crude oil products consisting of a coil of heavy pipe running through a furnace so that the oil passing through it is subject to high temperature.

Cracking temperature: the temperature (350 °C; 660°F) at which the rate of thermal decomposition of crude oil constituents becomes significant.

Criteria air contaminants (CAC): emissions of various air pollutants that affect our health and contribute to air pollution problems such as smog.

Crude assay: a procedure for determining the general distillation characteristics (e.g., distillation profile and other quality information of crude oil.

Crude oil: see Crude oil.

Crude oil refining: an integrated sequence of unit processes that results in the production of a variety of products.

Cryogenic plant: processing plant capable of producing liquid natural gas products, including ethane, at very low operating temperatures.

Cryogenics: the study, production and use of low temperatures.

Crystalline: a solid material whose constituent atoms, molecules, or ions are arranged in an orderly repeating pattern extending in all three spatial dimensions.

Cumulative effects: changes to the environment caused by an activity in combination with other past, present, and reasonably foreseeable human activities.

Cumulative production: production of crude oil, heavy oil, extra heavy oil, or tar sand bitumen to date.

Cut point: the boiling-temperature division between distillation fractions of crude oil.

Cuttings: the rock chippings cut from the formation by the drill bit, and brought to the surface with the mud. Used by geologists to obtain formation data.

Cyclic steam injection: the alternating injection of steam and production of oil with condensed steam from the same well or wells.

Cyclic steam stimulation (CSS): injecting steam into a well in a heavy-oil reservoir which introduces heat and thins the oil, allowing it to flow through the same well.

Deactivation: reduction in catalyst activity by the deposition of contaminants (e.g., coke, metals) during a process.

Deasphaltened oil: the fraction of crude oil after the asphaltene constituents have been removed.

Deasphaltening: removal of a solid powdery asphaltene fraction from crude oil by the addition of the low-boiling liquid hydrocarbons such as *n*-pentane or *n*-heptane under ambient conditions.

Deasphalting: the removal of the asphaltene fraction from crude oil by the addition of a low-boiling hydrocarbon liquid such as *n*-pentane or *n*-heptane; more correctly the removal asphalt (tacky, semisolid) from crude oil (as occurs in a refinery asphalt plant) by the addition of liquid propane or liquid butane under pressure.

Dendrite (crystal): a crystal that grows in a snowflake pattern or a tree branching pattern.

Density: the mass (or weight) of a unit volume of any substance at a specified temperature; see also Specific gravity.

Derrick: the tower-like structure that houses most of the drilling controls.

Desalting: removal of mineral salts (mostly chlorides) from crude oils.

Desulfurization: the removal of sulfur or sulfur compounds from a feedstock.

Dew point pressure: pressure at which the first drop of liquid is formed, when it goes from the vapor phase to the two-phase region.

Dielectric: an insulating material in which electrons are bound and unable to freely move within a substrate. A nonconducting substance, i.e., an insulator.

Differential-strain analysis: measurement of thermal stress relaxation in a recently cut well.

Diluents: low-boiling liquids used to higher-boiling liquids so that they can flow through pipelines.

Dispersion: a measure of the convective mi fluids due to flow in a reservoir.

Directional well (deviated) well: a well drilled at an angle from the vertical by using a slanted drilling rig or by deflecting the drill bit; directional wells are used to drill multiple wells from a common drilling pad or to reach a subsurface location beneath land where drilling cannot be done.

DNA: abbreviation of *deoxyribonucleic acid*: a nucleic acid that contains the genetic instructions used in the development and functioning of all known living organisms and some viruses.

Dome: geological structure with a semi-spherical shape or relief.

Donor solvent process: a conversion process in which hydrogen donor solvent is used in place of or to augment hydrogen.

Downhole: a term used to describe tools, equipment, and instruments used in the wellbore, or conditions or techniques applying to the wellbore.

Downhole steam generator: a generator installed downhole in an oil well to which oxygen-rich air, fuel, and water are supplied for the purposes of generating steam for it into the reservoir. Its major advantage over a surface steam generating facility is the losses to the wellbore and surrounding formation are eliminated.

Downstream: when referring to the oil and gas industry, this term indicates the refining and marketing sectors of the industry. More generically, the term can be used to refer to any step further along in the process.

Downstream sector: the refining and marketing sector of the crude oil industry.

Drainage radius: distance from which fluids flow to the well, that is, the distance reached by the influence of disturbances caused by pressure drops.

DRAM (dynamic random access memory): a memory that stores each bit of data in a separate capacitor within an integrated circuit. Since real capacitors leak charge, the information eventually fades unless the capacitor charge is refreshed periodically. Because of this refresh requirement, it is a dynamic memory as opposed to SRAM and other types of static memory.

Drill cuttings: the small pieces of rock created as a drill bit moves through underground formations while drilling.

Drilling rig: a drilling unit that is not permanently fixed to the seabed, e.g., a drillship, a semi-submersible or a jack-up unit. Also means the derrick and its associated machinery.

Drill stem test (formation test): conventional formation test method; a method of formation testing. The basic drill stem test tool consists of a packer or packers, valves or ports that may be opened and closed from the surface, and two or more pressure-recording devices. The tool is lowered on the drill string to the zone to be tested. The packer or packers are set to isolate the zone from the drilling fluid column.

Dry gas equivalent to liquid (DGEL): volume of crude oil that because of its heat rate is equivalent to the volume of dry gas.

Dry gas: natural gas containing negligible amounts of hydrocarbons heavier than methane. Dry gas is also obtained from the processing complexes.

Dry hole: any exploratory or development well that does not find commercial quantities of hydrocarbons.

Dubbs cracking: an older continuous, liquid-phase thermal cracking process formerly used.

Dykstra–Parsons coefficient: an index of reservoir heterogeneity arising from permeability variation and stratification.

Ebullated bed: a process in which the catalyst bed is in a suspended state in the reactor by means of a feedstock recirculation pump which pumps the feedstock upward at sufficient speed to expand the catalyst bed at approximately 35% above the settled level.

Elasticity: a material property that allows deformation under stress and reformation when stress is released.

Electron: subatomic particle that carries negative electric charge. The number of electrons in an atom and their energy levels determine many of the electrical properties of material. The electron is not known to have substructure; that is, it is not known to be made up of smaller particles.

Electron beam lithography: the practice of scanning a beam of electrons in a patterned fashion across a surface covered with a film (called the resist) ("exposing" the resist) and of selectively removing either exposed or nonexposed regions of the resist ("developing"). The purpose is to create very small structures in the resist that can subsequently be transferred into another material for a number of purposes, for example, for the creation of very small electronic devices.

Energy: a scalar physical quantity that describes the amount of work that can be performed by a force. Several different forms of energy exist, including kinetic, potential, thermal, gravitational, sound, light, elastic, and electromagnetic energy.

Entrained bed: a bed of solid particles suspended in a fluid (liquid or gas) at such a rate that some of the solid is carried over (entrained) by the fluid.

Environmetnal footprint: the impact on the environment, in terms of resource use, waste generation and changes to the physical environment.

Ethyl alcohol (ethanol or grain alcohol): an inflammable organic compound (C_2H_5OH) formed during fermentation of sugars; used as an intoxicant and as a fuel.

Evaporites: sedimentary formations consisting primarily of salt, anhydrite or gypsum, as a result of evaporation in coastal waters.

Expanding clays: clays that expand or swell on contact with water, e.g., montmorillonite.

Exploratory well: a well that is drilled without detailed knowledge of the underlying rock structure in order to find hydrocarbons whose exploitation is economically profitable.

Extra heavy oil: crude oil with relatively high fractions of heavy components, high specific gravity (low API density) and high viscosity at reservoir conditions..

Facies: one or more layers of rock that differs from other layers in composition, age or content.

FAST: fracture-assisted steam flood technology.

Faujasite: a naturally occurring silica–alumina (SiO_2–Al_2O_3) mineral.

Fault: fractured surface of geological strata along which there has been differential movement. Fluid saturation: Portion of the pore space occupied by a specific fluid; oil, gas and water may exist.

FCC: fluid catalytic cracking.

Feedstock: crude oil as it is fed to the refinery; a refinery product that is used as the raw material for another process; the term is also generally applied to raw materials used in other industrial processes.

Ferroelectricity: a material property that is characterized by natural electric polarizability that can be altered by an external electric field. The term is used in analogy to ferromagnetism, in which a material exhibits a permanent magnetic moment.

Filtration: a mechanical or physical operation which is used for the separation of solids from fluids (liquids or gases) by interposing a medium through which the fluid can pass, but the solids (or at least part of the solids) in the fluid are retained.

Fixed bed: a stationary bed (of catalyst) to accomplish a process (see Fluid bed).

Flaring/Venting: the controlled burning (flare) or release (vent) of natural gas that can't be processed for sale or use because of technical or economic reasons.

Flexicoking: a modification of the fluid coking process insofar as the process also includes a gasifier adjoining the burner/regenerator to convert excess coke to a clean fuel gas.

Flow line: pipe, usually buried, through which oil or gas travels from the well to a processing facility.

Flue gas: gas from the combustion of fuel, the heating value of which has been substantially spent and which is, therefore, discarded to the flue or stack.

Fluid: a reservoir gas or liquid.

Fluid bed: the use of an agitated bed of inert granular material to accomplish a process in which the agitated bed resembles the motion of a fluid; a bed (of catalyst) that is agitated by an upward passing gas in such a manner that the particles of the bed simulate the movement of a fluid and has the characteristics associated with a true liquid. See Fixed bed.

Fluid catalytic cracking: cracking in the presence of a fluidized bed of catalyst.

Fluid coking: a continuous fluidized solids process that cracks feed thermally over a bed of coke particles.

Fluid injection: the injection of gases or liquids into a reservoir to force oil toward and into producing wells

Formation: an interval of rock with distinguishable geologic characteristics.

Formation damage: the reduction in permeability in reservoir rock due to the infiltration of drilling or treating fluids into the area adjacent to the wellbore.

Formation pressure: the pressure at the bottom of a well when it is shut in at the wellhead.

Formation resistance factor (F): ratio between the resistance of rock saturated 100 percent with brine divided by the resistance of the saturating water.

Formation volume factor (B): the factor that relates the volume unit of the fluid in the reservoir with the surface volume. There are volume factors for oil, gas, in both phases, and for water. A sample may be directly measured, calculated or obtained through empirical correlations.

Formation water (interstitial water): salt water underlying gas and oil in the formation; water found in the pore spaces of a rock and might not have been present when the rock was formed. See Connate water.

Fossil fuel resources: a gaseous, liquid, or solid fuel material formed in the ground by chemical and physical changes (diagenesis) in plant and animal residues over geological time; natural gas, crude oil, coal, and oil shale.

Fractional composition: the composition of crude oil as determined by fractionation (separation) methods.

Fracturing: a method of breaking down a formation by pumping fluid at very high pressure. The objective is to increase production rates from a reservoir.

FRAM (Ferroelectric RAM (random access memory): a memory that uses a ferroelectric layer rather than a dielectric layer to achieve nonvolatility. A nonvolatile memory will retain the stored information even if it is not constantly supplied with electric power.

Free associated gas: natural gas that overlies and is in contact with the crude oil of the reservoir. It may be gas cap.

Fullerene: a family of molecules composed entirely of carbon, in the form of a hollow sphere, ellipsoid, tube, or plane. Spherical fullerenes are also called buckyballs, and cylindrical ones are called carbon nanotubes or buckytubes. Graphene is an example of a planar fullerene sheet.

Gas cap: a part of a hydrocarbon reservoir at the top that will produce only gas.

Gas compressibility ratio (*Z*): the ratio between an actual gas volume and an ideal gas volume. This is a dimensional amount that usually varies between 0.7 and 1.2.

Gas field: a field containing natural gas but no crude oil.

Gasification: a process to partially oxidize any hydrocarbon, typically heavy residues, to a mixture of hydrogen and carbon monoxide; can be used to produce hydrogen and various energy byproducts.

Gas injection: the process whereby separated associated gas is pumped back into a reservoir for conservation purposes or to maintain the reservoir pressure.

Gas lift: artificial production system that is used to raise the well fluid by injecting gas down the well through tubing, or through the tubing-casing annulus.

Gasohol: a term for motor vehicle fuel comprising between 80-90% unleaded gasoline and 10%–20% ethanol (see also Ethyl alcohol).

Gas-oil ratio (GOR): ratio of reservoir gas production to oil production, measured at atmospheric pressure.

Gas-to-liquids (GTL): the conversion of natural gas to a liquid form so that it can be transported easily; typically, the liquid is converted back to natural gas prior to consumption.

Geological province: a region of large dimensions characterized by similar geological and development histories.

Glycol dehydrator: field equipment used to remove water from natural gas by using triethylene glycol or diethylene glycol.

Graben: dip or depression formed by tectonic processes, limited by normal type faults.

Graphene: a single sheet of trigonally bonded (sp^2) carbon atoms in a hexagonal structure.

Gravitational segregation: reservoir driving mechanism in which the fluids tend to separate according to their specific gravities. For example, since oil is heavier than water it tends to move toward the lower part of the reservoir in a water injection project.

Gravity: see API gravity.

Gravity drainage: the movement of oil in a reservoir that results from the force of gravity.

Gravity segregation: partial separation of fluids in a reservoir caused by the gravity force acting on differences in density.
Gravity-stable displacement: the displacement of oil from a reservoir by a fluid of a different density, where the density difference is utilized to prevent gravity segregation of the injected fluid.
Greenhouse effect: warming of the earth due to entrapment of the energy of the Sun by the atmosphere.
Greenhouse gases (GHGs): a type of gas that contributes to the greenhouse effect by absorbing infrared radiation.
Greenhouse gas intensity (GHG intensity): the average emission rate of a given greenhouse gas from a specific source. For example: greenhouse gases released per barrel of production.
Guard bed: a bed of disposal adsorbent used to protect process catalysts from contamination by feedstock constituents.
Handling efficiency shrinkage factor (HESF): this is a fraction of natural gas that is derived from considering self-consumption and the lack of capacity to handle such. It is obtained from the gas-handling statistics of the final period in the area corresponding to the field being studied.
HCPV: hydrocarbon pore volume.
Hearn method: a method used in reservoir simulation for calculating a pseudo relative permeability curve that reflects reservoir stratification.
Heat value: the amount of heat released per unit of mass, or per unit of volume, when a substance is completely burned. The heat power of solid and liquid fuels is expressed in calories per gram or in BTU per pound. For gases, this parameter is generally expressed in kilocalories per cubic meter or in BTU per cubic foot.
Heavy ends: the highest boiling portion of a crude oil fraction; see also Light ends.
Heteroatom compounds: chemical compounds which contain nitrogen and/or oxygen and/or sulfur and /or metals bound within their molecular structure(s).
Heterogeneity: lack of uniformity in reservoir properties such as permeability.
Higgins–Leighton model: stream tube computer model used to simulate waterflood.
Horizontal drilling: drilling a well that deviates from the vertical and travels horizontally through a producing layer.
Horst: bock of the earth's crust rising between two faults; the opposite of a graben.
Huff-and-puff: a cyclic enhanced oil recovery method in which steam or gas is injected into a production well; after a short shut-in period, oil and the injected fluid are produced through the same well.
Hydration: the association of molecules of water with a substance.
Hydraulic fracturing: the opening of fractures in a reservoir by high-pressure, high-volume injection of liquids through an injection well which causes the surrounding rocks to crack and allows natural gas or oil to be produced from tight formations.; also known as fracking.
Hydrocarbon index: an amount of hydrocarbons contained in a reservoir per unit area.
Hydrocarbon compounds: chemical compounds containing only carbon and hydrogen.

Hydrocarbons: chemical compounds fully constituted by hydrogen and carbon.

Hydrocracking: a catalytic high-pressure high-temperature process for the conversion of crude oil feedstocks in the presence of fresh and recycled hydrogen; carbon-carbon bonds are cleaved in addition to the removal of heteroatomic species.

Hydrocracking catalyst: a catalyst used for hydrocracking which typically contains separate hydrogenation and cracking functions.

Hydrodenitrogenation: the removal of nitrogen by hydrotreating.

Hydrodesulfurization: the removal of sulfur by hydrotreating.

Hydrogen addition: an upgrading process in the presence of hydrogen, e.g., hydrocracking; see Hydrogenation.

Hydrogenation: the chemical addition of hydrogen to a material. In nondestructive hydrogenation, hydrogen is added to a molecule only if, and where, unsaturation with respect to hydrogen exists.

Hydrogen transfer: the transfer of inherent hydrogen within the feedstock constituents and products during processing.

Hydrophilicity/Hydrophilic: the tendency of a molecule to be solvated by water. Also refers to a physical property of a molecule that can bond with water.

Hydrophobicity/Hydrophobe: the physical property of a molecule (known as a hydrophobe) that is repelled from a mass of water.

Hydroprocessing: a term often equally applied to hydrotreating and to hydrocracking; also, often collectively applied to both.

Hydrotreating: the removal of heteroatomic (nitrogen, oxygen, and sulfur) species by treatment of a feedstock or product at relatively low temperatures in the presence of hydrogen.

Hydrovisbreaking: a noncatalytic process, conducted under similar conditions to visbreaking, which involves treatment with hydrogen to reduce the viscosity of the feedstock and produce more stable products than is possible with visbreaking.

Immiscible: two or more fluids that do not have complete mutual solubility and co-exist as separate phases.

Immiscible carbon dioxide displacement: injection of carbon dioxide into an oil reservoir to effect oil displacement under conditions in which miscibility with reservoir oil is not obtained; see Carbon dioxide augmented waterflooding.

Immiscible displacement: a displacement of oil by a fluid (gas or water) that is conducted under conditions so that interfaces exist between the driving fluid and the oil.

Impurities and plant liquefiables shrinkage factor (IPLSF): it is the fraction obtained by considering the nonhydrocarbon gas impurities (sulfur, carbon dioxide, nitrogen compounds, etc.) contained in the sour gas, in addition to shrinkage caused by the generation of liquids in gas processing plant.

Impurities shrinkage factor (ISF): it is the fraction that results from considering the nonhydrocarbon gas impurities (sulfur, carbon dioxide, nitrogen compounds, etc.) contained in the sour gas. It is obtained from the operation statistics of the last annual period of the gas processing complex (GPC) that processes the production of the field analyzed.

Incompatibility: the immiscibility of crude oil products and also of different crude oils which is often reflected in the formation of a separate phase after mixing and/or storage.

Initial boiling point: the recorded temperature when the first drop of liquid falls from the end of the condenser.

Initial vapor pressure: the vapor pressure of a liquid of a specified temperature and zero per cent evaporated.

Injection profile: the vertical flow rate distribution of fluid flowing from the wellbore into a reservoir.

Injection well: a well in an oil field used for injecting fluids into a reservoir.

Injectivity: the relative ease with which a fluid is injected into a porous rock.

Instability: the inability of a crude oil product to exist for periods of time without change to the product.

Integrity: maintenance of a slug or bank at its preferred composition without too much dispersion or mixing.

Interface: the thin surface area separating two immiscible fluids that are in contact with each other.

Interfacial film: a thin layer of material at the interface between two fluids which differs in composition from the bulk fluids.

Interfacial tension: the strength of the film separating two immiscible fluids, e.g., oil and water or microemulsion and oil; measured in dynes (force) per centimeter or milli-dynes per centimeter.

Interfacial viscosity: the viscosity of the interfacial film between two immiscible liquids.

Interference testing: a type of pressure transient test in which pressure is measured over time in a closed-in well while nearby wells are produced; flow and communication between wells can sometimes be deduced from an interference test.

Interphase mass transfer: the net transfer of chemical compounds between two or more phases.

Interstitial water (formation water): salt water underlying gas and oil in the formation; water found in the pore spaces of a rock and might not have been present when the rock was formed. See Connate water.

Jacket: the lower section, or legs, of an offshore platform.

Kaolinite: a clay mineral formed by hydrothermal activity at the time of rock formation or by chemical weathering of rock with high feldspar content; usually associated with intrusive granite rock with high feldspar content.

Kata-condensed aromatic compounds: compounds based on linear condensed aromatic hydrocarbon systems, e.g., anthracene and naphthacene (tetracene).

Kerogen: a complex carbonaceous (organic) material that occurs in sedimentary rock and shale; generally insoluble in common organic solvents; produces hydrocarbons when subjected to a heat.

K-factor: see Characterization factor.

Kinematic viscosity: the ratio of viscosity to density, both measured at the same temperature.

Kinetic Energy: the extra energy which it possesses due to its motion. It is defined as the work needed to accelerate a body of a given mass from rest to its current velocity.

LASER: abbreviation of "Light Amplification of Stimulated Emission Radiation"; a device that emits light (electromagnetic radiation) through a process called stimulated emission. Laser light is usually spatially coherent, which means that the light either is emitted in a narrow, low-divergence beam, or can be converted into one with the help of optical components such as lenses.

Lewis acid: a chemical species that can accept an electron pair from a base.

Lewis base: a chemical species that can donate an electron pair.

Light crude oil: the specific gravity of the oil is more than 25° API, but less than or equal to 38°. See Medium crude oil.

Light ends: the lower boiling components of a mixture of hydrocarbons; see also Heavy ends, Light hydrocarbons.

Limolite: fine grain sedimentary rock that is transported by water. The granulometrics ranges from fine sand to clay.

Limestone: calcium carbonate-rich sedimentary rocks in which oil or gas reservoirs are often found.

Liquefied natural gas (LNG): oilfield or naturally occurring gas, chiefly methane, liquefied for transportation.

Liquefied petroleum gas (LPG): light hydrocarbon material, gaseous at atmospheric temperature and pressure, held in the liquid state by pressure to facilitate storage, transport and handling. Commercial liquefied gas consists essentially of either propane or butane, or mixtures thereof.

Liquid crystal display (LCD): an electronically modulated optical device shaped into a thin, flat panel made up of color or monochrome pixels filled with liquid crystals. It is arrayed in front of a light source (backlight) or reflector and is often used in battery-powered electronic devices because it uses very small amounts of electric power.

Lithography (Photolithography): method of fabrication of integrated circuits and microelectromechanical systems that uses alternating steps of material deposition and removal. The process selectively removes parts of a thin film or the bulk of a substrate. It uses light to transfer a geometric pattern from a photo mask to a light-sensitive chemical photo resist on the substrate. A series of chemical treatments then engrave the exposure pattern into the material underneath the photo resist.

Lithology: the geological characteristics of the reservoir rock.

Lorenz coefficient: a permeability heterogeneity factor.

Magnetic memory: storage of information on magnetized material such as magnetic tape. Magnetic storage uses different patterns of magnetization in a magnetizable material to store data and is a form of nonvolatile memory (a memory that will retain the stored information even if it is not constantly supplied with electric power.) An example of a magnetic memory is a computer hard disk drive.

Maltenes (malthenes): that fraction of crude oil that is soluble in, for example, pentane or heptane; deasphaltened oil; also the term arbitrarily assigned to the pentane-soluble portion of crude oil that is relatively high boiling (>300°C, 760 mm) (see also Petrolenes).

Materials characterization: the process(es) used for analyzing the structure and properties of a material.

Materials science: an interdisciplinary field involving the properties of matter and its applications to several areas of science and engineering. Materials Science investigates the relationship between the structure of materials at atomic or molecular scales and their macroscopic properties. It includes elements of applied physics and chemistry, as well as chemical, mechanical, civil and electrical engineering.

Marx–Langenheim model: mathematical equations for calculating heat transfer in a hot water or steam flood.

Medium crude oil: liquid crude oil with a density between that of light and heavy crude oil; the specific gravity of the oil is more than 20° API, but less than or equal to 25°. See Light crude oil.

Melting point: the temperature point in which a material transitions from solid to liquid.

MEOR: microbial enhanced oil recovery.

Mesoporous: a material possessing pores with at least one dimension between 2 nm to 50 nm

Metagenesis: the alteration of organic matter during the formation of crude oil that may involve temperatures above 200 °C (390°F); see also Catagenesis and Diagenesis.

Metamorphic: group of rocks resulting from the transformation that commonly takes place at great depths due to pressure and temperature. The original rocks may be sedimentary, igneous or metamorphic.

Methane: the principal constituent of natural gas; the simplest hydrocarbon molecule, containing one carbon atom and four hydrogen atoms.

Metric tonne: equivalent to 1000 kilos, 2204.61 lbs; 7.5 barrels.

Mica: a complex aluminum silicate mineral that is transparent, tough, flexible, and elastic.

Micellar fluid (surfactant slug): an aqueous mixture of surfactants, co-surfactants, salts, and hydrocarbons. The term micellar is derived from the word micelle, which is a submicroscopic aggregate of surfactant molecules and associated fluid.

Micelle: an aggregation of surfactant molecules dispersed in a liquid.

Microcarbon residue: the carbon residue determined using a thermogravimetric method. See also Carbon residue.

Microemulsion: a stable, finely dispersed mixture of oil, water, and chemicals (surfactants and alcohols).

Microemulsion or micellar/emulsion flooding: an augmented waterflooding technique in which a surfactant system is injected in order to enhance oil displacement toward producing wells.

Microorganisms: animals or plants of microscopic size, such as bacteria.

Microscopic displacement efficiency: the efficiency with which an oil displacement process removes the oil from individual pores in the rock.

Mid-boiling point: the temperature at which approximately 50% of a material has distilled under specific conditions.

Middle distillates: medium-density refined crude oil products, including kerosene, stove oil, jet fuel and light fuel oil.

Middle-phase microemulsion: a microemulsion phase containing a high concentration of both oil and water that, when viewed in a test tube, resides in the middle with the oil phase above it and the water phase below it.

Midstream: the processing, storage and transportation sector of the crude oil industry.

Migration (primary) the movement of hydrocarbons (oil and natural gas) from mature, organic-rich source rocks to a point where the oil and gas can collect as droplets or as a continuous phase of liquid hydrocarbon.

Migration (secondary): the movement of the hydrocarbons as a single, continuous fluid phase through water-saturated rocks, fractures, or faults followed by accumulation of the oil and gas in sediments (traps) from which further migration is prevented.

Millipede memory (IBM): a nonvolatile memory developed by IBM that uses nano-imprints to code information and atomic force sensing to decode information. It promises a data density of more than 1 terabit per square inch (1 gigabit per square millimeter), about 4 times the density of magnetic storage available today.

Mineral hydrocarbons: crude oil hydrocarbons, considered *mineral* because they come from the earth rather than from plants or animals.

Mineral oil: the older term for crude oil; the term was introduced in the nineteenth century as a means of differentiating crude oil (rock oil) from whale oil which, at the time, was the predominant illuminant for oil lamps.

Minerals: naturally occurring inorganic solids with well-defined crystalline structures.

Mineral seal oil: a distillate fraction boiling between kerosene and gas oil.

Mine tailings: process water remaining after tar sand mining and stored in settling basins called tailings ponds.

Minimum miscibility pressure (MMP): see Miscibility.

Miscibility: an equilibrium condition, achieved after mixing two or more fluids, which is characterized by the absence of interfaces between the fluids: (1) *first-contact miscibility:* miscibility in the usual sense, whereby two fluids can be mixed in all proportions without any interfaces forming. Example: at room temperature and pressure, ethyl alcohol and water are first -contact miscible. (2) *multiple-contact miscibility (dynamic miscibility):* miscibility that is developed by repeated enrichment of one fluid phase with components from a second fluid phase with which it comes into contact. (3) *minimum miscibility* pressure: the minimum pressure above which two fluids become miscible at a given temperature, or can become miscible, by dynamic processes.

MMBOE: Million Barrels Oil Equivalent.

MMcf: Millions of cubic feet per day (of gas).

Mobility: a measure of the ease with which a fluid moves through reservoir rock; the ratio of rock permeability to apparent fluid viscosity.
Mobility buffer: the bank that protects a chemical slug from water invasion and dilution and assures mobility control.
Mobility control: ensuring that the mobility of the displacing fluid or bank is equal to or less than that of the displaced fluid or bank.
Mobility ratio: ratio of mobility of an injection fluid to mobility of fluid being displaced.
Modified alkaline flooding: the addition of a co-surfactant and/or polymer to the alkaline flooding process.
Molecular sieve: a synthetic zeolite mineral having pores of uniform size; it is capable of separating molecules, on the basis of their size, structure, or both, by absorption or sieving.
Molecule: a unit of two or more atoms held together by covalent bonds (a form of chemical bonding that is characterized by the sharing of pairs of electrons between atoms, or between atoms and other covalent bonds)
Mud (drilling mud): fluid circulated down the drill pipe and up the annulus during drilling to remove cuttings, cool and lubricate the bit, and maintain desired pressure in the well.
Muskeg: a water-soaked layer of decaying plant material, three-to-ten feet thick, found on top of the overburden.
Nanobiosensor: a device that combines a biological indicator with an electrical, mechanical, or chemical sensing system on the nanoscale.
Nanobiotechnology: biotechnology at the nanoscale; it includes the application of the tools and processes of nanotechnology to study and manipulate biological systems.
Nanocluster: a noncovalently or covalently bound group of atoms or molecules whose largest overall dimension is typically in the nanoscale.
Nanofiber: a flexible nanorod.
Nanocrystal: a single crystalline material that has one dimension on the order of 100 nm.
Nanomaterial: a material that exhibits distinct properties when studied on the order of less than 100 nm.
Nanometer (nm): a unit of length in the metric system, equal to one billionth of a meter (i.e., 10^{-9} m or one millionth of a millimeter).
Nanoparticle: a particle with a size on the order of 1-100 nm.
Nanoplate: a nano-object with one external dimension in the nanoscale and the two other external dimensions significantly larger mass of dry nanoparticles
Nanopore: a small pore in an electrically insulating membrane that can be used as a single-molecule detector.
Nanoporous: a material possessing pores with at least one dimension in the nanoscale.
Nanorod: a nano-object with two similar external dimensions in the nanoscale and the third dimension significantly larger than the other two external dimensions.
Nanoscale: a term is used to refer to objects with dimensions on the order 1–100 nm.
Nanostructure: nanoscale structure.

Nanotechnology: the study of materials and properties on the order of 1-100nm.

Nanowhisker: a nanoscale structure that consists of brushes attached along a common spine.

Nanowire: a wire of diameter on the order of a nanometer.

Naphtha: any of various volatile, often flammable, liquid hydrocarbon mixtures used chiefly as solvents and diluents boiling below 200 °C (390°F); used as a blend stock for gasoline manufacture.

Native asphalt: see Bitumen.

Natural gas: mixture of hydrocarbons existing in reservoirs in the gaseous phase or in solution in the oil, which remains in the gaseous phase under atmospheric conditions. It may contain some impurities or nonhydrocarbon substances (hydrogen sulfide, nitrogen or carbon dioxide).

Natural gas liquids (NGLs): the hydrocarbon liquids that condense during the processing of hydrocarbon gases that are produced from oil or gas reservoir; see also Natural gasoline.

Natural gasoline: a mixture of liquid hydrocarbons extracted from natural gas suitable for blending with refinery gasoline.

Net thickness: the thickness resulting from subtracting the portions that have no possibilities of producing hydrocarbon from the total thickness.

Nonassociated gas: the natural gas found in reservoirs that do not contain crude oil at the original pressure and temperature conditions.

Nonionic surfactant: a surfactant molecule containing no ionic charge.

Nonvolatile memory: a memory that will retain the stored information even if it is not constantly supplied with electric power.

Normal fault: the result of the downward displacement of one of the blocks from the horizontal. The angle is generally between 25° and 60° and it is recognized by the absence of part of the stratigraphic column.

Nucleation: a site in which a phase transition begins and grows outward from. Some examples of phases that may form via nucleation in liquids are gaseous bubbles, crystals, or glassy regions.

Observation wells: wells that are completed and equipped to measure reservoir conditions and/or sample reservoir fluids, rather than to inject or produce reservoir fluids.

Oil: typically, conventional crude oil that exists in the liquid phase in reservoirs and remains as such under original pressure and temperature conditions. Small amounts of nonhydrocarbon substances may be included. Also, the term *oil* is used in the context of this book as a generic term to include heavy oil, extra heavy oil, and tar sand bitumen and it is not intended to be used as a means for definition of these resources.

Optics: the study of the behavior and properties of light including its interactions with matter and its detection by instruments.

Optimum salinity: the salinity at which a middle-phase microemulsion containing equal concentrations of oil and water results from the mixture of a micellar fluid (surfactant slug) with oil.

Optoelectronics: the study and application of electronic devices that source, detect and control light (including invisible forms of radiation such as gamma rays, X-rays, ultraviolet and infrared, in addition to visible light). Optoelectronic devices are electrical-to-optical or optical-to-electrical transducers, or instruments that use such devices in their operation.

Optoelectronics: the study and application of electronic devices that source, detect and control light (including invisible forms of radiation such as gamma rays, X-rays, ultraviolet and infrared, in addition to visible light). Optoelectronic devices are electrical-to-optical or optical-to-electrical transducers, or instruments that use such devices in their operation.

Organic sedimentary rocks: rocks containing organic material such as residues of plant and animal remains/decay.

Original gas volume in place: amount of gas that is estimated to exist initially in the reservoir and that is confined by geologic and fluid boundaries, which may be expressed at reservoir or atmospheric conditions.

Original oil volume in place: amount of crude oil that is estimated to exist initially in the reservoir and that is confined by geologic and fluid boundaries, which may be expressed at reservoir or atmospheric conditions.

Original pressure: pressure prevailing in a reservoir that has never been produced. It is the pressure measured by a discovery well in a producing structure.

Ozone: ground-level ozone is a colorless gas that forms just above the earth's surface.

Particulate matter: the microscopic solid or liquid particles that remain suspended in the air for some time.

PDMS (Polydimethylsiloxane): an inorganic polymer used in nanotechnology applications such as nanoimprint and soft lithography. It is the most widely used silicon-based organic polymer.

Permeability: the capacity of a reservoir rock to transmit fluids; rock property for permitting a fluid pass; a factor that indicates whether a reservoir has producing characteristics or not.

Petroleum (crude oil): a naturally occurring mixture of gaseous, liquid, and solid hydrocarbon compounds usually found trapped deep underground beneath impermeable cap rock and above a lower dome of sedimentary rock such as shale; most crude oil reservoirs occur in sedimentary rocks of marine, deltaic, or estuarine origin.

Phase behavior: the tendency of a fluid system to form phases as a result of changing temperature, pressure, or the bulk composition of the fluids or of individual fluid phases.

Phase diagram: a graph of phase behavior. In chemical flooding a graph showing the relative volume of oil, brine, and sometimes one or more micro emulsion phases. In carbon dioxide flooding, conditions for formation of various liquid, vapor, and solid phases.

Phase properties: types of fluids, compositions, densities, viscosities, and relative amounts of oil, microemulsion, or solvent, and water formed when a micellar fluid (surfactant slug) or miscible solvent (e.g., CO_2) is mixed with oil.

Phase separation: the formation of a separate phase that is usually the prelude to coke formation during a thermal process; the formation of a separate phase as a result of the instability/incompatibility of crude oil and crude oil products.

Photolithography: a process used to transfer patterns and make layered materials using photosensitive substances and selective light exposure.

Physical limit: the limit of the reservoir defined by any geological structures (faults, unconformities, change of facies, crests and bases of formations, etc.), caused by contact between fluids or by the reduction, to critical porosity, of permeability limits, or the compound effect of these parameters.

Physical properties: the properties of a material that can be measured without changing the composition of the material; examples of physical properties are color, density and boiling temperature.

Pilot project: project that is being executed in a small representative sector of a reservoir where tests performed are similar to those that will be implemented throughout the reservoir. The purpose is to gather information and/or obtain results that could be used to generalize an exploitation strategy in the oil field.

PINA analysis: a method of analysis for paraffins, *iso*-paraffins, naphthenes, and aromatics.

Pinnacle reef: a conical formation, higher than it is wide, usually composed of limestone, in which hydrocarbons might be trapped.

PIONA analysis: a method of analysis for paraffins, *iso*-paraffins, olefins, naphthenes, and aromatics.

Pitch: the nonvolatile, brown to black, semi-solid to solid viscous product from the destructive distillation of many bituminous or other organic materials, especially coal.

Plant liquefiables shrinkage factor (PLSF): the fraction arising from considering the liquefiables obtained in transportation to the processing complexes.

Polarization: a property of waves that describes the orientation of their oscillations.

Polymer clay: deformable composite of polyvinyl chloride (PVC) that can be manipulated similar to a clay. This usually does not contain any actual clay.

PONA analysis: a method of analysis for paraffins (P), olefins (O), naphthenes (N), and aromatics (A).

Pool: a natural underground reservoir containing an accumulation of crude oil.

Pore volume: total volume of all pores and fractures in a reservoir or part of a reservoir; also applied to catalyst samples.

Porosity: the ratio between the pore volume existing in a rock and the total rock volume; a measure of rock's storage capacity; the percentage of rock volume available to contain water or other fluid.

Pour point: the lowest temperature at which oil will pour or flow when it is chilled without disturbance under definite conditions.

Primary structure: the first level of ordered structuring of matter above disorder.

Quadrillion: 1×10^{15}

Quantum dot: a semiconductor in which the electron propagation is confined in three dimensions (differing from quantum wires in which propagation is controlled in two dimensions, and quantum wells in which propagation is controlled in a single direction).

Quench: the sudden cooling of hot material discharging from a thermal reactor.

Ramsbottom carbon residue: see Carbon residue

Raw materials: minerals extracted from the earth prior to any refining or treating.

Refinery: a series of integrated unit processes by which crude oil can be converted to a slate of useful (salable) products.

Refinery gas: a gas (or a gaseous mixture) produced as a result of refining operations.

Refining: the process(es) by which crude oil is distilled and/or converted by application of a physical and chemical processes to form a variety of products are generated.

Reformulated fuels: gasoline, diesel or other fuels which have been modified to reflect environmental concerns, performance standards, government regulations, customer preferences or new technologies.

Regeneration: the reactivation of a catalyst by burning off the coke deposits.

Regenerator: a reactor for catalyst reactivation.

Regression: geological term used to define the elevation of one part of the continent over sea level, as a result of the ascent of the continent or the lowering of the sea level.

Relative permeability: the capacity of a fluid, such as water, gas or oil, to flow through a rock when it is saturated with two or more fluids. The value of the permeability of a saturated rock with two or more fluids is different to the permeability value of the same rock saturated with just one fluid.

Residuum: the nonvolatile fraction (b.p. > 570 °C, > 1050°F) remaining after processing or distillation of crude oil.

Resins: the portion of the maltenes (the nonasphaltene part of crude oil) that is adsorbed by a surface-active material such as clay or alumina; the fraction of deasphaltened oil that is insoluble in liquid propane but soluble in n-heptane.

Resistance factor: a measure of resistance to flow of a polymer solution relative to the resistance to flow of water.

Resource: total volume of hydrocarbons existing in subsurface rocks; also known as original in-situ volume.

Saline groundwater (brackish water): deep groundwater that is high in dissolved salt and unsuitable for domestic or agricultural uses.

Salinity: the concentration of salt in water.

Sand: a course granular mineral mainly comprising quartz grains that is derived from the chemical and physical weathering of rocks rich in quartz, notably sandstone and granite.

Sand face: the cylindrical wall of the wellbore through which the fluids must flow to or from the reservoir.

Sandstone: a sedimentary rock formed by compaction and cementation of sand grains; can be classified according to the mineral composition of the sand and cement.

SARA analysis: a method of fractionation by which crude oil is separated into saturates, aromatics, resins, and asphaltene fractions.

Saturation: the ratio of the volume of a single fluid in the pores to pore volume, expressed as a percent and applied to water, oil, or gas separately; the sum of the saturations of each fluid in a pore volume is 100 percent.
Saturation pressure: pressure at which the first gas bubble is formed, when it goes from the liquid phase to the two-phase region.
Saybolt Furol viscosity: the time, in seconds (Saybolt Furol Seconds, SFS), for 60 ml of fluid to flow through a capillary tube in a Saybolt Furol viscometer at specified temperatures between 70 and 210° F; the method is appropriate for high-viscosity oils such as transmission, gear, and heavy fuel oils.
Saybolt Universal viscosity: the time, in seconds (Saybolt Universal Seconds, SUS), for 60 mL of fluid to flow through a capillary tube in a Saybolt Universal viscometer at a given temperature.
Scanning force microscope: a very high-resolution microscope that uses a microcantilever to scan the surface of a substrate. This microscope can image and scan surface features on the order of less than a nanometer. Same as Atomic Force Microscope.
Scanning tunneling microscope: a widely used instrument for viewing surfaces at the atomic level. It provides a three-dimensional profiles of viewed surfaces, which is very useful for characterizing surface roughness, observing surface defects, and determining the size and conformation of molecules and aggregates on the surface.
Secondary structure: the second level of ordered structuring of matter above disorder.
Sediment: an insoluble solid formed as a result of the storage instability and/or the thermal instability of crude oil and crude oil products.
Sedimentary: formed by or from deposits of sediments, especially from sand grains or silts transported from their source and deposited in water, as sandstone and shale; or from calcareous remains of organisms, as limestone.
Sedimentary basin: a geographical area, such as the Western Canada Sedimentary Basin, in which much of the rock is sedimentary (as opposed to igneous or metamorphic) and therefore likely to contain hydrocarbons.
Sedimentary strata: typically consist of mixtures of clay, silt, sand, organic matter, and various minerals; formed by or from deposits of sediments, especially from sand grains or silts transported from their source and deposited in water, such as sandstone and shale; or from calcareous remains of organisms, such as limestone.
Seismic section: seismic profile that uses the reflection of seismic waves to determine the geological subsurface.
Semiconductor: a material that exhibits electrical conductivity properties between those of a conductor and an insulator. The conductivity of a semiconductor material can be varied under an external electrical field. Devices made from semiconductor materials are the foundation of modern electronics, including radio, computers, telephones, and many other devices. Semiconductor devices include the transistor, many kinds of diodes including the light-emitting diode, the silicon controlled rectifier, and digital and analog integrated circuits. Solar

photovoltaic panels are large semiconductor devices that directly convert light energy into electrical energy.

Self-assembly: processes in which a disordered system of pre-existing components forms an organized structure or pattern as a consequence of specific, local interactions among the components, without external direction.

Separation: the process of separating liquid and gas hydrocarbons and water. This is typically accomplished in a pressure vessel at the surface, but newer technologies allow separation to occur in the wellbore under certain conditions.

Shale: rock formed from clay.

Shutdown: a production hiatus during which the platform ceases to produce while essential maintenance work is undertaken.

Slime: a name used for crude oil in ancient texts.

Sludge: a semisolid to solid product which results from the storage instability and/or the thermal instability of crude oil and crude oil products.

Solar cell: semiconductor or organic device used to harness solar energy and convert into electrical energy.

Solution gas (associated gas): natural gas that is found with crude oil in underground reservoirs. When the oil comes to the surface, the gas expands and comes out of the solution.

Sonic log: a well log based on the time required for sound to travel through rock, useful in deter- mining porosity.

Sonic log: a well log based on the time required for sound to travel through rock, useful in determining porosity.

Sour crude oil: crude oil containing an abnormally large amount of sulfur compounds; see also Sweet crude oil.

Sour gas: natural gas at the wellhead may contain hydrogen sulphide (H_2S), a toxic compound. Natural gas that contains more than one per cent of H_2S is called sour gas.

Spacing: optimum distance between hydrocarbon producing wells in a field or reservoir.

Specific gravity: an intensive property of the matter that is related to the mass of a substance and its volume through the coefficient between these two quantities. It is expressed in grams per cubic centimeter or in pounds per gallon.

Spent catalyst: catalyst that has lost much of its activity due to the deposition of coke and metals.

Spills: spills include accidental release of crude oil, produced water or other hydrocarbon products from well sites, batteries or storage tanks. These spills can affect land, vegetation, water bodies and groundwater.

Spud-in: the operation of drilling the first part of a new well.

Standard conditions: the reference amounts for pressure and temperature. In the English system, it is 14.73 pounds per square inch for the pressure and 60°F for temperature.

Steam distillation: distillation in which vaporization of the volatile constituents is effected at a lower temperature by introduction of steam (open steam) directly into the charge.

Strata: layers including the solid iron-rich inner core, molten outer core, mantle, and crust of the earth.
Stratigraphy: part of geology that studies the origin, composition, distribution and succession of rock strata.
Stripper well: a well that produces (strips from the reservoir) oil or gas.
Structural nose: a term used in structural geology to define a geometric form protruding from a main body.
Sucker rod pumping system: a method of artificial lift in which a subsurface pump located at or near the bottom of the well and connected to a string of sucker rods is used to lift the well fluid to the surface.
Super-light oil: the specific gravity is more than 38° API.
Supramolecule: an ordered array of molecules, held together through noncovalent interactions, which exhibits at least a primary structure..
Surface active material: a chemical compound, molecule, or aggregate of molecules with physical properties that cause it to adsorb at the interface between *two* immiscible liquids, resulting in a reduction of interfacial tension or the formation of a microemulsion.
Surfactant: a type of chemical, characterized as one that reduces interfacial resistance to mixing between oil and water or changes the degree to which water wets reservoir rock.
Suspended well: a well that has been capped off temporarily.
Suspension: a heterogeneous fluid containing solid particles that are sufficiently large for sedimentation. An example of a suspension would be sand in water. The suspended particles are visible under a microscope and will settle over time if left undisturbed. This distinguishes a suspension from a colloid in which the suspended particles are smaller and do not settle.
Surface area: a measure of exposed area of na object.
Surface-to-volume ratio: the ratio of exposed surface area to volume of the particle. In nanotechnology, very high surface to volume ratio is the enabler for many nanoscale properties.
Sweep efficiency: the ratio of the pore volume of reservoir rock contacted by injected fluids to the total pore volume of reservoir rock in the project area. *(See also* horizontal sweep efficiency *and* vertical sweep efficiency.)
Sweet crude oil: crude oil containing little sulfur; see also Sour crude oil.
Sweetening plant: industrial plant used to treat gaseous mixtures and light crude oil fractions in order to eliminate undesirable or corrosive sulfur compounds to improve their color, odor and stability.
Swept zone: the volume of rock that is effectively swept by injected fluids.
Synthetic crude oil (syncrude): a hydrocarbon product produced by the conversion of coal, oil shale, or tar sand bitumen that resembles conventional crude oil; can be refined in a crude oil refinery.
Tar: the volatile, brown to black, oily, viscous product from the destructive distillation of many bituminous or other organic materials, especially coal; a name used for crude oil in ancient texts.

Tar sand (bituminous sand): another name for oil sand. A sandstone deposit that is saturated with bitumen. The viscous hydrocarbonaceous mixtures found in northern Alberta have historically been referred to (incorrectly) as tar, pitch or asphalt; see Bituminous sand.

Tcf: trillion Cubic Feet (of gas).

Tertiary structure: the third level of ordered structuring of matter above disorder.

Thermal coke: the carbonaceous residue formed as a result of a noncatalytic thermal process; the Conradson carbon residue; the Ramsbottom carbon residue.

Thermal cracking: a process that decomposes, rearranges, or combines hydrocarbon molecules by the application of heat, without the aid of catalysts.

Thermal process: any refining process that utilizes heat, without the aid of a catalyst.

Top down: a process that progresses from larger units to smaller units.

Trace element: those elements that occur at very low levels in a given system.

Transport liquefiables shrinkage factor (TLSF): the fraction obtained by considering the liquefiables obtained in transportation to the processing complexes.

Trap: a stratigraphic or structural feature that ensures the juxtaposition of reservoir and seal such that hydrocarbons remain trapped in the subsurface, rather than escaping (due to their natural buoyancy) and being lost; the geological geometry that permits the concentration of hydrocarbons; a sediment in which oil and gas accumulate from which further migration is prevented.

Triaxial borehole seismic survey: a technique for detecting the orientation of hydraulically induced fractures, wherein a tool holding three mutually seismic detectors is clamped in the borehole during fracturing; fracture orientation is deduced through analysis of the detected microseismic perpendicular events that are generated by the fracturing process.

Trillion: 1×10^{12}

True boiling point (True boiling range): the boiling point (boiling range) of a crude oil fraction or a crude oil product under standard conditions of temperature and pressure.

Tube-and-tank cracking: an older liquid-phase thermal cracking process.

Ultimate analysis: elemental analysis.

Ultra-violet light: electromagnetic radiation with a wavelength shorter than that of visible light, but longer than X-rays, in the range 10nm to 400nm, and energies from 3 to 124 eV. It is so named because the spectrum consists of electromagnetic waves with frequencies higher than those that humans identify as the color violet.

Universal viscosity: see Saybolt Universal viscosity.

Upper-phase microemulsion: a microemulsion phase containing a high concentration of oil that, when viewed in a test tube, resides on top of a water phase.

Upstream: the companies that explore for, develop and produce crude oil resources are known as the upstream sector of the crude oil industry.

Viscosity: a measure of the ability of a liquid to flow or a measure of its resistance to flow; the force required to move a plane surface of area 1 meter^2 over another parallel plane surface 1 meter away at a rate of 1 meter/sec when both surfaces are immersed in the fluid.

Volatile organic compounds (VOCs): gases and vapors, such as benzene, released by crude oil refineries, petrochemical plants, plastics manufacturing and the distribution and use of gasoline; VOCs include carcinogens and chemicals that react with sunlight and nitrogen oxides to form ground-level ozone, a component of smog.

Volumetric sweep: the fraction of the total reservoir volume within a flood pattern that is effectively contacted by injected fluids.

VSP: vertical seismic profiling, a method of conducting seismic surveys in the borehole for detailed subsurface information.

Well abandonment: the final activity in the operation of a well when it is permanently closed under safety and environment preservation conditions.

Wellbore: a hole drilled or bored into the earth, usually cased with metal pipe, for the production of gas or oil.

Well completion: the complete outfitting of an oil well for either oil production or fluid injection; also the technique used to control fluid communication with the reservoir.

Wellhead: that portion of an oil well above the surface of the ground.

Well logs: the information concerning subsurface formations obtained by means of electric, acoustic and radioactive tools inserted in the wells. The log also includes information about drilling and the analysis of mud and cuts, cores and formation tests.

Wells-to-wheels: the full product life cycle is considered—from production (wells) to the use of the fuel in a vehicle (wheels). Wells-to-wheels analysis can be used to assess total life cycle greenhouse gas emissions from production to combustion of different crude oils.

Wet gas: mixture of hydrocarbons obtained from processing natural gas from which nonhydrocarbon impurities or compounds have been eliminated, and whose content of components that are heavier than methane is such that it can be commercially processed.

Zeolite: a crystalline aluminosilicate used as a catalyst and having a particular chemical and physical structure.

Index

D

F

Q

R

S